Indium Phosphide and Related Materials: Processing, Technology, and Devices

Related Titles from Artech House

Indium Phosphide and Related Materials: Processing, Technology, and Devices

Avishay Katz
Editor

Artech House
Boston • London

Library of Congress Cataloging-in-Publication Data

Indium phosphide and related materials: processing, technology, and devices
  / Avishay Katz, editor.
       p. cm.
    Includes bibliographical references and index.
    ISBN 0-89006-512-8
    1. Electronics–Materials. 2. Indium phosphide.
    3. Semiconductors. I. Katz, Avishay.
    TK7870.I44   1991   91-29592
    621.3815'028–dc20   CIP

© 1992   ARTECH HOUSE, INC.
685 Canton Street
Norwood, MA 02062

International Standard Book Number: 0-89006-512-8
Library of Congress Catalog Card Number: 91-29592

10 9 8 7 6 5 4 3 2 1

Contributors

C.R. Abernathy, AT&T Bell Laboratories, Murray Hill, NJ 07974, USA

W.A. Anderson, Center for Electronic and Electro-optic Materials, State University of New York at Buffalo, Amherst, NY 14260, USA

E.K. Byrne, AT&T Bell Laboratories, Murray Hill, NJ 07974, USA

U.K. Chakrabarti, AT&T Bell Laboratories, Murray Hill, NJ 07974, USA

T.J. Coutts, Solar Energy Research Institute, Golden, CO 80401, USA

N.K. Dutta, AT&T Bell Laboratories, Murray Hill, NJ 07974, USA

T.A. Gessert, Solar Energy Research Institute, Golden, CO 80401, USA

M. Geva, AT&T Bell Laboratories, Breinigsville, PA 18031, USA

T.R. Hayes, AT&T Bell Laboratories, Murray Hill, NJ 07974, USA

B. Jalali, AT&T Bell Laboratories, Murray Hill, NJ 07974, USA

K.L. Jiao, Center for Electronic and Electro-optic Materials, State University of New York at Buffalo, Amherst, NY 14260, USA

A. Katz, AT&T Bell Laboratories, Murray Hill, NJ 07974, USA

R.N. Nottenburg, AT&T Bell Laboratories, Murray Hill, NJ 07974, USA

S.J. Pearton, AT&T Bell Laboratories, Murray Hill, NJ 07974, USA

D. Schmitz, AIXTRON GmbH, 5100 Aachen, FRG

V. Swaminathan, AT&T Bell Laboratories, Breinigsville, PA 18031, USA

Contents

Preface

The issues involved in processing high-quality, InP-based electronic and photonic devices have been widely explored during the last decade. Sophisticated applications of InP-based devices, such as those used in highly-reliable, high-speed, long-distance, fiber-optics communication systems, are placing a greater demand on the quality requirements of the material growth, analytical characterization techniques, and device processing concepts. It is essential to include these three elements in the manufacturing sequence in order to produce devices of the required quality.

Achieving improved control of the current technology, gaining a better understanding of the fundamental phenomena, improving the success and failure mode analysis, and addressing the even more complicated future challenges associated with InP-based photonic and electronic device technology have been the major concerns of the InP-based technology research and development community.

The amount of effort and resources invested in this technology during recent years reflects the important role that InP and related semiconductor materials have in manufacturing both the analog and digital high-frequency active devices. It is this intensive and thorough work that has made the rapid transition from the fundamental research phase to the design of engineering products possible.

This book brings together both material and process issues as well as circuit aspects of current InP-based devices. It provides an overview of this exciting field by compiling reviews on selected essential topics associated with InP-based device technology.

It is our hope that this compilation will provide readers with an introduction to the subject and generate further interest in and study of the available publications that address in greater detail specific issues of InP and related material technology.

This book is organized in the following way: Chapter 1 provides an introduction to the basic physical properties, bonding, and crystal structure of InP and its related semiconductor alloys. Chapter 2 contains a summary of the secondary ion mass spectrometry (SIMS) analysis technique, which is essential in carrying out a surface analysis of these materials. Chapter 3 describes the deep-level transient spectroscopy

(DLTS) technique, commonly used to define the material electrical properties. Chapters 4, 5, and 6 address material growth aspects, and discuss metalorganic vapor phase epitaxy (MOVPE), the metalorganic chemical vapor deposition (MOCVD), and gas-source molecular beam epitaxy (GSMBE) techniques, respectively. Process technologies are further pursued in the following chapters. Chapter 7 discusses ion beam processing; Chapter 8 covers dry etching technique; Chapter 9 deals with ohmic contact technology; and Chapter 10 outlines dielectric deposition techniques. At this point the discussion moves into the realm of devices, and examples of some device families are provided. Chapter 11 covers an optical device, the InGaAsP quantum well laser; Chapter 12 discusses an electronic device, the heterostructure bipolar transistor; and Chapter 13 examines solar cells.

Finally, I wish to thank most warmly the authors who have contributed their talent, skills, and time in writing the various chapters included in this book. I would also like to acknowledge the support of AT&T Bell Laboratories management in making this project possible.

Chapter 1
Properties of InP and Related Materials

V. Swaminathan
AT&T Bell Laboratories

1.1 INTRODUCTION

In recent years there has been considerable interest in indium phosphide (InP) and related alloys because of their applications in many electronic and photonic devices. The phenomenal progress made over the last decade in the preparation and processing of InP-related materials, together with the development of highly reliable devices, has pushed the frontiers of the optoelectronics technology to very high levels of sophistication. The ternary alloy, GaInAs, and the quaternary alloy, GaInAsP, grown lattice matched to InP substrates, are the materials of choice for making light sources and detectors for the present-day high-data-rate, long-haul, fiber-optic communication systems. High-speed electronic devices, such as heterojunction bipolar transistors and high-electron mobility transistors, are being realized from GaInAs/InP and GaInAs/AlInAs heterostructures. Novel devices, such as resonant tunneling transistors, are also gaining importance, and the GaInAs/AlInAs system is ideal for these devices because of the small electron mass in the barrier (AlInAs) and the large conduction band offset. Because of these myriad applications, InP-related materials have become an important class of III-V semiconductors [1]. In this chapter we will describe the bonding, crystal structure, band structure, and physical properties of InP and some of the alloy semiconductors based on it.

1.2 BONDING AND CRYSTAL STRUCTURE

InP crystallizes in the cubic zinc blende (sphalerite) structure, which consists of two interpenetrating fcc lattices, one shifted by $a/4$ [111] relative to the other fcc lattice

(a being the length of the fcc cube edge). The two fcc lattices are occupied by two different atoms, for example, In and P. The cubic unit cell of a sphalerite structure is shown in Figure 1.1. If the coordinates of the group III atoms are 000, $0\frac{1}{2}\frac{1}{2}$, $\frac{1}{2}0$ $\frac{1}{2}$, $\frac{1}{2}\frac{1}{2}0$, that of the group V atoms are $\frac{1}{4}\frac{1}{4}\frac{1}{4}$, $\frac{1}{4}\frac{3}{4}\frac{3}{4}$, $\frac{3}{4}\frac{1}{4}\frac{3}{4}$, $\frac{3}{4}\frac{3}{4}\frac{1}{4}$. Each group III atom has four nearest neighbor group V atoms, 12 next-nearest neighbor group III atoms, and so on. The sphalerite structure does not have inversion symmetry because the arrangement of atoms along the body diagonal, the order AB..AB..AB (where dots represent vacant sites), is not invariant under inversion.

Since each fcc lattice has four atoms per unit cell, there are four molecules of the compound AB per unit cell in the sphalerite structure. In the unit cell, the atoms are arranged such that for each group III atom there are four equally distant group V atoms arranged at the corners of a regular tetrahedron, and vice versa. The tetrahedral arrangement is the result of the covalent bonding of the eight valence electrons per molecule in sp^3 hybridized orbitals. The bonding is not, however, entirely covalent since there is some charge transfer between the two types of atoms, giving rise to a partial ionic character in the bonding. For InP, 42% of the bonding is estimated to be ionic [2.].

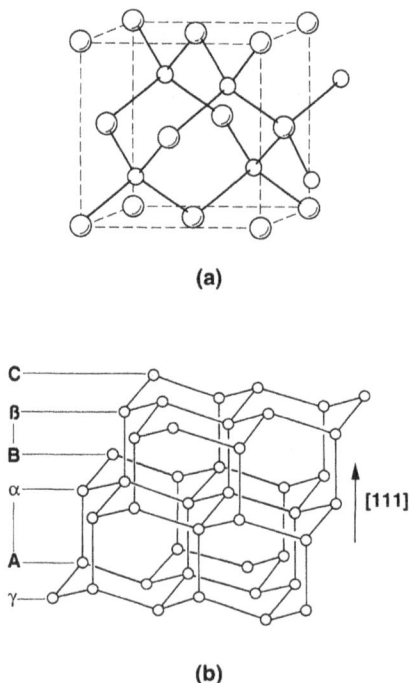

(a)

(b)

Figure 1.1 (a) The sphalerite structure; (b) stacking of (111) planes in the sphalerite structure. ABC . . . is the stacking order in one fcc lattice and $\alpha\beta\gamma$ is the stacking order in the other.

The stacking of {111} planes in the cubic semiconductors (Fig. 1.1) follows the ABCABC stacking of the closely packed {111} planes in the fcc lattice. However, the sphalerite structure is relatively empty in that the fraction of the total volume filled by spheres is only 0.34 compared to 0.74 of a closely packed structure. Because of the presence of two types of atoms, there is also a polarity in the stacking of the {111} planes—that is, [111] ≠ [$\overline{1}\overline{1}\overline{1}$].

1.3 ENERGY BAND STRUCTURE

The calculated energy band structure of InP is shown in Figure 1.2 [3]. Both the valence band maxima and the lowest conduction band minimum occur at $k = 0$, the Γ point. That is, InP is a direct band gap semiconductor. Higher conduction band minima occur in the $\langle 100 \rangle$ (Δ) and $\langle 111 \rangle$ (Λ) directions, as can be seen in Figure 1.2. The L and X minima are, respectively, 0.40 eV and 0.7 eV above the Γ_1 minimum [3]. The valence band maxima are derived from the p-like orbitals p_x, p_y, p_z, which remain degenerate under the tetrahedral group of the zinc blende lattice. The

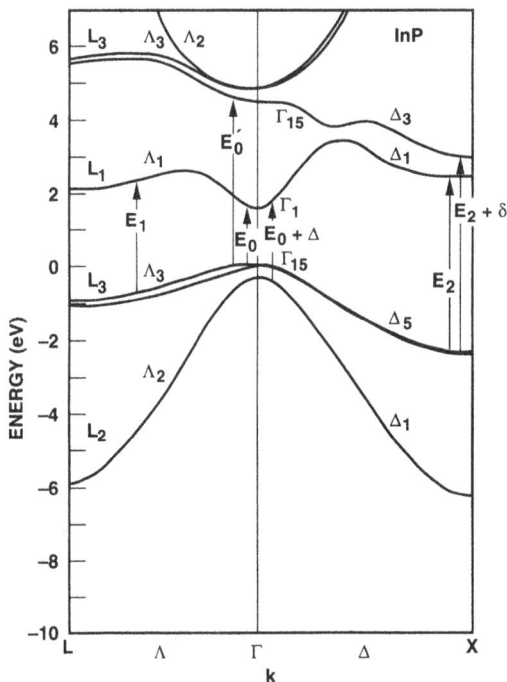

Figure 1.2 Calculated energy band structure of InP in two principal directions in the Brillouin Zone [3]. The various critical point energies are indicated.

representation of the valence band is Γ_{15}. Including electron spin, there are six states altogether. These six states are further split by spin-orbit interaction into a fourfold, degenerate $J = 3/2$ state and a twofold degenerate $J = 1/2$ state, the $J = 1/2$ band being the lowest band. The value of the spin-orbit splitting for a few III-V compounds is given in Table 1.1 [2]. The $J = 3/2$ bands are called the light hole and heavy hole bands. These two bands can be labeled by the azimuthal quantum number of J relative to the k axis, $m_J = 3/2$ or $1/2$. If uniaxial strain is applied to the semiconductor, the shear component of the strain further splits the degeneracy of the heavy and light hole bands [4]. This is particularly evident in heterostructures due to lattice-mismatch strain [5–8]. The conduction band is derived from s-like orbitals and denoted by the representation Γ_1. It is very much like the mirror image of the split-off spin-orbit valence band. The various critical point energies are also indicated in Figure 1.2. These critical points satisfy the condition $\nabla_k E(k) = 0$.

Table 1.1
Spin-Orbit Splittings [2]

Compound	Δ_0 (eV)
GaP	0.127
GaAs	0.34
GaSb	0.80
InP	0.11
InAs	0.38
InSb	0.82

1.4 ENERGY BAND STRUCTURE OF ALLOYS

It has been found experimentally that in many semiconductor alloys the energy gap varies nonlinearly with composition. For an $A_x B_{1-x} C$ compound, the energy gap can be expressed in the form [9]

$$E_0(x) = a + bx + cx^2 \tag{1.1}$$

where c is called the bowing parameter and is four times the deviation of $E_0(x)$ from linearity at $x = .05$. The other two parameters, a and b, are determined by the values of E_0 observed in the pure binary compounds. The bowing parameter is the sum of two terms, an intrinsic bowing and a short-range (within one unit cell) extrinsic bowing resulting from the aperiodic crystal potential caused by disorder in the alloys [10,11].

1.5 HETEROJUNCTIONS

Many of the InP-based optoelectronics devices contain both homojunctions and heterojunctions in the same structure. The abrupt change in the energy band structure at the heterointerface leads to discontinuities, or offsets, in the conduction and valence band edges. Such offsets act as potential steps in addition to the pure electrostatic potential variation that may be present. The uniqueness of the heterostructure devices results from their ability to control the flow and distribution of their electrons and holes via the band offsets.

Heterojunctions can be classified by their band lineup. The types of band offsets that occur at abrupt semiconductor heterojunctions are illustrated in Figure 1.3. In the straddling band lineup shown in Figure 1.3a, the band offsets in both the conduction and valence bands act as potential barriers and keep electrons and holes in the smaller bandgap semiconductors. Heterostructures that have straddling lineups are called type I heterostructures (e.g., InP-GaInAsP heterojunctions). Some heterojunctions have staggered (Fig. 1.3b) or broken-gap (Fig. 1.3c) band lineups. These are called type II heterostructures. In strained layers of $Ga_xIn_{1-x}As$ grown on InP, type I to type II transition was found to occur for $x > 0.8$ [12]. In this situation the electrons and holes are spatially separated with electrons in the conduction band of InP and holes in the valence band of $Ga_xIn_{1-x}As$.

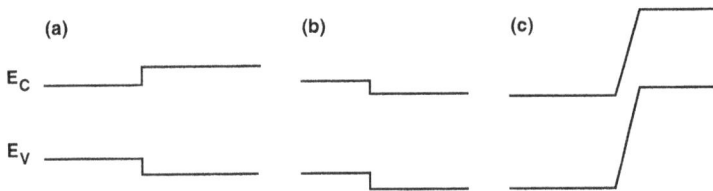

Figure 1.3 Schematic bandgap diagrams showing the different heterojunction band lineup: (a) straddling lineup, (b) staggered lineup; and (c) broken-gap lineup.

Theories of Band Offsets

The parameter of utmost importance in heterostructures for device applications is the band edge offset. The band edge offsets in the conduction and valence bands are related to the difference in the bandgap of the two components of the heterojunction as

$$\Delta E_V + \Delta E_C = \Delta E_g \tag{1.2}$$

Thus, if one of the offsets is calculated, the other is easily obtained because

ΔE_g is reasonably well known. The calculation of ΔE_V, for example, requires knowledge of the absolute energies of the valence band edges of the semiconductors. Further, when the two semiconductors are brought into contact with each other, the charge redistribution in the interface region and the associated electronic interface dipole effects may be significant. Given these two facts, ΔE_V may be written as

$$\Delta E_V = (E_{V_0}^A - E_{V_0}^B) - \delta E_V \qquad (1.3)$$

where $E_{V_0}^A$ and $E_{V_0}^B$ are the energies with respect to the vacuum level of the valence band maximum of the two semiconductors A and B, respectively, and δE_V is the energy shift in the band edges at the interface caused by the formation of the interface dipole. Apart from charge redistribution at the interface, the interface dipole effect may also be affected by the presence of localized interface defects, particularly in mismatched heterostructures [13].

As shown schematically in Figure 1.4, ΔE_C is equal to the difference in electron affinities of A and B. This is the *electron affinity rule* (EAR) suggested by Anderson [14] to calculate the band offset:

$$\Delta E_C = \chi_A - \chi_B \qquad (1.4)$$

ΔE_V is obtained from Equation (1.2), neglecting contribution of δE_V.

From Figure 1.4 the difference in the ionization energies of A and B is related to the valence band offset:

$$\Delta E_V = \phi_B - \phi_A = E_{V_0}^B - E_{V_0}^A \qquad (1.5)$$

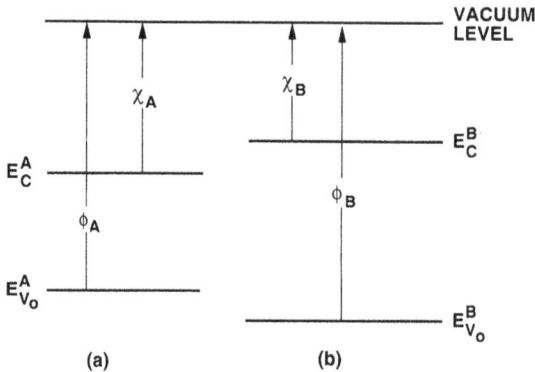

Figure 1.4 Schematic band diagram showing the differences in electron affinity and ionization potential of semiconductors A and B.

where ϕ_A and ϕ_B are the ionization energies of A and B. In the Harrison atomic orbital (HAO) model [15], the ionization energies are calculated using linear combination of atomic orbitals. Once again δE_V is neglected in determining ΔE_V.

Tersoff [16] has proposed the "midgap lineup" model in which a midgap level is assigned to each semiconductor. When two semiconductors form a heterojunction, their midgap levels are lined up. Good agreement has been obtained between offsets calculated using this model and determined experimentally for many III-V semiconductors on Ge and Si.

Ruan and Ching [17] have proposed an effective dipole theory for band lineups. They considered the penetration of the effective-mass electrons from the material with higher valence band top into the other, which results in an effective charge transfer across the heterointerface and the formation of an effective dipole. The offsets predicted by this model differ from experimental values by ~0.1 eV on average for 30 heterojunctions.

Hybertsen [18,19] considered the effect of strain on the band offset in the nominally lattice-matched (100) $Ga_{0.47}In_{0.53}As/InP$ system. The ideal interface as sketched in Figure 1.5a has a large (6%) interface strain because of the differences in bond lengths of InAs, $Ga_{0.47}In_{0.53}P$, and InP. Total energy minimization calculations for 6% strain gave for the valence band offset a value of 410 + 50 meV, close to the experimental value (see Table 1.2). If strain is neglected, a value of 290 meV was obtained. Hybertsen also found that the offset is unchanged when local mixing at the interface (as shown in Figure 1.5b) reduces the strain, giving the minimum energy at 3% strain. Table 1.2 lists the valence band offsets for the InP-based heterostructures.

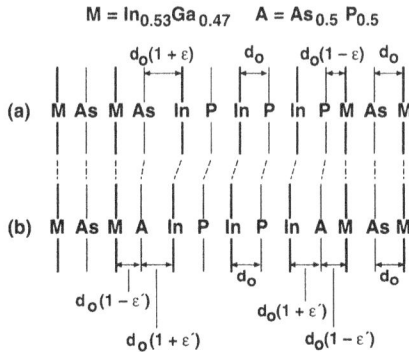

Figure 1.5 A schematic diagram of the nominally lattice-method sequence of planes in the (100) $Ga_{0.47}In_{0.53}As/Inp$ heterojunction showing interface strain due to differences in bond lengths of InAs, InP, and $Ga_{0.47}In_{0.53}P$ (a). Chemical intermixing leading to strain relief at the interface (b) [18,19].

Table 1.2

Calculated and Experimental Valence Band Offsets (eV) for GaAs, InP-Based Heterostructures
as Compiled in Ruan and Ching [17]

Heterostructure	Experiment	EAR	HAO	Tersoff's Model	Effective Dipole Model[d]	Total Energy Minimization
Ge/InP	0.64	0.90	0.64	0.58	0.67	
Ge/GaAs	0.25–0.65	0.70	0.41	0.52	0.51	
Si/InP	0.57	0.58	0.24	0.40	0.36	
Si/GaAs	0.05	0.38	0.03	0.34	0.22	
GaAs/AlAs	0.19–0.50	0.52	0.04	0.35	0.36	
GaAs/InAs	0.17	0.19	0.32	0.20	0.16	
GaAs/ZnSe	0.96–1.10	1.31	1.05	—	0.95	
InP/CdS	1.63	1.40	1.36	—	0.97	
InSb/InP	—	0.87			0.71	
InSb/GaAs	—	0.67			0.53	
InAs/InP		0.39			0.29	
GaAs/InP	—	0.20			0.13	
$Ga_{0.47}In_{0.53}As$/InP	0.346 ± 0.01[a] 0.39[b]					0.41
$Al_{0.48}In_{0.52}As$/InP	0.36 0.46[c]					0.25
$Ga_{0.47}In_{0.53}As$/ $Al_{0.48}In_{0.52}As$	0.13, 0.20 0.21[c]					0.17

(a) D.V. Lang, *Heterojunction Band Discontinuities*, ed. F. Capasso and G. Margaritondo, Amsterdam: North-Holland, 1987, p. 377.
(b) S.R. Forrest, *Heterojunction Band Discontinuities*, ed. F. Capasso and G. Margaritondo, Amsterdam: North-Holland, 1987, p. 311.
(c) Compiled in Hybertson [19].
(d) Ruan and Ching [17].
(e) Hybertson [18,19].

1.6 ALLOY SEMICONDUCTORS

Let us consider two III-V compounds, *AC* and *BC*, which form a cation alloy: $A_xB_{1-x}C$ (e.g., $Ga_xIn_{1-x}As$). Similarly, compounds *AC* and *AD* form an anion alloy: AC_xD_{1-x} (e.g., $InAs_xP_{1-x}$). Mixed compounds of this type are called ternary alloys. If the two compounds do not share a cation or an anion, one gets a quaternary alloy: for example, $Ga_xIn_{1-x}As_y$.

Figure 1.6 shows the lattice parameter versus energy gap at room temperature

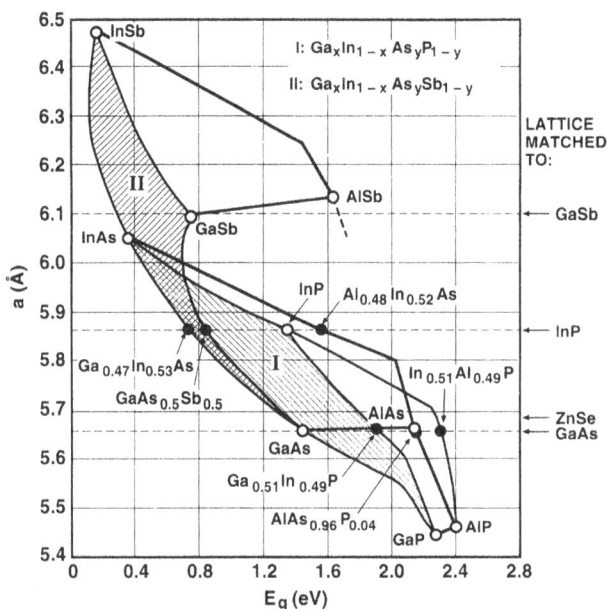

Figure 1.6 Lattice parameter *versus* energy gap at room temperature for various III-V compounds and their alloys [20].

for various III-V compounds and their alloys [20]. The figure also shows the composition of the lattice-matched ternary alloys and the corresponding substrates on which they are grown. The availability of good-quality substrates dictates to a large extent which of the lattice-matched heterostructures would be of use for device structures. At present, technologically important III-V heterostructures are limited to those grown on GaAs and InP substrates.

The shaded region I represents the quaternary alloy $Ga_xIn_{1-x}As_yP_{1-y}$, which can be grown lattice matched on GaAs or InP substrates. When GaAs is used as the substrate, the lattice-matched GaInAsP alloy covers the energy range 1.42–1.91 eV, more or less the same energy range as that covered by the ternary alloy AlGaAs. The GaInAsP alloy grown on InP substrate spans the energy range 0.75–1.35 eV between the bandgap of $Ga_{0.47}In_{0.53}As$ (0.75 eV) and that of InP (1.35 eV). The ternary alloy $Al_xIn_{1-x}As$ is lattice matched to InP at $x \sim 0.48$ and has a direct energy gap of 1.45 eV at room temperature. The $Al_xGa_yIn_{1-x-y}As$ alloy covers the spectral range 0.75–1.45 eV and is lattice matched in InP substrate for $x + y$ 1 $\sim$ 0.47. This alloy, containing only one group V atom, covers a wider spectral range than $Ga_xIn_{1-x}As_yP_{1-y}$ and may be advantageous for many heterostructure devices.

1.7 MATERIAL PARAMETERS OF ALLOY SEMICONDUCTORS

In deriving many physical parameters of the alloys, when specific experimental data are unavailable, a linear interpolation scheme is generally adopted using the values of the related binary compounds. For a ternary compound, $A_xB_{1-x}C$, the parameter P is derived from

$$P(A_xB_{1-x}C) = xP_{AC} + (1 - x)P_{BC} \tag{1.6}$$

Material parameters such as lattice constant vary linearly with composition in accordance with Equation (1.6).

The parameter of a quaternary compound such as $A_xB_{1-x}C_yD_{1-y}$ can be obtained from the respective values of the four binaries, AC, AD, BC and BD, in accordance with

$$P(A_xB_{1-x}C_yD_{1-y}) = xyP_{AC} + x(1 - y)P_{AD}$$
$$+ (1 - x)yP_{BC} + (1 - x)(1 - y)P_{BD} \tag{1.7}$$

If the parameters for the ternary alloys $A_xB_{1-x}C$, $A_xB_{1-x}D$, AC_yD_{1-y}, and BC_yD_{1-y} are available, then Equation (1.7) can be modified to give [21]

$$P(A_xB_{1-x}C_yD_{1-y})$$
$$= \frac{x(1 - x)[(1 - y)P_{ABD} + yP_{ABC}] + y(1 - y)[xP_{ACD} + (1 - x)P_{BCD}]}{x(1 - x) + y(1 - y)} \tag{1.8}$$

The ternary parameters, P_{ABD} and so on, may include a quadratic dependence given by Equation (1.1). This interpolation equation reduces to the average of the four ternary parameters at $x = y = 0.5$.

In the remainder of this section we have summarized the material properties of the InP-related alloys. The values for these properties, unless otherwise noted, are taken from the Landolt-Bornstein series [22,23].

1.7.1 Lattice Properties

Table 1.3 lists the lattice constant, density, melting point, and Debye temperature of a few of the III-V compounds. The lattice constant of ternary III-V alloys generally varies linearly with composition (Vegard's law). This is illustrated in Figure 1.7 for the $Ga_xIn_{1-x}As$ alloy [24]. Lattice-matched composition for growth on InP substrates occurs at $x \tilde{\ } 0.468$.

Table 1.3
Lattice Constant, Density, Melting Point, and Debye Temperature of III-V Compounds

Compound	Lattice Constant (A)	Density (gm cm^{-3})	Melting Point (K)	Debye Temperature[a] (K)
AlP	5.467	2.40	2823	588
AlAs	5.660	3.70	2013	417
AlSb	6.136	4.26	1338	292
GaP	5.4512	4.138	1740	456
GaAs	5.6532	5.3161	1513	344
GaSb	6.0959	5.6137	985	266
InP	5.8687	4.81	1335	321
InAs	6.0583	5.667	1215	249
InSb	6.4794	5.7747	800	203

(a) M.G. Holland, *Semiconductors and Semimetals, Vol. 2:3,* ed. R.K. Willardson and A.C. Beer, New York: Academic Press, 1966, p. 3.

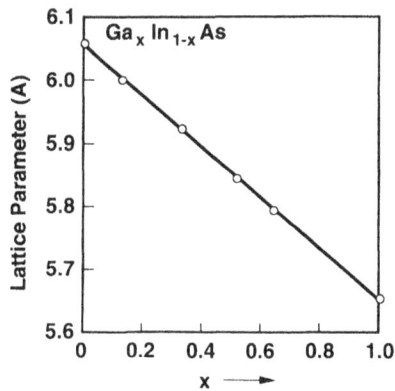

Figure 1.7 Lattice parameter of $Ga_xIn_{1-x}As$ *versus* x at 300K [24].

Vegard's law can also be assumed for the quaternary alloys. For the $Ga_xIn_{1-x}As_yP_{1-y}$ alloys, Vegard's law is given by

$$a(x,y) = 0.1896y - 0.4175x + 0.0124xy + 5.8687 \tag{1.9}$$

$$= a_{InP}\left[1 + \left(\frac{\Delta a}{a}\right)_0\right]$$

The quaternary alloys grown lattice matched on InP substrates can be considered part of a pseudo binary system between $Ga_{0.47}In_{0.53}As$ and InP. The lattice matching relation between x and y can be expressed as

$$x \cong 0.47y \ (0 < y < 1.0)$$

or using Equation (1.7) more rigidly as

$$x = \frac{0.1896y}{0.4175 - 0.012\,y} \ (0 < y < 1.0) \tag{1.10}$$

$(\Delta a/a)_0$ in Equation (1.9) is the relaxed mismatch related to the measured $(\Delta a/a)_\perp$ for the case of thin epitaxial layers [25]:

$$(\Delta a/a)_\perp = \frac{(1 + \nu)}{(1 - \nu)}(\Delta a/a)_0 \tag{1.11}$$

For {100} crystals according to Equation (1.29), Equation (1.11) reduces to

$$\left(\frac{\Delta a}{a}\right)_{100} = \frac{S_{11} - S_{12}}{S_{11} + S_{12}}\left(\frac{\Delta a}{a}\right)_0 \tag{1.12}$$

where S_{ij} is the elastic compliance (see Sec. 1.7.4). Using the S_{ij} values in Table 1.9, we obtain

$$\left(\frac{\Delta a}{a}\right)_{100} = 2.125\,(\Delta a/a)_0 \tag{1.13}$$

1.7.2 Band Structure

The parameters pertinent to the electronic band structure are summarized in Tables 1.4 and 1.5. The direct and indirect energy gaps, their temperature dependences,

Table 1.4
Energy Band Structure

Parameter		AlP	AlAs	AlSb	GaP	GaAs	GaSb	InP	InAs	InSb
Direct Energy Gap (eV)	$\Gamma_{15} - \Gamma_1$	3.62 (77K)	3.14	2.22	2.78	1.424	0.70	1.34	0.356	0.180
Indirect Energy Gap (eV)	$\Gamma_{15} - X_1$	2.45	2.14[a]	1.63[a]	2.268[a]	1.804	1.25 (10K)	2.04		
Temperature Dependence of Direct and Indirect Energy Gap ($\times 10^{-4}$ eV K^{-1})	dE_{d}/dT dE_{ind}/dT	−3.6	−5.2 −4.0	−3.5	−4.5 −5.2	−3.9 −2.4[b]	−3.7	−2.9 −3.7[b]	−3.5	−2.8
Pressure Dependence of Minimum Energy Gap ($\times 10^{-6}$ eV^{-1})	dE/dP			−1.5	−1.6	12.0	14.7	8.8	10.6	15.9
Indirect Energy Gap (eV)	$\Gamma_{15} - L_1$					1.81 (110K)	0.81	1.74		

(a) Minimum situated at the Δ axes near the boundary: $k = (0.903, 0, 0)$ for AlAs; $k = (0.95, 0, 0)$ for GaP.
(b) Temperature dependence of the energy separation between Γ and X minima.

Table 1.5
Critical Point Energies (eV)

Critical Points	AlP	AlAs	AlSb	GaP	GaAs	GaSb	InP	InAs	InSb
E_0' ($\Gamma_{15v} - \Gamma_{15c}$)		4.34	3.7 (77K)	4.8 (80K)	4.49 (4.2K)	3.2 (80K)	4.8 (77K)	4.5 (77K)	3.16
E_1 ($\Lambda_{3v} - \Lambda_{1c}$)			2.78	3.7 (80K)	2.90		3.15	2.5	1.88
$E_1 + \Delta$ ($\Lambda_{4v} - \Lambda_{1c}$)			3.18	3.9 (80K)	3.17		3.30 (77K)	2.75	2.38
E_2 ($X_{5v} - X_{1c}$)		4.54					5.04	4.72	4.08
$E_2 + \delta$ ($X_{5v} - X_{3c}$)		4.89					5.6	5.3	4.6 (110K)

and the pressure dependence of the minimum energy gap for the binary compounds
are given in Table 1.4. For AlP, AlAs, AlSb, and GaP, the conduction band minimum is located at the Δ axes near the Brillouin zone boundary. For AlAs, the
k value of the conduction band minimum is 0.087 k_{max} from the X point where k_{max}
$= 2\pi/a$, a being the lattice constant. For GaP the corresponding value is 0.043 k_{max}
from the X point. The valence band structure of all the compounds listed in the tables
is very similar. The valence band maximum is located at $k = 0$, and it is characterized by fourfold degenerate heavy hole and light hole bands, and a twofold degenerate spin-orbit band separated by the energy Δ_0 given in Table 1.1.

The temperature dependence of the bandgap is generally expressed by [26]

$$E_g(T) = E_g(0) - \frac{\alpha T^2}{(T + \beta)} \tag{1.14}$$

where $E_g(0)$ is the energy gap at 0K, and α and β are constants, β having a value
close to the Debye temperature. Table 1.6 gives the α and β values for obtaining

Table 1.6

Temperature Dependence of the Minimum Energy Gap. Parameters α and β as Defined in Equation
(1.14) are Obtained from Casey and Panish [27]

Compound	E_g (0) (eV)	α (X 10^{-4} eV K^{-1})	β
AlP	2.52	3.18	588
AlAs	2.239	6.0	408
AlSb	1.687	4.97	213
GaP	2.338	5.771	372
GaAs	1.519	5.405	204
GaSb	0.810	3.78	94
InP	1.421	3.63	162
InAs	0.420	2.50	75
InSb	0.236	2.99	140
$Ga_{0.47}In_{0.53}As$/InP	0.822	4.5	327[a,b]

(a) T.P. Pearsall, L. Eaves, and J.C. Portal, "Photoluminescence and Impurity Concentration in
$Ga_xIn_{1-x}As_yP_{1-y}$, Alloys Lattice Matched to InP," Vol. 54 (2), 1983, pp. 1037–1047. The α and β
values are found to be applicable for the entire composition range.
(b) The linear temperature coefficient obtained from absorption measurement is -3.48×10^{-4} eV/K.
E. Zielinski, H. Schweizer, K. Streubel, H. Eisek, and G. Weimann, "Excitonic Transitions and Exciton
Damping Processes in InGaAs/InP," J. Appl. Phys. Vol. 59 (6), 1986, pp. 2196–2204.

E_g below room temperature [27]. Near and above room temperature, $E_g(T)$ varies linearly with temperature. The values of the linear temperature coefficient of the energy gap are given in Table 1.4. For $Ga_{0.47}In_{0.53}As$ and $Ga_xIn_{1-x}As_yP_{1-y}$ lattice matched to InP, the linear temperature coefficient of the energy gap have been determined to be -0.348 meV K^{-1} [28] and -0.4 meV K^{-1} [29], respectively.

Table 1.5 gives the critical point energies of the different binary III-V compounds. These critical point energies are indicated in Figure 1.2. The temperature dependence of the critical point energies also follows Equation (1.14), with different α and β values.

For the alloys, a parameter of interest is the variation of E_g with composition. Besides the variation of the minimum energy gap, the variation of the energy separation between the different conduction band minima is also of interest. The composition dependence of the energy gap is often represented by Equation (1.1). Table 1.7 gives E_g versus composition for some important ternary alloy semiconductors [20,22,23].

Table 1.7
Compositional Dependence of the Energy Gap in III-V Ternary Alloy Semiconductors at 300K

Alloy	Direct Energy Gap (E_Γ)	Indirect Energy Gap	
		E_x	E_L
$Al_xIn_{1-x}P$	$1.34 + 2.23x$	$2.24 + 0.18x$	
$Al_xIn_{1-x}As$	$0.36 + 2.35x + 0.24x^2$	$1.8 + 0.4x$	
$Ga_xIn_{1-x}P$	$1.34 + 0.511x + 0.6043x^2$		
	$(0.49 < x < 0.55,$ VPE layer)		
$Ga_xIn_{1-x}As$	$0.356 + 0.7x + 0.4x^2$		
InP_xAs_{1-x}	$0.356 + 0.675x + 0.32x^2$		

1.7.2.1 $Ga_xIn_{1-x}As$

The calculated compositional dependence of the Γ, X, and L energy gaps for the ternary alloy $Ga_xIn_{1-x}As$ is shown in Figure 1.8 [30]. The variation of the Γ gap at 2K in a narrow range of composition for unstrained $Ga_xIn_{1-x}As$ layers is shown in Figure 1.9 [31]. The solid line in Figure 1.9 represents the expression

$$E_g(2K) = 0.4105 + 0.6337x + 0.475x^2 \qquad (1.15)$$

The bowing parameter given by Equation (1.1) is 0.475. At $x \sim 0.468$ (lattice matched

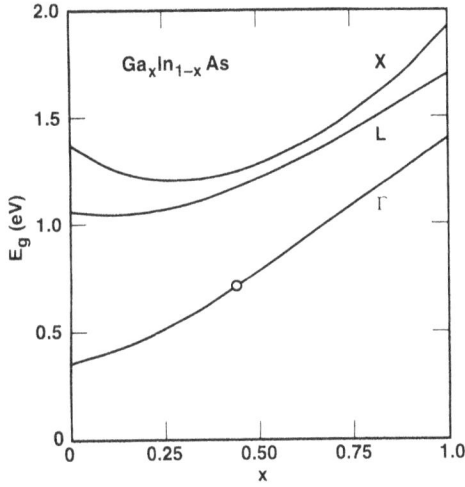

Figure 1.8 Calculated compositional dependence of the Γ, X, and L energy gaps in room temperature in $Ga_xIn_{1-x}As$ [30].

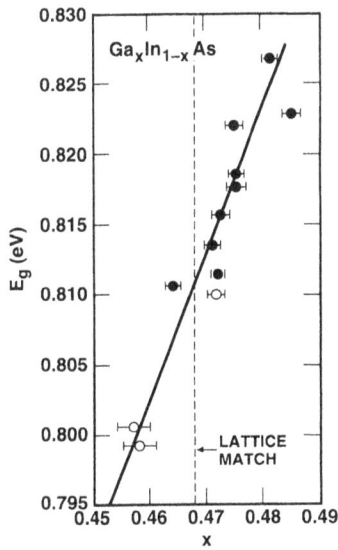

Figure 1.9 Energy gap at 2K versus composition for $Ga_xIn_{1-x}As$ near the lattice-matched compositions on InP substrates. Full circles: LPE layers, open circles: VPE layers. Solid line is given by $E_g = 0.4105 + 0.6337x + 0.475\, x^2$ [31].

to InP), Equation (1.15) gives $E_g(2K) = 0.811$ eV. Recent measurements [8,33] have given a value for $E_g(0) = 0.803$ eV. Using this value and the $E_g(0)$ values of GaAs and InAs gives a bowing parameter of 0.523 eV. Away from the lattice-matched composition, considerable shift in E_g from the calculated value can occur caused by built-in heteroepitaxial elastic strain [5–8] (see Sec. 1.7.4).

1.7.2.2 $Al_xIn_{1-x}P$

The compositional dependence of the Γ and X energy gaps is given in Table 1.7. The direct (Γ) to indirect (X) energy gap crossover occurs at $x = 0.44$.

1.7.2.3 $Al_xIn_{1-x}As$

The variation of E_g versus composition for $Al_xIn_{1-x}As$ is given in Figure 1.10. The bandgap changes at $x = 0.68$ from a direct gap (Γ) in InAs to an indirect gap (X) in AlAs. The compositional dependence of Γ is represented by Equation (1.1) with the bowing parameter ~ 0.24 eV. Near compositions ($x \sim 0.45$–0.54) lattice matched to InP ($x \sim 0.477$), the bandgap (not corrected for lattice-mismatch strain) is expressed by a linear relation [34]:

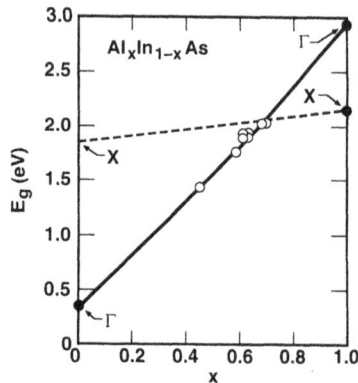

Figure 1.10 Γ and X energy gaps for $Al_xIn_{1-x}As$.

$$E_g(300\text{K}) = 0.357 + 2.29x \qquad (1.16)$$

$$E_g(4\text{K}) = 0.447 + 2.22x$$

Equation (16) gives $E_g(300\text{K}) = 1.45$ eV.

1.7.2.4 Quaternary Alloys

For determining the bandgap of the quaternary alloy, Moon *et al.* [35] have used the following equation:

$$E_g(A_xB_{1-x}C_yD_{1-y}) = xE_{ACD} + (1 - x)E_{BCD} - \Delta \qquad (1.17)$$

where Δ is the bowing parameter term for the quaternary alloy and is given by

$$\Delta = x(1 - x)[(1 - y)c_{ABD} + yc_{ABC}]$$
$$+ y(1 - y)[xc_{ACD} + (1 - x)c_{BCD}] \qquad (1.18)$$

where c_{ABD} and so on are the ternary bowing parameters of Equation (1.1). For $x = y = 0.5$, Δ is one-eighth of the sum of the four ternary bowing parameters, which is twice as large as the quaternary bowing parameter given by Equation (1.8). A comparison of the two interpolation schemes given by Equations (1.8) and (1.17) with measured values of the bandgap in $Ga_xIn_{1-x}As_yP_{1-y}$ alloys showed that both schemes give comparable errors [21]. However, for calculation of the bandgap, Equation (1.17), which is the quaternary version of Equation (1.1), is preferred since it has some theoretical basis. Using the bandgaps and lattice constants of the binary compounds, Glisson *et al.* [21] have derived the lowest energy bandgap and lattice constant of several quaternary alloys of the III-V semiconductors at room temperature. For calculating the bandgap, they used Equation (1.17). The unknown bowing parameters of the ternary semiconductors were obtained by extrapolating from the available experimental data [9]. When the experimental c values were plotted against the difference in the lattice constants of the two end point binary compounds, a definite trend toward larger c with larger lattice constant difference was noted. Glisson *et al.* [21] obtained the quaternary lattice constants from the binary values using Equation (1.8). The lattice constants of the ternaries are assumed to follow the linear relation given by Equation (1.6). The energy bandgap and lattice constant contours of several InP quaternary alloys as derived by Glisson *et al.* are reproduced in Figures 1.11 and 1.12.

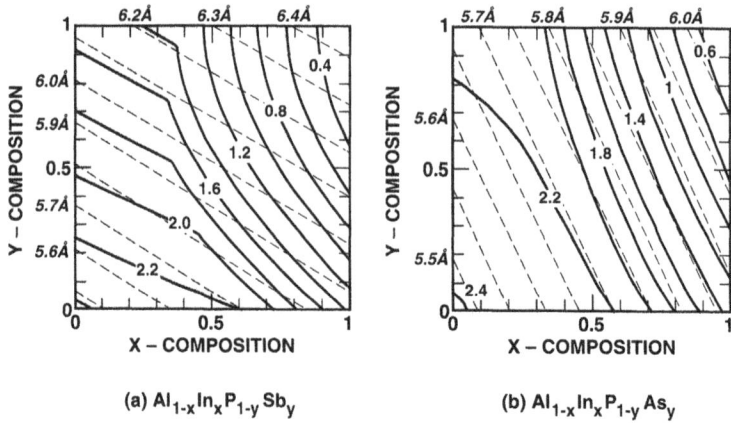

Figure 1.11 Energy bandgap and lattice constant contours at 300K for (a) $Al_{1-x}In_xP_{1-y}Sb_y$ and (b) $Al_{1-x}In_xP_{1-y}As_y$ alloys. The solid curves are the energy gap contours, and the dashed curves are lattice constant contours. The shaded region shows the compositional range over which the material has an indirect bandgap [21].

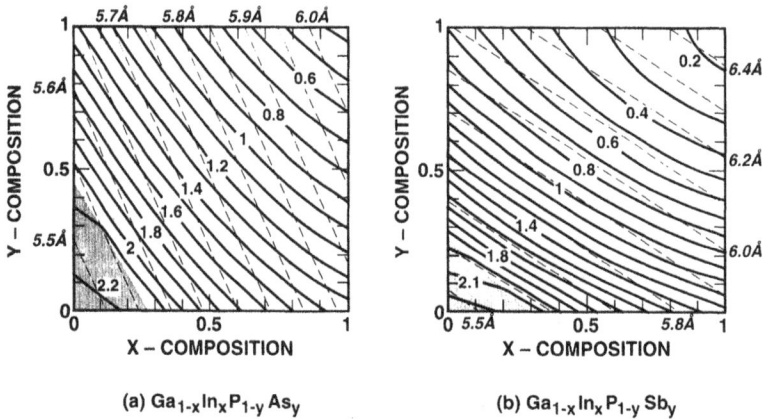

Figure 1.12 Energy bandgap and lattice constant contours at 300K for (a) $Ga_{1-x}In_xAs_yP_{1-y}$ and (b) $Ga_{1-x}In_xP_{1-y}Sb_y$ alloys. The solid curves are the energy gap contours, and the dashed curves are lattice constant contours. The shaded region shows the compositional range over which the material has an indirect bandgap [21].

1.7.2.5 $Ga_xIn_{1-x}As_yP_{1-y}$ on InP Substrate

GaInAsP lattice matched to InP has a direct gap over the entire range of alloy compositions. By fitting photoluminescence and electroreflectance data at room temperature to Equation (1.1), the following expression for E_g has been obtained [36]:

$$E_g(y) = 1.35 - 0.775y + 0.149y^2 \qquad (1.19)$$

Pearsall [36] calculated the bowing parameter for the lattice-matched compositions, for which the only contribution to bowing is the extrinsic bowing caused by alloy disorder since the intrinsic bowing parameter is, in principle, zero. The disorder bowing parameter calculated according to the prescription in Van Vechton and Bergstresser [9] is given by the following expression:

$$c(y) = 0.219y - 0.149y^2 \qquad (1.20)$$

Equation (1.20) gives the bowing parameter for all transitions involving the valence and conduction bands. Figure 1.13 shows the experimental and calculated $E_g(y)$.

For the general case of arbitrary x and y, Kuphal [37] has given the following expression for E_g:

$$E_g(x,y) = 1.35 + 0.668x - 1.068y + 0.758x^2 \qquad (1.21)$$
$$+ 0.078y^2 - 0.069xy - 0.322x^2y + 0.03xy^2$$

This general expression for E_g closely reproduces Equation (1.19) for lattice-matched alloy. The composition parameters x and y can be obtained from the E_g measurement (e.g., photoluminescence) and the lattice mismatch $\Delta a/a$ measurement (e.g., X-ray rocking curve) using Equations (1.21) and (1.9).

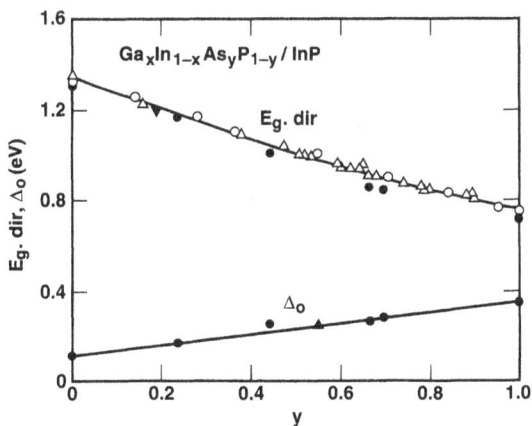

Figure 1.13 Direct energy gap $E_{g\mathrm{dir}}$ and spin orbit splitting Δ_0 *versus* composition in $Ga_xIn_{1-x}As_yP_{1-y}/$ InP at room temperature. The data points represent several experimental sources, and the solid lines are calculated [36].

Figure 1.13 also shows the compositional dependence of the spin-orbit splitting $\Delta(y)$ energy. The curve shows slight upward bowing, and the bowing itself is very small [36]. For $Ga_{0.47}In_{0.53}As$, $\Delta(y)$ is 0.35 eV compared to 0.11 eV for InP.

1.7.2.6 $Al_xGa_yIn_{1-x-y}As$ on InP Substrate

This quaternary alloy can be considered a pseudobinary of $Al_{0.48}In_{0.52}As$ and $Ga_{0.47}In_{0.53}As$. For lattice matching to InP, $x + y = 0.47$. An expression for E_g similar to Equation (1.21) can be obtained and is given by

$$E_g(x,y) = 0.36 + 2.093x + 0.629y + 0.577x^2 \qquad (1.22)$$
$$+ 0.436y^2 + 1.013xy - 2.0xy(1 - x - y)$$

For the lattice matched alloy,

$$E_g(z) = 0.76 + 0.49z + 0.20z^2 \qquad (1.23)$$

where $z = x/0.48$ [38].

1.7.3 Effective Mass

Table 1.8 lists the effective masses at conduction and valence band edges. Semiconductors that have conduction band minimum at $\Gamma = 0$ are characterized by one mass only. Those having minimum near the X point are characterized by longitudinal and transverse effective masses, m_l and m_t. Because of lack of inversion symmetry, the degenerate bands at the X point are split into a higher lying X_3 band and a lower lying X_1 band. As a result, these semiconductors also exhibit what is known as a *camel's back* structure in the conduction band [39]. For the indirect bandgap semiconductors, the density of states effective mass is given by $\nu^{2/3} (m_l m_t^2)^{1/3}$ where ν is the number of equivalent conduction band minima. When the minimum occurs at the X-point $\nu = 3$. The valence band is characterized by three effective masses corresponding to the heavy hole, light hole, and spin-orbit bands. The density of states effective mass for holes is given by $(m_{hh}^{3/2} + m_{lh}^{3/2})^{2/3}$.

The conduction band effective mass is also a function of doping. When the doping level increases, the bottom of the conduction band becomes filled and the electron gas becomes degenerate. Since the nonparabolicity of the band becomes significant as the Fermi level moves up in the band, the electron mass increases with doping level for carrier concentrations larger than 10^{18} cm^{-3}. Figure 1.14 shows the dependence of electron effective mass on carrier density for InP [40].

Table 1.8
Effective Mass Parameters

Compound	Conduction Band (units of m_o)			Valence Band (units of m_o)			
	m_l	m_t	m_e	m_{hh}	m_{lh}	m_{hso}	m_h^b
AlP				0.63	0.20	0.29	0.70
AlAs	1.5	0.19	0.79[a]	0.76	0.15	0.24	0.80
AlSb	1.64	0.23	0.92[a]	0.94	0.14	0.29	0.98
GaP	7.25	0.313	1.86[c]	0.54	0.16	0.24	
GaAs			0.067[d]	0.49	0.08	0.15	
GaSb			0.044	0.34	0.044	0.13	
InP			0.075[d]	0.56	0.12	0.12	
InAs			0.024	0.37	0.025	0.14	
InSb			0.014	0.39	0.016	0.47	

(a) For three equivalent minima, m_e is given by $3^{2/3} (m_l m_t^2)^{1/3}$
(b) $m_h = (m_{hh}^{3/2} + m_{lh}^{3/2})^{2/3}$
(c) Because of the presence of a camel's back conduction band structure and extremely high nonparabolicity, there is considerable scatter in the effective mass values reported in the literature.
(d) The concentration dependence of the effective mass for InP is shown in Figure 1.14.

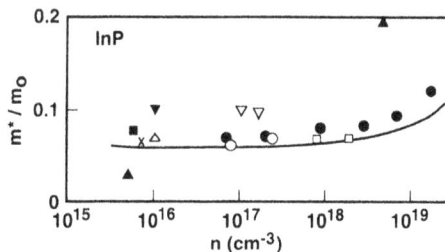

Figure 1.14 The electron effective mass *versus* carrier concentration in InP. The data points represent experimental results of several authors. The solid line is a calculated curve [40].

Effective Mass in Ternary and Quaternary Alloys

According to the $k \cdot p$ approximation [41], the effective mass of the carrier, m^*, is given by

$$\frac{1}{m^*} = \frac{1}{m} + \frac{2P^2}{3\hbar^2}\left(\frac{2}{E_g} + \frac{1}{E_g + \Delta}\right) \qquad (1.24)$$

where P is the matrix element connecting the conduction band and the light hole and spin-orbit split valence bands. It has been found in many ternary alloys that the effective mass calculated under the assumption that P^2 varied linearly between the values for the binary compound does not agree well with the experimental values. The origin of this discrepancy has been attributed to disorder-induced mixing of the conduction and valence bands at the Γ point [42]. The effect of this mixing is to reduce P^2, and thus to increase the effective mass above that which would be obtained by the $k \cdot p$ calculation. Figures 1.15 and 1.16 show the electron effective mass for $Ga_xIn_{1-x}As$ [43] and $Al_xIn_{1-x}As$ [44], respectively. For $Ga_{0.47}In_{0.53}As$, the heavy hole and light hole masses have been determined to be $0.465m_0$ and $0.05m_0$, respectively [45]. Figures 1.17 and 1.18 show the electron effective mass and light hole effective mass for lattice-matched $Ga_xIn_{1-x}As_yP_{1-y}$, respectively [36]. The heavy

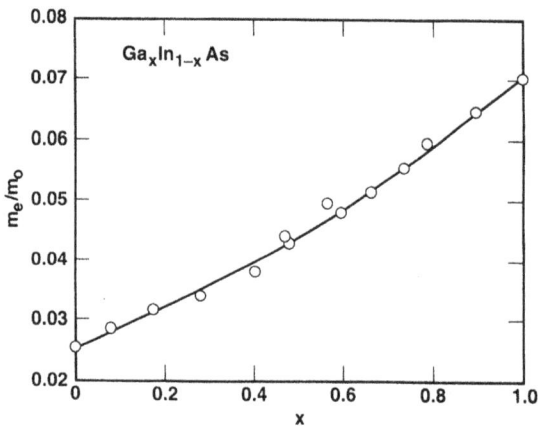

Figure 1.15 Electron effective mass *versus* composition in $Ga_xIn_{1-x}As$ [43].

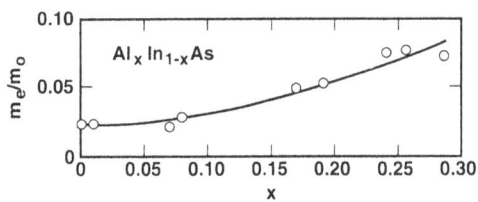

Figure 1.16 Electron effective mass in $Al_x In_{1-x} As$ [44].

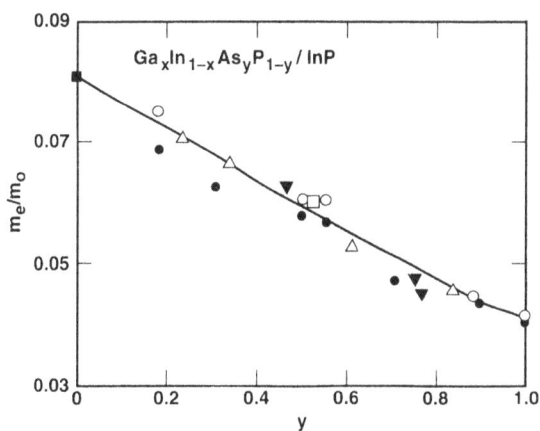

Figure 1.17 Electron effective mass *versus* composition in $Ga_x In_{1-x} As_y P_{1-y}/InP$ [36].

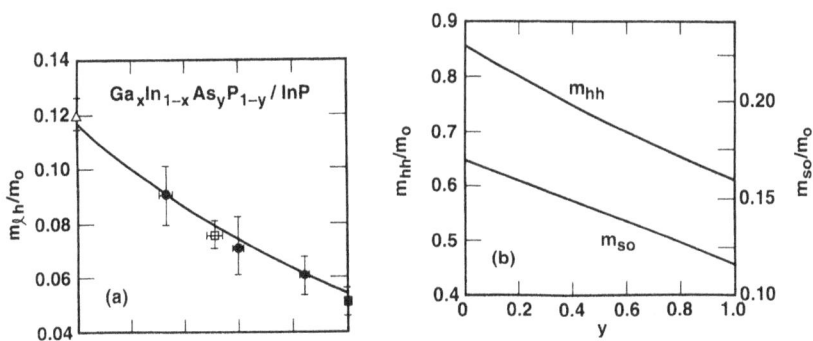

Figure 1.18 (a) Light hole effective mass and (b) heavy hole effective mass and effective mass of the spin-orbit split-off valence band *versus* composition in $Ga_x In_{1-x} Al_y P_{1-y}$ on InP substrates [36].

hole and split-off band masses can be obtained from the values of the binary compounds using Equation (1.7).

1.7.4 Elastic Properties

The elastic compliances S_{ij} in units of 10^{-12} cm^2 dyne^{-1} are listed in Table 1.9. For a cubic crystal, there are three independent elastic constants: S_{11}, S_{12}, and S_{44}. The elastic compliance relates strain to stress:

$$\epsilon_i = \sum_j S_{ij}\, \sigma_j \qquad (1.25)$$

where the strain ε_i and stress σ_j are the components of second-rank tensors and S_{ijs} are components of a fourth-rank tensor. Equation (1.25) can be expressed in its covariant form:

$$\sigma_i = \sum_j C_{ij}\, \epsilon_j \qquad (1.26)$$

where C_{ijs} are the elastic stiffness constants. The compliance tensor is the reciprocal of the stiffness tensor. For a cubic crystal, C_{ijs} are related to S_{ijs} according to [46]

Table 1.9
Elastic Compliances in Units of 10^{-12} cm^2 dyne^{-1}

Compound	S_{11}	S_{12}	S_{44}
AlP	1.090	−0.350	1.630
AlAs	1.070	−0.320	1.840
AlSb	1.696	−0.562	2.453
GaP	0.973	−0.298	1.419
GaAs	1.176	−0.365	1.684
GaSb	1.582	−0.495	2.314
InP	1.650	−0.594	2.170
InAs	1.945	−0.685	2.525
InSb	2.443	−0.863	3.311

$$C_{11} = \frac{S_{11} + S_{12}}{S}$$

$$C_{12} = -\frac{S_{12}}{S} \qquad (1.27)$$

$$C_{44} = \frac{1}{S_{44}}$$

where

$$S = (S_{11} - S_{12})(S_{11} + 2S_{12})$$

From the point of practical utility, the elastic constant of interest is Young's modulus E. Another elastic property of interest is Poisson's ratio ν. In the cubic crystals, both E and ν are anisotropic. The expressions for E and ν for an arbitrary crystallographic direction in a cubic crystal are given by Brantley [47]:

$$\frac{1}{E} = S_{11} - 2\left(S_{11} - S_{12} - \frac{1}{2}S_{44}\right)(\ell_1^2\ell_2^2 + \ell_2^2\ell_3^2 + \ell_1^2\ell_3^2)$$

$$\nu = -\frac{S_{12} + \left(S_{11} - S_{12} - \frac{1}{2}S_{44}\right)(\ell_1^2 m_1^2 + \ell_2^2 m_2^2 + \ell_3^2 m_3^2)}{S_{11} - 2\left(S_{11} - S_{12} - \frac{1}{2}S_{44}\right)(\ell_1^2\ell_2^2 + \ell_2^2\ell_3^2 + \ell_1^2\ell_3^2)} \qquad (1.28)$$

where ℓ is the longitudinal stress axis and m is an orthogonal direction to ℓ and the ℓ_s and m_s are the direction cosines for ℓ and m with respect to the cube axes. For the $\langle 100 \rangle$ axes,

$$\frac{1}{E} = S_{11}$$

$$\nu = -\frac{S_{12}}{S_{11}} \qquad (1.29)$$

For the biaxial plane stress conditions associated with thin films, often encountered in the case of heterostructures, longitudinal stresses and strains parallel to the film substrate are related by $E/(1 - \nu)$. The expressions for this composite elastic constant are easily obtained from Equation (1.28).

The bulk modulus for zinc blende type crystals is given by

$$B = \frac{1}{3(S_{11} + 2S_{12})} \tag{1.30}$$

For $\langle 100 \rangle$ axes, $B = E/3 \, (1 - 2\nu)$. Knowing the density ρ and C_{ij}, one can obtain the sound velocity v using the relation

$$v = (C_{ij}/\rho)^{1/2} \tag{1.31}$$

Deformation Potential

Application of an external stress has a profound effect on the band structure. Hydrostatic stress causes a shift of the energy states. Uniaxial stress lowers the symmetry of the lattice and splits some degenerate states. In cubic semiconductors, the fourfold degeneracy of the valence band is lifted, resulting in a $J = 3/2$, $m_j = 3/2$ and $J = 3/2$, $m_j = 1/2$ heavy hole and light hole bands [4]. The shift of E_g with hydrostatic stress and the splitting of the valence band δE_v are given by the deformation potential constants. The values of these constants are listed in Table 1.10. Variation of direct energy gap with hydrostatic pressure p is described by constant a:

$$a = -\frac{1}{3}(C_{11} + 2C_{12})\frac{dE_g}{dp} \tag{1.32}$$

The splitting of the valence band under [100] stress is described by the deformation potential constant b and the splitting under [111] stress by constant d, according to the relation

$$\delta E_v = b(S_{11} - S_{12})\sigma$$

$$\delta E_v = \left(\frac{d}{2\sqrt{3}}\right)S_{44}\sigma \tag{1.33}$$

where σ is the applied uniaxial stress.

Table 1.10
Deformation Potential Constants

	a *(eV)*			
Compound	*Direct Gap*	*Indirect Gap*	b (eV)	d (eV)
AlP				
AlAs				
AlSb	−5.9	2.2	−1.35	−4.3
GaP	−9.6		−1.65	−4.5
GaAs	−9.77		−1.70	−4.55
GaSb	−8.28		−2.0	−4.7
InP	−6.35		−2.0	−5.0
InAs	−6.0		−1.8	−3.6
InSb	−7.7		−2.0	−4.9

Heteroepitaxial Strain and Band Gap

One of the aims in growing device quality heterostructures is to reduce as much as possible the lattice mismatch between the epitaxial layer and the substrate. Because the layers are usually much thinner than the substrate, the strain caused by lattice mismatch can be accommodated elastically or plastically by the generation of dislocations when the thickness of the layers exceed a certain critical thickness [48]. Both types of strain have a profound influence on device characteristics. Misfit dislocations may play a crucial role in determining the recombination velocity of minority carriers at heterointerfaces [49]. Elastic strain alters the bandgap of the semiconductor, and, as a result, the electro-optical properties of the devices can be affected [50].

In the elastic regime, for coherent growth (in-plane lattice parameter of the epitaxial layer equals that of the substrate) the lattice mismatch is accommodated by the tetragonal distortion of the layer [25]. For this situation, the lattice parameter of the unstrained epitaxial layer is given by Equation (1.11). When $(\text{DELTA}a/a)_\perp$ is positive (the layer has a larger lattice parameter than the substrate), there is in-plane compressive strain in the layer and tensile strain normal to the interface. When $(\text{DELTA}a/a)_\perp$ is negative, there is in-plane tension and compression normal to the layer. The hydrostatic component of the biaxial strain shifts the bandgap while the shear component lifts the degeneracy of the valence band as sketched in Figure 1.19. For biaxial compression, the valence band closer to the conduction band is the heavy

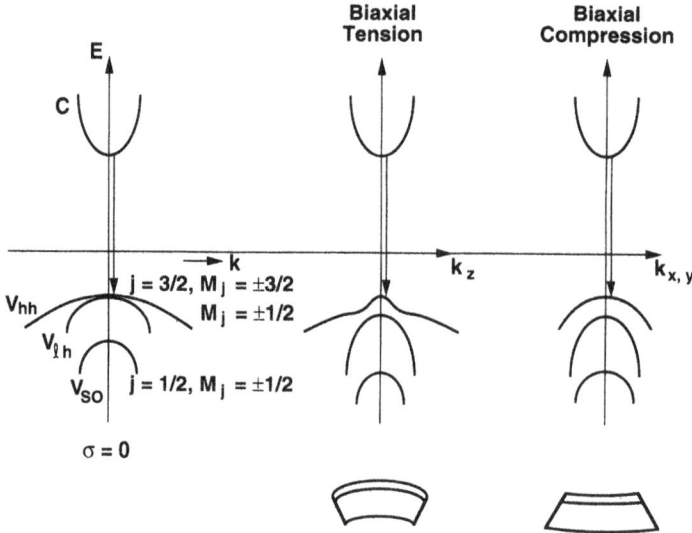

Figure 1.19 Valence band structure of cubic semiconductors under biaxial strain. The fourfold degenerate valence band is split into two twofold degenerate bands. The valence band closest to the conduction band is different depending on the nature of the strain. Wafer bowing resulting from biaxial strain is also shown.

hole band, whereas the light hole band is closer for biaxial tension. A similar situation is realized for externally applied uniaxial strain of tension and compression. The wafer bowing resulting from the biaxial strain is also shown in Figure 1.19.

Asai and Oe [5] derived the transition energies to the $J = 3/2$, $m_j = 3/2$ and $J = 3/2$, $m_j = 1/2$ valence bands using the stress-dependent Hamiltonian [4], and they are as follows:

$$E_{3/2,3/2} = \left(-2a\frac{C_{11} - C_{12}}{C_{11}} + b\frac{C_{11} + 2C_{12}}{C_{11}}\right)\varepsilon \qquad (1.34)$$

$$E_{3/2,1/2} = \left(-2a\frac{C_{11} - C_{12}}{C_{11}} - b\frac{C_{11} + 2C_{12}}{C_{11}}\right)\varepsilon \qquad (1.35)$$

where ε is the mismatch strain, taken to be positive for compression.

From Equations (1.34) and (1.35) it follows that for in-plane compressive strain (positive mismatch), the lowest energy optical transition involves the $m_j = 3/2$ valence band, and for in-plane tensile strain (negative mismatch), the $m_j = 1/2$ valence band. Figure 1.20 shows the band gap, derived from photoluminescence measurements, of $Ga_xIn_{1-x}As$ grown on InP for lattice mismatch varying from -0.2% to

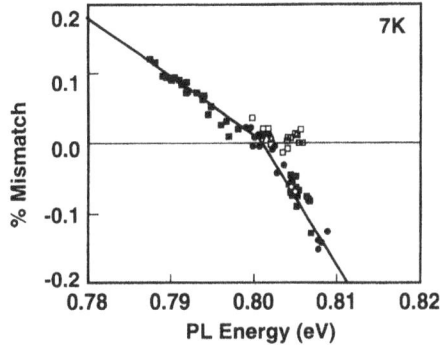

Figure 1.20 Photoluminescence peak energy at 7K of $Ga_xIn_{1-x}As/InP$ as a function of lattice mismatch. The effect of biaxial strain on peak energy is explained on the basis of Equations (1.34) and (1.35) [8].

+0.2% [8]. The data clearly illustrates the strain effect on bandgap. There is good agreement between the experimental points and the solid lines derived from Equations (1.34) and (1.35) using the extrapolated values of elastic and deformation potential constants for $Ga_{0.47}In_{0.53}As$ from the binary values.

1.7.5 Optical Properties

Table 1.11 lists the optical properties-refractive index near the bandgap wavelength, is temperature dependence, the static and high-frequency dielectric constants, and the frequencies of the zone center LO and TO phonons. The optical absorption of solids can be described either by the complex index of refraction $n + ik$ or by the complex dielectric function $\in + i\in_2$, where n, k, $\in_1$, and $\in_2$ are all real functions of the frequency ω.

$$\in_1 = n^2 - k^2, \qquad \in_2 = 2nk \qquad (1.36)$$

The functions $\in_1(\omega)$ and $\in_2(\omega)$ are connected by the Kramers-Kroenig relations. The absorption coefficient α is related to k by

$$\alpha = \frac{2\omega k}{c} \qquad (1.37)$$

where k is the imaginary part of the refractive index and c is the extinction coefficient. One quantity that is measured in experiments is reflectance. For normal incidence it is given by

Table 1.11
Optical Properties

Compound	Refractive Index n Near E_g	$\dfrac{1}{n}\dfrac{dn}{dT} \sim (K^{-1})$	Dielectric Constant[a]		Wave Number of Raman Phonon (cm^{-1})	
			$\in(0)$	$\in(\infty)$	LO	TO
AlP	3.03	3.5×10^{-5}	9.8	7.54		
AlAs	3.18	4.6×10^{-5}	10.06	8.16	404.1	360.9
AlSb	3.4	3.5×10^{-5b}	12.04	10.24	339.6	318.8
GaP	3.45	2.5×10^{-5b}	11.1	9.08	403.0	367.3
GaAs	3.65	4.5×10^{-5}	12.91	10.9	291.9	268.6
GaSb	3.82	8.2×10^{-5}	15.69	14.44	233.0	224.0
InP	3.41	2.7×10^{-5}	12.61	9.61	345.0	303.7
InAs	3.52	6.5×10^{-5b}	15.15	12.25	238.6	217.3
InSb	4.00	1.2×10^{-4}	17.7	15.68	190.8	179.8

(a) In units of $\in_0$, the permittivity of free space 8.85×10^{-14} F cm^{-1}.
(b) D.K. Ghosh, L.K. Samanta, and G.C. Bhar, *Infrared Phys.*, Vol. 26, 1986, p. 111.

$$R = \frac{(n-1)^2 + k^2}{(n+1)^2 + k^2} \tag{1.38}$$

When $k = 0$, that is, in the transparent range, then

$$R = \frac{(n-1)^2}{(n+1)^2} \tag{1.39}$$

In the limit $\omega \to 0$, $\in(0)$, the static dielectric constant, is given by the relation [2]

$$\in(0) = 1 + \left(\frac{\hbar\omega_p}{E_h}\right)^2 A \tag{1.40}$$

where A is a number close to unity, ω_p is the free electron plasma frequency, and E_h is the homopolar energy gap. The free electron plasma frequency is given by

$$\omega_p^2 = \frac{4\pi n e^2}{m} \tag{1.41}$$

where m is the free electron mass, and $n = 8$ electrons per diatomic unit cell volume. The *static* and *high* frequency dielectric constants are related by

$$\in(0) = \in(\infty) + \frac{4\pi N e_T^2}{\omega_{TO}^2 M} \tag{1.42}$$

where M is the reduced mass, e_T the effective charge, ω_{TO} the TO phonon frequency, and N the number of ion pairs. Note from Table 1.11 that both $\in(0)$ and $\in(\infty)$ show an inverse dependence on E_g, decreasing with increasing E_g.

In the study of transport and optical properties of polar semiconductors, the coupling between the electron and LO phonons becomes important. The electron-LO phonon interaction is given by the well-known Frohlich coupling constant [51]:

$$\alpha_F = \frac{e^2}{4\pi\in_0} \left(\frac{m^*}{2\hbar^3 \omega_{LO}}\right)^{1/2} \left[\frac{1}{\in(\infty)} - \frac{1}{\in(0)}\right] \tag{1.43}$$

where ω_{LO} is the LO phonon frequency. Equation (43) shows that α_F depends on the ionic polarizability of the crystal related to $\in(\infty)$ and $\in(0)$.

Values of k and n over a wide range of photon energies for many III-V semiconductors have been compiled by Seraphin and Bennett [52].

Phonons in Ternary and Quaternary Alloys

In ternary or quaternary alloys, one can expect phonon frequencies due to the different binary compounds. Such multiple-mode behavior has been observed in the Raman scattering and infrared reflectivity measurements in GaInAs [53,54] and GaInAsP [55,56] alloys. Figure 1.21 shows the phonon frequencies in lattice-matched $Ga_x In_{1-x} As_y P_{1-y}$ as a function of y [56]. The low-frequency (220–260 cm^{-1}) phonons are assigned to GaInAs modes. At $y = 1$, they are separated into InAs-like and GaAs-like vibrations. The high-frequency (300–360 cm^{-1}) band appears for all compositions except for $y = 1$, and they are attributed to InP and GaP vibrations.

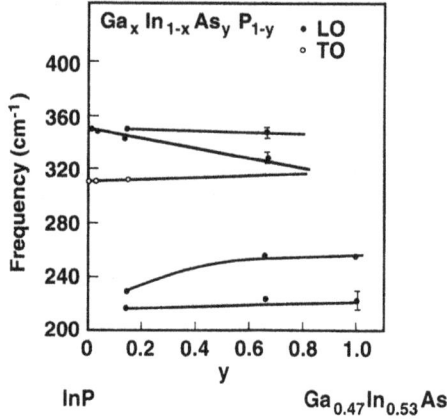

Figure 1.21 Measured phonon frequencies of various Raman modes in $Ga_xIn_{1-x}As_yP_{1-y}$ as a function of y. The solid lines emphasize the experimental trends [56].

Refractive Index

One of the important factors in the design of heterostructure lasers and optoelectronic devices is the refractive index, n. Often, knowledge of n as a function of photon energy near and below the bandgap is required. Several models have been proposed to obtain n as a function of photon energy and composition in ternary and quaternary alloys. In many models the starting point for calculating n in alloys is the semiempirical, single-effective oscillator model proposed by Wemple and DiDomenico [57] to analyze refractive index dispersion in many covalent, ionic, and amorphous materials. According to this model, n is given at photon energy E:

$$n^2 - 1 = E_dE_0/(E_0^2 - E^2) \tag{1.44}$$

where E_0 is the single oscillator energy and E_d is the dispersion energy, a measure of the strength of interband optical transitions. The single oscillator parameters E_0 and E_d of the binaries can be interpolated according to Equation (1.6) or (1.7) to yield n of ternary or quaternary alloys. However, description of n near the band edge, the region of most interest, using this model appears to be inadequate even for binary compounds. Afromowitz [58] proposed a modified single oscillator model to explain dispersion of n near the bandgap. In this model, n^2 is expressed as

$$n^2 - 1 = \frac{E_d}{E_0} + \frac{E_dE^2}{E_0^3} + \frac{\eta E^4}{\pi} \ln\left(\frac{2E_0^2 - E_g^2 - E^2}{E_g^2 - E^2}\right) \tag{1.45}$$

where

$$\eta = \frac{\pi E_d}{2E_0^3(E_0^2 - E_g^2)} \tag{1.46}$$

The index describing the group velocity, $\bar{n}$, which is related to the longitudinal mode spacing in lasers, is given by

$$\bar{n} = n + E\,dn/dE \tag{1.47}$$

and can be obtained from Equation (1.45).

Buus and Adams [59] used Equation (1.45) to calculate n as a function of E and y in $Ga_x In_{1-x} As_y P_{1-y}$ lattice matched to InP. The values of E_0 and E_d were obtained from their respective values in the binary compounds using Equation (1.7), and are given by

$$E_0(y) = 3.391 - 1.652y + 0.863y^2$$
$$E_d(y) = 28.91 - 9.278y + 5.626y^2 \tag{1.48}$$

Equation (1.48), together with Equation (1.19) for $E_g(y)$, can be substituted in Equations (1.45) and (1.46) to obtain n as a function of E and y.

Burkhard et al. [60] have measured n for $Ga_x In_{1-x} As_y P_{1-y}$ lattice matched to InP as a function of y and E for $E > E_g$. They fitted their experimental results using the interpolation given by Equation (1.7) and using the quantity $(\in - 1)/(\in + 2)$ as the parameter. Then n was obtained taking $n^2 \approx \in$.

Adachi [6] proposed a generalized model of dielectric constants of semiconductors based on models of the intraband transitions. According to this model, the real part of the dielectric constant $\in$ is expressed as a function of photon energy E as

$$\in(E) = A_0 \left\{ f(x) + \frac{1}{2} \left(\frac{E_g}{E_g + \Delta_0} \right)^{3/2} f(x_0) \right\} + B_0 \tag{1.49}$$

where A_0 and B_0 are constants obtained by fitting experimental data to Equation (1.49), and

$$f(x) = x^{-2}(2 - (1 + x)^{1/2} - (1 - x)^{1/2})$$
$$x = E/E_g \tag{1.50}$$
$$x_0 = \frac{E}{E_g + \Delta_0}$$

Figure 1.22 shows the experimental values of $\in (\approx n^2)$ for $Ga_x In_{1-x} As_y P_{1-y}$ as a function of wavelength [62] for several compositions. The dispersion curves calculated based on Equations (1.49) and (1.50) are also shown in Figure 1.22. The agreement with experimental curves is quite good over most of the wavelength range. Bacher et al. [63] used the data from Burkhard et al. [60] and Chandra et al. [62] to obtain the spectral dependence of refractive index of $Ga_{0.47}In_{0.53}As/InP$ in the energy range 0.5–1.5 eV, as shown in Figure 1.23. The dashed curve in Figure 1.23 represents an estimate by the authors.

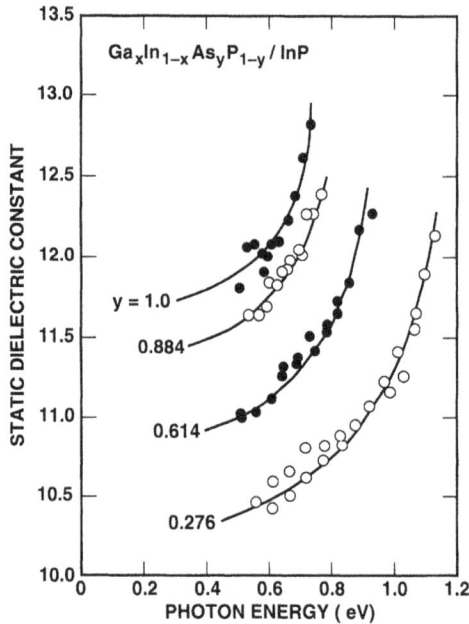

Figure 1.22 Dielectric constant *versus* photon energy for $Ga_x In_{1-x} As_y P_{1-y}/InP$ at various compositions [51].

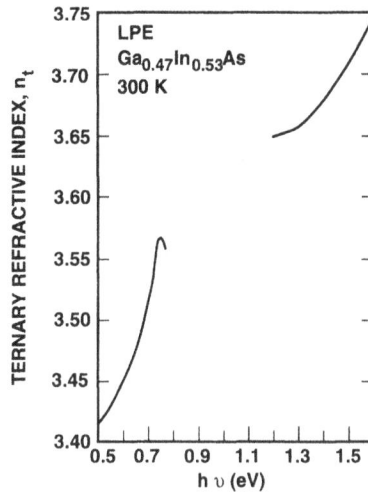

Figure 1.23 Refractive index for $Ga_{0.47}In_{0.53}As$ *versus* photon energy at 300K as reported in Chandra *et al.* [62] for $h\nu < 0.76$ eV and in Burkhard *et al.* [60] for $h\nu > 1.2$ eV. The dashed curve is an estimate of intermediate energies [63].

1.7.6 Thermal Resistivity

Thermal resistivity of semiconductors is an important parameter in the design of power-dissipating devices such as semiconductor lasers. It is also important in calculating the figure of merit of thermoelectric devices (e.g., Peltier devices). The thermal resistivity values near room temperature for the binary III-V compounds are given in Table 1.12. For the ternary and quaternary alloys, the thermal resistivity follows a quadratic relation the same as E_g. For the ternary alloy $A_xB_{1-x}C$, it can be written as

$$W(x) = xW_{AC} + (1 - x)W_{BC} + c_{A-B}x(1 - x) \tag{1.51}$$

where W_{AC} and W_{BC} are the thermal resistivities of the binary compounds. Adachi [64] has obtained the bowing parameter c_{A-B} for several ternary alloys, and each is given in Table 1.13. Note that c_{As-P} values of InAsP and GaAsP alloys are considerably smaller than c_{In-Ga} of InGaAs, and that $c_{Ga-Al} \approx c_{As-P}$. The mean atomic weight of In-Ga is about two times larger than that of As-P, which is nearly the same as that of Al-Ga. The anharmonic contribution to thermal resistivity resulting from mass difference between the atoms is supposed to be responsible for the differences and similarities in the c values.

Table 1.12

Coefficient of Thermal Expansion and Thermal Conductivity of III-V Compounds
Near Room Temperature

Compound	Coefficient of Thermal Expansion $(10^{-6}/°C)$	Thermal[a] Conductivity $(W\ cm^{-1}\ K^{-1})$
AlP	4.5	0.9
AlAs	4.9	0.8
AlSb	4.0	0.57
GaP	4.5	0.77
GaAs	6.86	0.46
GaSb	7.75	0.39
InP	4.75	0.68
InAs	4.52	0.273
InSb	5.37	0.166

(a) M.G. Holland, *Semiconductors and Semimetals,* Vol. 2, ed. R.K. Willardson and A.C. Beer, New York: Academic Press, 1966, p. 3.

Table 1.13

The Alloy Disorder Bowing Parameter c_{A-B} Denoting the Deviation from Linearity of the Thermal Resistivity of Ternary Alloys. The Bowing Parameter is Obtained by Fitting Experimental Data to Equation (1.51) [64]

Ternary Alloy	C_{A-B} $(W^{-1}\ deg\ cm)$
InGaAs	72
InAsP	30
GaAsP	20
AlGaAs	30
InGaP	72[est]
InGaSb	72[est]

The thermal resistivity of a quaternary alloy $A_xB_{1-x}C_yD_{1-y}$ can be expressed in the same way as Equation (1.51) but with two bowing parameters, c_{A-B} and c_{C-D}. The thermal resistivity of $Ga_xIn_{1-x}As_yP_{1-y}$ lattice matched to InP calculated using $c_{In-Ga} = 72$ W^{-1} K cm and $c_{As-P} \simeq 25$ W^{-1} K cm is shown in Figure 1.24 [64]. The thermal resistivity increases with increasing As content and reaches a maximum value of 24 W^{-1} K cm at $y \sim 0.75$.

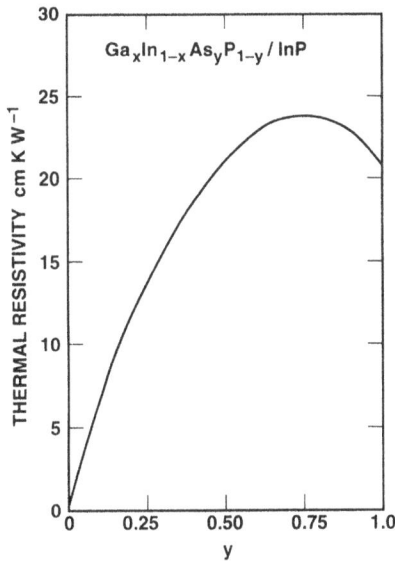

Figure 1.24 Thermal resistivity *versus* composition calculated for $Ga_xIn_{1-x}As_yP_{1-y}$/InP [64].

1.7.7 Thermal Expansion Coefficient

In double heterostructures consisting of compositionally different layers, differences in the thermal expansion coefficient of the layers can generate elastic stresses during cooling from the growth temperature to room temperature or during thermal cycling of heterostructure devices. The values of the thermal expansion coefficients near 300K for the binary compounds are given in Table 1.12. For InP, the temperature dependence of the expansion coefficient below [65] and above [66] room temperature has been measured and the results are shown in Figure 1.25. In Figure 1.26, the thermal expansion coefficient of $Ga_xIn_{1-x}As_yP_{1-y}$ alloy lattice matched to InP or GaAs obtained from linear interpolation is shown together with experimental data at selected y values [67]. Pietsch and Marlow [68] found that the thermal expansion coefficient of the relaxed lattice is about 30% smaller than the values shown in Figure 1.26.

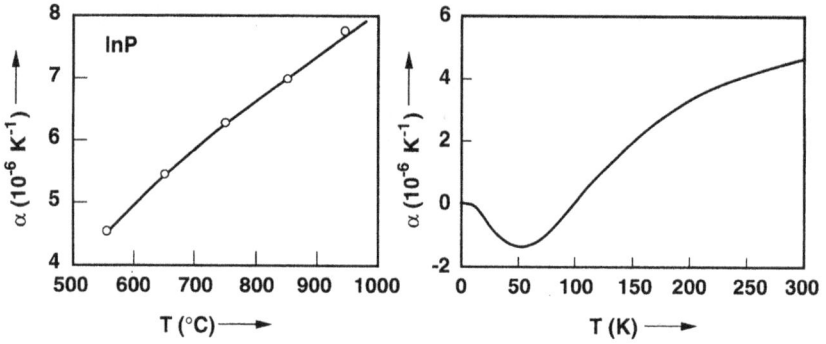

Figure 1.25 Coefficient of thermal expansion of InP above (a) and below (b) room temperature [65, 66].

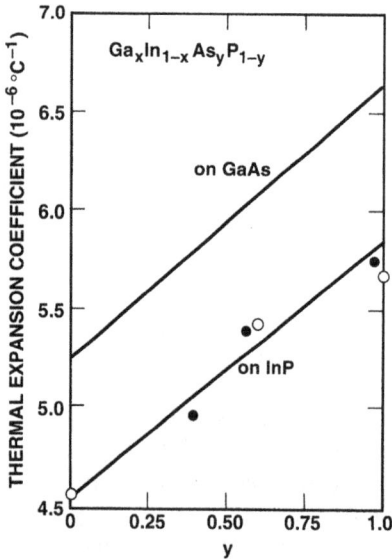

Figure 1.26 Thermal expansion coefficient of $Ga_xIn_{1-x}Al_yP_{1-y}$ lattice matched to InP, and GaAs as a function of composition [67].

1.8 SUMMARY

In this chapter we briefly reviewed the bonding, crystal structure, and band structure of InP and related alloys. We summarized the various physical properties of InP and other III-V compounds, and outlined the interpolation procedures for obtaining the values of the physical properties of alloys from the corresponding binary values when experimental data are not available. We also discussed the properties of GaInAs, AlInAs, GaInAsP, and AlGaInAs alloys when lattice matched to InP.

ACKNOWLEDGMENTS

I acknowledge the support and encouragement of A.C. Adams, W.A. Gault, and P.J. Anthony during the preparation of this manuscript.

REFERENCES

1. V. Swaminathan and A.T. Macrander, *Materials Aspects of GaAs and InP Based Structures*, New Jersey: Prentice Hall, 1991.
2. J.C. Phillips, *Bonds and Bands in Semiconductors*, New York: Academic Press, 1973.
3. M.L. Cohen and J.R. Chelikowsky, *Electronic Structure and Optical Properties of Semiconductors*, Berlin: Springer-Verlag, 1988.
4. G.L. Bir and G.E. Pikus, *Symmetry and Strain Induced Effects in Semiconductors*, New York: John Wiley & Sons, 1974.
5. H. Asai and K. Oe, *J. Appl. Phys.*, Vol. 54, 1983, p. 2052.
6. C.P. Kuo, S.K. Vong, R.M. Cohen, and G.B. Stringfellow, *J. Appl. Phys.*, Vol. 57, 1985, p. 5428.
7. A.T. Macrander and V. Swaminathan, *J. Electrochem. Soc.*, Vol. 134, 1987, p. 1247.
8. I.C. Bassignana, C.J. Miner, and N. Puetz, *J. Appl. Phys.*, Vol. 65, 1989, p. 4299.
9. J.A. Van Vechten and T.K. Bergstresser, *Phys. Rev. B* Vol. 1, 1970, p. 3351.
10. J.C. Phillips and J.A. Van Vechten, *Phys. Rev. Lett.*, Vol. 22, 1969, p. 705.
11. J.A. Van Vechten, *Phys. Rev.*, Vol. 187, 1969, p. 1007.
12. D. Gershoni, H. Temkin, J.M. Vandenberg, S.N.G. Chu, R.A. Hamm, and M.B. Panish, *Phys. Rev. Lett.*, Vol. 60, 1988, p. 448.
13. C. Tejedor and F. Flores, *J. Phys.*, Vol. C11, 1975, p. L19.
14. R.L. Anderson, *Solid State Electron.*, Vol. 5, 1962, p. 341.
15. W.A. Harrison, *J. Vac. Sci. Technol.*, Vol. 14, 1977, p. 1017.
16. J. Tersoff, *Phys. Rev. Lett.*, Vol. 52, 1984, p.465; *Phys. Rev. B*, Vol. 32, 1985, p. 3968.
17. Y.C. Ruan and W.Y. Ching, *J. Appl. Phys.*, Vol. 62, 1987, p. 2885.
18. M. Hybertsen, *Phys. Rev. Lett.*, Vol. 64, 1990, p. 555.
19. M. Hybertsen, *J. Vac. Sci. Technol. B*, Vol. 8, 1990, p. 773.
20. Landolt-Bornstein, "Numerical Data and Functional Relationships" in *Science and Technology*, Vol. 22: Semiconductors subvol. a, ed. O. Madelung and M. Schulz, New York: Springer, 1987.
21. T.H. Glisson, J.R. Hauser, M.A. Littlejohn, and C.K. Williams, *J. Electron. Mat.*, to be published.
22. Landolt-Bornstein, "Numerical Data and Functional Relationships," in *Science and Technology*, Vol. 17: Semiconductors subvol. a, ed. O. Madelung, M. Schulz, and H. Weiss, New York: Springer, 1982.

23. Landolt-Bornstein, "Numerical Data and Functional Relationships" in *Science and Technology,* Vol. 17: Semiconductors subvol. d, ed. O. Madelung, M. Schulz, and H. Weiss, New York: Springer, 1984.

24. E.F. Hocking, I. Kudman, T.E. Seidel, C.M. Schmalz, and E.F. Steigmeier, *J. Appl. Physics,* Vol. 37, 1966, p. 879.

25. J. Hornstra and W.J. Bartels, *J. Cryst. Growth,* Vol. 44, 1978, p. 513.

26. Y.P. Varshni, *Physica,* Vol. 34, 1967, p. 149.

27. H.C. Casey, Jr. and M.B. Panish, *Heterostructure Lasers Part B: Materials and Operating Characteristics,* New York: Academic Press, 1978, p. 9.

28. E. Zielinski, H. Schweizer, K. Streubel, H. Eisele, and G. Weimann, *J. Appl. Phys.,* Vol. 59, 1986, p. 2196.

29. Y. Yamazoe, T. Nishino, and Y. Hamakawa, *IEEE J. Quantum. Elec.,* Vol. QE17, 1981, p. 139.

30. W. Porod and D.K. Ferry, *Phys. Rev.,* Vol. 27B, 1983, p. 2587.

31. K.H. Goetz, D. Bimberg, H. Jurgensen, J. Selders, A.V. Solomonov, G.F. Glinskii, and M. Razhegi, *J. Appl. Phys.,* Vol. 54, 1983, p. 4543.

32. M.R. Lorenz and A. Onton, *Proc. 10th Int. Conf. Semiconductor Phys.,* Cambridge/Mass, USAEC, New York, 1970, p. 444.

33. E.H. Reihlen, D. Birkedal, T.Y. Wang, and G. Stringfellow, *J. Appl. Phys.,* Vol. 68, 1990, p. 1750.

34. B. Wakefield, M.A.G. Halliwell, T. Kerr, D.A. Andrews, G.J. Davies, and D.R. Wood, *Appl. Phys. Lett.,* Vol. 44, 1984, p. 341.

35. R.L. Moon, G.A. Antypas and L.W. James, *J. Electron. Mater.,* Vol. 3, 1974, p. 635.

36. T.P. Pearsall, ed., *GaInAsP Alloy Semiconductors,* New York, 1982, p. 295.

37. E. Kuphal, *J. Cryst. Growth,* Vol. 67, 1984, p. 441.

38. D. Olego, T.Y. Chang, E. Silberg, E.A. Caridi, and A. Pinczuk, *Appl. Phys. Lett,* Vol. 41, 1982, p. 476.

39. D. Bimberg, M.S. Skolnick, and L.M. Sander, *Phys. Rev. B,* Vol. 19, 1979, p. 2231.

40. F.P. Kesamanly, D.N. Nasledov, A. Ya. Nashelskii, and V.A. Skripkin, *Sov. Phys. Semicond.,* Vol. 2, 1969, p. 1221.

41. E.O. Kane, Chapter 4A in *Handbook on Semiconductors,* Vol. 1, ed. W. Paul, Amsterdam: North Holland, 1982.

42. O. Berolo, J.C. Woolley, and J.A. Van Vechten, *Phys. Rev. B,* Vol. 8, 1973, p. 3794.

43. M.B. Thomas and J.C. Woolley, *Can. J. Phys.,* Vol. 49, 1971, p. 2052.

44. E.E. Matyas and A.G. Karoza, *Phys. Stat. Solidi,* Vol. 111b, 1982, p. K45.

45. K. Alavi, R.L. Aggrawal, and S.H. Groves, *Phys. Rev. B,* Vol. 21, 1980, p. 1311.

46. J.F Nye, *Physical Properties of Crystals,* Oxford: Clarendon Press, 1957.

47. W.A. Brantley, *J. Appl. Phys.,* Vol. 44, 1973, p. 534.

48. J.W. Matthews and A.E. Blakeslee, *J. Cryst. Growth,* Vol. 27, 1974, p. 118.

49. P.M. Petroff, R.A. Logan, and A. Savage, *Phys. Rev. Lett,* Vol. 44, 1980, p. 289.

50. V. Swaminathan, P. Parayanthal, and R.L. Hartman, *Appl. Phys. Lett,* Vol. 52, 1988, p. 1461.

51. J.T. DeVreese, "Polarons" in *Ionic Crystals and Polar Semiconductors,* Amsterdam: North Holland, 1972.

52. B.O. Seraphin and H.E. Bennett, *Semiconductors and Semimetals,* ed. R.K. Willardson and A.C. Beer, New York: Academic Press, 1967, p. 499.

53. K. Kakimoto and T. Katoda, *Appl. Phys. Lett.,* Vol. 40, 1982, p. 826.

54. T.P. Pearsall, R. Carles, and J.C. Portal, *Appl. Phys. Lett.,* Vol. 42, 1983, p. 436.

55. A. Pinczuk, J.M. Worlock, R.E. Nahory, and M.A. Pollack, *Appl. Phys. Lett.,* Vol. 33, 1978, p. 461.

56. R.K. Soni, S.C. Abbi, K.P. Jain, M. Balkanski, S. Slempkes, and J.C. Benchimol, *J. Appl. Phys.,* Vol. 59, 1986, p. 2184.

57. S.H. Wemple and M. DiDomenico, Jr., *Phys. Rev. B,* Vol. 3, 1971, p. 1338.

58. M.A. Afromowitz, *Solid State Commun.,* Vol. 15, 1974, p. 59.

59. J. Buus and M.J. Adams, *Proc. IEEE Solid State Elect. Dev.,* Vol. 3, 1979, p. 189.

60. H. Burkhard, H.W. Dinges, and E. Kuphal, *J. Appl. Phys.,* Vol. 53, 1982, p. 655.

61. S. Adachi, *J. Appl. Phys.,* Vol. 53, 1982, p. 5863.

62. P. Chandra, L.A. Coldren and K.E. Strege, *Electron Lett.,* Vol. 17, 1981, p. 6.

63. F.R. Bacher, J.S. Blakemore, J.J. Ebner, and J.R. Arthur, *Phys. Rev. B,* Vol. 37, 1988, p. 2551.

64. S. Adachi, *J. Appl. Phys.,* Vol. 54, 1983, p. 1844.

65. T. Soma, J. Satoh, and H. Matsuo, *Solid. State. Comm.,* Vol. 42, 1982, p. 889.

66. V.M. Glazov, K. Darletov, A. Ya. Nashelskii, and M.M. Mamedov, *Zh. Fiz. Khim.,* Vol. 51, 1977, p. 2558. Taken from Landolt-Bornstein [22].

67. R. Bisaro, P. Mevenda, and T.P. Pearsall, *Appl. Phys. Lett.,* Vol. 34, 1979, p. 100.

68. U. Pietsch and D. Marlow, *Phys. Stat. Solid.,* Vol. 93a, 1986, p. 143.

Chapter 2
SIMS Analysis of InP and Related Materials

M. Geva
AT&T Bell Laboratories

2.1 INTRODUCTION

Most III-V devices are based on layered structures where strict requirements are imposed upon properties such as material composition, interface quality, and surface characteristics, which may crucially affect the device performance and reliability. It is, therefore, essential for the material grower and device engineer to have good material characterization throughout the device processing. They need to know the actual material stoichiometry in each layer, dopants and impurities concentration level within and between layers, dopant and impurity diffusion profiles across interfaces, the extent of oxides and debris on the surface, and other properties that may affect the electronic and photonic properties of the material.

The need to resolve variations in material composition within layers, which may be as thin as 10Å, to accurately follow a dopant concentration profile across sharp interfaces, and to distinguish surface from bulk composition implies that the diagnostic probe must have high depth resolution. Analytical techniques possessing such a quality fall under the category of surface analysis.

A variety of surface analytical techniques have been developed in the past three decades, and new and improved ones are being developed continually. In all surface analytical techniques, a beam of particles hits a surface and particles emitted from the surface are analyzed to reveal information about its properties, primarily the composition. The incident or probing beam consists of photons, electrons, ions, or atoms, and the same is true for the particles emitted from the surface. The analyzed properties of the emitted particles are mainly energy and mass. Table 2.1 lists the major surface analytical techniques, indicating the type of the incident and emitted particles, the analyzed property of the emitted particles, the elemental range and

Table 2.1
Major Surface Analytical Techniques.

Analytical Techniques	Incident Particles	Emitted (Analyzed) Particles	Analyzed Properties	Elemental Range	Elemental Detection Limit (%)	Depth Resolution (Å)	Lateral Resolution	Reference
X-ray Photoelectron Spectroscopy (XPS)[a]	X-ray photons (of a particular wavelength)	Photoelectrons	Kinetic energy (binding energy)[b]	>Li	0.01–1 (atomic)	5–20	0.15–1 mm[c]	[1–4]
Auger Electron Spectroscopy (AES)[d]	Electrons	Auger electrons	Kinetic energy[e]	>Li	0.1–1 (atomic)	5–20	<0.1 μm	[4–7]
Ion Scattering Spectroscopy (ISS)	Ions (monoenergetic)	Reflected ions	Kinetic energy[f]	All elements	0.01–1 (atomic)	<5	10 μm	[8–11]
Rutherford Backscattering Spectrometry (RBS)	Ions (usually He+), 0.5–4 MeV	Backscattered ions	Kinetic energy[g]	>Li[h]	1–10 for low Z, 10–100 ppm for high Z elements[i]	20–500	1 mm	[12–14]
Secondary Ion Mass Spectrometry (SIMS)	Ions	Secondary ions	Mass	All elements and isotopes[j]	1 ppm–1 ppb	<50	0.1–1 μm	[15–17]

REMARKS:

[a] Also called *electron spectroscopy for chemical analysis* (ESCA).
[b] Reveals elemental composition and chemical information. Semiquantitative.
[c] In some instruments, 30 μm.
[d] Also, scanning auger microprobe (SAM).
[e] Reveals elemental composition and some chemical information. Semiquantitative.
[f] Energy loss due to collisions with surface atoms. Reveals elemental composition.
[g] Reveals composition, depth distribution of elements, crystallographic information.
[h] H can be detected in a different mode of operation.
[i] Mass resolution is poor for elements much heavier than probe ions; sensitivity is poor for elements lighter than the matrix elements.
[j] High mass resolution.

detection limit, and depth and lateral resolution. It also lists reference numbers for the publications containing information regarding each technique.

The depth resolution depends upon such properties as the depth of penetration of the incident particles, the interaction of the incident beam with the substrate's atoms, and the depth from which emitted particles can emerge with their relevant characteristics (such as energy) unaltered. By its nature, the surface analytical technique reveals the composition of the top surface region only. However, by repeatedly sputtering the surface layer while analyzing, a composition depth profile of the material can be achieved with a high depth resolution.

Since the surface analytical techniques reveal the surface composition of the sample, it is important that the top layer should not be contaminated by residual species from the chamber's atmosphere. The rate of deposition of gas species on the sample surface at a pressure of 10^{-6} torr is roughly a monolayer per second, as can be found from the kinetic theory of gases (assuming a sticking coefficient of one, namely, that each impinging particle sticks to the surface). In order to avoid interference of background particles with the analysis of the surface, the monolayer deposition time should be much longer than the analysis time. This can be achieved by lowering the pressure in the analysis chamber to the 10^{-9} to 10^{-11} torr range, where a monolayer deposition time extends from roughly fifteen minutes to more than a day. This pressure range is termed ultrahigh vacuum (UHV). Thus, all surface analytical techniques, with the exception of Rutherford back scattering (RBS), require a UHV environment in the analysis chamber.

Although the various surface analytical techniques are complementing each other in many ways, and all of them are used, to varying degrees, in the characterization of semiconductor materials, including InP-based structures, secondary ion mass spectrometry (SIMS) assumes a leading role because of its high elemental sensitivity and excellent depth resolution. It is the only technique that can detect trace-level dopant and impurity elements, and measure their concentration in the structure while maintaining high depth resolution. The rest of this chapter will, therefore, be devoted to SIMS analysis. This chapter does not deal with the fine details of the technique, which can be found in the vast published literature (see, for instance, [15–21]). Rather, it is intended to provide the reader with a general understanding of the technique, its capabilities as well as its limitations as a diagnostic tool. Some typical examples of SIMS analysis relating to InP-based and other semiconductor materials are given as illustration. For the benefit of readers who are interested in greater detail, representative publications are referenced at the beginning of some sections.

2.2 FUNDAMENTALS OF SECONDARY ION MASS SPECTROMETRY

2.2.1 Description of Technique

Figure 2.1 shows a schematic of the SIMS instrument. The sample is placed in a UHV chamber, where a primary ion beam, of 0.5 to 20 keV impact energy, is focused on its surface and sputters material off the outermost top region. A small

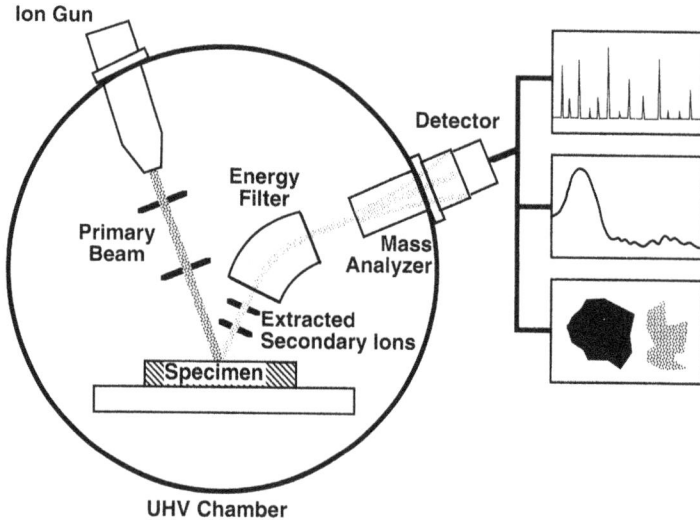

Figure 2.1 Schematic view of SIMS instrument.

fraction, typically 10^{-1} to 10^{-4} of the sputtered material from the sample, is in an ionized state (positive and negative ions). These secondary ions are extracted and focused by an optical system passing through an energy filter. The ions are then mass analyzed, revealing the material composition of the top layer. SIMS analysis can be presented in various ways. By recording the detected secondary ion current *versus* mass, a mass spectrum is generated. By recording the secondary ion current corresponding to a single mass, or sequentially to several elements, while sputtering the surface, a profile of signal intensity *versus* sputter time is produced. It can be converted rather easily to a profile of concentration *versus* depth, or a concentration depth profile, as it is commonly called. By recording the intensity of a signal corresponding to a single mass peak as a function of position on the sample's surface, an elemental mapping, or image, of the surface is produced.

2.2.2 The SIMS Hardware

The SIMS hardware [16] is composed of a UHV specimen chamber, with one or more ion guns and a mass analyzer attached to it. An airlock or sample introduction chamber is connected to the specimen chamber with a gate valve separating them. Most SIMS chambers also include an electron gun, which is used during the analysis of an insulating sample to prevent charging of the surface. Another essential component is an energy filter, which limits the energy band of the extracted secondary ions that enter the mass spectrometer. As discussed in Section 2.3.6.1, the energy filter is also used to reduce mass interference caused by molecular ions. The two main components in the SIMS hardware, the ion gun and the mass analyzer, are discussed in greater detail below.

2.2.2.1 Ion Gun

Three major types of ion guns are presently employed in SIMS instruments [16,22,23]: duoplasmatron, cesium, and liquid metal guns. All guns consist of an ion source and an electrostatic optical system that focuses and deflects the beam. Most guns also include a mass filter that removes neutral and ionic species different from the intended ions. The impact energy of the primary (incident) ions in SIMS instruments is typically in the range of 1 to 20 keV.

In the duoplasmatron gun, a low-pressure (10^{-4}–10^{-3} torr) gas (O_2, N_2, Ar, Xe) is introduced into a hollow cathode, where an electric arc is maintained between the cathode and an anode in front of it. The plasma is confined to the axis by means of an intermediate conical electrode and an axial magnetic field. Positive or negative ions are extracted from the plasma through an extraction aperture, and they then pass through the primary beam optical system. The most commonly employed primary ions from duoplasmatron guns are O_2^+, O^-, and Ar^+. Of these, the O_2^+ beam is the most frequently applied in SIMS analyses, especially for the analysis of electropositive species (see Section 2.2.3.2). The O^- beam is used primarily for analyzing insulating samples since it can drastically reduce the electrostatic charging of the sample [24].

The cesium ion gun consists of a Cs reservoir (elemental or compound), which when heated releases Cs atoms (vapor). The Cs atoms are thermally ionized upon collision with a hot tungsten surface. The ions are extracted through an aperture in the ionizing tungsten plate into the optical system. The thermal ionization produces a very pure ion beam with a very narrow thermal spread of the order of 0.1 eV. The Cs primary beam is particularly beneficial in the analysis of electronegative species, such as O, F, and S (see Section 2.2.3.2).

The liquid metal ion source is usually composed of a metal needle (typically tungsten) with a sharp tip at one end. The needle is deposited with, or a portion of its rod is embedded in a reservoir of, a metal with a low-melting temperature, such as Ga, In, Sn, and Au. It is maintained at a high enough temperature so that a liquid film of the source metal forms on its surface. The liquid film produces a very sharp tip at the needle apex, and a stable, jet-like vapor protrusion forms there. A high electric field maintained between the needle tip and an extraction electrode causes field ionization of the evaporated atoms. The extracted ions then pass through the gun's optical system. The very small emitting area and the high local electric field there produce a high-brightness ion source and a highly collimated ion beam. Current densities of over 1 A/cm^2, and beam diameters of less than 500Å can be achieved.

Most liquid metal ion guns use a Ga source, and are used for high-spatial-resolution SIMS, primarily for static SIMS (see Section 2.2.4.1).

2.2.2.2 Mass Analyzer

The main types of mass filters presently employed in SIMS instruments for the analysis of secondary ions are [16]: the quadrupole, the magnetic sector, and the time of flight (TOF) mass spectrometers. The first two are the most commonly used.

The quadrupole mass filter consists of four smooth metal rods with an RF electric field and a superimposed DC field applied to them, resulting in stable trajectories (i.e., passing through the exit aperture) only for ions in a mass-to-charge range of ΔM around the mass M. The values of the RF and of the DC voltages, and their ratio determine the transmitted mass value and the mass resolution, defined as $M/\Delta M$, where M is the mass corresponding to the center (top) of the peak, and ΔM is the width at 10% of the peak height. Typical mass resolution values are in the range of 60 to 600. In many quadrupole mass spectrometers, there is an additional set of short four rods placed in front of the main rods, with only RF voltage applied to them. They act as a prefilter. For optimal performance, the incoming ions should have a kinetic-energy spread of about 10 to 15 eV or less. This is achieved by placing an energy filter in front of the mass analyzer, which brings about an additional advantage of preventing a line of sight between the sample and the detector, thus eliminating the detection of neutrals. The quadrupole mass spectrometer is the most compact and the least expensive of the mass analyzers. It is also fairly insensitive to some topographical variations on the analyzed surface, such as minor height differences. Since mass switching involves only a change in the voltages applied to the spectrometer's rods, the change can be done very quickly. Although mass ranges of over 1000 amu can be easily achieved, most spectrometers exhibit reduced transmission at the high mass range. The transmission of quadrupole mass spectrometers is limited typically to 1%, and the mass resolution to about 1000. In most SIMS instruments equipped with a quadrupole mass spectrometer, it is possible to change the specimen angle relative to the primary beam and to the spectrometer, thus enhancing the analytical capabilities of the instrument.

In SIMS instruments with a magnetic sector mass spectrometer, the sample is typically biased 3–6 kV relative to the secondary ions extraction optics, and the extracted ions pass through an optical system, an energy filter, and the magnetic sector mass spectrometer. In the magnetic sector, the ions follow a circular trajectory with the radius preportional to $(MV/e)^{1/2}/B$, where M and e are the ion's mass and charge, respectively, V is the bias voltage, and B is the magnetic field intensity. By adjusting the spectrometer's entrance and exit slits, the mass resolution can be varied. In most cases, it can be increased from about 200 to 10,000, and in some cases, to over 25,000. However, increasing the mass resolution means, of course, decreasing the transmission of the spectrometer. The high mass resolution allows resolving some mass interferences, as discussed in Section 2.3.6. The transmission of the magnetic sector mass spectrometer is typically higher by about an order of magnitude than that of a quadruple mass spectrometer. The mass range, however, is limited in most cases to about 500 amu. Since the sample is biased several kilovolts relative to the extraction optics, its surface should be smooth and flat to avoid artifacts caused by local electric field variations. The mass switching may take a longer time than in a quadrupole mass spectrometer because of possible hysteresis effects. In most instruments with a magnetic sector mass analyzer, the sample position and orientation are fixed. However, the primary beam incidence angle changes with the beam impact energy.

The TOF mass spectrometer is an energy-focusing, electrostatic sector, or linear analyzer. It comes in conjunction with a pulsed ion beam (TOS-SIMS) [16], or a pulsed laser beam (known as laser ion mass analyzer (LIMA) or laser ion mass spectrometry (LIMS) [25]. A short ion beam (or laser beam) pulse of 10 ps to 10 ns duration, is focused on a surface and sputters or vaporizes material off a small region on it. About 0.1% to 1% of the sputtered particles are in an ionized state. The ionized species are extracted by a potential difference of a few kilovolts into a linear or curved drift tube, and are detected at its end. Since the velocity of particles of the same energy is inversely proportional to the square root of the mass, the arrival time of the ions increases with mass. The mass resolution of the TOF spectrometer depends on the pulse length, the accelerating bias, and the details of the drift tube. It is uniform over a large mass range, which can exceed 10,000. The advantages of the TOF spectrometer are its broad energy pass, high transmission, the parallel detection of all the masses (within the spectrometer range), and the high mass resolution. The disadvantage is the pulsed operation with a relatively low duty cycle. The TOF-SIMS is particularly advantageous for static SIMS (see Section 2.2.4), where the detailed atomic and molecular composition of the surface can be gathered at a high spatial resolution. The high detection sensitivity of the spectrometer can be increased even further by post-ionizing the emitted neutrals. This can also make the technique more quantitative, since it would be less vulnerable to matrix effects. This applies, of course, to all types of SIMS. Post-ionization is currently being studied extensively [26–29].

2.2.3 The SIMS Process

SIMS analysis involves a number of processes. Although a great deal of information has been gathered and various models have been proposed, most processes are not yet fully understood. In this section we will briefly review the main SIMS processes [15–17,22,23,30–33].

2.2.3.1 The Sputtering Process

The sputtering of the sample's surface by the impinging primary ions is the basic process in SIMS analysis. Figure 2.2 gives a schematic representation of the sputtering process. The primary ions collide with surface atoms and transfer energy to them. A small fraction of the primary ions are scattered back from the surface, while most lose their energy in a succession of collisions with surface atoms. The surface atoms that acquire sufficient energy in the collision are displaced. They are called *recoils*. Many of the recoils, in turn, undergo series of collisions with neighboring atoms, transferring their energy and producing new recoils. This process is called *collision cascade*. As a result of the collision cascade, some primary ions are implanted into the sample, and surface atoms are repositioned. Some are knocked deeper into the sample, while others move sideward or upward, causing a mixing effect [34]. A typical mixing range is of a few tens of angstroms. Some recoils at the surface region acquire enough upward directed momentum to overcome the potential barrier and are sputtered off the surface. The escape depth of the sputtered species increases with the impact energy of the primary ions, and it depends upon parameters such as matrix and surface composition, primary beam species, and angle of incidence. In a typical SIMS analysis, the escape depth is of the order of a few to ten angstroms. Outward-moving particles that originate from greater depths are likely to suffer collisions before reaching the surface. The average number of sputtered particles per incident primary ion is called the *sputtering yield, S. S* is typically of the

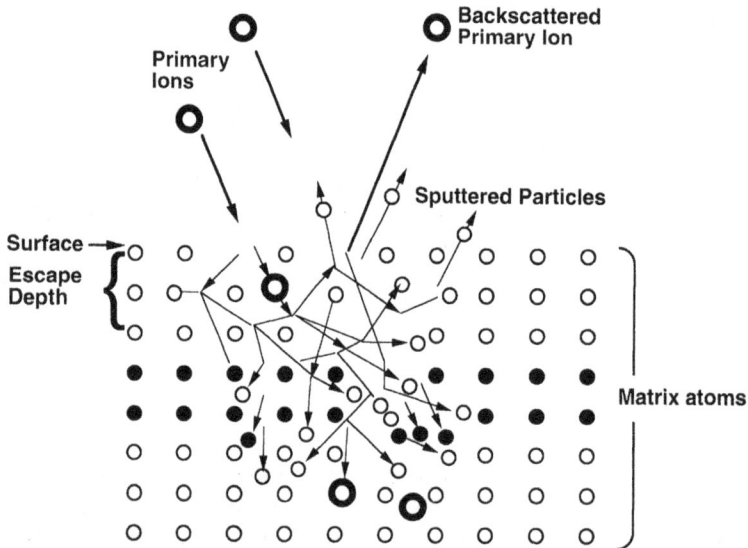

Figure 2.2 Schematic presentation of the sputtering process.

order of a few sputtered atoms per incident primary ion. S is a function of the surface composition and of the primary beam species, current density, impact energy, and incidence angle relative to the surface normal. The energy distribution curve of the sputtered particles peaks at a low energy of several electron-volts, and then decreases slowly toward higher energies. The actual distribution profile depends on the sputtered species and on parameters such as those listed above for S.

The intensity cross-section of the sputtering ion beam is typically of a gaussian profile. In order to achieve a homogeneous distribution of the sputtering ions across the region of interest, therefore, it is necessary to raster the beam, with the raster dimension at least several times larger than the beam diameter.

If the material is sputtered homogeneously across the surface, the surface morphology is preserved. However, in many cases the sputtering is spatially nonuniform, resulting in surface structuring such as cone formation. Surface structuring is commonly seen, for instance, in sputtered Au, Al, and other metal surfaces. This effect depends on parameters similar to those mentioned above for S. Common ways to reduce surface structuring include the selection of optimal primary beam species, energy, and incidence angle. A very effective way of minimizing it is to bombard the surface from more than one angle, either by using two ion guns or by rotating the sample during sputtering.

In many crystalline surfaces, such as those of semiconductor wafers, almost no surface structuring occurs under normal SIMS conditions, and the sputtered surface is eroded uniformly and remains smooth. The rastered ion beam sputters, then, a flat-bottomed crater, whose depth can be measured subsequent to the analysis by means of a micro-profilometer, for instance. The sputter rate can thus be converted to a depth scale when analyzing a homogeneous (single-matrix) wafer, since it is proportional to the crater depth. One should be careful, though, in a case of layered structures (e.g., InGaAs/InP), since the sputter rate usually changes with matrix composition. If the relative rates are known for the various layer compositions, then the conversion of sputtering time into depth can be accomplished rather easily.

In a matrix composed of more than one element, such as InP, the sputter rate may be different for each constituent. However, the preferential sputtering reaches an equilibrium state very fast, and, consequently, the sputtered material has the same composition as the bulk [35]. Thus, the composition of the sputtered material *versus* time, in the case of a matrix composed of more than one element, exhibits a transient behavior at the very top surface region. In a layered structure, the equilibrium has to be re-established at each interface. This can cause some artifacts at the interface regions when profiling through a layered structure.

2.2.3.2 Ionization Yield

Only a fraction of the sputtered species are in an ionized state. Since the charged particles are the ones that are analyzed in the SIMS, the secondary ion emission from sputtered surfaces is a particulary important SIMS process.

The ionization yield for a certain type of secondary ions (such as singly charged, positive secondary ions of a particular species) sputtered from the surface is defined as the number of such secondary ions emitted per incident primary ion. For monoelemental matrices, yield values can vary over as much as six orders of magnitude, depending on the surface composition and structure, the analyzed ion, and the primary ion beam species, angle of incidence, and impact energy. In matrices composed of more than one element, such as InP and InGaAs, the situation is further complicated by the effect of species of one type on the ionization yield of the others. This effect, which can cause variations of several orders of magnitude in the ionization yield of an element when changing the matrix, is called matrix effect. Thus, the ionization yield for In in InGaAs may be quite different than its yield in InGaP. The same is true for the ionization yields of dopant or impurity species, which, beside their dependence on the primary-beam properties, depend upon matrix effects and surface composition.

It has been observed [17,45] that there is a correlation between the ionization yield of positive secondary ions and the degree of electropositivity (or the reciprocal of the ionization potential), and between the yield for negative secondary ions and the degree of electronegativity (or electron affinity).

The primary ion beam species can have a marked effect on the ionization yield of the analyzed species [16,17,36]. The yield of some positive secondary ions, such as Mg^+, Al^+, or Ti^+, can be enhanced by more than three orders of magnitude when replacing an Ar^+ primary beam by an O_2^+ beam. A similar effect is found in the case of negative secondary ions, such as O^- and F^-, when switching from an Ar^+ primary beam to a Cs^+ beam. Although the exact mechanism causing this effect is not known, it is believed that the implanted O_2^+ or Cs^+ ions change the chemistry of the top layer, affecting the work function, electron affinity, and other properties of the surface region, which can strongly influence the probability for the sputtered particle to be in an ionized state. In the case of impurity elements with a higher affinity than the matrix, for instance, the implanted Cs atoms probably lower the electron affinity of the surface matrix even further, causing an enhancement in the negative ion yield of the impurities. Due to their strong effect on the ionization yield of positive and negative secondary ions, O_2^+ and Cs^+ primary beams are presently used routinely in SIMS analyses, and only in special cases are other types of primary beams used, such as Ar^+, Xe^+, or Ga^+.

As discussed in Section 2.2.3.1, when the surface is bombarded by the primary ion beam, incident ions are implanted in the matrix and collision cascades cause mixing at the top region. Thus, the top layer, from which the secondary ions are emitted, has a different structure and chemistry than the bulk material below it. This "mixed" top region remains present as the top layer is being eroded. This has two major consequences: (a) A certain sputtering should occur prior to reaching a steady-state situation, where the structure and composition of the top layer of a certain matrix remain stable. Normally, a layer of a few tens of angstroms has to be removed. The original top surface, therefore, represents a transient region. This is further enhanced by the impurities, which are almost always found at the surface and strongly affect the ionization yield. However, by depositing on the surface a thin layer (a few tens of angstroms or more) of the same matrix, the original surface does not exhibit a transient behavior anymore [15]. (b) Due to the damage and displacement caused by the collision cascade, the secondary ion yield is normally not affected by the crystallographic orientation and structure of the matrix.

2.2.3.3 Energy Distribution of the Secondary Ions

While the energy distribution of secondary monoatomic ions is similar to that of the uncharged sputtered atoms, the distribution curve of secondary molecular ions is narrower, shifted towards lower energy. The higher the number of atoms in the sputtered molecular (or cluster) ion, the narrower is its energy distribution curve. Such a trend is demonstrated in Figure 2.3. As discussed in Section 2.3.6.1, this has an important application in reducing mass interference caused by molecular ions.

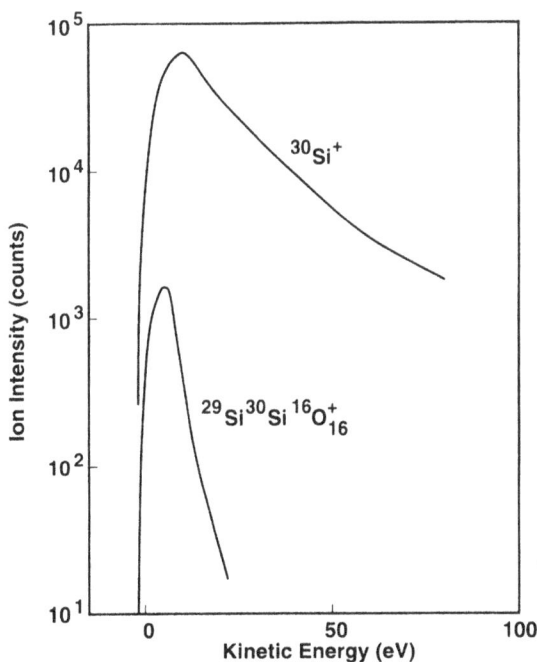

Figure 2.3 Energy distribution curves of sputtered monoatomic ($^{30}Si^+$) and triatomic ($^{29}Si^{30}Si^{16}O^+$) secondary ions. (From Charles Evans & Associates)

2.2.3.4 Quantification of SIMS Profiles

The large variation (several orders of magnitude) in ionization yield values for different elements in a matrix and their strong dependence on matrix composition, primary beam species and parameters, and other factors make the quantification of the SIMS data (i.e., the conversion of secondary ion current into atomic concentration) a very involved issue [16,17,37,38]. This difficulty is compounded by the complexity of the sputtering and ionization processes and the lack of a thorough understanding of these processes based on first principles modeling. Consequently, in order to accurately quantify the SIMS profiles, it is necessary to use standards. The ionization yield of dopant species in a matrix was found to vary linearly with the dopant's concentration up to a concentration of about 1% (atomic), where the matrix chemistry starts to be affected. This property provides a relatively simple and accurate way to quantify dopant concentrations in SIMS analysis using standards containing dopant elements at a known concentration in a matrix similar to that of the sample. A brief description of the procedure is given below.

Let the impurity or dopant concentration in a matrix be ρ_d, and its measured secondary ion current or count rate be I_d, then

$$I_d = K_d \rho_d$$

where K_d is a constant of proportionality containing matrix and instrumental effects, such as sputtering, ionization yield, and spectrometer transmission. Similarly, the measured secondary ion current of a matrix element (such as P or In in InP), I_m, is given by

$$I_m = K_m \rho_m$$

where ρ_m is the element's concentration in the matrix.

By dividing I_d by I_m and rearranging terms, ρ_d can be expressed as

$$\rho_d = \left(\frac{K_m}{K_d} \rho_m \right) \frac{I_d}{I_m} = \text{RSF} \, (d,m) \, \frac{I_d}{d_m}$$

where RSF (d,m) is the relative sensitivity factor of the dopant element, d, in respect to the matrix element, m. RSF (d,m) is a constant for the dopant element at concentrations under 1% in a particular matrix. If the matrix element has more than one isotope, then I_m is the secondary ion current of the isotope being monitored. If the dopant element has more than one isotope present in the matrix, with the monitored isotope having a relative abundance f_i, then the concentration of the dopant element is given by

$$\rho_d = \text{RSF}\,\frac{(d_i,m)}{f_i} \cdot \frac{I_d}{I_m}$$

where RSF (d_i,m) is the RSF of the monitored isotope of the dopant element in relation to the matrix element (or isotope), m.

The RSF value for a particular dopant element in a particular matrix is found either by analyzing a similar matrix with a known concentration of the dopant element, or by analyzing the matrix with the dopant element (isotope) implanted into it at a known fluence.

2.2.4 Static and Dynamic SIMS

By its nature, SIMS analysis involves a sputtering of the sample surface. However, based on the sputter rate, we distinguish between two main modes of analysis: *static* SIMS [16,23,39,40] and *dynamic* SIMS [16,17,35].

2.2.4.1 Static SIMS

If the sputter rate is such that a monolayer sputtering time is much longer than the analysis time, then the analysis falls under the category of static SIMS. In static SIMS the current density of the primary beam on the sample surface should be below 5 nA/cm^{-2}. Static SIMS analysis is usually given in a form of spectra or elemental images of the surface. Since the top atomic layers are not sputtered away, no equilibrium is reached and, therefore, static SIMS provides qualitative rather than quantitative information. Although static SIMS can be used for qualitative studies of surface composition and contaminations on III-V wafers and devices, its main applications are in other fields, such as metallurgy, polymers, and biology.

2.2.4.2 Dynamic SIMS

With an increased sputtering rate, the surface of the sample is being continuously eroded. Dynamic SIMS is, therefore, a destructive technique. However, the increased sputtering rate produces higher flux of secondary ions, thus increasing the SIMS sensitivity. By sequentially removing thin surface layers and following certain mass peaks, dynamic SIMS can also provide concentration depth profiles for the analyzed elements in the sample. The depth resolution of such depth profiles depends on various parameters, such as the flatness of the crater sputtered by the rastered beam, the primary beam impact energy and incidence angle, the primary beam species, and the analyzed elements and matrix. In some cases, depth resolution of less than 10Å can be achieved. Due to its superb sensitivity and high depth resolution and its ability to perform a quantitative concentration analysis, dynamic SIMS is by far the major analysis mode applied to the analysis of III-V materials and structures.

2.2.5 SIMS Modes of Operation

Two main modes of operation are employed in SIMS instruments in order to provide spatial resolution in the material analysis: the ion microphobe and the ion microscope [16].

2.2.5.1 The Ion Microprobe

In the ion microprobe mode, the SIMS operates in an analogous way to the electron microprobe or the scanning electronic microscope (SEM). The primary ion beam, focused to a fine spot, is rastered across the surface, while the spectrometer output, tuned for a specific mass, is synchronically displayed. In all instruments it is possible to gate the detected signal, (i.e., to output only the counts due to secondary ions sputtered from a central region of the rastered area) either by electronic or optical means. The gating size is adjustable, and it is typically 10% to 50% of the raster dimension. In some instruments it is possible to store the spectrometer's output for each mass peak, synchronously with the rastered beam, in a computer memory matrix, thus mapping the secondary ion counts across the surface. In static SIMS mode, the counts in each matrix element of the computer memory keep accumulating or averaging and produce an elemental imaging of the surface. In the depth profiling mode, however, a new memory matrix is generated for each cycle, thus forming a three-dimensional memory matrix in the course of the analysis. Subsequent to the analysis, it is possible to reconstruct an elemental depth profile from a selected set of matrix pixels that corresponds to a subregion of the analyzed area. Using this method, it is not necessary to know a priori the exact position of the feature of interest (which may be buried), as is the case for a simple gating. It is only necessary to know its approximate location on the sample's surface so that it is included in the analyzed area. Another advantage is that, in some cases, it is possible to view each cycle (two-dimensional memory matrix) separately, construct a variety of cross-sections, or even reconstruct a three-dimensional elemental profile from the three-dimensional memory matrix. The spatial resolution in the microprobe mode depends on the spot size of the primary beam. Duoplasmatron and cesium guns can typically produce beams with a spot size down to several microns. However, in some cases a sub-micron spatial resolution is attainable, and when using a liquid metal gun the spot size may even be reduced below 0.1 μm. One should realize, however, that by reducing the beam size the primary beam current is reduced, and, consequently, the secondary ion emission may be reduced as well, causing the signal-to-noise ratio to decrease.

2.2.5.2 *The Ion Microscope*

In the ion microscope mode, the SIMS operates in an analogous way to an optical microscope, where the primary ion beam illuminates the surface and causes emission of secondary ions. The secondary ions from the central region are extracted and pass through the secondary ion optics and through energy and mass filters. The transmitted ions form an enlarged image of the source on a collecting plate at the focal plane. This plate is typically a phosphorous screen or an image intensifier. A spatial resolution of about 1 μm can be achieved in this mode by the imaging optics, while the illuminating primary beam can be broad and intense. The ion microscope mode is typically found in magnetic sector and TOF-based SIMS instruments. By employing apertures in the optical system, the signal from the central region of the restered area is optically gated. Instead of being imaged on a position-sensitive plate, the total secondary ions current, transmitted from the imaged region, can be collected by a pulse counting detector or by a Faraday cup. Thus, the SIMS can be used in this mode for routine depth profiles where the spatial resolution is determined by the dimension of the imaged area.

The advantage of the ion microscope mode is that the primary beam current can be relatively high since the spatial resolution is determined by the imaging optics. The analysis time is, therefore, reduced and the signal-to-noise ratio is increased.

2.3 FEATURES OF SIMS ANALYSIS

In this section, the main features of SIMS analysis are briefly discussed: sensitivity, detection limit, dynamic range, and depth, lateral, and mass resolution [15–17,31,35].

2.3.1 Sensitivity

The SIMS sensitivity for a specific element in a specific matrix depends on the element's secondary ion flex which is monitored by the detector. The sensitivity, therefore, depends upon the ionization yield of the element in the matrix, and, hence, on the matrix, the analyzed element, and the analytical conditions such as the primary ion beam species, energy, and angle of incidence. It also depends upon the primary beam current density and on the spectrometer's transmission. For comparative purposes, the SIMS sensitivity values are inversely related to the RSF values [17].

2.3.2 Detection Limit

The SIMS detection limit for a specific element in a matrix is the lowest concentration of that element that can be detected in a matrix by the SIMS in the course of the analysis. It is determined by the SIMS sensitivity for the element in the matrix and by the background signal. To optimize the detection limit, the background contributions should be minimized. One common contribution is due to mass interference (i.e., a signal due to other species) such as molecular ions, which have a similar mass-to-charge ratio. Examples for mass interference are ^{31}PH and ^{32}S in InP, or ^{27}AlH and ^{28}Si in AlGaAs (see Section 2.3.6). Another source of background signal can be secondary ions of the analyzed element resputtered from surfaces at the vicinity of the sample, such as the extraction lens (which may have been deposited previously by a sputtered material). This contribution is often referred to as *instrument memory effect*. The resputtering is done primarily by back-scattered primary ions or by sputtered species with a high kinetic energy. Another source of high background signal may be secondary ions originating out of the flat region of the crater's bottom, such as the edge. To avoid this stray signal, the crater dimension must be much larger than the beam spot size, and the signal should be gated so that only secondary ions emitted from the central region of the crater would be collected (see Section 2.3.4.2). Species of the chamber atmosphere, like O_2, H_2, H_2O, and hydrocarbons, can contribute to the background signal when analyzing the related elements in a matrix. This contribution can be reduced by improving the vacuum and lowering the partial pressure of such species. A common way to improve the detection limit in this case is to increase the primary beam current density over the probed region. This, however, leads to a higher sputter rate. Another source for a background signal may be impurities in the primary ion beam. Most ion guns are equipped with a mass filter to minimize this effect.

2.3.3 Dynamic Range

The SIMS dynamic range for the concentration (or count rate) of an element in a given depth profile of a matrix is the difference between the element's highest measured concentration (count rate) value and the lowest, or the background value, expressed in decades or in orders of magnitude. Dynamic range usually refers to a profile of an ion-implanted element. A large dynamic range implies high sensitivity and low detection limit. The linear response range of a typical pulse-counting detector used in SIMS instruments is limited to under 10^6 counts per second. This limitation may be overcome if the detection at the high end can be transferred, without losing linearity, to a Faraday cup (current measurement). However, this is generally not done, and the limiting factor in most cases is the background signal due to memory (resputtering) and mixing effects or to secondary ions extracted from regions off the crater's center, such as the edges.

2.3.4 Depth Resolution

Ideally, the atomic layers were sputtered sequentially by the primary ion beam, without affecting the layers underneath, and probed one by one. The depth resolution would then be of the order of 1Å. In reality, of course, it is not so. The main factors that reduce the depth resolution in SIMS analysis are mixing effects [15–17,35,41–43] and the shape of the sputtered crater [15–17]. In addition, the combination of sputter rate and the time interval between successive data acquisition cycles can affect the depth resolution. Simply, if a layer of 100Å is sputtered between successive cycles, the profile depth resolution cannot be better than this value. As a result of the limited depth resolution, an abrupt, monolayer-thick interface would appear broadened in the SIMS depth profile. As a measure for the depth resolution, it is possible to use, for instance, the depth difference between the 84% and 16% levels ($1/e^2$ of the peak value), or the 90% and 10% levels of the signal of a relevant element in the profile across the interface.

2.3.4.1 Mixing Effects

As discussed in Section 2.2.3.1, surface atoms are driven deeper into the bulk when hit by impinging primary ions. These displaced atoms are termed *primary recoils* or *knock-ons*. Matrix atoms displaced because of collisions with recoils or implanted primary ions are called *secondary*, or *cascade*, recoils. While the average displacement of primary recoils has an inward pointing component, the displacement of the secondary recoils is more isotropic. The recoil displacement is the main cause of mixing that occurs during SIMS analysis (another cause may be due to enhanced diffusion of various types).

The range of the mixing depends on the primary ion beam species, impact energy, angle of incidence, and the matrix composition and structure. In general, the mixing range increases with the primary beam impact energy, and decreases with the angle of incidence (relative to the surface normal). Thus, for a high-depth-resolution profile, it is desirable to use a primary ion beam with low impact energy that hits the surface at an oblique angle.

2.3.4.2 The Sputtered Crater

A necessary condition for high depth resolution is that the material sputtered off the region from which the secondary ions are extracted be removed evenly. This ensures that at each instant the surface represents a single depth. To achieve such a flat region, the primary beam is kept rastered during the analysis, and the raster dimension is large compared to the beam's spot size. The crater must be large enough so that the collected secondary ions emerge only from a region within the flat portion of the crater's bottom. Any contribution from the crater's edge or adjacent regions must be avoided. This is done by limiting the extraction to only those secondary ions sputtered from the central region of the crater's bottom. This technique is called *gating*. Two main types of gating are used in SIMS instruments: electronic (i.e., the extraction or detection is done only when the rastered beam is over the central region) and optical (i.e., only the central region is imaged by the secondary ions optics). A flat, well-defined bottom in the crater is also important for an accurate depth measurement subsequent to the analysis, which is used in the conversion of a sputtering-time scale into a depth scale. This measurement is usually done by means of a profilometer.

2.3.5 Spatial Resolution

In the microscope mode (see Section 2.2.5.2), the spatial resolution of the SIMS analysis is determined by the secondary ion optics. When using the electron multiplier (or Faraday cup) as a collector for secondary ions transmitted from the entire imaged area, the resolution is determined by the diameter of the imaged surface area. A typical diameter is in the 10 to 100-μm range. If a position-sensitive detector is used for the collection of the transmitted ions, a spatial resolution of the order of 1 μm can be achieved.

In the microprobe mode (see Section 2.2.5.1), the spatial resolution is determined by the spot size of the primary ion beam. Some recent O_2^+ and Cs^+ guns can produce beams with a minimal spot size of the order of 1 μm. Liquid metal guns (such as a Ga^+) produce much brighter beams with spot sizes as small as 100Å. However, at these dimensions lateral mixing can become a factor.

2.3.6 Mass Resolution

The dynamic range in SIMS analysis is often limited by mass interference (i.e., overlapping of the peak of interest by peaks due to other species). Mass interference has several causes, including different isotopes with the same mass number, as in the case of ^{54}Cr and ^{54}Fe, and species at different ionization states which have the same mass-to-charge ratio, such as $^{56}Fe^{++}$ and $^{28}Si^+$. However, the most common sources of mass interference are molecular or cluster ions: $^{16}O_2$ or ^{31}PH and ^{32}S, $^{16}O_3$ and ^{48}Ti, ^{27}AlH and ^{28}Si, or $^{27}Al^{16}O_3$ and ^{75}As.

Two main methods are employed in SIMS instruments to reduce and even eliminate mass interference: energy filtering and high mass resolution.

2.3.6.1 Energy Filtering

As discussed in Section 2.2.3.3, the energy distribution curve of secondary molecular ions is narrower than that of secondary monoatomic ions. The narrowing effect is particularly evident in the energy distribution curves of triatomic or larger molecular ions. If the mass interference is caused by such molecular ions, shifting the energy band of the secondary ions that enter the spectrometer to higher energies can reduce, and in some cases even eliminate, the mass interference. This would improve, of course, the detection limit for the species of interest [44]. Shifting the energy band of the analyzed secondary ions is accomplished by the SIMS energy filter, through which pass the extracted secondary ions prior to entering the mass analyzer. An example of reducing mass interference due to molecular ions by energy filtering is shown in Figure 2.4. Unfortunately, this method is not effective in reducing mass interference due to diatomic molecular ions or to multiply charged ions and isotopes of the same mass.

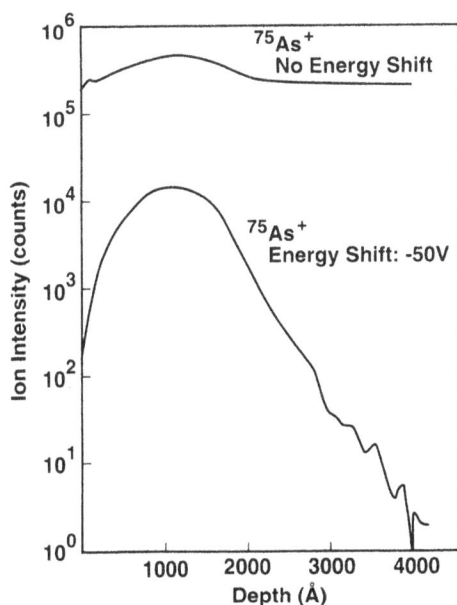

Figure 2.4 SIMS depth profiles of ^{75}As implant in Si. The upper profile was obtained with an energy band of 0 to 150V of the transmitted secondary ions, where mass interference by $^{29}Si^{30}Si^{16}O^+$ causes poor detection limit. The lower profile was obtained with the energy band shifted 50V toward higher energies, resulting in the elimination of the mass interference. (From Charles Evans & Associates)

2.3.6.2 High Mass Resolution

SIMS instruments equipped with a magnetic sector or TOF mass analyzer, can resort to another method of reducing or eliminating mass interference—high mass resolution. In most cases, a mass resolution $(M/\Delta M)$ of several thousand is required to resolve the peak of interest, not only from interference caused by a molecular ion, but also that by another isotope or a multiply charged ion. For example, the mass difference of ^{31}PH and ^{32}S is 0.00951 amu, requiring a mass resolution of about 3365 to resolve them. A mass resolution of 959 is needed to resolve ^{14}N^{+} from ^{28}Si^{++}, and a mass resolution of about 10,000 to resolve ^{48}Ca from ^{48}Ti. These mass resolution values are all within the range of such instruments. However, not all interferences can be resolved, since the peaks in some cases are too close. The mass resolution required to distinguish ^{54}Fe from ^{54}Cr, for instance, is about 73,000. In most instruments with a magnetic sector mass analyzer, a mass resolution of 10,000 can be easily achieved, and in some instruments can exceed 25,000. However, the higher the mass resolution, the lower the transmission of the spectrometer. Figure 2.5 shows a mass spectrum of a Si-doped $Al_{0.11}Ga_{0.89}As$ wafer, obtained on a magnetic sector SIMS, at mass resolution of about 4000. It shows the ^{28}Si and ^{27}AlH peaks well resolved.

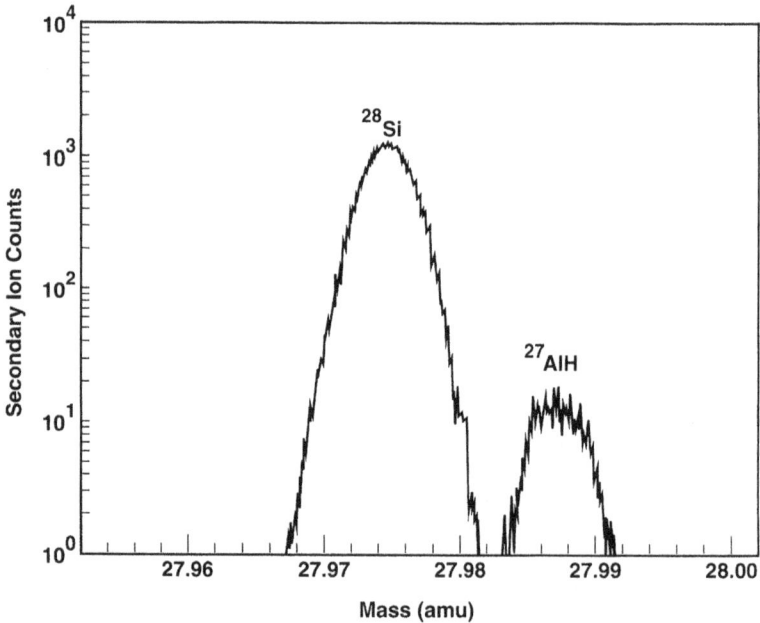

Figure 2.5 High-mass-resolution mass spectrum of a Si-doped $Al_{0.11}Ga_{0.89}As$ wafer, obtained on a magnetic-sector SIMS, with $M/\Delta M \simeq 4000$.

2.4 SIMS APPLICATIONS TO InP AND OTHER III-V STRUCTURES

The very high elemental sensitivity of the SIMS is, of course, the main advantage in using it in material characterization of InP and other structures. However, other features of SIMS analysis, such as depth resolution, spatial resolution, mass resolution, *et cetera*, may be of importance in various applications; in some cases, optimizing them could result in reduced sensitivity. In this section, some representative SIMS applications in the analysis of InP and other III-V structures are reviewed.

2.4.1 Surface Contamination Analysis

Analysis of surfaces is usually done in the static-SIMS mode of analysis, where spectra or elemental images of the surface provide a qualitative or comparative measure of levels of impurities, oxides, *et cetera*. Figure 2.6 shows a high lateral resolution static-SIMS elemental mapping of As on an InP surface. The image indicates some As accumulation in small regions on the surface, which was probably deposited during a processing step.

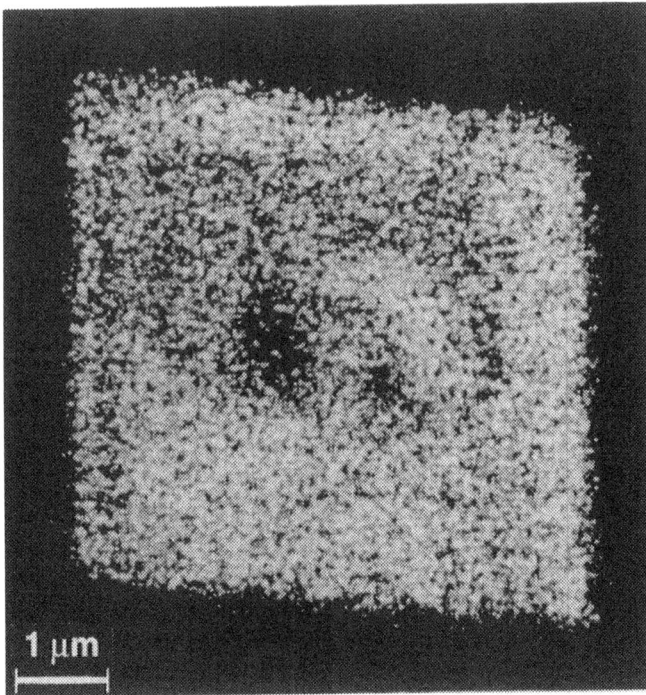

Figure 2.6 High-spatial-resolution As mapping of an InP surface, obtained in the static SIMS mode. The dark regions are rich in As. (From M. Schumacher, CAMECA)

2.4.2 Bulk Impurity Analysis

Bulk analysis refers to depth profile analysis of a wafer or a grown layer, usually intended to determine the concentration level of dopant and impurity elements in the matrix. In bulk analysis the main objective usually is to maximize the sensitivity, whereas depth and lateral resolution are not essential. Therefore, the primary beam current density is often increased to achieve lower detection limits. Figure 2.7 shows a SIMS bulk analysis of two $Al_{0.30}Ga_{0.70}As$ epilayers grown by a metalorganic chemical vapor deposition (MOCVD) under different conditions. In Figure 2.7(a), the concentration levels of O and C in the grown layer are 3–4 $\times$ 10^{17} cm^{-3} and 8–9 $\times$ 10^{18} cm^{-3}, respectively, while in Figure 2.7(b) the O level is about 2.5 $\times$ 10^{16} cm^{-3}, and the C concentration is below the instrument detection limit of about 1.5 $\times$ 10^{16} cm^{-3}. The quantification of the impurities' concentration was done by analyzing, under the same analytical conditions, O and C implants, of known fluencies, in GaAs and AlGaAs wafers. The depth scale was found by measuring the depth of the sputtered craters subsequent to the analysis.

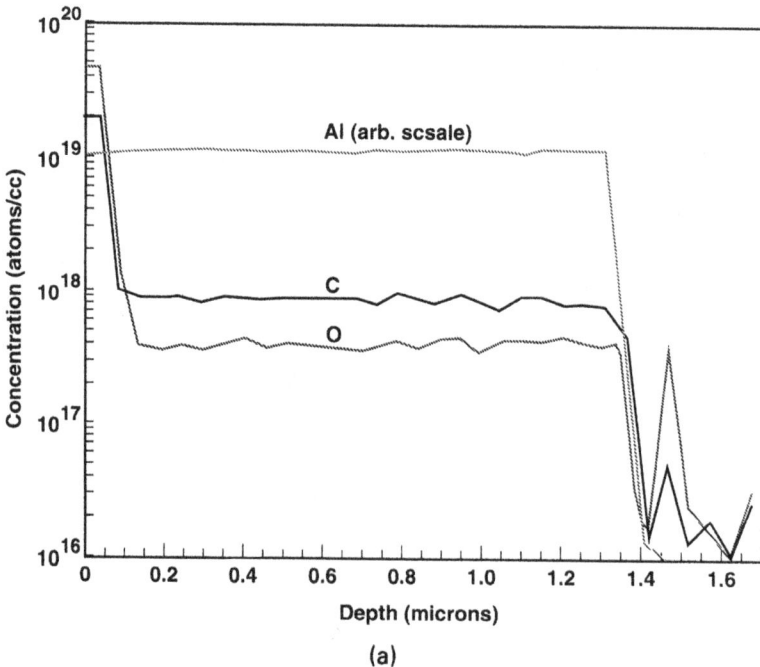

(a)

Figure 2.7 SIMS bulk analysis of C and O in MOCVD-grown $Al_{0.3}Ga_{0.7}As/GaAs$. The profiled epilayers were grown under conditions for high C and O concentration level (a), and for low levels (b). (Sample grown by K.D.C. Trapp and J.L. Zilko, AT&T Bell Laboratories)

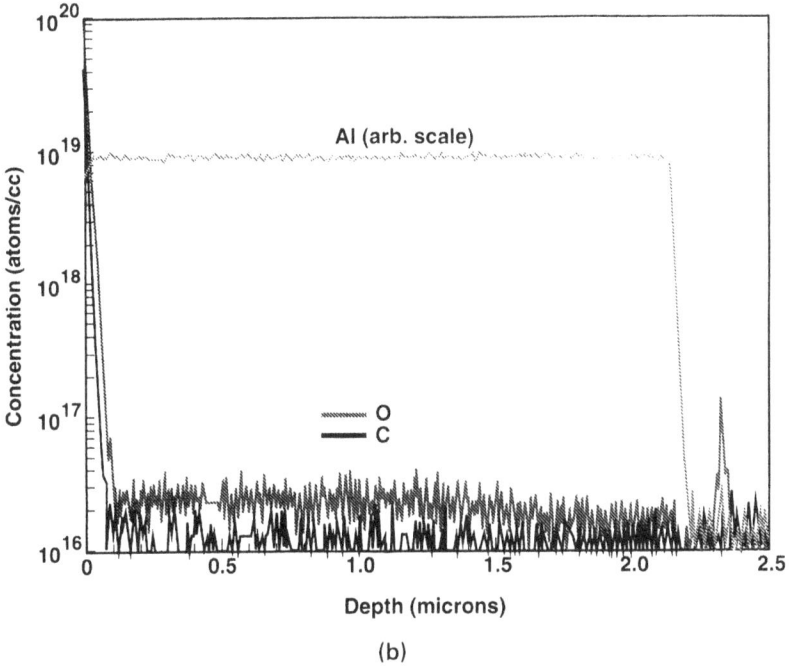

Figure 2.7 Continued

2.4.3 Layered Structures

Layered structures require special attention to two issues: (a) When sputtering across an interface from one layer to the next, the matrix may change, and with it the ionization yields for the analyzed elements and the sputtering rate may change as well. It is, therefore, essential to quantify the concentrations for each type of layer separately. It is also necessary to measure the sputter rate for each type of matrix in order to convert correctly the sputtering time into a depth scale. (b) In order to resolve impurities that may be present at an interface, or dopant diffusion across it, the SIMS depth profiling must be done at a high depth resolution. Since increased depth resolution may imply decreased sensitivity in some cases, the analytical parameters should be adjusted so that the sensitivity and depth resolution are optimized.

Figure 2.8 shows a SIMS depth profile of a layered structure grown in a metalorganic molecular beam epitaxy (MOMBE) system. The structure consists of three repetitions of 250Å InP/2300Å InGaAs layers, grown on an InP substrate. Each of the InGaAs layers contains a 750Å thick, Be-doped region. The profile shows the Be level in the three doped regions. The sharp interfaces of these regions imply that the Be did not diffuse. The profile also indicates an increase in the Be base level in the structure, corresponding to the increase in the Be doping level, and Be gettering in all InP/InGaAs interfaces.

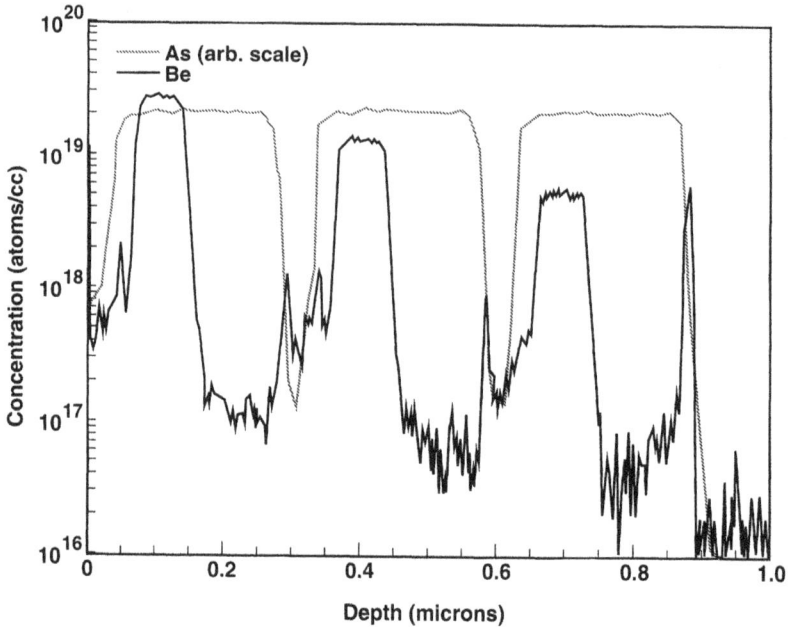

Figure 2.8 SIMS depth profile of a MOMBE-grown structure, consisting of 3 × [InP(250Å)/In-GaAs(2300Å)]/InP, with a 750Å-thick Be-doped region within each InGaAs layer. (Sample grown by D. Ritter, AT&T Bell Laboratories)

2.4.4 High Depth Resolution Depth Profiling

The thin layers in a multi-quantum well (MQW) structure impose even more stringent requirements on the depth resolution of the SIMS. Figure 2.9 shows a SIMS depth profile of a MOCVD-grown MQW laser structure. The structure includes a Si doped InP buffer layer, grown on a InP substrate, followed by a Si doped InGaAsP layer, an undoped InGaAs/InGaAsP MQW region, a Zn doped InGaAsP layer, and a Zn doped InP cap layer. The analysis was done using a Cs^+ primary beam of 3 keV impact energy and detecting $(MCs)^+$ molecular ions, where M is the analyzed element (Zn, Si, *et cetera*). The profile indicates sharp interfaces and no dopant diffusion into the MQW active region. To quantify this profile, the RSF and sputter rate values were measured for each type of matrix under the same analytical conditions.

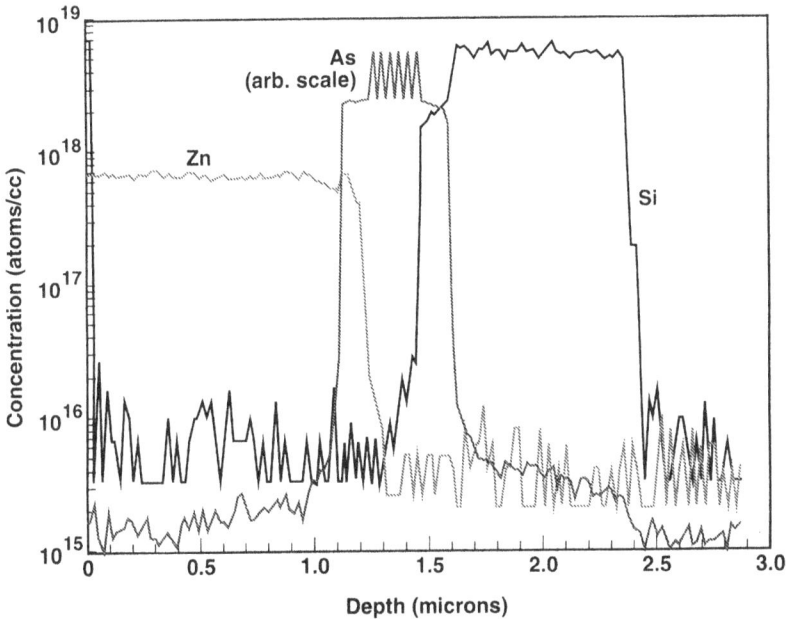

Figure 2.9 High-depth-resolution SIMS depth profile of an InP-based MQW laser structure. (Sample grown by V.R. McCrary, AT&T Bell Laboratories)

The highest depth resolution is required in the analysis of δ-doped layers, where the dopant element is confined to a thin region of only several atomic monolayers [46]. The analysis is done with a low impact energy beam, of 1 to 2 keV, and at an oblique angle of incidence.

2.4.5 Imaging Depth Profile

Figure 2.10 shows SIMS imaging depth profiles of a Be doped GaAs wafer, one as grown and one annealed. The analysis was done in the microprobe mode, forming a three-dimensional concentration image (see Section 2.2.5.1). The figure shows both concentration depth profiles and cross-section images of the Be in the wafers. Be outdiffusion is clearly seen in the annealed wafer. Cross-section images in other planes can be easily generated from the stored data in the computer memory. While the images give an overall view of the dopant distribution, the depth profile provides a quantitative measure of the dopant concentration as a function of depth in the structure.

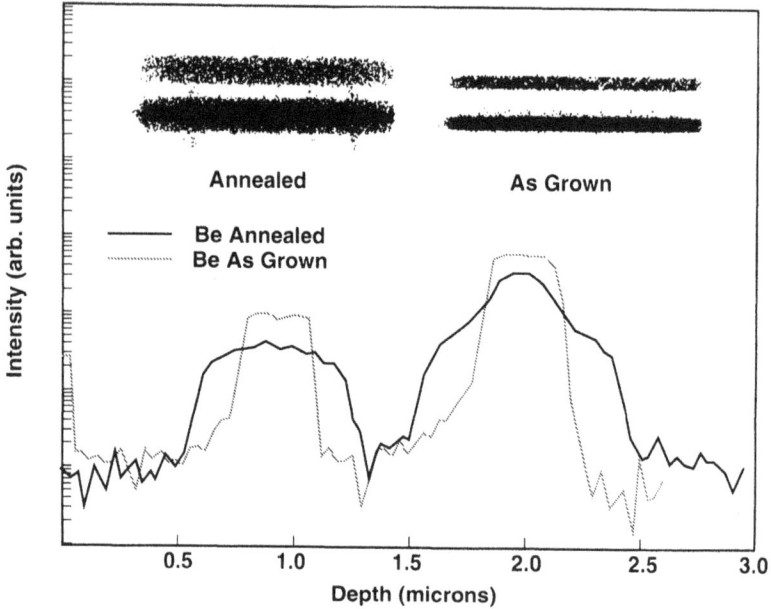

Figure 2.10 SIMS three-dimensional imaging profile of As-grown and annealed Be-doped GaAs wafers. The concentration depth profiles are shown in the lower portion of the figure, and vertical cross-section Be images are shown in the upper portion. Be outdiffusion is clearly seen in the annealed sample. (From Charles Evans & Associates)

2.4.6 Metal-Semiconductor Interface

Metallization is a crucial processing step in the fabrication of most III-V and other devices. A key issue is often the effect of annealing on the metal-semiconductor interface. Diffusion across the interface can result in the formation of alloys and intermetallic compounds. Since the ionization yields usually change appreciably when sputtering through the metal layers into the semiconductor, the quantification of the concentration profiles in the transient region of the interface may be complicated [47]. As is the case with other layered structures, the RSF and the sputtering rate values should be measured for each type of matrix.

Figure 2.11 depicts SIMS depth profiles of P in WSi_x/InP structures, as grown and annealed at different temperatures for 10 seconds. While the P outdiffusion increases with the annealing temperature, the W and Si profile (not shown) do not exhibit apparent diffusion into the substrate.

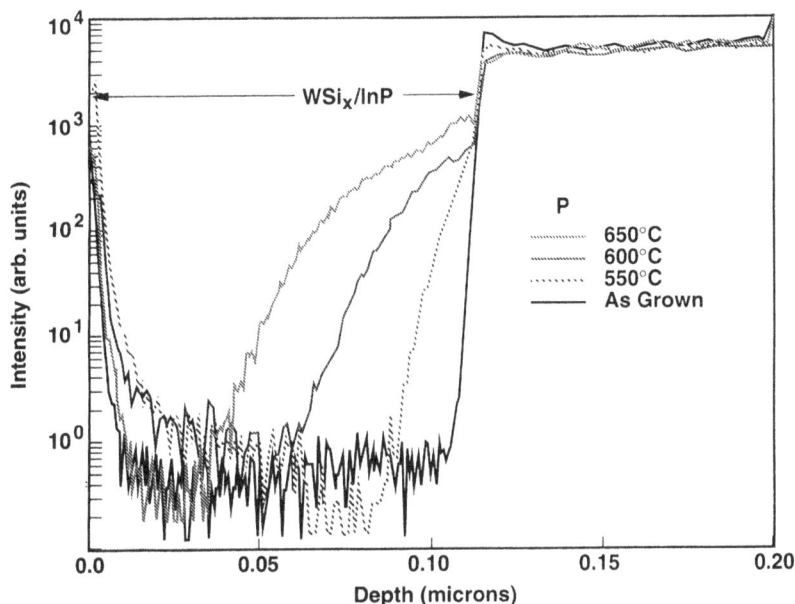

Figure 2.11 Superimposed SIMS depth profiles of P in WSi$_x$/InP structures, As-grown and rapid thermal annealed, at various temperatures for 10 sec. The W and Si profiles (not shown for clarity) did not show appreciable change after annealing. (Samples from A. Katz, AT&T Bell Laboratories)

REFERENCES

1. W.M. Briggs and M.J. Parker, in *Methods of Surface Analysis,* ed. A.W. Czanderna, New York: Elsevier, 1975, pp. 103–158.

2. S.H. Hercules and D.M. Hercules, in *Characterization of Solid Surfaces,* ed. P.F. Kane and G.B. Larrabee, New York: Plenum Press, 1978, pp. 307–336.

3. A.B. Chrisie, in *Methods of Surface Analysis,* ed. J.N. Walls, New York: Cambridge Univ. Press, 1989, pp. 127–168.

4. D. Briggs and M.P. Seah, eds., *Practical Surface Analysis by Auger and X-ray Photoelectron Spectroscopy,* New York: John Wiley & Sons, 1983.

5. C.C. Chang, in *Characterization of Solid Surfaces,* ed. P.F. Kane and G.B. Larrabee, New York: Plenum Press, 1978, pp. 509–521.

6. A. Joshi, L.E. Davis, and P.W. Palmberg, in *Methods of Surface Analysis,* ed. A.W. Azanderna, New York: Elsevier, 1975, pp. 159–222.

7. H.E. Bishop, in *Methods of Surface Analysis,* ed. J.M. Walls, New York: Cambridge Univ. Press, 1989, pp. 87–126.

8. T.M. Buck, in *Methods of Surface Analysis,* ed. A.W. Czanderna, New York: Elsevier, 1975, pp. 75–102.

9. E.N. Haeussler, *Surface and Interface Analysis,* Vol. 2, 1980, p. 134.

10. W.L. Baun, *Surface and Interface Analysis,* Vol. 3, 1981, p. 243.

11. D.G. Armour, in *Methods of Surface Analysis*, ed. J.M. Walls, New York: Cambridge Univ. Press, 1989, pp. 263–298.

12. L.C. Feldman, J.W. Mayer, and S.T. Picraux, *Materials Analysis by Ion Channeling*, New York: Academic Press, 1982.

13. L.C. Feldman, in *Methods of Surface Characterization*, Vol. 2, ed. A.W. Czanderna, New York: Plenum Press, 1991, pp. 311–360.

14. W.A. Grant, in *Methods of Surface Analysis*, ed. J.M. Walls, New York: Cambridge Univ. Press, 1989, pp. 299–337.

15. P. Williams, in *Applied Atomic Collision Physics*, Vol. 4, ed. S. Datz, New York: Academic Press, 1983, pp. 327–377.

16. A Benninghoven, F.G. Rüdenauer, and H.W. Werner, *Secondary Ion Mass Spectrometry*, New York: John Wiley & Sons, 1987.

17. R.G. Wilson, F.A. Stevie, and C.W. Magee, *Secondary Ion Mass Spectrometry*, New York: John Wiley & Sons, 1989.

18. A. Benninghoven, J. Okano, R. Shimizu, and H.W. Werner, eds., *Secondary Ion Mass Spectrometry, SIMS IV*, New York: Springer-Verlag, 1984.

19. A. Benninghoven, R.J. Colton, D.S. Simons, and H.W. Werner, eds., *Secondary Ion Mass Spectrometry, SIMS V*, New York: Springer-Verlag, 1986.

20. A. Benninghoven, A.M. Huber, and H.W. Werner, eds., *Secondary Ion Mass Spectrometry, SIMS VI*, New York: John Wiley & Sons, 1988.

21. A. Benninghoven, C.A. Evans, K.D. McKeegan, H.A. Storms, and H.W. Werner, eds., *Secondary Ion Mass Spectrometry, SIMS VII*, New York: John Wiley & Sons, 1990.

22. R. Smith and J.M. Walls, in *Methods of Surface Analysis*, ed. J.M. Walls, New York: Cambridge Univ. Press, 1989, pp. 20–56.

23. J.C. Vickerman, in *Methods of Surface Analysis*, ed. J.M. Walls, New York: Cambridge Univ. Press, 1989, pp. 169–215.

24. C.A. Andersen, H.J. Roden, and C.F. Robinson, *J. Appl. Phys.*, Vol. 40, 1969, p. 3419.

25. L.A. Heimbrook, "Laser Photoionization and Desorption Surface Analysis Techniques," *Proceedings of SPIE*, Vol. 1208, 1990, pp. 52–62.

26. R. Jede, in *Secondary Ion Mass Spectrometry, SIMS VII*, ed. A. Benninghoven, C.A. Evans, K.D. McKeegan, H.A. Storms, and H.W. Werner, New York: John Wiley & Sons, 1990, pp.169–177.

27. W.H. Christie and J.D. Fassett, in *Secondary Ion Mass Spectrometry, SIMS VII*, ed. A. Benninghoven, C.A. Evans, K.D. McKeegan, H.A. Storms, and H.W. Werner, New York: John Wiley & Sons, 1990, pp. 179–183.

28. R.W. Odom, in *Secondary Ion Mass Spectrometry, SIMS VII*, ed. A. Benninghoven, C.A. Evans, K.D. McKeegan, H.A. Storms, and H.W. Werner, New York: John Wiley & Sons, 1990, pp. 185–187.

29. C. Becker, in *Secondary Ion Mass Spectrometry, SIMS VII*, ed. A. Benninghoven, C.A. Evans, K.D. McKeegan, H.A. Storms, and H.W. Werner, New York: John Wiley & Sons, 1990, pp. 189–191.

30. K. Wittmaak, in *Inelastic Ion-Surface Collisions*, ed. N.H. Tolk, J.C. Tully, W. Heiland, and C.W. White, New York: Academic Press, 1977, pp. 153–199.

31. J.A. McHugh, in *Methods of Surface Analysis*, New York: Elsevier Sci. Pub., 1975, pp. 223–278.

32. D.E. Harrison, Jr., *Rad. Effects*, Vol. 70, 1983, p. 1.

33. H.H. Andersen, in *Sputtering by Particle Bombardment I*, ed. R. Behrisch, New York: Springer-Verlag, 1981, p. 72.

34. S. Hofmann, *Surface Interface Analysis*, Vol. 2, 1980, p. 143.

35. D.E. Sykes, in *Methods of Surface Analysis*, ed. J.M. Walls, New York: Cambridge Univ. Press, 1989, pp. 216–262.

36. H.A. Storms, K.F. Brown, and J.D. Stein, *Anal. Chem.*, Vol 49, 1977, p. 2023.

37. P.R. Boudewijn and H.W. Werner, in *Secondary Ion Mass Spectrometry, SIMS V*, ed. A. Benninghoven, R.J. Colton, D.S. Simons, and H.W. Werner, New York: Springer-Verlag, 1986, pp. 270–278.

38. H.W. Werner, *Surface and Interface Analysis*, Vol. 2, 1980, p. 56.

39. A. Benninghoven, *Surface Science*, Vol. 53, 1975, p. 596.

40. A. Benninghoven, in *Chemistry and Physics of Solid Surfaces*, ed. R. Vanselow and S.Y. Tong, New York: Springer-Verlag, 1976, pp. 207–233.

41. W. Vandervost, H.E. Maes, and R.F. DeKeesmaecker, *J. Appl. Phys.*, Vol. 56, 1984, p. 1425.

42. W. Vandervost and F.R. Shepherd, *J. Vac. Sci. Technol. A*, Vol. 5, 1987, p. 313.

43. C.W. Magee and R.E. Honig, *Surface Interface Analysis*, Vol. 4, 1982, p. 35.

44. R.J. Blattner and C.A. Evans, *Scanning Electron Microscopy*, Vol. 4, 1980, p. 55.

45. V. Swaminathan, M. Geva, and W.C. Dautremont-Smith, unpublished.

46. E.F. Schubert, *J. Vac. Sci. Technol. A*, Vol. 8, 1990, p. 2980.

47. M.L. Yu and W. Reuter, *J. Vac. Sci. Technol.*, Vol. 1, 1980, p. 36.

Chapter 3
Deep Levels in InP and Related Materials

W.A. Anderson and K.L. Jiao
State University of New York at Buffalo

3.1 INTRODUCTION

InP is becoming a semiconductor material of great importance because of its unique combination of electrical and optical properties. InP may be formed by many growth techniques. Bulk growth techniques include liquid encapsulated Czochralski (LEC), single crystal Bridgman technique, and float zone (FZ) process. Thin film epitaxy crystal growth techniques include liquid phase epitaxy (LPE), molecular beam epitaxy (MBE), metalorganic chemical vapor deposition (MOCVD or MOVPE), and combined MOCVD and MBE (MOMBE, CBE). The many alloy forms, including InGaAs or InGaAsP, provide a variety of energy gaps and carrier mobilities. Increased interest in electro-optics and photonics places InP and its alloys at the forefront of materials research. International conferences on InP and related materials are now held on an annual basis with the most recent located in Cardiff, Wales in April 1991.

Most electro-optic applications require defect-free material. Material growth and device fabrication must be conducted to minimize the introduction of deep levels. Several techniques exist for evaluating deep levels, including deep-level transient spectroscopy (DLTS) and photoluminescence (PL), and photoreflectance (PR) spectroscopy. This chapter will focus on DLTS evaluation of InP and related materials. First, we will review early studies, and then explore later work on more specialized materials and impurity effects. InP is unique, compared to Si or GaAs, because of its tolerance to radiation. We will address this issue as well.

We will also present a useful compilation of reported deep levels, including their respective causes. We will make a comparison of various results reported by the many research teams, and look at significant differences in trap characteristics based upon growth technique, dopant, type of junction, and post fabrication conditions. Many good agreements are evident, along with contradictions.

3.2 THEORY OF DEEP-LEVEL TRANSIENT SPECTROSCOPY AND VARIATIONS OF THE BASIC TECHNIQUE

DLTS is a powerful tool for investigating the deep levels in semiconductor materials. However, the principles behind it are not very complicated. It is simply a combination of the transient capacitance (current or charge) measurements at a scanning temperature using the rate-window sampling technique. The development of DLTS theory may be described by a flowing block diagram shown in Figure 3.1. The theory of the generation-recombination statistics presented by Hall [1] and Shockley and Read [2], plus the *p-n* junction theory, could be viewed as the basestones; the setup of the capacitance transient technology serves as the walls; and the introduction of the rate-window concept with the double boxcar sampling method [3] finish the roof of the DLTS architecture.

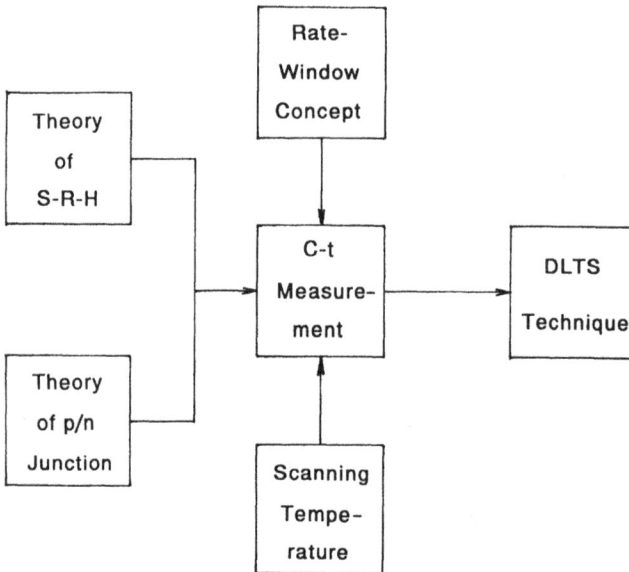

Figure 3.1 Block diagram of the formation of the DLTS technique.

3.2.1 Indirect Generation-Recombination Statistics

In nonequilibrium semiconductors, generation or recombination can occur either directly between conduction band and valence band (intrinsic generation-recombination) or indirectly by virtue of deep energy levels (extrinsic generation-recombination), which are called generation-recombination centers (*G-R centers*) or traps. For simplicity, assuming only one single energy level to exist in the bandgap, the generation-recombination dynamics can be divided into four basic steps or processes as shown in Figure 3.2, where a) is during the process and b) represents process completion. The arrows indicate direction of electron motion. Step I is the capture of an electron from the conduction band, E_c; step II is electron emission back to E_c; step III is the capture of a hole from the valence band, E_v; and IV is hole emission back to E_v. Step I plus Step III completes recombination process, while step II followed by IV will finish a generation cycle. Such a single energy level is called a G-R center. If step I is followed by II or step III is followed by IV, the trapping event will occur instead of recombination or generation, and the level will be named a trap center or level. Whether an energy level behaves like a G-R center or a trap depends upon its relative position with respect to the Fermi level, the sample temperature, and the capture cross section. Usually, only the trap properties of a level can be investigated using DLTS measurements.

Three important properties are involved with a trap level, namely, the trap density, N_T, the activation energy, E_A, and the capture cross section, σ. Through

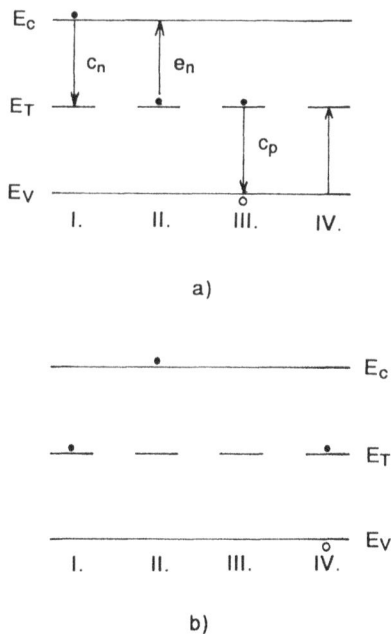

Figure 3.2 Diagram showing generation-recombination dynamics: a) during process; b) after process. After W. Shockley and W.T. Read, "Statistics of the Recombinations of Holes and Electrons," *Phys. Rev.*, Vol. 87, 1952, pp. 835–842.

some mathematical analysis, the following equations can be derived to describe an electron trap level:

$$e_n = (g\,\sigma_n\,N_C\,v_{th})\exp - (E_A/kT) \tag{3.1}$$

where e_n is the electron emission rate, g is the degeneracy factor, N_c is the effective density of states in the conduction band, and v_{th} is the thermal velocity. The term e_n is almost exponential with temperature except for the precoefficients v_{th} and N_c, which contain a T^2 dependence. For a capture process, there is the expression

$$c_n = \sigma_n\,v_{th}\,n \tag{3.2}$$

where c_n is the electron capture rate and n is the free electron concentration. Parallel expressions can be derived for the hole trap levels.

3.2.2 Capacitance Transient Measurements

Whenever an electron or hole is captured or emitted, the occupation of the trap level will be changed, and the rate of change is given by

$$dn_T/dT = dp/dt - dn/dt = (c_n n + e_p)(N_T - n_T) - (c_p p + e_n)n_T \tag{3.3}$$

Equation (3.3) can be simplified in the reverse biased space charge region and if the electron emission is predominant to give

$$dn_T/dT = -e_n\,n_T \tag{3.4}$$

The initial conditions are

$$n_T(t) = \begin{cases} N_T & \text{at } t = 0 \\ 0 & \text{at } t = \infty \end{cases} \tag{3.5}$$

where N_T is the total trap density. Therefore, the time dependence of n_T becomes

$$n_T(t) = N_T \exp(-e_n t) \tag{3.6}$$

From the well known p-n junction theory, the junction capacitance can be written as

$$C = A \sqrt{\frac{q\varepsilon_s \varepsilon_o N_{scr}}{2(V_{bi} - V)}} = C_o \qquad (3.7)$$

where A is the junction area, V_{bi} is the built-in voltage, V is the applied voltage, and $N_{scr} = N_D - n_T$ is the total ionized impurity concentration. Thus, the change in occupied trap level can be mirrored in the change of capacitance:

$$C = C_o \sqrt{1 - \frac{n_T(t)}{N_D}} \qquad (3.8)$$

under the assumption that $N_D \gg n_T$.

Figure 3.3 shows the capacitance transient measurement as well as band diagrams before, during, and after the pulse. It can be seen that majority and minority

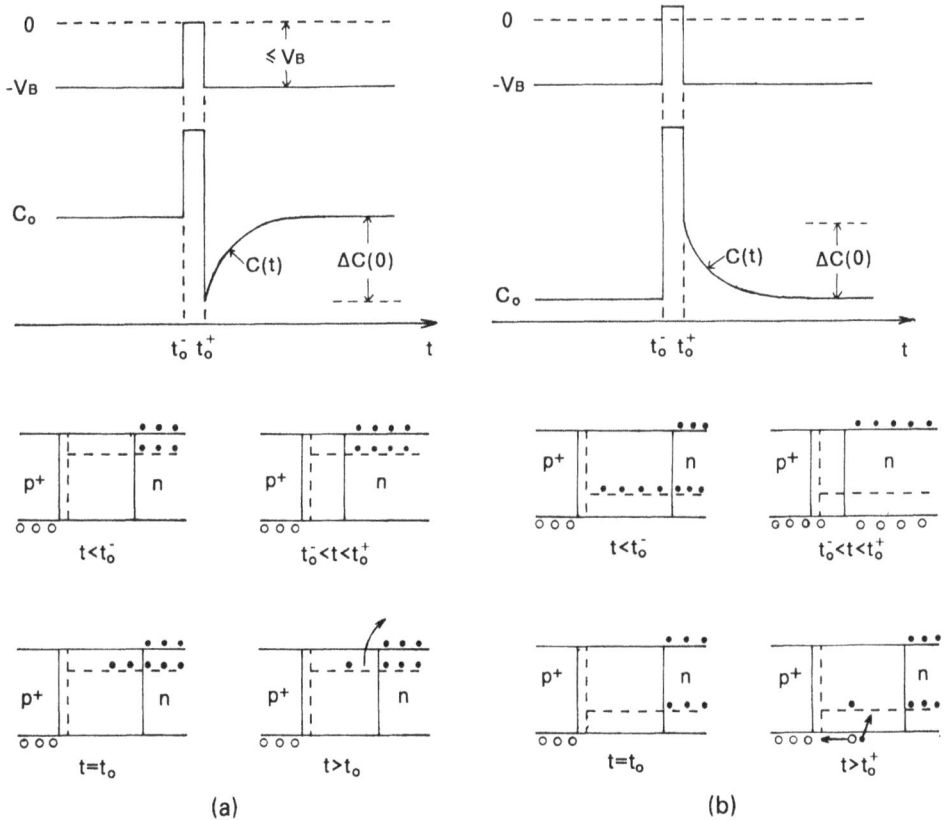

Figure 3.3 Capacitance transient: a) from a majority carrier trap, b) from a minority carrier trap. After C. Barnes, "Short Course on Deep Level Transient Spectroscopy," Materials Research Society, Boston, Nov. 1985.

carrier traps give the opposite contribution to the capacitance transient, so distinguishing them is possible by this method.

3.2.3 DLTS Theory

From the above discussion, it can be seen that if a series of capacitance transient measurements are conducted at different temperatures, the properties of a trap level are obtainable, although such a task would be in a single-shot form rather than a spectrum form. Hence, it is very time consuming and less aesthetic. It was the introduction of the rate-window concept with the double boxcar sampling technique, presented by Lang in 1974, that made automated data acquisition available. Figure 3.4 illustrates the idea of such a creative introduction. It can be seen that for a repetitive C-t test with a varied decay time constant at different temperatures, the data sampling through a given rate window (t_1, t_2) can lead to a waveform output of the relative change of capacitance (current or charge). This waveform can be called the DLTS output or DLTS spectrum if using temperature as the X-axis. For a regular trap level, the amplitude of the DLTS output tends to be zero if the sample temperature is much lower or higher, but will reach a maximum at a certain temperature (T_{max}). The emission rate can be derived as follows:

$$\frac{d\Delta C}{dT}\bigg|_{T=T_{max}} = \frac{e\Delta C}{de_n}\frac{de_n}{dT} = \frac{de_n}{dT}(-t_1 e^{-e_n t_1} + t_2 e^{-e_n t_2}) \tag{3.9}$$

$$e_n(T_{max}) = \frac{\ln(t_2/t_1)}{t_2 - t_1} \tag{3.10}$$

By varying the rate window, several data points of T_{max} can be obtained. Therefore, the Arrhenius plot of $\ln(e_n/T^2)$ versus $1/T$ can be made, and the slope will lead to the solution of the activation energy, E_A, if the emission follows the exponential form in Equation (1). The trap density, derived from the maximum DLTS output, C_{max}, is given by

$$N_T = \frac{2N_D \, r^{r/(r-1)}}{1-r}\frac{\Delta C}{C_o} \tag{3.11}$$

where $r = t_2/t_1$. Usually, the approximation $N_T = 2 N_D \Delta C/C_o$ is used, which holds if r is very large.

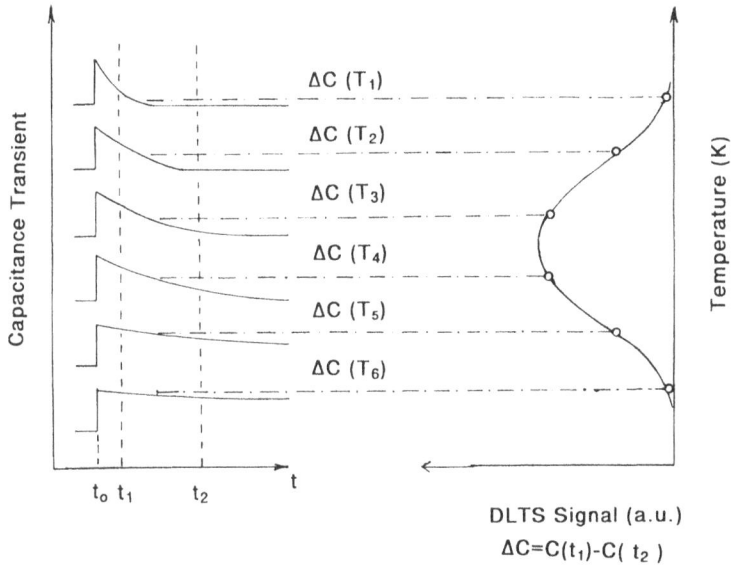

Figure 3.4 Introduction of the rate-window concept and formation of the DLTS signal. After D.V. Lang, "Deep-Level Transient Spectroscopy: A New Method to Characterize Traps in Semiconductors," *J. Appl. Phys.*, Vol. 45, 1974, pp. 3023–3032.

3.2.4 Variation from the Basic Technique

Since the first presentation of the DLTS technique using the double boxcar approach by Lang, many variations or alternatives have been released. It is not easy to give a complete classification and comparison because of the many variations. Types of instrumentation include boxcar DLTS, lock-in amplifier DLTS [4,5], correlation DLTS [5,6], and computer DLTS [7–12], which can digitize the DLTS waveform so that only one temperature sweep is needed for data analysis. Dependent upon different transient signals, there are capacitance DLTS, current DLTS, and charge DLTS. By changing the pulse mode, several refinements of the basic boxcar technique have been developed. Typically, the double-correlation DLTS (DDLTS) method [13] uses two pulses instead of one so that the trap concentration profiles can be obtained by varying the observation windows, the pulse amplitudes, or the dc reverse basis.

A contrast to the conventional constant voltage DLTS (CV-DLTS), is the constant capacitance DLTS (CC-DLTS) [14–21]. As the name implies, the capacitance will be held constant during the carrier emission measurement by dynamically varying the applied voltage during the transient response through a feedback path [17]. It is the time-varying voltage, rather than capacitance, that gives the trap information, resulting in the advantage that the width of the depletion region remains constant so that the arbitrary N_T can be tested without the restriction of $N_T < N_D$. CC-DLTS is also good for trap concentration profiling, especially when combined with DDLTS.

Other sources, such as light or electron beam (e.g., in a scanning electron microscope), can be used to excite minority carriers in place of the electrical filling pulses, yielding the so-called optical DLTS (O-DLTS) [22–25] and scanning DLTS (S-DLTS) [26–29]. Alternatively, a variable-energy light can replace the temperature sweep whether the sample is pulsed optically or electrically to form different O-DLTS themes. However, the emission rate, e_n, must be changed into $e_n + e_n^o$, where $e_n^o = \sigma_n^o \, \phi$ is the optical electron emission rate, σ_n^o is the optical electron capture cross section, and ϕ is the photon flux density. O-DLTS can determine the optical properties, such as σ_n^o, and is unique in detecting multilevel traps, such as the double-donor nature of sulfur in silicon [30]. In addition, O-DLTS offers the ability to generate and modulate minority carriers without the need of a p-n junction, for example, in semiinsulating materials. The photo-induced current transient spectroscopy (PITS or PICTS) [31–36] is, in a broad sense, a pure form of DLTS since the photocurrent is to be measured instead of an electrical signal. The results of PICTS will not be discussed here, however.

In summary, DLTS has become the most commonly used deep-level characterization technique in the area of semiconductor research. The more sophisticated theories for nonexponential emission, for nonuniform doping concentration, *et cetera*, have also been developed.

3.3 EARLY WORK ON DEEP LEVELS IN InP

In 1978, Majerfeld *et al.* [37] reported on two studies of deep levels in InP in which great detail was not given. Most early work was conducted on material grown by bulk techniques or vapor phase epitaxy, as was the case here, and used Au-Schottky contact, although some work involved in Al- or Ti-Au-Schottky. Majerfeld reported traps with activation energies at 0.43, 0.59, and 0.63 eV. Other data are also given in Table 3.1 for electron traps and Table 3.2 for hole traps.

Wada *et al.* [38] studied deep levels using the Au-Schottky contact on n-bulk by LEC, vapor phase epitaxy (VPE), and LPE materials. No traps were found in LPE-grown InP, but many in the other two cases given in Table 3.1. Traps A, B, E, and F were seen in both LEC and VPE material, whereas C, D, G, and H were only seen in LEC material. They concluded there was a 0.39-eV separation in energy between L-G bands from an argument based on electron emission and capture.

Bremond *et al.* [39] investigated n- and p-type LEC and VPE layers grown on n^+ or SI InP substrates using an Au-Schottky contact. Deep traps were identified as being related to Fe and Cr dopants. Significant quantities of both electron and hole traps were detected. Their analysis identified the IE1 trap due to the one electron Fe^{2+} trap state, and the IE3 trap due to the Cr^{2+} acceptor level. Similarly, Levinson

Table 3.1
Summary of Electron Traps in InP for Early Work

Identity	Activation Energy $E_A(eV)$	Capture Cross-Section σ (cm^2)	Trap Density N_T (cm^{-3})	Reference	Reason
E2	0.11	3×10^{-17}		[40]	
H	0.11	1.4×10^{-16}		[38]	
G	0.12	1.8×10^{-18}		[38]	
E10	0.12		5×10^{11}	[46]	MOCVD
F	0.15	1.6×10^{-18}		[38]	
E9	0.16		9.3×10^{11}	[46]	
IE6	0.17	6×10^{-18}		[39]	VPE-n
V	0.18	1×10^{-15}		[41]	VPE
IE5	0.20	2.5×10^{-13}		[39]	Bulk-n
E7	0.22			[46]	
E1	0.23	4×10^{-19}		[47]	MOCVD
IE4	0.24	7×10^{-16}		[39]	Bulk-n&n:Fe
E8	0.24		1.7×10^{12}	[46]	
I1	0.24			[44,45]	Fe-doped bulk
E5	0.27				
E2	0.29	5×10^{-20}		[47]	
E6	0.30		3.3×10^{13}	[46]	
W	0.30	8×10^{-19}		[41]	VPE
E1	0.30	$>4.0 \times 10^{-15}$	1.0×10^{15}	[43]	Zn
E2	0.35	3.8×10^{-15}	8×10^{14}	[43]	Excess P
E7	0.37			[39]	
MOE1	0.37	3×10^{-18}	3×10^{13}	[44]	Undoped n
E3	0.38	3×10^{-20}		[47]	
U	0.40			[41]	VPE
IE3	0.40	5×10^{-15}		[39]	Many cases; Cr^{2+} on cation (substitution)
T	0.43	8×10^{-16}		[41]	VPE
E4	0.43		2.8×10^{13}	[46]	
E	0.43	3.4×10^{-15}	6.4×10^{13}	[37][b]	Found in all cases and coupled to conduction band minimum
				[38]	
IE2	0.45	5×10^{-16}		[39]	Bulk-n, VPE n/n^+
E3	0.45		8×10^{10}	[46]	
C	0.47	1.3×10^{-14}		[38]	
F1	0.48	6×10^{-19}		[45]	MOCVD:Fe[a]

Table 3.1

Continued

Identity	Activation Energy $E_A(eV)$	Capture Cross-Section σ (cm^2)	Trap Density N_T (cm^{-3})	Reference	Reason
E3*	0.49	1×10^{-20}	3×10^{14}	[43]	In-vac or P-INTER
MOE2	0.53	6×10^{-20}	1.5×10^{14}	[44]	In-vac or P-INTER
S	0.53	$>2 \times 10^{-15}$		[41]	VPE, Co
V	0.54	$>5 \times 10^{-15}$		[41]	VPE
B	0.55	1×10^{-14}		[38]	
E1	0.56		1.7×10^{14}	[46]	MOCVD, Fe Substitute
R	0.58	$>1 \times 10^{-15}$		[41]	VPE, Fe substitute on In
D	0.59	3×10^{-14}	2×10^{13}	[37][b]	One bulk, indirect L-MINIMA
D	0.59	3.5×10^{-11}		[38]	
E4	0.59	3×10^{-15}	4.6×10^{13}	[43]	LEC-bulk
MOE3	0.60	9×10^{-19}	6.7×10^{14}	[44]	
E4	0.61	4×10^{-20}		[47]	
IE1	0.63	3×10^{-14}		[39]	Many cases; one electron trap state Fe^{2+} on cation site
A	0.63	5.5×10^{14}		[38]	
E9	0.64			[39]	
A	0.63		1.2×10^{14}	[37][b]	Bulk + VPE, indirect L-MINIMA
E2	0.66		1×10^{13}	[46]	
Q	0.70	$>2 \times 10^{-15}$		[41]	VPE
F2	0.78	1×10^{-20}		[45]	MOCVD:Fe[a]

*Main trap

(a) Donor-Fe complex center (F_1) and Fe-related defect (F_2) due to Fe-out diffusion from substitute.
(b) LEC and VPE show traps. LPE does not.
[37]: Au-Schottky on n-bulk and VEP-epi
[38]: Au-Schottky on n-bulk by LEC, VPE, and LPE
[39]: Au-Schottky on n- and p-type LEC or VPE material
[40]: Al-Schottky on undoped (100) LEC-InP
[41]: Ti-Au Schottky on VPE-InP (also seen in LEC bulk)
[43]: Au-Schottky on n-type by LEC or p^+/n by Zn diffusion
[44]: Au-Schottky on MOCVD-grown
[46]: Au-Schottky on MOCVD-grown
[47]: Au-Schottky on MOCVD-grown

Table 3.2
Summary of Hole Traps in InP for Early Work

Identity	Activation Energy E_A (eV)	Capture Cross Section σ (cm²)	Trap Density N_T (cm⁻³)	Reference	Reason
IH1	0.70			[39]	Bulk p:Fe
H2	0.64		8×10^{14}	[43]	LPE, Zn-doped
IH2	0.48	1×10^{-13}		[39]	Bulk p-undoped
IH3	0.36			[39]	Bulk p-undoped
H1	0.27		2×10^{14}	[43]	LPE, Zn-doped

[39]: LEC or VPE (n and p), Au-Schottky
[62,63]: n^+/p(Zn) and p^+/n(Si) (also seen as a minority trap in n-InP after irradiation)
[64]: n^+/p diffused, Zn-doped (4.5E15-1E18)

---

et al. [40] studied undoped (100) LEC-grown material using an Al-Schottky contact. He and others explored effects of electron irradiation which introduced levels at 0.14, 0.24, and 0.41 eV below the conduction band. An original level at $E_c - 0.46$ eV was detected. A separate section will address effects of radiation on InP. Tapster [41] used a Ti-Au-Schottky contact to study VPE-grown InP. Many levels were detected that correspond to those reported by others. In addition, trap *R* was attributed to substitutional Fe on an In site, and trap *S* to Co. From the same group, Skolnick *et al.* [42] studies Co-doped InP to reveal an electron trap at $E_c - 0.53$ eV and a hole trap at $E_v + 0.50$ eV. The former was attributed to $Co^{2+} \rightarrow Co^{3+} + e$, and the latter to $Co^{4+} \rightarrow Co^{3+} + h$ from Co substitution for In.

McAfee *et al.* [43] investigated Zn-diffusion or an Au-Schottky barrier on LEC-grown, (100) n-type InP. Electron traps at $E_c - 0.30$ eV, $E_c - 0.35$ eV, $E_c - 0.49$ eV, and $E_c - 0.59$ eV were respectively attributed to Zn diffusion, combined effects of Zn and P, excess P, and unknown reasons. Hole traps were seen at $E_v + 0.27$ eV and $E_v + 0.64$ eV.

InP formed by MOCVD was studied by Ogura *et al.* [44,45], Nicholas *et al.* [46], and Pudensi *et al.* [47] using Au-Schottky structures. Ogura reported three levels with $E_A = 0.37$, 0.53, and 0.60 eV for the epitaxial layer, as well as $E_A = 0.37$ and 0.59 eV for the bulk LEC-grown material. The two common traps probably originate with the substrate, whereas the 0.53 eV contribution was speculated to be due to a defect introduced by phosphorous overpressure. Pudensi reported electron levels with $E_A = 0.23$, 0.29, 0.38, and 0.61 eV, two of which agree very well with those found by Ogura [45]. The 0.61-eV signal was attributed to Fe. Many levels were reported by Nicholas, most of which did not correspond to the ones reported by Ogura. This points out the dependence of defects in InP upon even minor changes in growth conditions.

It is clear that proper growth of InP can be accomplished without the introduction of deep levels. This will be even more clear in the next two sections of this chapter. Table 3.1 shows the multitude of electron traps that have been detected. Fewer hole traps have been reported, which may be due in part to less effort spent in the search for hole traps. Yet more recent growth of InP has produced defect-free material such as that done by LPE [38]. Defects have been attributed to Fe-doping [42], In-vacancy or P-interstitial [42], Cr-substitution [39], Co [41], and Fe-substitution [43].

3.4 MORE RECENT WORK ON DEEP LEVELS IN InP AND RELATED MATERIALS

Since 1984, more specialized studies have been conducted on deep levels in InP and related materials. Effects of specific dopants or impurities have been explored. Studies have been made on InP grown by LPE, MOCVD, LEC, low pressure MOCVD (LPMOCVD), and VPE. Effects of lattice mismatch, strain, rapid thermal processing, and electrothermal stress have been explored. Finally, ternary and quaternary forms have been studied to some extent and will be reported herein. A summary of some of these works is given in Table 3.3.

Tin and Barnes [48] studied Au-oxide-InP where the InP was grown by LEC. Deep levels were seen on as-grown material at $E_c - 0.59$, 0.54, and 0.45 eV. Application of an electrothermal stress at 390K and $-2V$ added levels at $E_c - 0.35$ and 0.22 eV. These deep levels were attributed to P-interstitial or In-vacancy ($E_c - 0.59$ eV), oxygen ($E_c - 0.54$ eV), P-vacancy ($E_c - 0.45$ eV), or recombination-enhanced defect reaction of the stress-related pair. Kim et al. [49] studied a similar structure and observed three deep levels with two at similar energies as reported by Tin. These levels were attributed to the insulating (I)-layer since DLTS peaks shifted with fill pulse amplitude [50]. They also studied effects of rapid thermal processing, which added two new levels attributed to a P-vacancy. InP grown by MOCVD was studied by Yamamoto et al. [51] using a Au/anodic-oxide/InP structure. Traps were seen with activation energy $E_A = 0.80$, 0.44, and 0.24 eV for companion LEC-grown InP and MOCVD material grown at temperature $T \leq 500°C$. No traps were seen in MOCVD material grown at $T \geq 600°C$. The 0.44 eV level was attributed to a P-vacancy or In-interstitial. A study of MBE-grown material by Iliadis et al. [52] used an $Al/SiO_2/InP$ structure to reveal three traps. Traps at $E_c - 0.57$ eV and $E_c - 0.52$ eV were attributed to low growth temperature, causing antisite defects, interstitials, or vacancies. They attributed the $E_c - 0.43$ level to a P-vacancy: Fe

Table 3.3
More Recent Data on Deep Levels in InP and Related Materials

Material	Device Structure	Special Conditions	Trap Level (eV)	Capture Cross Section (cm^2)	Reference
(100)n-InP by LEC	Au/oxide/InP	Original	$E_c - 0.59$	3×10^{-16}	[48]
			$E_c - 0.54$	8×10^{-17}	
			$E_c - 0.45$	3×10^{-17}	
		Electrothermal	$E_c - 0.35$	8×10^{-19}	
		stress	$E_c - 0.22$	4×10^{-20}	
n-InP by LEC	M/I/InP	Due to I	$E_c - 0.79$	8.4×10^{-13}	[49]
		Due to I	$E_c - 0.58$	8.4×10^{-16}	
		Due to I	$E_c - 0.55$	1.7×10^{-15}	
		Due to RTA	$E_c - 0.43$	8.8×10^{-14}	
		Due to RTA	$E_c - 0.35$	8.4×10^{-14}	
n-InP by MOCVD	Au/oxide/InP	Phos. deficient	$E_c - 0.80$		[51]
		Low temp.	$E_c - 0.44$		
			$E_c - 0.24$		
n-InP by MBE	Al/SiO$_2$/InP		$E_c - 0.57$	7×10^{-11}	[52]
			$E_c - 0.52$	1.3×10^{-14}	
		Fe impurity	$E_c - 0.43$	2.6×10^{-11}	
InGaAsP/InP by VPE	p^+/n	$1.6 \times 10^{16}/cm^3$	$E_v + 0.18$	5×10^{-16}	[53]
		$1.6 \times 10^{16}/cm^3$	$E_c - 0.60$	2×10^{-16}	
		All samples	$E_c - 0.60$	1×10^{-14}	
In$_{0.53}$Ga$_{0.47}$As:Zn by LPE	p-InGaAs/n-InP	PtTi diffusion and RTA	$E_v + 0.89$[a]	3×10^{-14}	[54]
			$E_c - 0.61$[b]		
			$E_c - 0.45$[b]		
			$E_c - 0.35$[b]		
			$E_c - 0.30$[b]		
InP:Zn by LEC	Yb/p-InP Schottky		$E_v + 0.56$	3×10^{-18}	[55]
			$E_v + 0.22$	9×10^{-19}	
			$E_v + 0.12$	1.4×10^{-17}	
			$E_c - 0.40$	9×10^{-18}	
			$E_v + 0.25$	4×10^{-21}	
n-InP by LEC	Pd/oxide/InP		$E_c - 0.20$	3.8×10^{-13}	[56]
n-InP by MOCVD	Pd/oxide/InP		$E_c - 0.375$	4.0×10^{-16}	
p-InP by MOCVD	n^+/P		$E_c - 0.38$	2.0×10^{-14}	
			$E_v + 0.93$	1.4×10^{-13}	
InP:Mn by OMVPE	p-InP:Mn/n^+-InP:S	Mn-doped	$E_v + 0.25$	3.4×10^{-18}	[57]
n-GaInAs/InP by MBE	Au/oxide/GaInAs	O-related	$E_c - 0.33$	1×10^{-16}	[58]

Table 3.3
Continued

Material	Device Structure	Special Conditions	Trap Level (eV)	Capture Cross Section (cm²)	Reference
$In_{0.5}(Ga_{1-x}Al_x)_{0.5}P$:Zn by LPMOCVD	Au-Schottky	Zn-doped, $x = 1$	$E_v + 0.42$		[59]
		Zn-doped, $x = 0.4$	$E_v + 0.23$		
		Zn-doped, $x = 0$	$E_v + 0.04$		
$In_y(Ga_{0.3}Al_{0.7})_{1-y}As$:Si by LPMOVPE	Schottky contact	Role of In	$E_c - (0.46-0.49)$		[60]
			$E_c - (0.35-0.38)$		
			$E_c - (0.20-0.24)$		
$Ga_xIn_{1-x}P$ by MOCVD	Au/p-GaInP	Lattice mismatch	$E_v + 0.84$	9×10^{-12}	[61]

(a) Seen only for T > 500°C.
(b) Seen in all cases.

impurity complex. It is clear that different growth techniques and the use of different oxides in metal insulator semiconductor (MIS) devices can greatly influence deep-level formation.

We next consider other studies on InP, but now using different structures or effects of other dopants and impurities. Macrander and Johnston [53] used p^+-n diodes made from VPE-grown InGaAsP on n-InP. All samples gave an electron trap at $E_c - 0.60$ eV because of interaction with the Γ-conduction band minimum. Another trap at the same level was due to interaction with the L-conduction band minimum caused by richness in phosphorus for a carrier density of 1×10^{16} cm^{-3}. This same material gave a unique hole trap with an activation energy of 0.18 eV. Another study by Jiao et al. [54] involved LPE-grown p-InGaAs on n-InP. The Pt:Ti ohmic contact was treated by rapid thermal anneal (RTA) at different temperatures in order to evaluate influence on electrical properties. All samples showed electron traps at $E_A = 0.61, 0.45, 0.35,$ and 0.30 eV. A unique hole trap was seen for RTA > 500°C giving $E_A = 0.89$ eV, $\sigma = 3 \times 10^{-14}$ cm^{-2}, and $N_T = 1-3 \times 10^{-14}$ cm^{-3}. Most work has been conducted on n-InP, whereas Singh and Anderson et al. [55] studied Zn-doped bulk InP using a Schottky junction. Hole traps were seen at $E_v + 0.56$

eV, $E_v + 0.25$ eV, $E_v + 0.22$ eV, and $E_v + 0.12$ eV, with an electron trap at $E_c - 0.40$ eV. The trap at $E_v + 0.25$ eV was due to the interface and not the bulk. Other studies [56] showed an electron trap at $E_c - 0.20$ eV for a Pd/oxide/bulk n-InP, $E_c - 0.375$ eV for a Pd/oxide/epitaxial n-InP, and $E_c - 0.38$ eV for an n^+/p-epitaxial InP grown by MOCVD. The sample structures are shown in Figure 3.5 and the different DLTS profiles in Figure 3.6. The dotted lines refer to minority carrier traps and the solid ones refer to majority carrier traps. The first electron trap was also seen by Bremond [39], whereas the others were seen by McAfee [43] and attributed to excess P during MOCVD growth. Huang and Wessels [57] studied MOVPE-grown p-InP:Mn/n^+-InP:S. A hole trap at $E_v + 0.25$ eV was attributed to the Mn on an In site, with hole emission from Mn^{+3} to the valence band.

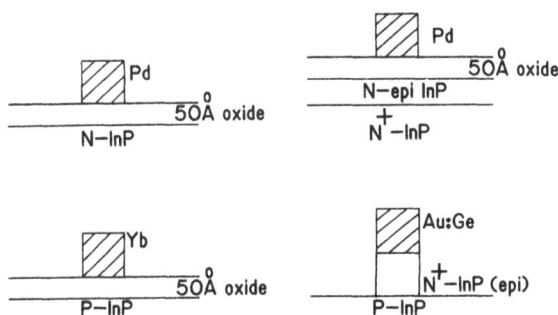

Figure 3.5 Several structures for DLTS studies of InP.

The final discussion in this section considers ternary and quaternary semiconductors related to InP. Loualiche *et al.* [58] studied MBE-grown GaInAs/InP using a structure of Au/oxide/GaInAs. An oxygen-related trap was seen at $E_c - 0.33$ eV in the MBE-grown material and in the companion LPE material. A Au-Schottky contact was used in a study of LPMOCVD-grown $In_{0.5}(Ga_{1-x}Al_x)_{0.5}P$ with Zn doping [59]. A hole trap was seen with dependence upon x giving levels at $E_v + 0.42$ eV, $E_v + 0.23$ eV, and $E_v + 0.04$ eV, respectively, for $x = 1.0, 0.4$, and 0.0. Pann *et al.* [60] used a Schottky contact to study the influence of In on LPMOVPE-grown, Si-doped $In_y(Ga_{0.3}Al_{0.7})_{1-y}As$. Indium was shown to modify the DX-center by reducing E_A when In is near the Si donor. MOCVD-grown p-GaInP was studied using a Au-Schottky contact [58]. Lattice mismatch was deemed the cause of a hole trap at $E_v - 0.84$ eV because of strain relaxation.

An example of the work first mentioned by Yamasaki *et al.* [50] is shown in Figure 3.7 for a Pd/oxide/n-InP diode [56]. This represents a bulk trap since bulk traps have a discrete energy causing peaks to occur at the same temperature for different pulse voltages. Surface traps have an energy distribution of surface states causing both shape and peak location to change with pulse voltage.

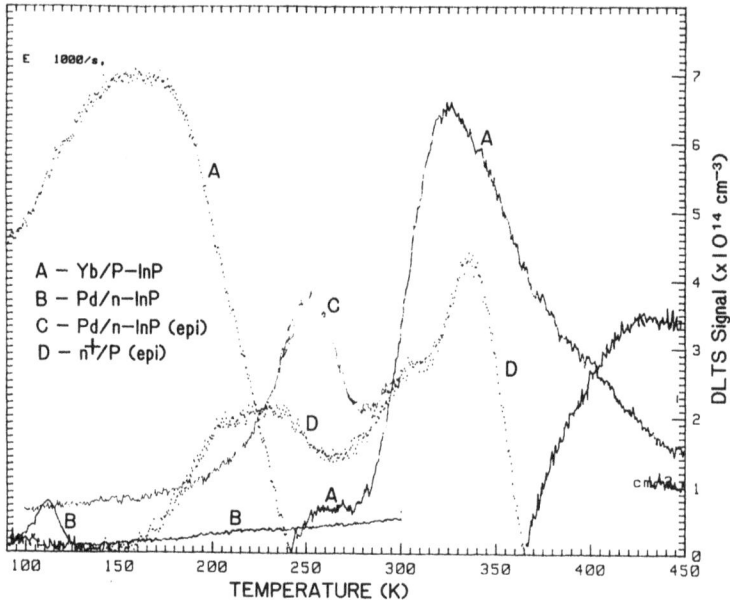

Figure 3.6 DLTS signal for four InP cases [56].

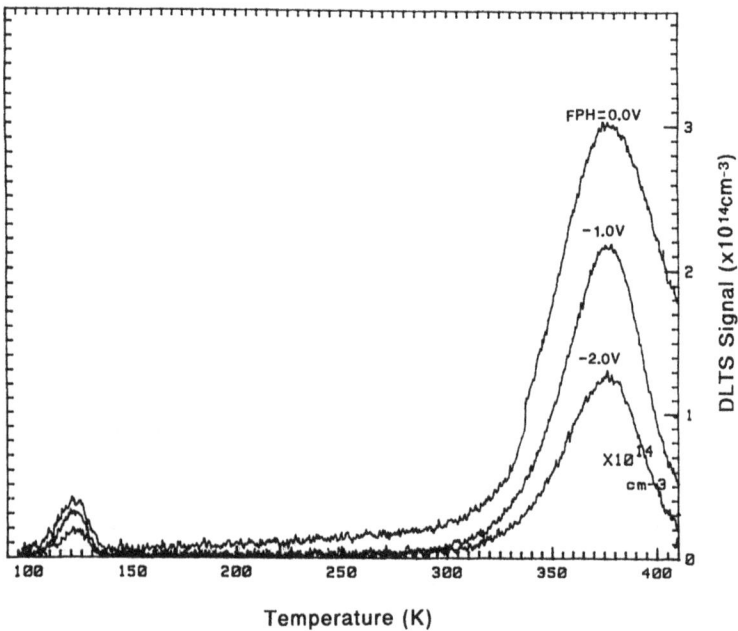

Figure 3.7 DLTS signal for a bulk trap in InP with different values of fill pulse voltage (FPH).

3.5 RADIATION EFFECTS IN InP

Indium phosphide has received extensive study regarding its performance in a radiation environment [62–74]. Much of this interest stems from initial work indicating that InP is far superior to Si or GaAs in a radiation environment. A summary of these results is given in Table 3.4. Yamaguchi and coworkers have perhaps published more on this topic than others [62–67].

They studied p-type InP with a report of 1.0 MeV electron irradiation-induced electron traps at activation energies of E_A = 0.15, 0.17, 0.22, 0.30, and 0.55 eV, with hole traps at E_A = 0.15, 0.22, 0.2, 0.37, and 0.52 eV. Among these traps, most significant were hole traps at E_A = 0.37 and 0.52 eV, and an electron trap at E_A = 0.19 eV. They attribute the radiation resistance of InP to the migration energy of In- and P-displaced atoms in InP, which is favorable compared to GaAs. The 0.37-eV hole trap is attributed to a Frenkel pair, $V_p \rightarrow P_i$, and the 0.52-eV hole trap to a P_n-Zn complex. The 0.37-eV level can be annealed at room temperature at a rate of $\alpha(p)^{2/3}$ [66]. The forward bias or light injection annealing rate in p-InP is six to seven orders above that in n-InP [64]. Irradiation in the presence of light or forward bias will significantly reduce the rate of degradation [65]. The major trap in n-InP due to e^--irradiation is found at E_A = 0.38 eV, which corresponds to a similar electron trap in p-InP. This work has focused upon n^+/p or p^+/n junctions, which then avoids some of the confusion about contributions from the surface.

Sibille $et\ al.$ [68,69] reported on 1.0 MeV electron irradiation of p-InP using a Ti-Au Schottky contact. They reported hole traps with E_A = 0.17, 0.33, 0.37, and 0.54 eV, in good agreement with Yamaguchi, although capture cross section differed in some cases by an order of magnitude. Their initial InP, grown by LPE on a (100) p^+-InP substrate had a preirradiation hole trap close to the 0.17-eV level. Sibille observed a significant reduction in radiation damage by annealing at 470K for 10 min. They also identified the 0.37-eV level as a point defect and the 0.52-eV level as a point defect-impurity complex.

Benton $et\ al.$ [70] studied n-InP using a p^+/n structure and 1.5 MeV electron irradiation. Electron traps were detected having E_A = 0.09, 0.12, 0.22, 0.37, 0.64, and 0.79 eV, with the 0.79-eV trap as the main defect. Yamaguchi [62] did not report such a wide variety of trap levels in n-InP. Levinson $et\ al.$ [40] studied similar material with an Al-Schottky barrier and reported similar traps (except for the 0.79-eV main trap) as well as electron traps at E_A = 0.11 and 0.17 eV prior to irradiation. The absence of the 0.79-eV level in the Schottky device may indicate Zn diffusion during p/n junction formation as the cause. Minority carrier injection-enhanced annealing was seen with recovery at an E_A = 1.3 eV at -6V or 0V bias, or 0.42 eV at 20 mA bias. A study by Suski and Bourgorn [71] on n-InP using a Au-Schottky barrier and 2.0 MeV electron irradiation resulted in electron traps at E_A = 0.14, 0.24, and 0.41 eV, with a 0.46-eV level prior to irradiation. Similar and other levels were found by Benton as well. Anderson $et\ al.$ [72] studied 1.0 MeV electron irradiation of p-InP using a Yb-Schottky structure. Figures 3.8 and 3.9 show preirradiation DLTS signal and activation plots, respectively, whereas Figures 3.10 and

Table 3.4
Summary of Radiation Effects in InP

Substrate	Device	Irradiation Condition	Trap Energy (eV)	Capture Cross Section (cm^2)	Reference
n-InP	p^+/n	1.0 MeV-e^-	H4@E_A = 0.38	5×10^{-15}	[62]
p-InP	n^+/p	^{60}Co			[63]
p-InP	n^+/p	1.0 MeV-e^-	H4@0.37	5×10^{-15}	[64]
			E2@0.19		
p-InP	n^+/p	1.0 MeV-e^-	H4@0.37		[65,66]
			H5@0.52		
p-InP	n^+/p	1.0 MeV-e^-	E5@0.15		[67]
			EPX@0.17	3×10^{-17}	
			E6@0.22		
			E8@0.30		
			E9@0.55		
			H1@0.15	2.1×10^{-15}	
			H2@0.22	3×10^{-17}	
			H3@0.32	6×10^{-16}	
			H4@0.37	8×10^{-16}	
			H5@0.52	5×10^{-15}	
			H6@0.23	2.5×10^{-13}	
p-InP	Ti-Au Schottky	1.0 MeV-e^-	H1@E_a = 0.167	2.1×10^{-15}	[68, 69]
			H3@E_a = 0.330	2.4×10^{-15}	
			H4*@E_a = 0.371	1.7×10^{-15}	
			H5@E_a = 0.535	6.0×10^{-15}	
n-InP	p^+/n	1.5 MeV-e^-	E1@0.09		[70, 40]
			E3@0.12		
			E4@0.22		
			E7@0.37		
			E9@0.64		
			E10*@0.79	$>7.6 \times 10^{-17}$	
n-InP	Au-Schottky	As fabricated 2 MeV-e^-	0.46		[71]
			E1@0.14		
			E2@0.24	3×10^{-20}	
			E3@0.41		
p-InP	Yb-Schottky	As fabricated 1.0 MeV-e^-	H1@0.48	1.3×10^{-18}	[72]
			E1@0.13	9.0×10^{-20}	
			H1@0.50	2.6×10^{-18}	
			H2@0.38	3.0×10^{-20}	
			E1@0.031	1.1×10^{-21}	
n-InP	Au/Oxide/InP	935°C anneal only	E1@0.65	$1-3 \times 10^{-16}$	[73]

Table 3.4
Continued

Substrate	Device	Irradiation Condition	Trap Energy (eV)	Capture Cross Section (cm²)	Reference
		935°C anneal only	E2@0.52	$1\text{--}4 \times 10^{-15}$	
		150 KeV-Ar @5×10^{11} and $5 \times 10^{12}/cm^2$	E3@0.37	$1\text{--}3 \times 10^{-14}$	
			E4@0.25	1×10^{-15}	
			E5@0.19	1×10^{-15}	
			E6@0.17	1×10^{-15}	
n-InP	Au-Schottky	As fabricated	0.29		[74]
		As fabricated	0.41		
		As fabricated	0.63		
		200 KeV deuteron	0.40		

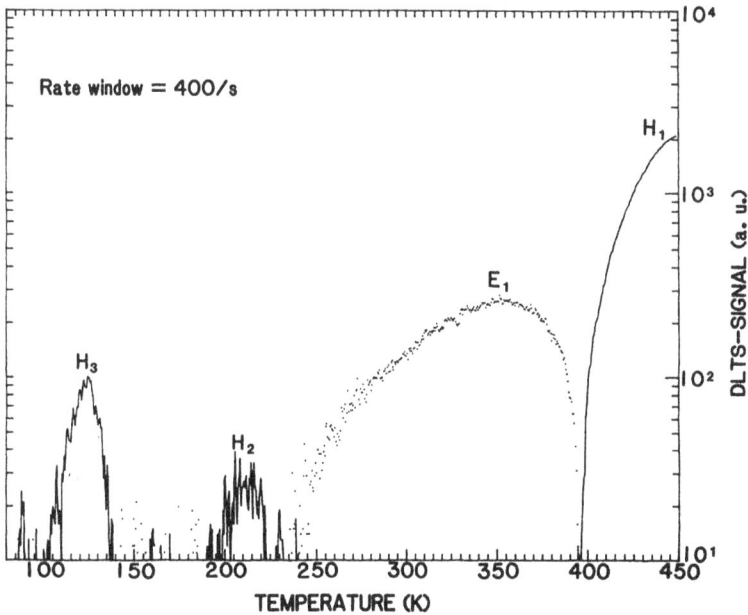

Figure 3.8 DLTS spectrum prior to irradiation showing three hole traps and one electron trap [56].

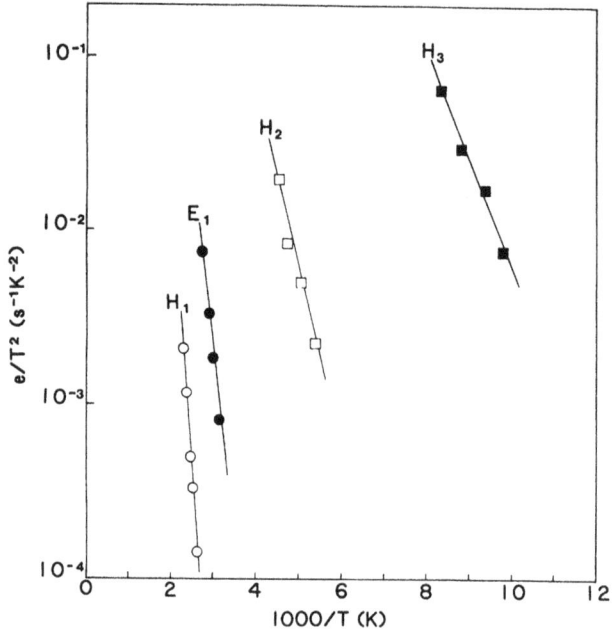

Figure 3.9 Activation energy plot for data of Figure 3.8 [59].

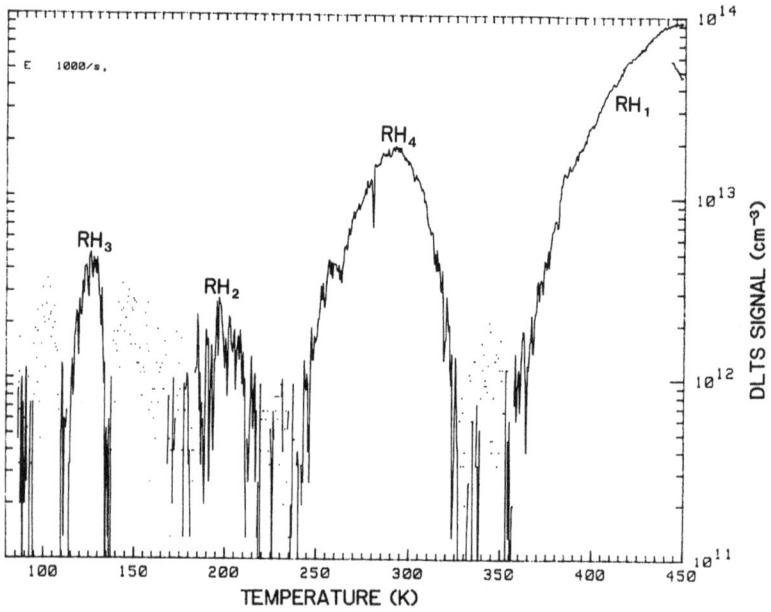

Figure 3.10 DLTS spectrum after 1.0-MeV electron irradiation showing a new hole trap [56].

3.11 show corresponding data after irradiation. The as-fabricated material had a major hole trap at 0.48 eV, previously seen by Bremond [39], and an electron trap at 0.13 eV, previously seen by Levinson [40], Wada [38], Bremond [39], and Nicholas [46]. After irradiation, a new hole trap at 0.38 eV and one similar to the original at 0.52 eV correspond closely to the ones seen by Sibille [68] and Yamaguchi [67], except for a different capture cross section. The original electron trap at 0.13 eV was now gone. An electron trap at 0.031 eV may be impurity or surface related.

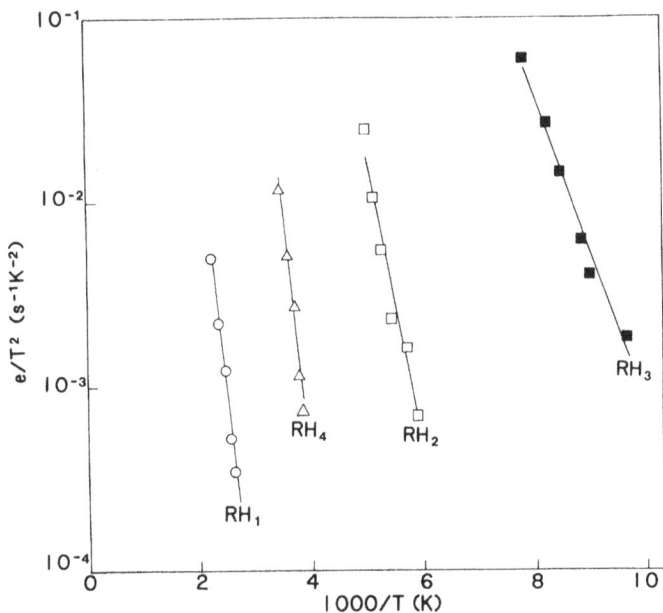

Figure 3.11 Activation energy plot for data of Figure 3.10 [56].

Luo *et al.* [73] studied 150 keV Ar irradiation of Au/oxide/*n*-InP devices. Annealing alone caused electron traps at 0.65 eV and 0.52 eV, previously seen by Bremond [39] and Ogura [44,45], respectively. After irradiation, electron traps appeared at 0.17, 0.19, 0.25, and 0.37 eV. Annealing was effective in removing the radiation-induced defects. Macrander *et al.* [74] studied 200 keV deuteron irradiation of a Au/*n*-InP structure. Electron traps were detected at 0.29, 0.41, and 0.63 eV prior to irradiation, and at 0.40 eV after irradiation. This new level is close to one seen after electron irradiation but not really close enough to say that it is the same effect.

Agreement among researchers as to effects of radiation in InP is quite good regarding trap levels, although some researchers detect more levels because of increased sensitivity setting on their instrumentation. Agreement on capture cross section is not as good and some do not report a value. Reproducibility on *p-n* junction devices is better than for Schottky-type devices. This brings to mind a number of variables that can enter into the final result:

(a) For *p-n* junction devices, the epitaxial top layer thickness can influence radiation effects in the junction region.

(b) For Schottky devices, thickness and stopping power of the metal can influence radiation damage. A junction at the surface is influenced by surface states and effects of the oxide, if it exists.

(c) Bias during DLTS measurement can greatly influence the result. This is evident from data in Table 3.5 which shows variation in trap data with bias during DLTS testing.

Table 3.5

Variation of the Parameters of Dominant Hole Trap with Quiescent Reverse Bias in Zn-Doped *p*-InP by DLTS-Using Schottky Diodes [72]

Reverse Bias (V)	$E_{T_p} - E_V$ (eV)	Capture Cross Section σ_p $(10^{-21} Cm^2)$	Trap Density N_{T_p} $10^{14} cm^{-3}$	W^a (μm)
3.0	0.56	4000	1.5	1.02
2.0	0.48	400	2.1	0.85
1.0	0.38	30	2.4	0.67
0.3	0.34	9	2.7	0.53
0.0	0.36	40	2.3	0.45
0.2[b]	0.25	4	0.9	0.33

(a) Space change layer width.
(b) Quiescent forward bias value.

3.6 DISCUSSION

Early work on deep levels in InP revealed a multitude of deep levels ranging from $E_c - 0.12$ eV to $E_c - 0.78$ eV for electron traps. In many cases, imperfections in the growth process were responsible for the defects. In other cases, impurities were the cause. Figures 3.12 and 3.13 record deep-level locations for electrons and holes

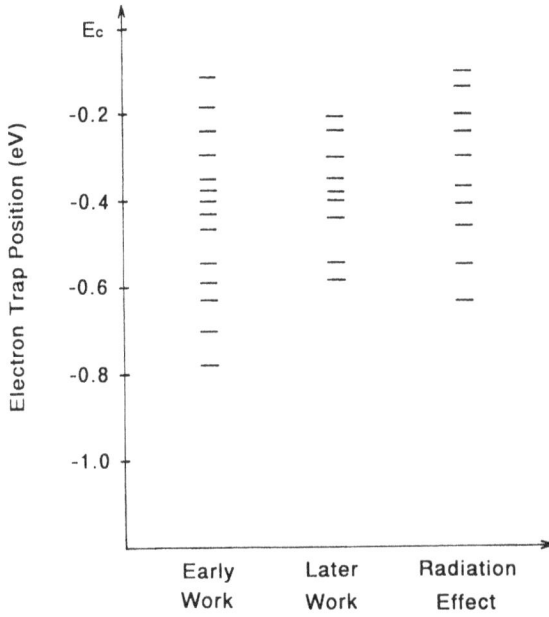

Figure 3.12 Energy level diagram showing positions of measured electron traps in InP.

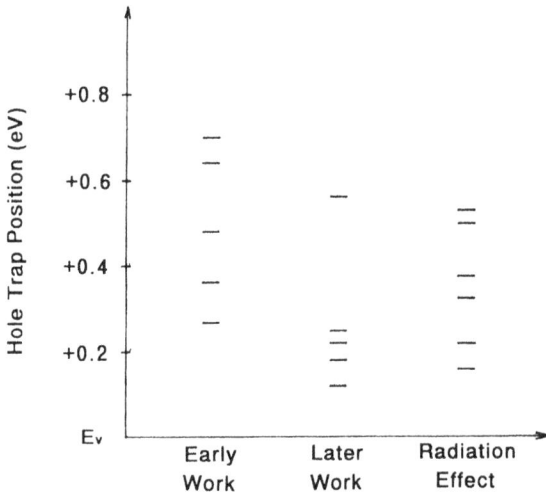

Figure 3.13 Energy level diagram showing positions of measured hole traps in InP.

from early work, later work, and radiation studies. Defects from the growth process were recorded at $E_c - 0.35$ eV, $E_c - 0.49$ eV, $E_c - 0.53$ eV, $E_v + 0.36$ eV, and $E_v + 0.48$ eV, usually caused by excess P as an interstitial or deficient In as a vacancy. Fe-doping has been identified as the cause of electron traps at $E_c - 0.24$ eV, $E_c - 0.56$ eV, $E_c - 0.58$ eV, $E_c - (0.61$ to 0.63 eV$)$, and $E_c - 0.78$ eV, and a hole trap at $E_v + 0.70$ eV. Zinc is a popular p-type dopant and accounts for an electron trap at $E_c - 0.24$ eV, and hole traps at $E_v + 0.27$ eV and $E_v + 0.64$ eV. Electron traps at $E_c - 0.40$ eV and $E_c - 0.53$ eV were caused by Cr and Co, respectively.

Later work examined other impurities, different growth methods, and various stressing effects. Several authors identified the $E_c - (0.43$ to $0.45)$ eV level as a result of a P-vacancy or In-interstitial. Some of this was found during MOCVD growth, whereas electrothermal stress or RTA could give the same contribution. The work on electrothermal stress revealed electron traps at $E_c - 0.54$ eV due to O, and $E_c - 0.59$ eV due to a P-interstitial or In-vacancy. Extra levels were seen because of the I-layer in MIS structures. Electron traps were seen at $E_c - 0.52$ eV and $E_c - 0.57$ eV due to antisite defects, interstitials, or vacancies. A hole trap at $E_v + 0.89$ eV was identified during RTA as caused by Ti-diffusion at high temperature. Mn on an In site contributed a hole trap at $E_v + 0.25$ eV. Defect-free material can be obtained using proper growth and impurity control.

Much attention has been given to InP as a radiation-tolerant material. Most significant 1.0 MeV electron radiation-induced traps in Zn-doped InP were seen at $E_v + 0.37$ eV as a Frenkel pair, $E_v + 0.52$ eV as a P-Zn complex, and at $E_c - 0.19$ eV. In n-InP, an electron trap at $E_c - 0.38$ eV is usually seen after 1.0 MeV electron irradiation. One was also seen at $E_c - 0.79$ eV after 1.5 MeV electron irradiation. Minority carrier injection-enhanced annealing was shown to be effective in removing the radiation damage. The annealing rate in p-InP was seen to be six to seven times faster than in n-InP. Yet, care must be taken during annealing to avoid introduction of other defects, since annealing alone may create deep levels.

Examination of Figures 3.12 and 3.13 shows good consistency in some deep-level positions from all three areas of examination (columns). Electron traps were repeatedly seen at $E_c - 0.24$ eV, $E_c - 0.30$ eV, $E_c - 0.37$ eV, $E_c - 0.40$ eV, and $E_c - 0.55$ eV. This consistency was not seen in hole trap positions. Several conclusions are possible as a result of careful examination of the references used herein:

(a) Schottky of MIS devices can give good DLTS data, but care must be taken to separate surface and bulk traps. Improper surface preparation can introduce traps near the surface that appear as deep levels.

(b) Diffused junctions can introduce added levels that were not originally in the substrate. For example, Zn-diffusion to form a p^+/n junction may introduce new levels.

(c) DLTS is a good technique for evaluating growth processes or fabrication processes. It is possible to precisely identify the cause of certain process-induced defects.

(d) Radiation effects in InP depend greatly on device design and radiation source. A surface junction (MIS or MS) may give different trap data because of the effect of the metal layer or surface-related defects. It is important to begin such a study with defect-free material.

ACKNOWLEDGMENT

We wish to acknowledge the patience of Maria Kuang in preparing the manuscript.

REFERENCES

1. W. Shockley, and W.T. Read, "Statistics of the Recombinations of Holes and Electrons," *Phys. Rev.*, Vol. 87, 1952, pp. 835–842.

2. R.N. Hall, "Electron-Hole Recombination in Germanium," *Phys. Rev.*, Vol. 87, 1952, p. 387.

3. D.V. Lang, "Deep-Level Transient Spectroscopy: A New Method to Characterize Traps in Semiconductors," *J. Appl. Phys.*, Vol. 45, 1974, pp. 3023–3032.

4. L.C. Kimerling, "New Developments in Defect Studies in Semiconductors," *IEEE Trans. Nucl. Sci.*, Vol. NS-23, 1976, pp. 1479–1505.

5. G.L. Miller, J.V. Ramirez, and D.A. Robinson, "A Correlation Method for Semiconductor Transient Signal Measurements," *J. Appl. Phy.*, Vol. 46, 1975, pp. 2638–2644.

6. K. Dmowski, and Z. Pioro, "Noise Properties of Analog Correlators with Exponentially Weighted Average," *Rev. Sci. Inst.*, Vol. 58, 1987, pp. 2185–2191.

7. E.E. Wagner, D. Hiller, and D.E. Mars, "Fast Digital Apparatus for Capacitance Transient Analysis," *Rev. Sci. Inst.*, Vol. 51, 1980, pp. 1205–1211.

8. M.D. Jack, R.C. Pack, and J. Henriksen, "A Computer-Controlled Deep-Level Transient Spectroscopy System for Semiconductor Process Control," *IEEE Trans. Elect. Dev.*, Vol. ED-27, 1980, pp. 2226–2231.

9. K. Asada, and T. Sugano, "Simple Microcomputer-Based Apparatus for Combined DLTS-C-V Measurement," *Rev. Sci. Inst.*, Vol. 53, 1982, pp. 1001–1006.

10. T.R. Jervis, W.M. Teter, T. Cole, and D. Dunlavy, "Deep Level Transient Spectroscopy Using CAMAC Components," *Rev. Sci. Inst.*, Vol. 53, 1982, pp. 1160–1162.

11. C.Y. Chang, W.C. Hsu, C.M. Uang, Y.K. Fang, and W.C. Liu, "A Simple and Low-Cost Personal Computer-Based Automatic Deep-Level Transient Spectroscopy System for Semiconductor Devices Analysis," *IEEE Trans. Inst. Meas.*, Vol. IM-33, 1984, pp. 259–263.

12. K. Holzlein, G. Pensl, M. Schulz and P. Stolz, "Fast Computer-Controlled Deep Level Transient Spectroscopy System for Versatile Applications in Semiconductors," *Rev. Sci. Inst.*, Vol. 57, 1986, pp. 1373–1377.

13. H. Lefevre, and M. Schulz, "Double Colleration Technique (DDLTS) for the Analysis of Deep Level Profiles in Semiconductors," *Appl. Phy.*, Vol. 12, 1977, pp. 45–53.

14. J.A. Pals, "Properties of Au, Pt, Pd and Rh Levels in Silicon Measured with a Constant Capacitance Technique," *Solid-State Electron.*, Vol. 17, 1974, pp. 1139–1145.

15. G. Goto, S. Yanagisawa, O. Wada, and H. Takanashi, "Determination of Deep-Level Energy and Density Profiles in Inhomogeneous Semiconductors," *Appl. Phys. Lett.*, Vol. 23, 1973, pp. 150–151.

16. N.M. Johnson, "Measurement of Semiconductor-Insulator Interface States by Constant-Capacitance, Deep-Level Transient Spectroscopy," *J. Vac. Sci. Tech.*, Vol. 21, 1982, pp. 303–314.

17. G.L. Miller, "A Feedback Method for Investigating Carrier Distribution in Semiconductors," *IEEE Trans. Elec. Dev.*, Vol. ED-19, 1972, pp. 1103–1108.

18. J.M. Horas, "Thermal Filling Effect on Constant Capacitance Transient Spectroscopy," *Phys. Stat. Sol.*, Vol. 69a, 1982, pp. K209–K213.

19. M.F. Li, and C.T. Sah, "New Technique of Capacitance-Voltage Measurements of Semiconductor Junctions," *Solid-State Elec.*, Vol. 25, 1982, pp. 95–99.

20. R.Y. DeJule, M.A. Haase, D.S. Ruby, and G.E. Stillman, "Constant Capacitance DLTS Circuit for Measuring High Purity Semiconductors," *Solid-State Elec.*, Vol. 28, 1985, pp. 639–641.

21. J.J. Shiau, A.L. Fahrenbruch, and G.H. Bube, "A Method to Improve the Speed and Sensitivity of Constant-Capacitance Voltage Transient Measurements," *Solid-State Elec.*, Vol. 30, 1987, pp. 513–518.

22. B. Monemar, and H.G. Grimmeiss, "Optical Characterization of Deep Energy Levels in Semiconductors," *Prog. Cryst. Growth Charact.*, Vol. 5, 1982, pp. 47–88.

23. H.G. Grimmeiss, "Deep Level Impurities in Semiconductors," in *Annual Review of Material Science,* ed. R.A. Huggins, R.H. Bube, and R.W. Robers, Palo Alto: Annual Reviews, 1977, pp. 341–376.

24. A. Chantre, G. Vicent, and D. Bois, "Deep-Level Optical Spectroscopy in GaAs," *Phys. Rev.*, Vol. B23, 1981, pp. 5335–5339.

25. P.M. Mooney, "Photo-Deep Level Transient Spectroscopy: A Technique to Study Deep Levels in Heavily Compensated Semiconductors," *J. Appl. Phys.*, Vol. 54, 1983, pp. 208–213.

26. P.M. Petroff, and D.V. Lang, "A New Spectroscopic Technique for Imaging the Spatial Distribution of Non–Radiative Defects in a Scanning Transmission Electron Microscope," *Appl. Phys. Lett.*, Vol. 31, 1977, pp. 60–62.

27. O. Breitenstein, "A Capacitance Meter of High Absolute Sensitivity Suitable for Scanning DLTS Application," *Phys. Stat. Sol.*, Vol. 71a, 1982, pp. 159–167.

28. J. Heydenreich, and O. Breitenstein, "Characterization of Defects in Semiconductors by Combined Application of SEM(EBIC) and SDLTS," *J. Microsc.*, Vol. 141, 1986, pp. 129–142.

29. K. Wada, K. Ikuta, J. Osaka, and N. Inoue, "Analysis of Scanning Deep Level Transient Spectroscopy," *Appl. Phys. Lett.*, Vol. 51, 1987, pp. 1617–1619.

30. C.T. Sah, L.L. Rosier, and L. Forbes, "Direct Observation of the Multiplicity of Impurity Charge States in Semiconductors from Low-Temperature High-Frequency Capacitance," *Appl. Phys. Lett.*, Vol. 15, 1969, pp. 316–318.

31. C. Hurtes, M. Boulou, A. Mitonneau, and D. Bois, "Deep-Level Spectroscopy in High-Resistivity Materials," *Appl. Phys. Lett.*, Vol. 32, 1978, pp. 821–823.

32. J.K. Rhee, and P. K. Bhattacharya, "Photoinduced Current Transient Spectroscopy of Semi-insulating InP:Fe and InP:Cr," *J. Appl. Phys.*, Vol. 53, 1982, pp. 4247–4249.

33. R.E. Kremer, M.C. Arikan, J.C. Abele, and J.S. Blakemore, "Transient Photoconductivity Measurements in Semi-insulating GaAs. I An Analog Approach," *J. Appl. Phys.*, Vol. 62, 1987, pp. 2424–2431.

34. J.C. Abele, R.E. Kremer, and J.S. Blakemore, "Transient Photoconductivity Measurements in Semi-insulating GaAs. II A Digital Approach," *J. Appl. Phys.*, Vol. 62, 1987, pp. 2432–2438.

35. O. Yoshie, and M. Kamihara, "Photo-Induced Current Transient Spectroscopy in High-Resistivity Bulk Materials," *Jpn. J. Appl. Phys.*, Vol. 22, 1983, pp. 621–635.

36. D.C. Look, "The Electrical and Photoelectronic Properties of Semi-Insulating GaAs," in Semi-conductors and Semimetals, ed. R.K. Willardson and A.C. Beer, Vol. 19, Orlando: Academic Press, 1983, pp. 75–170.

37. A. Majerfeld, O. Wada, and A.N.M.M. Choudhury, "Deep-Level Traps and the Conduction-Band Structure of InP," *Appl. Phys. Lett.*, Vol. 33, 1978, pp. 957–959.

38. O. Wada, A. Majerfeld, and A.N.M.M. Choudhury, "Interpretation of Deep-Level Traps with the Lowest and Upper Conduction Minima," *J. Appl. Phys.*, Vol. 51, 1980, pp. 423–432.

39. D. Bremond, A. Nouailhat, and G. Guillot, "Deep Impurity Levels in InP," *Proc. Int. Symp. GaAs & Related Compounds*, Japan, 1981, pp. 239–243.

40. M. Levinson, J.L. Benton, H. Temkin, and L.C. Kimerling, "Defect States in Electron Bombarded *n*-InP," *Appl. Phys. Lett.*, Vol. 40, No. 11, June 11, 1982, pp. 990–992.

41. P.R. Tapster, "Emission and Capture Measurements on Deep Levels in InP," *J. Phys. C: Solid State Phys.*, Vol. 16, 1983, pp. 4173–4180.

42. M.S. Skolnick, P.R. Tapster, P.J. Dean, R.G. Humphreys, B. Cockayne, W.R. MacEwan, and J.M. Noras, "Deep Levels in Co-Doped InP," *J. Phys. C: Solid State Phys.*, Vol. 15, 1982, pp. 3333–3358.

43. S.R. McAfee, F. Capasso, D.V. Lang, A. Hutchinson, and W.A. Bonner, "A Study of Deep Level in Bulk *n*-InP by Transient Spectroscopy," *J. Appl. Phys.*, Vol. 52, No. 10, Oct. 10, 1981, pp. 6158–6165.

44. M. Ogura, M. Mizuta, N. Hase, and H. Kukimoto, "Deep Levels in InP Grown by MOCVD," *Jpn. J. Appl. Phys.*, Vol. 22, 1983, pp. 658–662.

45. M. Ogura, M. Mizuta, N. Hase, and H. Kukimoto, "A Study of Deep Levels in MOCVD-Grown InP/Semi-Insulating InP Structure," *Jpn. J. Appl. Phys.*, Vol. 23, 1984, pp. 79–83.

46. D.J. Nicholas, D. Allsopp, B. Hamilton, and A.R. Peaker, "Characterization of MOCVD InP Grown from Different Adduct Sources," *J. Crystal Growth*, Vol. 68, 1984, pp. 326–333.

47. M.A.A. Pudensi, K. Mohammed, and J.L. Merz, "Effects of Growth Temperature on Optical and Deep Level Spectroscopy of High-Quality InP Grown by Metalorganic Chemical Vapor Deposition," *J. Appl. Phys.*, Vol. 57, No. 8, 1985, pp. 2788–2792.

48. C.C. Tin, and P.A. Barnes, "Thermal and Electric Field Induced Defects in InP Metal-Insulator-Semiconductor Structures," *Appl. Phys. Lett.*, Vol. 53, No. 20, Nov. 14, 1988, pp. 1940–1942.

49. E.K. Kim, H.Y. Cho, J.H. Yoon, and S.K. Min, "Deep Levels in Undoped Bulk InP after Rapid Thermal Annealing," *J. Appl. Phys.*, Vol. 68, No. 4, Aug. 1990, pp. 1665–1668.

50. K. Yamasaki, M. Yoshida, and T. Sugano, "Deep Level Transient Spectroscopy of Bulk Traps and Interface States in Si MOS Diodes," *Jpn. J. Appl. Phys.*, Vol. 18, 1979, pp. 113–122.

51. N. Yamamoto, K. Uwai, and K. Takahei, "An Electron Trap Related to Phosphorus Deficiency in High-Purity InP Grown by Metalorganic Chemical Vapor Deposition," *J. Appl. Phys.*, Vol. 65, No. 8, Apr. 15, 1989, pp. 3072–3075.

52. A.A. Iliadis, S.C. Laih, and E.A. Martin, "Deep Levels in *n*-InP Grown by Molecular Beam Epitaxy," *Appl. Phys. Lett.*, Vol. 54, No. 15, Apr. 10, 1989, pp. 1436–1438.

53. A.T. Macrander, and W.D. Johnston, Jr., "Deep Levels in InGaAsP/InP p^+n Diodes Grown by Vapor Phase Epitaxy," *J. Appl. Phys.*, Vol. 54, No. 2, Feb. 1983, pp. 806–813.

54. K.L. Jiao, A.L. Soltyka, and W.A. Anderson, "Hole Trap Level in Pt-Ti/*p*-InGaAs/*n*-InP Heterostructures Due to Rapid Thermal Processing," *Appl. Phys. Lett.*, Vol. 57, No. 18, Oct. 29, 1990, pp. 1913–1915.

55. A. Singh, and W.A. Anderson, "Deep-Level Transient Spectroscopy Studies of Near-Surface Hole and Electron Traps in Zn-Doped InP Using High Barrier Yb/p-InP Schottky Diodes," *J. Appl. Phys.*, Vol. 64, No. 8, Oct. 15, 1988, pp. 3999–4005.

56. A. Singh, W.A. Anderson, Y.S. Lee, and K. Jiao, "Deep Levels in InP by DLTS and TSCAP: Survey and Recent Data," *SPIE, Indium Phosphide and Related Materials for Advanced Electronic and Optical Devices*, Vol. 1144, 1989, pp. 61–68.

57. K. Huang, and B.W. Wessels, "Deep-Level Properties of Mn in InP," *J. Appl. Phys.* Vol. 67, No. 11, June 1990, pp. 6882–6885.

58. S. Loualiche, A. Gauneau, A. Le Corre, D. Lecrosnier, and H. L'Haridon, "Residual Defect Center in GaInAs/InP Films Grown by Molecular Beam Epitaxy," *Appl. Phys. Lett.,* Vol. 51, No. 17, Oct. 26, 1987, pp. 1361–1363.

59. C. Nozaki, and Y. Ohba, "A Deep Level in Zn-Doped InGaAlP," *J. Appl. Phys.,* Vol. 66, No. 11, Dec. 1, 1989, pp. 5394–5397.

60. L.S. Pann, M.A. Tischler, and P.M. Mooney, "Effect of Indium on the Properties of DX Centers in Si-Doped $In_y(Ga_{0.3}Al_{0.7})_{1-y}As$," *J. Appl. Phys.,* Vol. 68, No. 4, Aug. 15, 1990, pp. 1674–1681.

61. J.F. Chen, J.C. Chen, Y.S. Lee, Y.W. Choi, K. Xie, P.L. Liu, W.A. Anderson, and C.R. Wie, "Shallow Levels, Deep Levels and Electrical Characteristics in Zn-Doped GaInP/InP," *J. Appl. Phys.,* Vol. 67, No. 8, Apr. 15, 1990, pp. 3711–3716.

62. M. Yamaguchi, Y. Itoh, and K. Ando, "Room-Temperature Annealing of Radiation-Induced Defects in InP Solar Cells," *Appl. Phys. Lett.,* Vol. 45, No. 11, Dec. 1, 1984, pp. 1206–1208.

63. M. Yamaguchi, C. Uemura, and A. Yamamoto, "Radiation Damage in InP Single Crystals and Solar Cells," *J. Appl. Phys.,* Vol. 55, No. 6, March 15, 1984, pp. 1429–1436.

64. M. Yamaguchi, K. Ando, A. Yamamoto, and C. Uemura, "Injection-Enhanced Annealing of InP Solar-cell Radiation Damage," *J. Appl. Phys.,* Vol. 58, No. 1, July 1, 1985, pp. 568–574.

65. K. Ando, and M. Yamaguchi, "Radiation Resistance of InP Solar Cells Under Light Illumination," *Appl. Phys. Lett.,* Vol. 47, No. 8, Oct. 15, 1985, pp. 846–848.

66. M. Yamaguchi, and K. Ando, "Effects of Impurities on Radiation Damage in InP," *J. Appl. Phys.,* Vol. 60, No. 3, Aug. 1, 1986, pp. 935–940.

67. M. Yamaguchi, and K. Ando, "Mechanism for Radiation Resistance of InP Solar Cells," *J. Appl. Phys.,* Vol. 63, No. 11, June 1, 1988, pp. 5555–5562.

68. A. Sibille, and J.C. Bourgoin, "Electron Irradiation Induced Deep Levels in p-InP," *Appl. Phys. Lett.,* Vol. 41, No. 10, Nov. 15, 1982, pp. 956–957.

69. A. Sibille, "Origin of the Main Deep Electron Trap in Electron Irradiated InP," *Appl. Phys. Lett.,* Vol. 48, No. 9, March 3, 1986, pp. 593–595.

70. J.L. Benton, M. Levinson, A.T. Macrander, H. Temkin, and L.C. Kimerling, "Recombination Enhanced Defect Annealing in n-InP," *Appl. Phys. Lett.,* Vol. 45, No. 5, Sept. 1, 1984, pp. 566–568.

71. J. Suski, and J.C. Bourgoin, "Defects Induced by Electron Irradiation in InP," *J. Appl. Phys.,* Vol. 54, No. 5, May 1983, pp. 2852–2854.

72. W.A. Anderson, A. Singh, K. Jiao, and B. Lee, "Deep Levels and Radiation Effects in p-InP," *Photovoltaic Research and Technology Conference,* NASA-Lewis, Cleveland, Apr. 19–21, 1988.

73. J.K. Luo, T. Kimura, S. Yugo, and Y. Adachi, "Deep Levels in Argon-Implanted and Annealed Indium Phosphide," *Jap. J. Appl. Phys.,* Vol. 26, No. 4, April 1987, pp. 582–587.

74. A.T. Macrander, B. Schwartz, and M.W. Focht, "Deep and Shallow Levels in n-Type Indium Phosphide Irradiated with 200-keV Deuterons," *J. Appl. Phys.,* Vol. 55, No. 10, May 15, 1984, pp. 3595–3602.

Chapter 4
Low-Pressure MOVPE of InP-Based Compound Semiconductors

D. Schmitz
AIXTRON GmbH

4.1 INTRODUCTION

Since the first attempts at producing compound semiconductor materials from organyl complex compounds as precursors for the group III element, large numbers
of variations in the processing have been investigated to finally improve the material
to a quality applicable to industrial devices.

The accuracy and the easy control of the applied vapor precursor are the major
advantages of the metalorganic vapor phase epitaxy (MOVPE) as an inferior crystal-
growth technique. All precursors can be handled conveniently by flow controllers
and valves. The concentration of each reactant in the process can be adjusted directly
and independently of all other parameters. Thus, it is possible to grow extremely
complex alloy crystals by properly adjusting the gas phase compositions. Fast switching of constituents in the gas phase generates abrupt crystal interfaces between grown
layers. This is an essential feature for electronic effects exploited in device structures
such as quantum well lasers and high electron mobility transistor (HEMT). First
observations of two-dimensional electron gases and energy quantization have been
in molecular beam epitaxy (MBE)-grown structures. State-of-the-art MOVPE is capable of producing structures with properties similar to those of MBE.

It is simpler in MOVPE than in MBE to produce compounds containing elements difficult to handle in extremely low-pressure ambients, such as phosphorus.
This is, of course, especially valid for all InP-based compound semiconductors. In
the cold-wall MOVPE reactor, the volatile Al-containing precursor is less aggressive
and much easier to handle than in hydride vapor phase epitaxy (VPE), where the

use of highly reactive materials such as Al in the gas phase generates problems of interaction with the hot quartz walls of the reactor vessel [1].

During the past five years, MOVPE process has become established as a production tool for device structures in the InP- and GaAs-based material system. Optical communication devices, such as lasers containing GaInAsP alloys with emission wavelengths of 1.05, 1.3, and 1.55 μm, or InP/InGaAs structures for photodiodes, pin-photodetectors, and optoelectronic integrated circuits (OEICs) for data conversion and adaption to optical networks, are routinely produced using this technique.

Al-containing compounds based, like AlInAs, on InP have been mostly used in MBE-grown structures to substitute InP as a wide bandgap material not containing P. The band structure of GaInAs/AlInAs and InP/AlInAs heterostructures has several advantages, and thus MOVPE has also developed to the production of materials from the Al-Ga-In-As material system.

Materials from the Al-Ga-In-P family have recently caused more interest in visible light laser components in commercial applications (barcode readers, counters, bright spot pointers, optical data storage devices, *et cetera*). The advantages of MOVPE for producing In- and P-containing compounds led to this technique acquiring major importance in this field.

The growth of extremely mismatched crystalline layers makes it possible to a large extent to use the physical properties of the strained layers in device applications. It has been shown that strained layer or superlattice buffers reduce the influence of substrate crystal defects on device performance and longevity in devices such as lasers or HEMTs, where enhanced degradation due to high current densities or interface defects could be expected.

The application of low-pressure operation in the MOVPE process makes it possible to improve layer uniformity and interface abruptness by means of higher gas velocities. The application of precursors with extremely low vapor pressures is simplified, and the risk of parasitic prereactions is further decreased. The impact, compared to the atmospheric pressure MOVPE, is better control of the process, higher flexibility in the choice of precursors, and reduced gas consumption.

Presently, a large part of MOVPE activities is performed in low-pressure reactors. Examples of the performance and the state of the art of this technique will be given in this chapter.

One problem in transferring process parameters between laboratories was the technical realization of reactors, which was very different among users some years ago. The shape and size of the reaction chambers were particularly different. Thus, the comparison of results obtained in different publications was not possible in most cases, regardless of the equipment used. Another problem was the upscaling of processes to production scale by simple transfer of parameter sets to upscaled reactor cells.

This chapter reviews the MOVPE results that were obtained while using a standard-designed reactor for industrial application: a horizontal-flow reactor operating at low pressures (20–100 mbar) and at high gas velocities (1–2 m/s), which will be described in detail through the chapter.

4.2 THE LOW-PRESSURE MOVPE PROCESS

4.2.1 Basic Considerations on the MOVPE Process

This section briefly describes the fundamentals of the MOVPE growth technique and parameters that influence the growth of III-V compound semiconductors. Many insights have been generated in GaAs MOVPE studies. Assuming that the general growth mechanisms of GaAs and InP are very similar, the understanding of the knowledge that was gained through the numerous studies on GaAs growth can be basically transferred to the InP technology [2], and extended to GaInAs, GaInAsP, and other compound semiconductors.

As an example, the growth of a binary compound semiconductor from a metalorganic complex compound of the group III element and the hydride compound of the group V element is discussed here. The deposition reaction takes place at the heated substrate surface in a cold-wall reactor. Figure 4.1 shows a schematic of the MOVPE process in a horizontal-flow reactor.

The substrate is placed on a heated graphite susceptor in the reaction chamber, with its surface parallel to the gas flow transporting all precursors. A laminar flow field in the deposition zone is considered to provide optimum distribution of precursors in a stationary flow chamber. In contrast, when using a gas stream vertical to the substrate surface, the flow pattern resulting from convection in the temperature gradient above the susceptor and the initial velocity vector might not yield in the desired transport conditions for the reactants to the substrate surface.

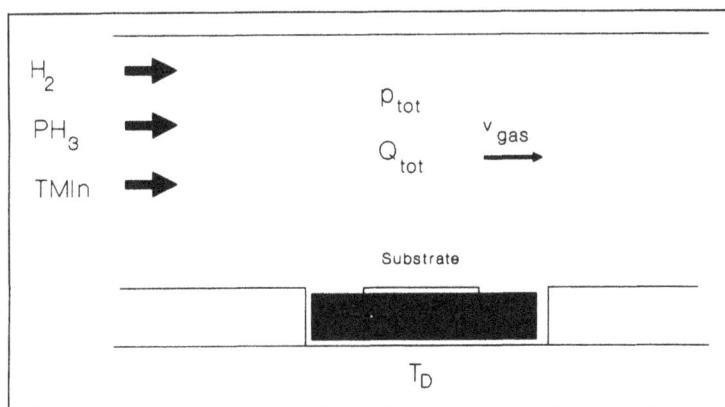

Figure 4.1 Principle of the MOVPE process in a cold wall reactor.

The susceptor can be heated either by RF or IR radiation from high-temperature lamps. Both methods are contactless and provide cold-reactor walls. The susceptor temperature is measured by a thermocouple placed in the center of the graphite block.

Pd-diffused hydrogen is used as a carrier gas to provide transport of the process reactants. All precursors are present as gases or vapors.

The gas velocity in the reactor v_g is defined by the gas flow rates into the reaction chamber and by the reactor pressure. The partial pressures of the precursors are controlled by the flow of each gaseous compound introduced into the reaction chamber.

The precursors used in MOVPE are volatile compounds of the required elements. Typically, the precursors for the group V elements are the hydrides, and for the group III element, the organic complexes. The latter are evaporated from thermostated bubbler cylinders by passing a carrier gas through it.

The following is a rather simple model of the essential steps in the epitaxial growth process:

(a) Injection of the precursors into the reactor.
(b) Gas phase transport of the precursors to the hot substrate surface, considering the concentration and velocity field of the reactor.
(c) Thermalization of the precursors catalyzed by the hot substrate surface, surface reactions, and formation of the crystalline material.
(d) Desorption of the reaction products.
(e) Transport of the byproducts to the exhaust.

The slowest of the steps in the sequence above limits the ratio of the total reaction and, therefore, the growth rate.

Considering a deposition reaction as follows:

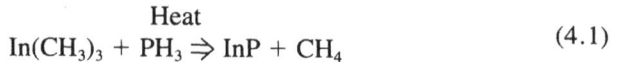

$$In(CH_3)_3 + PH_3 \overset{\text{Heat}}{\Rightarrow} InP + CH_4 \tag{4.1}$$

and taking into account that the reaction takes place in excess PH_3, it is clear that the deposition rate should be directly proportional to the partial pressure of the TMIn.

The relation of the growth rate to the susceptor temperature is basically divided into three temperature ranges:

(a) At lower temperatures, an exponential increase with $r \sim e^{(Ea/kT)}$, known as kinetically controlled regime, is observed. Here the surface adsorption and diffusion limit the growth rate.
(b) A temperature-independent range, where the gas phase diffusion limits the growth rate, occurs at intermediate temperatures.
(c) At very high temperatures a decrease in growth rate is observed. This could be attributed to enhanced prereactions by heat absorption in the gas phase.

A pronounced thermodynamically controlled regime, as observed in hydride VPE, does not occur because the process is not reversible.

Generally, the growth parameters for the crystals are chosen to have low temperature dependence in order to avoid the strong effects of nonuniform temperatures. From that point of view, the temperature range where the growth rate is diffusion-limited is preferred.

A more pronounced transition from the kinetically to the diffusion-controlled regime is observed at higher reactor pressures. This indicates an increase in surface diffusibility from pressure reduction.

4.2.2 Model of Diffusion-Controlled Deposition by MOVPE

An extended study of the kinetics of GaAs growth from metalorganic precursors and AsH$_3$ was done by Pütz [3] based on the results of GaAs growth in a horizontal-flow reactor at atmospheric and low pressure, and also in a UHV chamber. The latter process is known as Metal Organic Molecular Beam Epitaxy (MOMBE) or Chemical Beam Epitaxy (CBE). Considerations of the global chemical and physical processes for the growth of InP are comparable to those described in the study on GaAs. The model for the deposition in a diffusion-controlled regime is given below.

At a constant growth temperature, the dependence of the growth rate on the various parameters can be estimated according to a simplified model. Assuming the partial pressure of the group V precursor is sufficient at every place in the reaction chamber, the growth rate is proportional to the metalorganyl flux (I_{MO}) to the substrate surface. In a first approach, the transport from the gas phase carrying the precursors at a certain velocity (v_g) obeys a stagnant layer model as assumed for horizontal-flow processes [4].

The dependence of the thickness of the stagnant layer (δ) on the external parameters' reactor pressure (p_r) and gas velocity (v_g) can be described as:

$$\delta \sim (p_r \cdot v_g)^{-1/2} \qquad (4.2)$$

The diffusion flux (I_{MO}) to the substrate surface is described by Fick's law:

$$I_{MO} = D_{MO} \cdot d(p_{MO}/R \cdot T)/dy \qquad (4.3)$$

where D_{MO} is the diffusion coefficient of the metalorganyl in the carrier gas, p_{MO} is the partial pressure of the metalorganyl, R is the gas constant, and T the ambient temperature. The fact that the process is diffusion controlled allows the assumption that all arriving metalorganyl molecules are consumed at the substrate surface, and the concentration at the surface is 0. The proportionality of the growth rate with p_{MO}^0, representing the gas phase concentration of the organyl, is

$$r \sim I_{MO} \sim D_{MO} \cdot p_{MO}^0 / \delta \qquad (4.4)$$

and with the relation

$$D_{MO} \sim p_r^{-1} \qquad (4.5)$$

We finally obtain

$$r \sim p_{MO}^0 \cdot (v_g / p_r)^{1/2} \qquad (4.6)$$

which indicates the proportionality between the growth rate of a compound semi-conductor material in a horizontal-MOVPE-flow reaction chamber on the vapor pressure of the metalorganic compound, and the square root of the ratio of gas velocity and reactor pressure.

This is a relatively simple model, but its principal validity could be proven by experimental results from the growth of binary and the even more critical ternary and quaternary materials [2].

The influence of the reactor pressure on the growth rate can be seen in Equation (4.6). A decrease of the reactor pressure at constant partial pressure of the precursors increases the growth rate by enhancing the gas molecule diffusion. At the same time, the reduction of the reactor pressure increases the gas velocity if the total gas amount is kept constant, thus accelerating the exchange of the gas phase in the reaction chamber when changing the injected precursor composition.

The influence of the variation of the velocity v_g can be explained in an analogous way.

4.2.3 Process Precursors

The InP-based material system in general is described by the alloys that can be generated from the materials Al-Ga-In-As-P. The precursor materials preferably used in the MOVPE process have to meet several stringent conditions:

   (a) High purity.
   (b) Sufficient vapor pressure.
   (c) Stability at low temperatures.
   (d) Cracking temperatures in an appropriate low range.
   (e) No parasitic side reactions.
   (f) Safety from toxicity and other hazards.
   (g) Reasonable cost/efficiency ratio.

The precursors for the group V elements are generally AsH_3 and PH_3, gaseous compounds with vapor pressures of 14 and 35 bars, respectively, at room temperature. For these materials the amount of purity in manufacturing is very high. Since considerably high purities are required in the reactor, both gases are undiluted. Thus, relatively low operating pressure of the gas lines provides higher safety margins, protecting against mixtures such as hydrides with hydrogen, which are at 150 bars in the source cylinders. A second advantage is the reduced risk of impurities during blending. The compounds decompose in temperature ranges blow 500°C [5], so they are applicable to the MOVPE technique.

As group III precursors in most of the MOVPE applications, the methyl complexes of the different metals are used. The appropriate precursors for the described alloy system are listed below.

The state-of-the-art purity of these compounds is extremely high and reproducible. So they are very compatible with the high requirements in industrial production. The applicability of the process is excellent, since the high purity eliminates to a considerable degree the tendency to parasitic reactions with other precursors.

TMAl and TMGa are liquids at temperatures below 20°C, so evaporation is possible by passing carrier gas through a bubbler filled with the compound and held below room temperature. The vapor pressures are in the 10- to 200-mbar range, depending on the adjusted temperature. No heating of the bubblers and the lines is necessary.

TMIn is somewhat exceptional with respect to vapor pressure and handling because at room temperature this compound is a solid. The melting point of about 80°C appears too high to evaporate the compound from liquid phase without additional technical effort. The vapor pressure of about 1.8 mbar at 20°C is low but sufficient to sublime the material immediately from the solid phase at room temperature. Growth rates of up to 10 μm/h can be obtained when using the source in room temperature conditions. The solid nature was often expected to cause severe problems in the stability of the evaporation efficiency because of the change in the surface properties of the material. Proper handling of the operating parameters adjusted to this source allows the use of more than 95% of this source material to produce reproducible results for years.

Besides attempts to replace the group V hydrides with organic compounds and to develop precursors for the In with a liquid state at room temperature, the precursors listed in Table 4.1 have been established for the production of the compound semiconductors in the InP-based material system.

For n-type doping of low-pressure, MOVPE-grown layers, Si and S are applied with high reproducibility and low risk of memory effects in the reactor. For some applications, Se and Te are also used. The precursors for the dopants are SiH_4, H_2S, H_2Se, and organic compounds like diethyltellurium.

A very applicable p-type dopant is zinc from diethylzinc or dimethylzinc. Both metalorganic compounds are very easy to decompose and do not form resident compounds with other precursors. The stability and application in the processes will be described later. Other dopants like Mg or Be are less convenient because of high reactivity (Mg) or extreme toxicity (Be).

Table 4.1
Process Precursors

Element	Precursor	Formula	Vapor Pressure @ 17°C	Melting Point
Aluminum	Trimethyl Aluminum (TMAl)	$Al(C_2H_5)_3$	9.13 mbar dimeric	15.4°C
Gallium	Trimethyl Gallium (TMGa)	$Ga(C_2H_5)_3$	207.38 mbar monomeric	−15.8°C
Indium	Trimethyl Indium (TMIn)	$In(C_2H_5)_3$	1.76 mbar monomeric	88°C

4.3 EQUIPMENT DESIGN

The subject of this section is the design of low-pressure MOVPE reactors with horizontal-type growth chambers.

The basic functional groups of a reactor are

(a) The gas blending system.
(b) The reaction chamber.
(c) The reactor pressure control unit.
(d) Supervision and control electronics.
(e) Exhaust gas treatment.

From the construction point of view, the different functional groups must be held modular. This allows the design of equipment for a wide variety of applications.

The transfer of research and development equipment to production is often a limiting factor if, for a developed process, a completely new reactor arrangement is necessary. The following will describe a reactor concept that is applicable to different reactor cell types and is widely accepted in low-pressure MOVPE for research and development as well as production.

4.3.1 Design of the Gas Blending System

The precursors used in the MOVPE processes are gaseous compounds, such as AsH_3 and PH_3 and liquid, or even solid, materials, that have to be evaporated and transported in a carrier gas flux. The requirements on the gas blending system are

(a) Accurate and reproducible control of all gas fluxes.
(b) Fast transport of the gases into the deposition chamber.
(c) Fast composition changes of the gas phase to obtain sharp interfaces between grown layers.
(d) Simple, standard arrangements for easy matching to changed requirements.

Figure 4.2 shows a typical schematic of a MOVPE reactor designed for the growth of compounds containing Al-Ga-In-As-P.

All gas flow rates are adjusted by electronically operated mass flow controllers (MFC) in order to provide high accuracy and reproducibility. The flow controllers are held under gas flow continuously, even if not under process conditions, by the assembly of three-way valves at the inlet that toggle between purge and process gas. The valves in the blending system are chosen to ensure fail-safe operation.

Electronic pressure controllers (PC) are used in places where accurate definition of the gas pressures is necessary, such as at the metalorganyl source containers and in the gas dilution units. It is necessary to provide control of the pressure in the source containers of the metalorganyls, especially at low pressure operation of the reactor, because the evaporation of the compound is strongly dependent on the pressure.

Diluted gas sources using three MFCs and one PC allow the adjustment of a wide range of gas concentrations without changing source concentrations.

Rapid transport of the precursors into the reactor is provided by the use of a vent/run manifold. It consists of continuously purged valves, allowing a stabilized precursor flow to be added to a carrier gas stream constantly purged into the reactor or into a vent line, which is held at the same pressure as the reactor. Thus, a fast

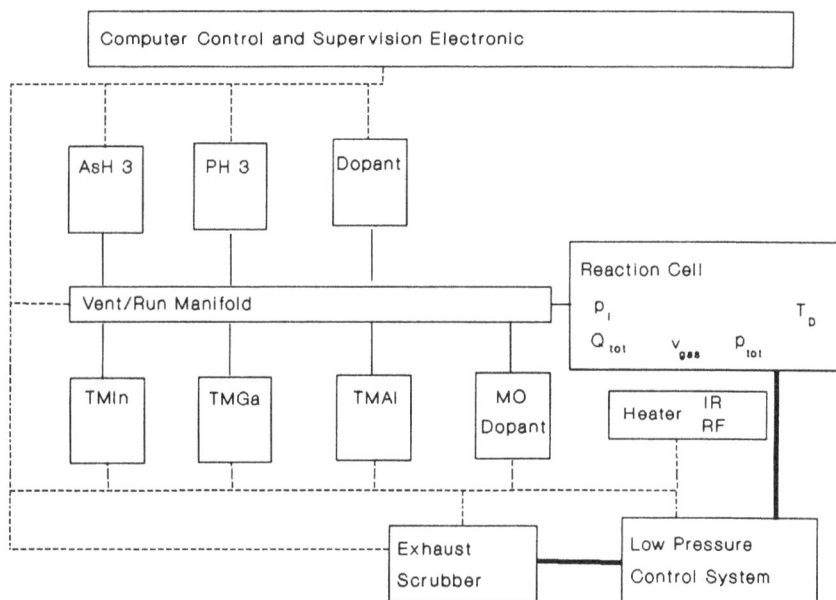

Figure 4.2 Schematic of a low-pressure MOVPE reactor.

exchange of the growth atmosphere in the reactor is possible by the use of high gas velocities in the transport gas lines (called *run lines*) and of minimized, zero dead volume valves in the vent/run manifold. With gas velocities of 1 to 3 m/s, gas phase switching times below 0.2s are possible.

The number of sources that have to be included in the gas blending system depends on the different material compositions and sequences that have to be grown in the reactor. If, for a certain precursor, multiple sources are included, the transition time between different concentrations in the reactor can be accelerated to the transition time of the valves, if necessary, for interface formation [6].

The reaction chamber is shown only schematically and will be described in more detail in Section 4.3.2, which deals with the design of the reaction chamber.

The constant pressure in the reactor must be established by an appropriate controlling assembly. A proven combination is a two-stage roughing pump equipped for hazardous gas applications with a controlled throttle valve in the reactor exhaust. For accurate measurement of the pressure, a capacitive gauge with sufficient resolution in the desired process pressure range is used.

4.3.2 Design of the Reaction Chamber

An example of a linear, horizontal process chamber is shown in Figure 4.3. This design is the most widely used standard for low-pressure MOVPE applications [7] of InP-based compound semiconductors.

The process chamber includes an exchangeable inner quartz tube, which is held in a cylindrically shaped outer protection tube and connected to the process gas inlet by a quartz ball fitting. This reactor has been described in various publications [2,7]. The shape of the growth chamber is rectangular in the deposition zone. The height is in the order of magnitude of the expected stagnant layer thickness to keep the gas phase concentrations as stable as possible over the deposition length.

The exchangeable growth chamber contains a graphite susceptor coated in an appropriate way to prevent outgassing from the porous material. The temperature of the susceptor is measured by thermocouples in quartz tubes that support the graphite block.

The susceptor is heated either by RF or IR radiation from high radiation density lamps. Both methods allow cold reactor walls and are applicable to the process pressure range from atmospheric pressure down to a value around 10 mbar. Below 10 mbar the high energy density in the RF field may cause a plasma in the growth atmosphere. In this case the uncontrolled discharge might cause undesired side reactions between precursors. The IR heater concept does not suffer from such a drawback and has been applied in processes even below 5 mbar.

As described previously, the diffusion-controlled regime of the deposition process is preferred, in general, as the parameter set for the growth of most materials.

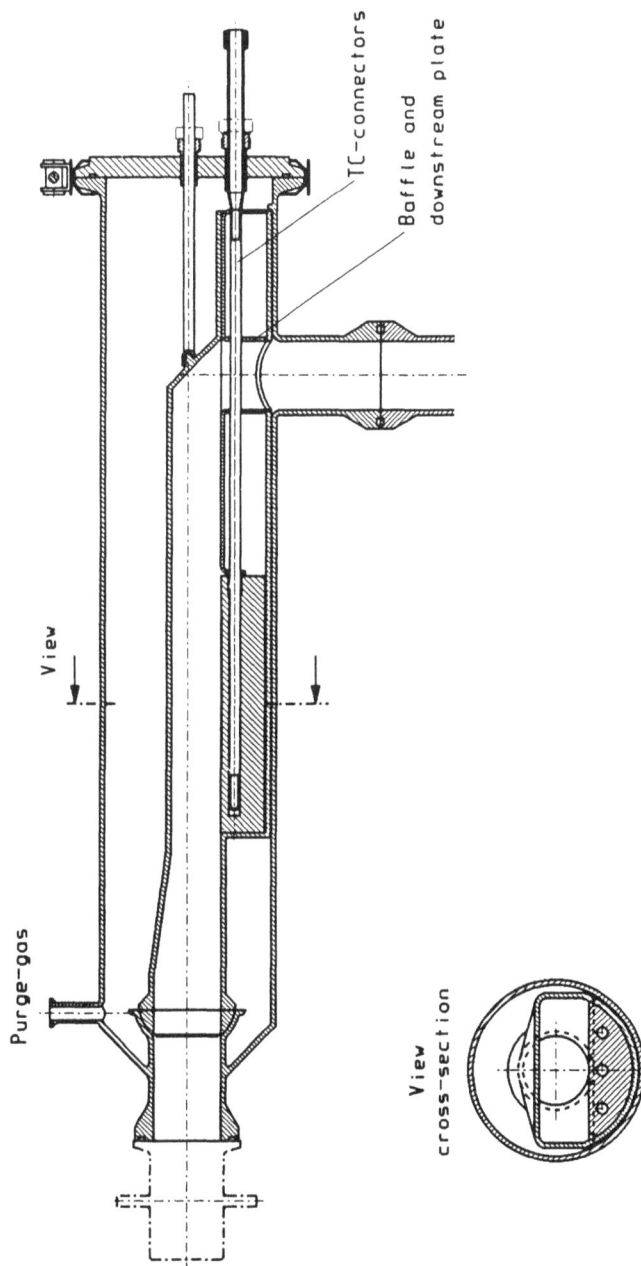

Figure 4.3 Schematic of a reaction cell for one 2-inch wafer.

Taking into account the considerations in Section 4.2, it can be concluded that under these conditions it is necessary to reduce the depletion of the gas phase along the flow direction of the gas in order to obtain homogeneous layers. This can be achieved by choosing high run gas flows so that the depletion of the gas phase is dominated by the horizontal transport velocity [4]. Typically for InP-based compound semiconductors, gas velocities between 1 and 3 m/s show optimum conditions for uniform layer growth.

Another approach for obtaining homogeneous layers with good results is to apply substrate rotation in horizontal gas flow reactors [8]. The revolving substrate is exposed to the different gas phase compositions during the growth, and depletion effects can be compensated in that way if revolution speed and growth rate match each other in an appropriate way. However, the technique used for the substrate rotation in the reactor described in Mircea *et al.* [8] requires special provisions, such as mechanical feedthroughs and heating techniques in the reactor cell, and cannot easily be included in an already existing reactor.

4.3.3 Substrate Rotation Technique

A new method for rotating susceptors without external moving parts has been introduced by P.J.M. Frijlink [9]: the *gas foil rotation* principle. In this technique, graphite wafer carrier plates are levitated on a gas foil generated through an arrangement of specially shaped grooves in the susceptor body. The shearing forces of the gas fed into the grooves apply a momentum to the floating carrier plate. This technique was first used in a planetary reactor for homogeneous growth of GaAs/AlGaAs structures on seven 2-inch wafers in one run. The arrangement of this reactor will be described more accurately later.

A gas foil rotation susceptor can easily be retrofitted to single-wafer reactor cells as described above by putting an additional gas path feeding into the susceptor. In this way, the benefits of rotating susceptors can be included in any gas phase reactor. The improvement in the results has been reported in detail [10].

A reactor for low-pressure MOVPE growth on five 2-inch wafers in one run is shown in Figure 4.4. The basic construction is similar to the reactor tube for the small-scale production described above.

The height of the growth chamber is adjusted to accommodate the low pressure and high gas velocity conditions. The outer shell is made of stainless steel with chambers for watercooling. Infrared heaters are included in the evacuated cell, separated from the growth chamber by a quartz plate. The growth chamber itself is an exchangeable quartz part containing the susceptor.

The susceptor of this reactor uses the gas foil rotation principle for each single substrate and the carrier plate for all five substrates. The carrier gas stream is injected

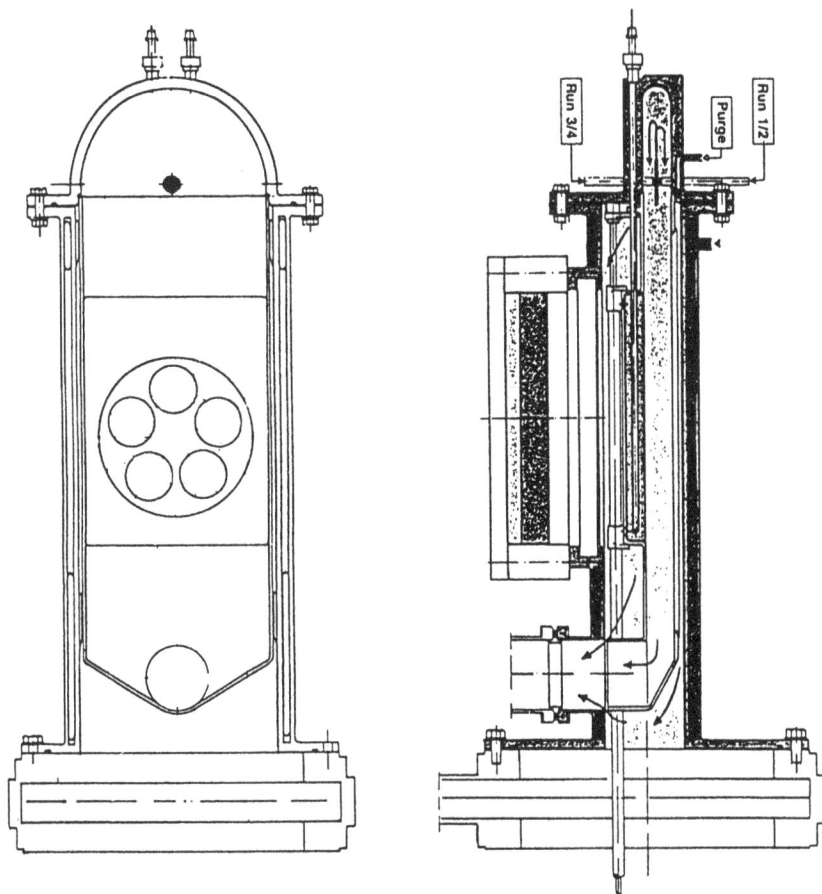

Figure 4.4 Schematic of a reaction cell for five 2-inch or three 3-inch wafers applying the gas foil rotation technique for a planetary motion of the substrates.

in parallel with the susceptor surface from the outside. In this way a parallel laminar flow of the reactants is established above the double axis rotating substrates.

This reaction chamber can replace the smaller one described above if required for large-scale production. Large-scale reactors applying this technology at atmospheric pressure have been described in the paper that introduced it in the AlGaAs/GaAs material system [9]. Results on both reactor types will be presented later.

The original planetary reactor cell principle is described in Figure 4.5. The reactant flow in this reactor is injected radial symmetrically from the center of the susceptor. The susceptor is also heated by an IR assembly similar to that described

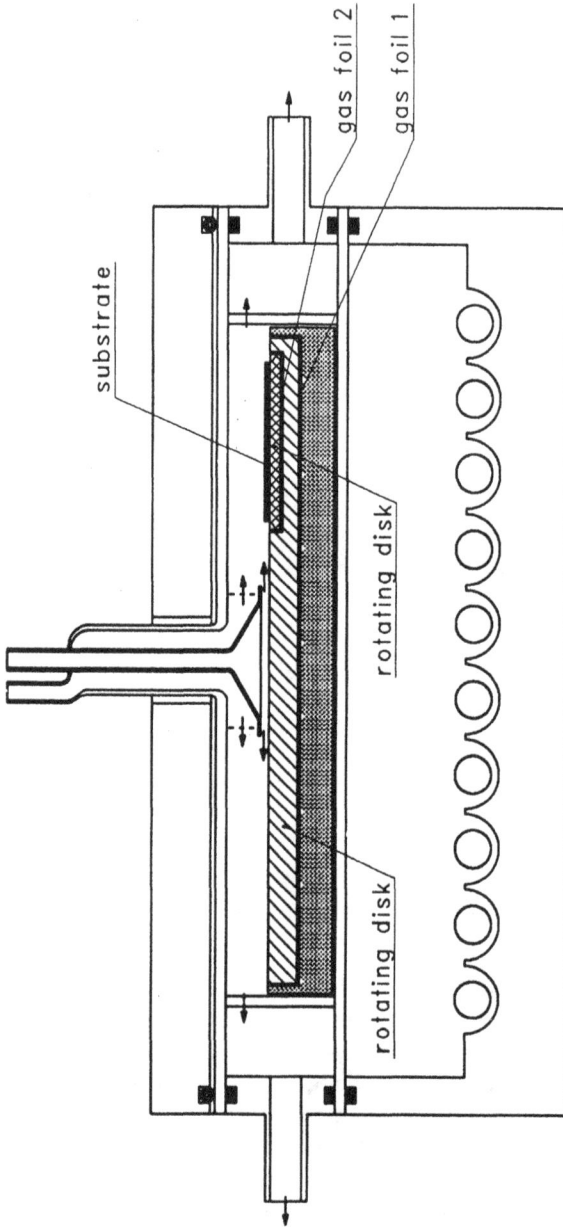

Figure 4.5 Cross-sectional view of the low-pressure planet reaction cell for seven 2-inch or five 3-inch wafers (published with kind permission of P.J.M. Frijlink [9]).

in the reactor for five 2-inch wafers. The excellent results of the growth of pseudomorphic HEMT device structures of GaInAs on GaAs in this reactor have already been reported [11]. This reactor design is now also used for commercial applications (e.g., AlGaInP visible laser structures) of the low-pressure MOVPE growth technique.

4.3.4 Electronic Safety and Control System

Important requirements of a producton tool such as a MOVPE reactor are reproducibility and reliability of results, but safety aspects also have to be considered since hazardous materials are handled. This has to be followed by a highly reliable Adesign of the controlling and supervising electronic equipment. The complete reactor must be continuously supervised by an appropriate set of sensors to ensure laboratory safety. A microprocessor-controlled operating system provides good reproducibility and the ability to perform complicated sequences for any structure preparation with the requested reliability.

State-of-the-art commercial MOVPE reactors have to provide all these features with a comfortable, interactive user interface to ensure safe and simple operation.

4.3.5 Exhaust Gas Treatment

Even though they have a highly effective reaction process, considerable numbers of the precursors, mainly group V hydrides, are released into the reactor exhaust. Laboratory safety and environmental issues require very thorough treatment of these toxic exhaust gases to ensure safety amid the complexity of semiconductor plants. To achieve this, high-efficiency exhaust gas scrubbers are used. Typically, the toxic gas concentrations in the exhaust after the scrubber treatment should be below the appropriate detection limit for each gas.

An example of such a scrubbing system for MOVPE and related processes is described by Luthardt *et al.* [12]. It does not require additional gases for the reaction and removes AsH_3 and PH_3 in a broad concentration range of up to 1% in carrier flow rates of about 50 l/min and down to the detection limit in precision hydride gas monitors (<5 ppb).

4.4 EXPERIMENTAL RESULTS AND DISCUSSION

Table 4.2 contains an overview of crystal alloys that can be grown lattice matched to InP using low-pressure MOVPE.

Table 4.2
Alloy Semiconductors Grown by Low-Pressure MOVPE

Material	Bandgap at 300 K Lattice Matched to InP	Applications
InP	1.35 eV	• Microwave devices • Detectors • Solar cells • HBTs
GaInAs	0.75 eV	• Detectors • HBTs • QW devices • HEMTs • MISFETs
AlInAs	1.46 eV	• HEMTs • HBTs
GaInAsP	1.55 μm = 0.8 eV 1.3 μm = 0.95 eV 1.1 μm = 1.13 eV 1.05 μm = 1.18 eV All compositions between InP and $Ga_{0.47}In_{0.53}As$ are possible	• Lasers • QW lasers • Waveguides • Modulators

4.4.1 Substrates and Preparation

Substrates for InP-based growth in low-pressure MOVPE are usually high-quality InP wafers cut from single crystals with low defect density. The growth surface, in most cases, is the exact oriented $\langle 100 \rangle$ or slightly misoriented (0.5° to 3° in $\langle 110 \rangle$ direction) crystal plane on which both species in the crystal, group III and V elements, are present.

Since the deposition reaction is not as close to the equilibrium as in other epitaxial processes (e.g., liquid phase epitaxy (LPE) or hydride VPE), MOVPE does not require enhanced surface step densities generated by specific substrate misorientation. The influence of different misorientations on physical properties of layer structures will be described later.

The relations of the growth behavior of different materials on the crystal orientation of the substrates play a role, especially when layers are deposited on structured substrates so that different crystal planes are exposed to the deposition process. Results of this issue will also be discussed later.

The substrate preparation prior to loading the reactor differs in many studies. A typical procedure for InP consists of the following steps:

(a) Clean in solvents (e.g., *propanol*).

(b) Rinse in solvent, prepare for polishing etchant (e.g., rinse in water).

(c) Etch in polishing etchant ($H_2SO_4:H_2O_2:H_2O$ 5:1:1, at 50°C $r \approx 0, 2$ μm/ min) and remove 2 to 3 μ.

(d) Remove etchant residues (rinse in water).

(e) Rinse in solvents or water.

(f) Dry.

State-of-the-art substrates may require less attention to surface polishing. In our opinion the most critical step in the preparation is the final drying. Good results can be obtained by using eccentric spinners.

4.4.2 Basic Process Parameters

It has been described that comparability between processes is strongly dependent on the reactor design. The reactors described in Section 4.3 represent the design most widely used for a large variety of materials.

The general parameter ranges adjusted for the above mentioned materials, according to literature, are

Deposition temperature: 550°–700°C
Reactor pressure: 20–100 mbar
Gas velocities: 1–3 m/s

Considering the preparation of thin layer structures with altering material combinations, it is desirable not to have parameter sets that are too different for the materials to be grown. The criteria for the choice of certain parameters and results are presented in the following sections.

4.4.3 Growth of InP-Based Materials

4.4.3.1 Growth of InP on InP

The quality of epitaxial InP plays an important role for all InP-based structures, because most of them include InP layers in their active regions as confinement or injector layers, or at least as buffer layers, for separation of substrate defects, which cannot be removed by the substrate preparation.

Deposition temperatures between 550°C and 650°C [13,14] are reported in the literature for the deposition of InP in low-pressure MOVPE. Extremely low and high values relate to precursor requirements. It is desirable to keep the temperature as low as possible to avoid perturbations from enhanced dopant diffusion in doped layer structures grown underneath.

It was shown in several investigations [14,15] that at temperatures between 600°C and 650°C layers with outstanding optical and electrical properties could be grown using the above-mentioned precursor combination at a reactor pressure of 20 mbar and high gas velocity conditions (v_{gas} = 140 to 200 cm/s). In Figure 4.6 the background carrier concentration and mobility in unintentionally doped InP layers is plotted against the inverse deposition temperature.

Above 640°C no variation of the electrical properties of the InP layers was observed. At lower temperatures (below 570°C) the material properties deteriorate, and strongly enhanced background carrier concentrations and increased compensation occur.

The InP growth rate is linearly dependent on the TMIn partial pressure calculated from the carrier gas flow passed through the source bubbler. The solid nature of TMIn does not generate saturation problems in the carrier gas stream and allows growth rates between 1 and 10 μm/h. This is proving the suitability of TMIn as a controllable source for the epitaxial growth.

The V/III ratio, primarily defined by the partial pressures of the precursors in the gas phase, shows no effects on the growth rate, pointing out that parasitic reactions in the gas phase are not present. The dependence of electrical properties on the V/III ratio is not drastic above 50 in InP grown at a temperature of 640°C [16], but is strongly related to the impurities in the precursors.

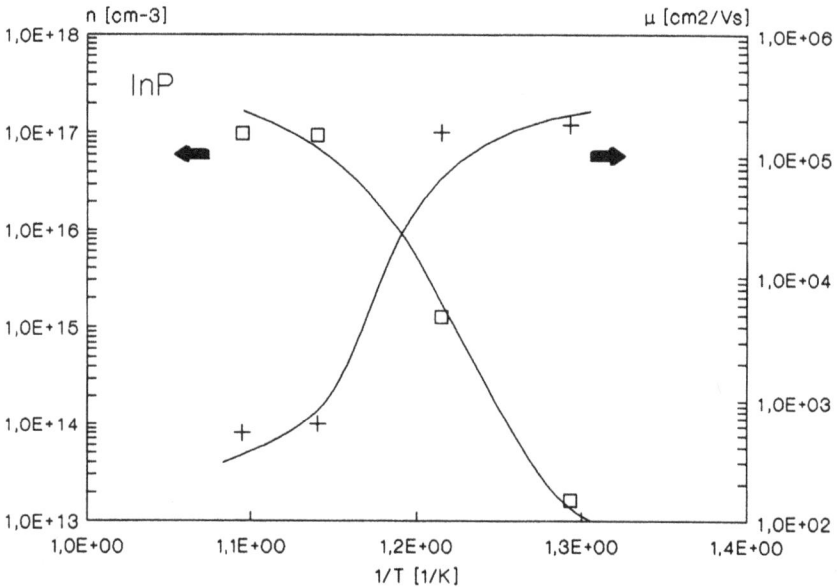

Figure 4.6 Background electron concentration and mobility in InP layers grown by low-pressure MOVPE, measured at 77K in dependence on inverse deposition temperature.

As already mentioned, the purity of the precursors is currently very good, so that the growth of high-purity material is possible. Figure 4.7 shows the dependence of electron mobility on the carrier concentration measured at liquid nitrogen temperature in undoped and intentionally n-doped InP, grown under the conditions mentioned at the end of Section 4.4.2. The solid line represents the theoretical value after Walukiewicz [17]. The dopants used for intentionally doped layers are S or Si from their hydride sources H_2S and SiH_4. The results presented here compare well to state-of-the-art data in the literature [2].

As can be clearly seen, the experimental values are close to the theoretical maximum curve for uncompensated material. Electron mobilities of 160,000 $cm^2/$ Vs measured at 77K can be obtained at background carrier concentrations around 10^{14} cm^{-3}. Comparing this with earlier data, it is obvious that there has been progress in material technologies.

Low-temperature photoluminescence measurements at 1.5K on high-purity InP layers show clearly resolved free excitonic transition peaks with full-width half-maximum (FWHM) typically below 0, 7 meV [15,18]. Additional signals originating from transitions at undesired acceptor impurities cannot be detected.

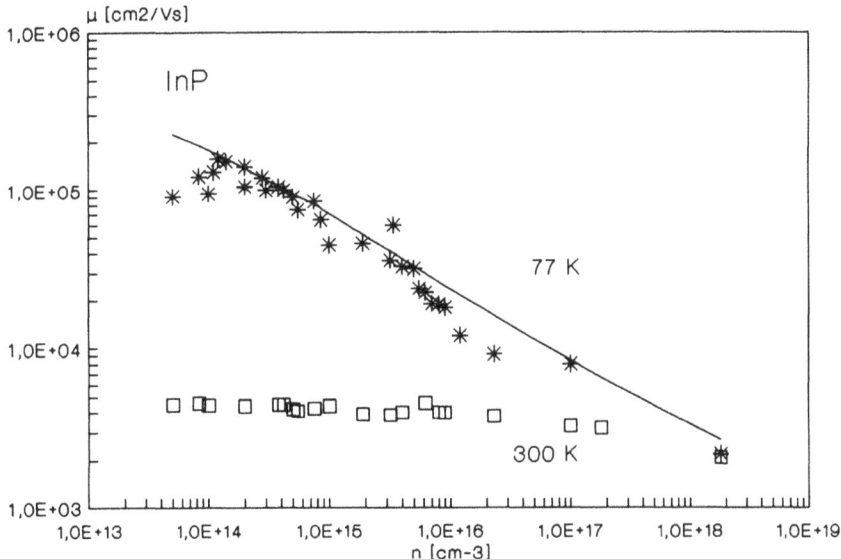

Figure 4.7 Electron mobility as a function of electron concentration in intentionally doped and unintentionally doped InP layers grown by low-pressure MOVPE. The solid line represents the 77K mobility at zero compensation ratio after Walukiewicz *et al.* [17].

Figure 4.8 shows a typical low-temperature PL spectrum of a high-purity InP layer. The spectrum clearly resolves the line of the free exciton at an energy of 1.4188 eV. Two donor-bound and one very low acceptor-bound exciton lines are also present in the spectrum. In comparison, the van der Pauw measurement of this specific layer yielded less than 10^{14} cm^{-3} free electrons, and a mobility of about 130,000 cm^2/Vs at 77K.

Doping of InP in low-pressure MOVPE with donors and acceptors has been studied extensively. Since a variety of effects occur in InP-based alloy semiconductors, the results of doping with respect to the different alloy compositions of InP and GaInAs will be described later.

The influence of flow dynamics and reactor geometry on the uniformity of film thickness and of the electrical properties of InP-based material layers over 2-inch wafer areas has been reported in various publications [2]. For most applications it was sufficient to have the uniformities below 5% over the described area. The use of InP-based materials for light-emitting devices with wavelength tunability according to the choice of layer thickness (quantum well (QW) and multi-quantum well (MQW) active regions in laser diodes [19]) has increased the requirements on the uniformities.

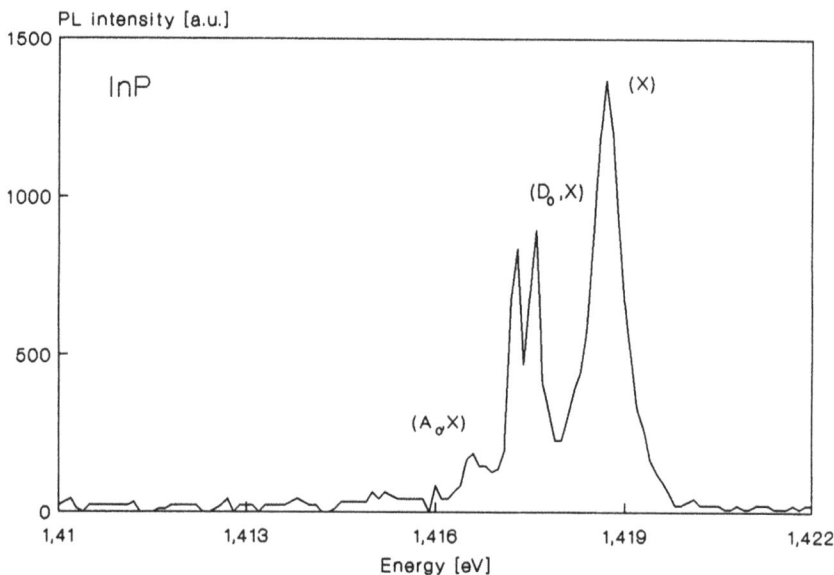

Figure 4.8 Low-temperature PL spectrum of a high-purity InP layer grown by low-pressure MOVPE. After Dr. F. Scholz, at 4. Physikalisches Institut Universität Stuttgart, Pfaffenwaldring 57, W-7000 Stuttgart 80, FRG, private communication [42].

The technique of wafer rotation to improve the uniformity of material properties, as mentioned before, has already been employed in specially constructed reactors. The newly developed *gas foil rotation* technique, used as a retrofit in commercially available MOVPE reactors, is a tool that could be applied to the entire InP-based material system. Figure 4.9 shows the distribution of film thickness on an InP layer grown in a reactor with gas foil rotation [10].

The film thickness variation was evaluated by mechanical measurement of surface steps with a profiler. The steps were prepared by selectively etching a grid into the InP layer deposited on a thin GaInAs buffer. The uniformity of film thickness is better than 1.5% over the entire wafer area.

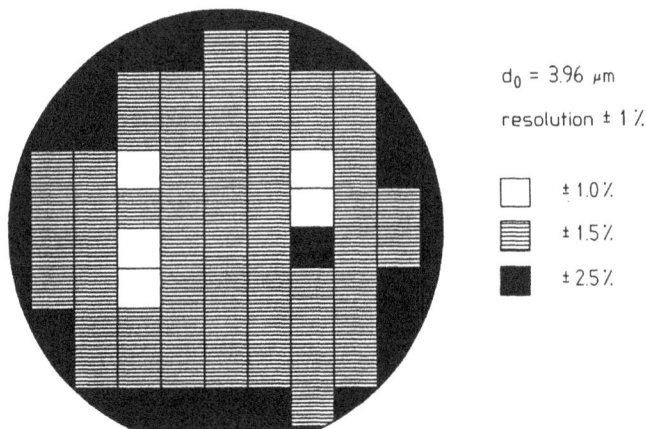

$d_0 = 3.96 \ \mu m$

resolution $\pm 1 \ \%$

☐ $\pm 1.0 \ \%$

▤ $\pm 1.5 \ \%$

■ $\pm 2.5 \ \%$

Figure 4.9 Uniformity of film thickness of an InP layer on a 2-inch wafer grown by low-pressure MOVPE applying the gas foil rotation technique.

4.4.3.2 Growth of GaInAs Lattice Matched to InP

GaInAs growth from TMIn, TMGa, and AsH$_3$ is done using low-pressure MOVPE in a variety of compositions [11,20] on various substrates. Particularly interesting is the composition lattice matched to InP. In addition to that, the growth of thin strained layers on GaAs and InP is being investigated because the variation of band structure caused by the strain in the crystal can be used to improve the properties of electronic devices [20,21].

At room temperature, GaInAs lattice matched to InP has a composition of 46.8% Ga and 53.2% In on the group III sublattice. Practically, variations of about 2% around this value are possible without features such as cracks or dislocations caused by lattice strain on the crystal surface. The crystal composition is controlled by the gas phase concentration of the group III precursors. No influence resulting from deposition temperature is observed at deposition temperatures higher than 600°C. The composition also does not show a dependence on the V/III ratio, giving evidence that undesired prereactions of the TMIn, as reported earlier, are strongly reduced in the low-pressure process.

The dependence of the solid crystal composition, represented by the lattice mismatch, on the TMIn/TMGa ratio in the gas phase is presented in Figure 4.10. The curve is determined for a constant V/III ratio and at a constant deposition temperature.

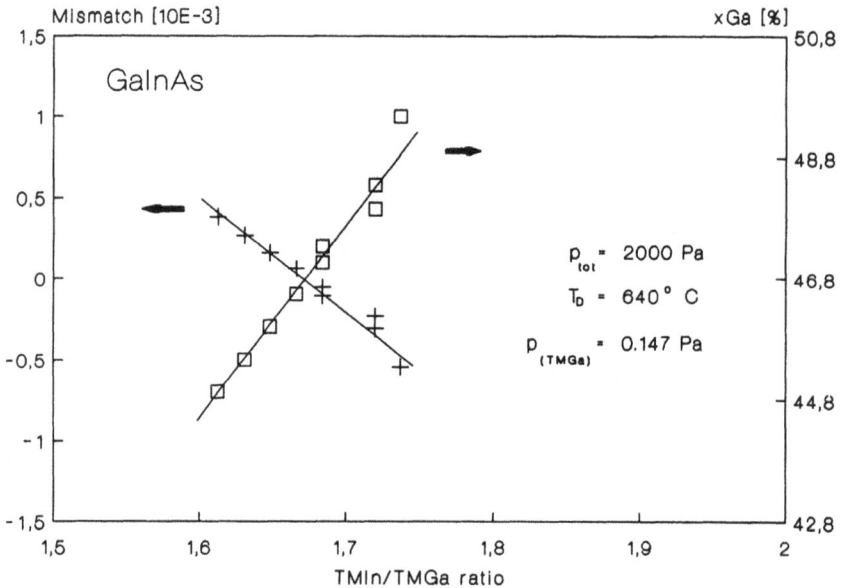

Figure 4.10 Lattice mismatch and Ga content of a GaInAs layer grown by low-pressure MOVPE on an InP substrate dependent on a TMGa/TMIn ratio in the gas phase.

It may be noted that the exact lattice matching of the GaInAs is at a ratio that is not identical with the In/Ga ratio in the solid crystal. This is due to the different deposition rates of the two binaries GaAs and InAs.

The variation of the above-described results with time of use of the TMIn bubbler is negligible. The reproducibility of lattice-matched GaInAs in the low-pressure process layers after consumption of 90% of a 100g cylinder required no flow adjustments of more than 5% to achieve optimum conditions.

The electronic properties of GaInAs films from low-pressure MOVPE processes using TMIn, TMGa, and AsH$_3$ as precursors have improved because the higher quality of the precursors available, and because of the better process control possible with advanced reactor technology. Figure 4.11 shows the variation of electron mobility in undoped and intentionally doped GaInAs films versus the electron concentration in the films.

As can be seen from these data, the background carrier concentration in MOVPE-grown samples is around 10^{15} cm^{-3}. It has been attributed to the purity level of the precursors. Experiments in which the V/III material ratio was reduced were carried out to determine the origin of the impurities. These experiments revealed the

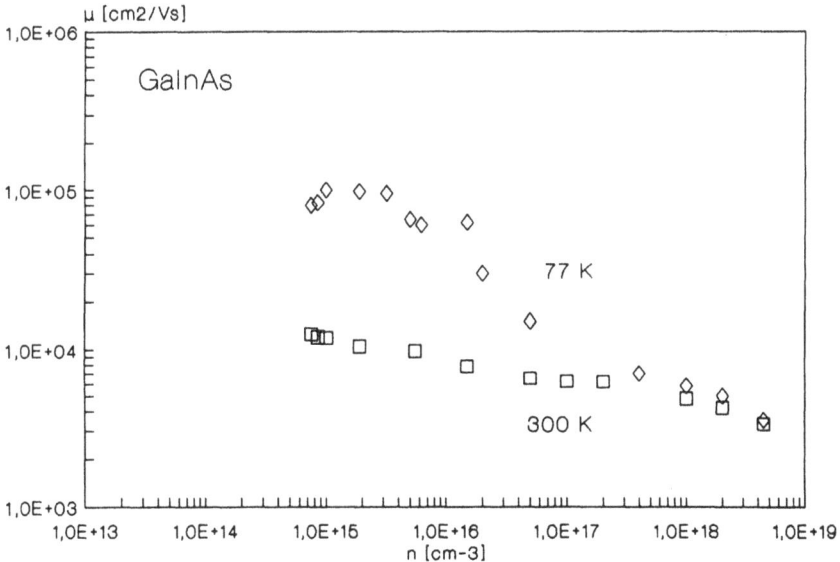

Figure 4.11 Electron mobility versus carrier concentration in intentionally doped and unintentionally doped GaInAs films grown by low-pressure MOVPE.

reduction of the impurities while lowering the V/III values. The electron mobility, however, was reduced as well, suggesting that the reduction was caused due to the enhanced compensation. The microwave reflectance, or CV measurement, is used to determine the carrier concentration at the surface of the GaInAs layer, where the conducting layer at the GaInAs/InP interface does not contribute to the measurement, and lower values in the range of 5×10^{14} cm^{-3} are measured. A stringent condition for the appearance of two dimensional electron gas (2DEG) effects is the formation of a sharp interface. The low-pressure MOVPE process with high gas velocity usually generates extremely sharp interfaces in material transitions. This will be described later.

The highest electron mobilities achieved in bulk GaInAs layers obtained in our laboratory are in the range of 13,000 cm^2/Vs at room temperature, and about 120,000 cm^2/Vs at 77K. The numbers compare well to the best reported in the literature [18].

Quantum and 2-DEG effects caused by the band discontinuities in heterostructures (e.g., InP/GaInAs) have gained more importance in the fabrication of new devices (lasers involving quantum wells or multi-quantum wells, HEMTs). The appearance of such effects is related to the properties of the interfaces between the different layers. Single InP/GaInAs transitions form 2DEGs at the interface to the InP in the GaInAs. By determination of the Hall resistance or the magnetoresistance of appropriate structures, it is possible to determine the carrier concentration and mobility. In AlGaAs/GaAs structures, electron mobilities up to 2×10^6 cm^2/Vs could be measured. Such high values are not expected in GaInAs because of the alloy scattering contribution that does not occur in the GaAs. However, it was already possible to achieve mobilities around 2×10^5 cm^2/Vs at sheet carrier concentrations of 10^{12} cm^{-2} [22]. More details on results of QW growth will be presented in Section 6.

Regarding the large area growth of GaInAs layers, it has been reported earlier that, comparable to the InP, the compositional and film thickness uniformity can be influenced by the gas velocity in the reactor. In this way the variation in film thickness can be adjusted to better than 3% parallel to the gas flow direction. At the same set of parameters, the lattice mismatch of the GaInAs film to the InP substrate varies less than 5×10^{-4} lattice constants. In the same way as described for the InP, rotating the wafer around the vertical axis applying the gas foil rotation, the uniformity of the layer thickness can be improved drastically.

The distributions shown in Figure 4.12 reveal a thickness uniformity better than 1% over most of the wafer area. The relaxed lattice mismatch variation of the grown film measured by X-ray diffraction measurement over the area of the 2-inch wafer, excluding a 4-mm rim, is less than 3×10^{-4} lattice constants.

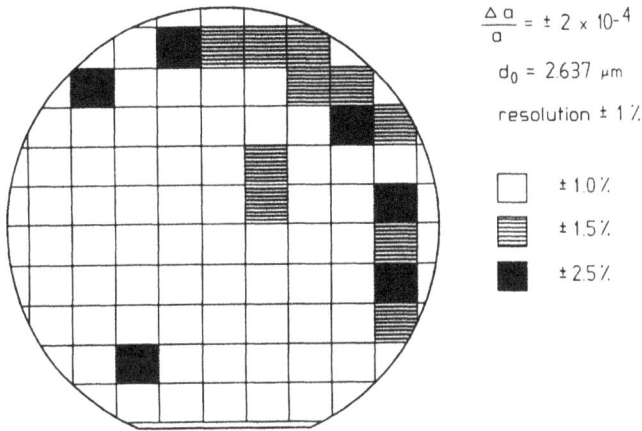

Figure 4.12 Film thickness distribution of a GaInAs layer grown on a 2-inch wafer grown by low-pressure MOVPE applying the gas foil rotation technique.

4.4.3.3 Growth of GaInAsP Lattice Matched to InP

GaInAsP is an alloy semiconductor consisting of a mixture of four different binary materials, GaP, GaAs, InP, and InAs. The condition of lattice matching to InP limits the possible composition range, and the bandgap represented by the emission wavelength spectrum can vary between 1.63 μm (GaInAs) and 0.879 μm (InP). Practically, compositions with an emission wavelength λ of 1.55, 1.3, 1.1, and 1.05 μm play the important role for light-emitting devices and waveguides in optical fiber communications. Just recently the materials have gained more importance for MQW laser devices.

An overview of the possible compositions of GaInAsP and the lattice constants, as well as emission wavelengths of the different compositions, has been given by Olsen and Zamerowski [23], which discusses the influence of gas phase parameters on the crystal composition in hydride chemical phase deposition (CVD) of Ga-In-As-P alloys.

The following will describe the basic parameters for low-pressure MOVPE of alloys in the described system. When growing GaInAsP from TMGa, TMIn, AsH$_3$, and PH$_3$, the partial pressures of all precursors, their different thermal stabilities, and the distribution coefficients for the lattice constituents between the gas phase and the solid determine the final composition of the crystal. Figures 4.13 and 4.14 describe the dependence of emission wavelength of various GaInAsP layers grown by low-pressure MOVPE on the TMIn/TMGa and AsH$_3$/PH$_3$ ratios in the gas phase as the determining factors of growth.

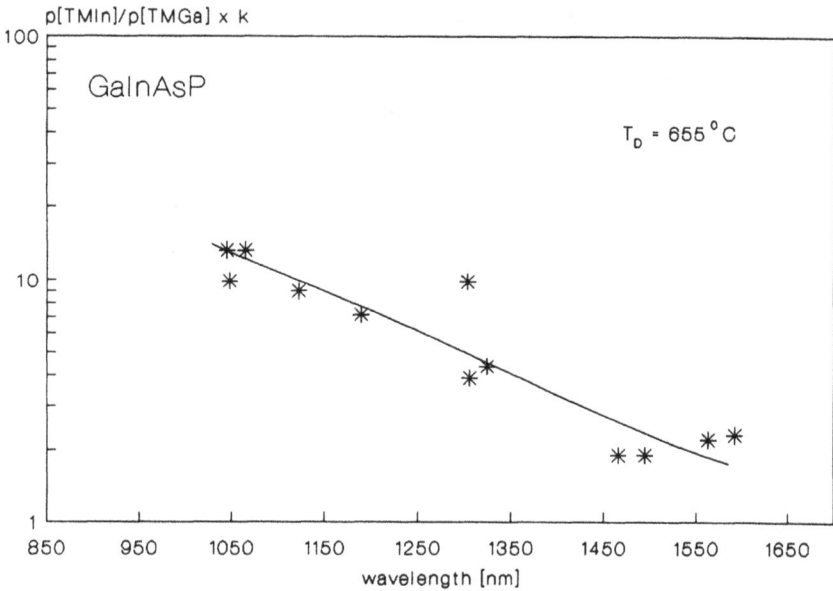

Figure 4.13 Room temperature emission wavelength of GaInAsP layers dependent on a TMIn/TMGa ratio in the gas phase.

The criteria for values included in these figures were lattice mismatched lower than 4×10^{-4} to the InP substrate and constant deposition temperature. The margins for the described composition range are the materials $Ga_{0.53}In_{0.47}As$, with the longest wavelength, and InP, with the shortest wavelength.

The slight curvatures in the semilogarithmic plots indicate that the crystal composition does not obey simple low-order functions of the gas phase composition. But in general it follows monotone regularities, so that an increased TMIn concentration in the gas phase results in a composition with higher bandgap, and an elevated AsH_3 pressure produces lower bandgap material. It is also important to note that the above correlation does not depend on the growth rate of the layers adjusted by the sum of the group III precursor pressures. Similar results have been obtained as a preparation for the fabrication of GRINSCH laser structures [24].

The electrical properties of GaInAsP layers of different compositions are characterized by the electron mobility in unintentionally doped layers. A plot of this parameter obtained from a variety of samples is given in Figure 4.15. The data points included are discriminated by the variation of lattice constant to InP below 4×10^{-4}.

For all compositions of materials, the maximum of the electron concentration was lower than 2×10^{15} cm^{-3}. This could be expected from the excellent values

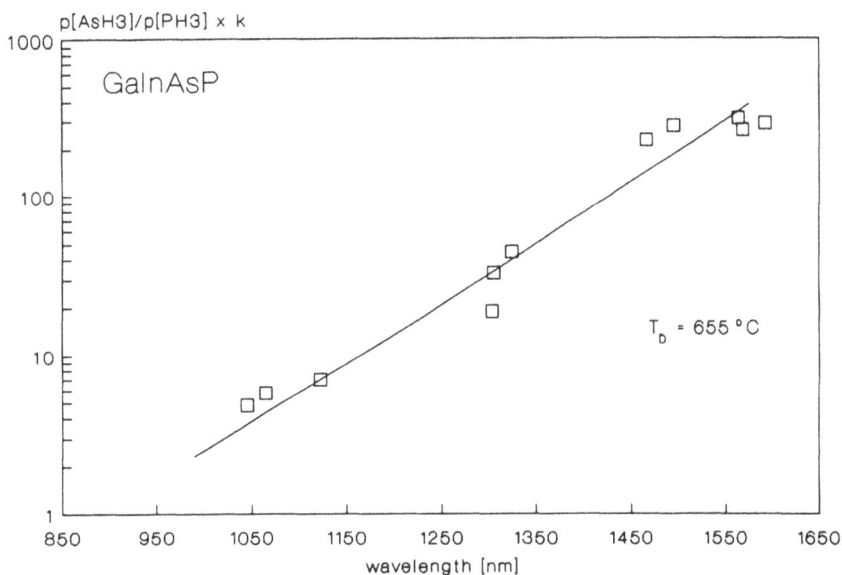

Figure 4.14 Room temperature emission wavelength of GaInAsP layers grown by low-pressure MOVPE dependent on an AsH₃/PH₃ ratio in the gas phase.

obtained from InP and GaInAs. The mobility at 77K was highest for InP and dropped with the introduction of alloy constituents. For GaInAs, the electron mobility at 77K is similar to InP at comparable carrier concentrations. Alloy scattering here shows effects at temperatures below 77K. For GaInAs alloys, the value drops with increased bandgap and seems to have a minimum at a composition corresponding to $\lambda = 1.3$ μm. Room temperature data show a monotone increase towards the composition of GaInAs.

Because GaInAsP is a material for a wide range of applications in optoelectronics, it is particularly interesting to have production techniques for large-area growth. The gas foil rotation technique has also found application for this material in single and multiwafer low-pressure MOVPE reactors, and it can produce remarkable results.

The following figures demonstrate the high productivity of this technique. Figure 4.16 shows the film thickness distribution of a GaInAsP layer ($\lambda = 1.289$ μm) grown on a 2-inch InP wafer.

The main part of the wafer shows thickness variations of $\pm 1.5\%$. This result compares well to the values obtained in InP. The achieved uniformity qualifies the material for the advanced homogeneity requirements of device structures for QW

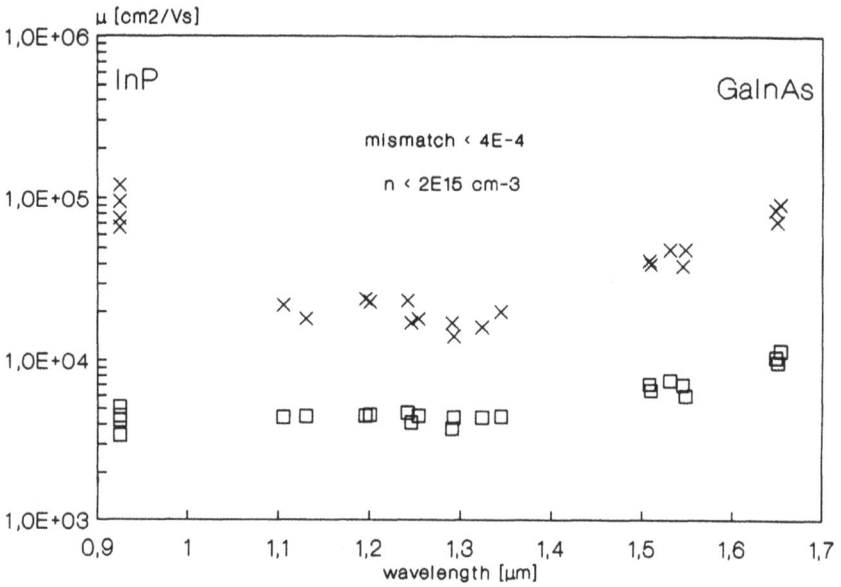

Figure 4.15 Electron mobility in unintentionally doped InP, GaInAsP, and GaInAs layer measured by the van der Pauw method at room temperature (squares) and liquid nitrogen temperature (crosses) plotted against the emission wavelength as reference for composition.

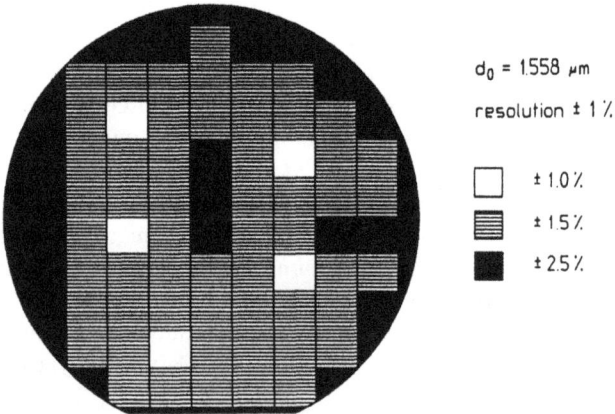

Figure 4.16 Film thickness distribution over a GaInAsP ($\lambda = 1{,}289\ \mu$m) layer on a 2-inch wafer grown by low-pressure MOVPE applying the gas foil rotation technique.

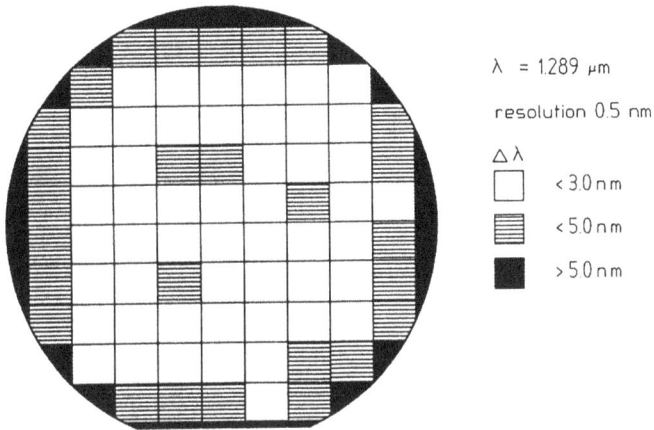

Figure 4.17 Emission wavelength distribution over a GaInAsP ($\lambda = 1,289$ μm) layer on a 2-inch wafer grown by low-pressure MOVPE applying the gas foil rotation technique.

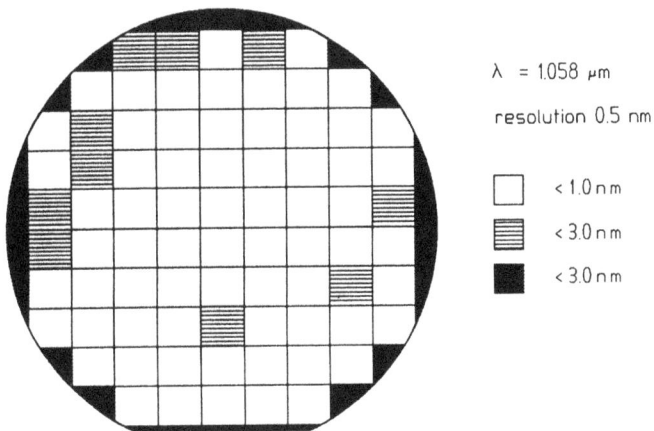

Figure 4.18 Emission wavelength distribution over a GaInAsP ($\lambda = 1,058$ μm) layer on a 2-inch wafer grown by low-pressure MOVPE applying the gas foil rotation technique.

(a)

(b)

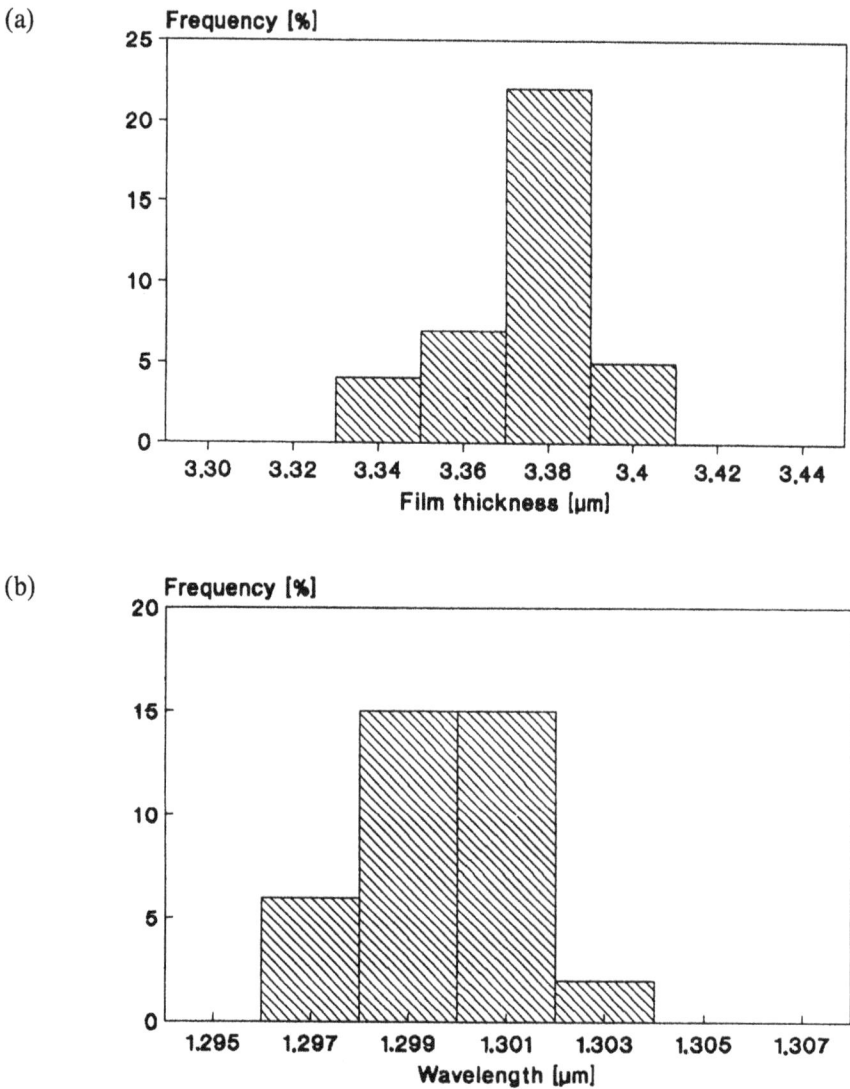

Figure 4.19 Histograms of film thickness (a) and emission wavelength (b) of a GaInAsP ($\lambda = 1.3 \ \mu$m) layer measured over five 2-inch wafers grown in one run in a reaction chamber applying planetary gas foil rotation in low-prssure MOVPE.

lasers or waveguides in which the performance depends strongly on the absolute thickness of layers.

On the other hand, the emission wavelength uniformity of the material is very important for the application as base material for devices, as mentioned above.

Figure 4.17 shows the distribution of the emission wavelength over the same layer measured at 300K in a Fourier transform PL spectrometer using the green line of an Ar laser. The major area of the wafer showed a wavelength variation below 3 nm, including the points within 5 nm of the average. Only a rim of 3 mm is out of range. For material with shorter wavelength, $\lambda = 1.05$ μm, this result could be further improved as demonstrated in Figure 4.18.

In this case the variation over the entire wafer was below 3 nm (except several points resulting from defects).

Finally, a result of GaInAsP layers grown in a multiwafer, low-pressure MOVPE reactor is presented in Figures 4.19a and b. The figures show the histograms of film thickness and emission wavelength measurements from a five-wafer load from the growth chamber described in Section 4.3.

The histograms show only minor variations in film thickness over the entire reactor load. The deviation of the measured film thickness is less than 2% over the area of the five wafers. Variations of the average emission wavelength are less than 3 nm over the grown layer area, excluding rim areas of the single wafers.

4.5 DOPING OF InP-BASED MATERIALS IN LOW-PRESSURE MOVPE

A basic condition for producing the desired device structures is the ability to control the majority carrier concentration in the grown layers by accurately incorporating dopants. The materials to be used as donor- or acceptor-type controlled impurity are to be selected carefully, primarily according to the properties they have in the matrix material to be grown and their effect on the band structure and band bending. The incorporated impurity atoms should be spatially stable, and thermally enhanced diffusion should be as low as possible. The activation of the dopants in the crystal should be high, preferably close to 100%. Memory effects due to the formation of volatile deposits in the reactor and saturation limitations should not occur.

A second stringent condition to be applied to a process is the availability of appropriate precursors. For the choice of the precursors, the important criteria are the same as for the precursors of the other constituents, such as low thermalization temperature, small affinity to prereactions with other components in the gas phase, and sufficient stability at lower temperatures and over longer times.

Dopants that have turned out to meet most of the above criteria are listed in Table 4.3.

Figure 4.20 shows the concentration of activated donors incorporated as S or Si atoms from SiH_4 and H_2S in alloys between InP and GaInAsP, measured by the Van der Pauw method at 300K.

Table 4.3

Dopants and Precursors

Element	Type	Precursor	State @ 17°C	Vapor Pressure @ 17°C
Silicon	Donor	SiH_4, diluted	Gaseous	Dependent on dilution typically 0.01% to 2%
		Si_2H_6	Gaseous	
Sulphur	Donor	H_2S diluted	Gaseous	Dependent on dilution typically 0.01% to 2%
Selenium	Donor	H_2Se diluted	Gaseous	Dependent on dilution typically 0.01% to 2%
Zinc	Acceptor	Dimethylzinc	Liquid	348.23 mbar
		Diethylzinc	Liquid	13.38 mbar
Cadmium	Acceptor	Dimethylcadmium	Liquid	31.9 mbar
Magnesium	Acceptor	Bis-cyclo-pentadienyl-magnesium	Solid	0.026 mbar

The results shown in this plot are obtained from layers grown at similar temperature, reactor pressure, and growth rate conditions. In all cases the carrier concentrations show a linear relation to the precursor partial pressures, yielding a doping range from 10^{15} to 10^{18} cm^{-3} with no indication of a saturation or limitation. From this point of view, the conditions for accurate controllability are given.

A definite difference between the incorporation rates of S and Si in InP is remarkable. The reduced distribution coefficient between gas phase concentration and solid might be due to the different incorporation mechanism of S and Si in the semiconductor material. While S is located on the group V sublattice, the donor Si has principally to be incorporated in the III sublattice. InP is generally grown at relatively high V/III ratios, so the high concentration of phosphorus might reduce the S uptake.

The question of local stability of an *n*-type dopant in low-pressure MOVPE-grown semiconductor material is the subject of Figure 4.21.

The figure shows a CV profile of an InP layer structure with two Si-doped InP layers on an Fe-doped semi-insulating substrate. The top layer is doped to 2×10^{17} cm^{-3}, the thicker layer below to 2×10^{18} cm^{-3}. An electromchemical etch profile is used to determine the interface sharpness between the doped layers and the absolute doping levels in the material.

The abruptness of the doping level transition at both interfaces is limited by the resolution of the measurement method. There is obviously no smear out of the dopant, due either to uncontrolled diffusion or to memory effects in the reactor.

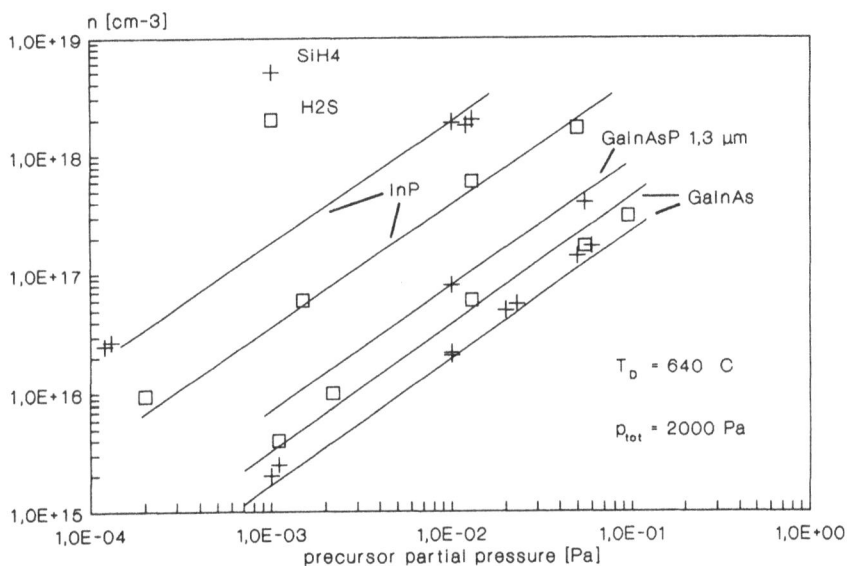

Figure 4.20 Electron concentration dependent on the partial pressure of various dopant precursors (H$_2$S and SiH$_4$) in low-pressure MOVPE-grown layers of the Ga-In-As-P alloy system measured by the van der Pauw method at room temperature.

Acceptor doping of InP and related materials faces several problems, especially at high doping levels. All acceptor dopants show a pronounced tendency to diffusion in the solid InP crystals. Depending on the element, InP reveals a saturation due to solubility limitations (e.g., for zinc at about 2 to 3 $\times$ 10^{18} cm^{-3}).

In comparison to the above-mentioned acceptor type dopants, zinc has finally turned out to be the most applicable material. Cadmium and magnesium, in contrast, have been considered to have lower diffusion coefficients [25]. Cadmium has a lower incorporation ratio in the solid InP at comparable vapor pressures, reducing its applicability in a production environment. Magnesium shows comparable results by means of diffusion in InP, but the controllability of the acceptor concentration in the crystal is more difficult as a result of the high reactivity of the precursor and an increased tendency to side reactions with trace impurities in other precursors and carrier gases.

In general, Zn has taken the main role of acceptor-type dopant for most applications. It has been found that the enhanced diffusibility of Zn is only valid for Zn atoms on interstitial sites. This does not appear at doping levels of below 5 $\times$ 10^{17} cm^{-3} under normal growth conditions at low pressures.

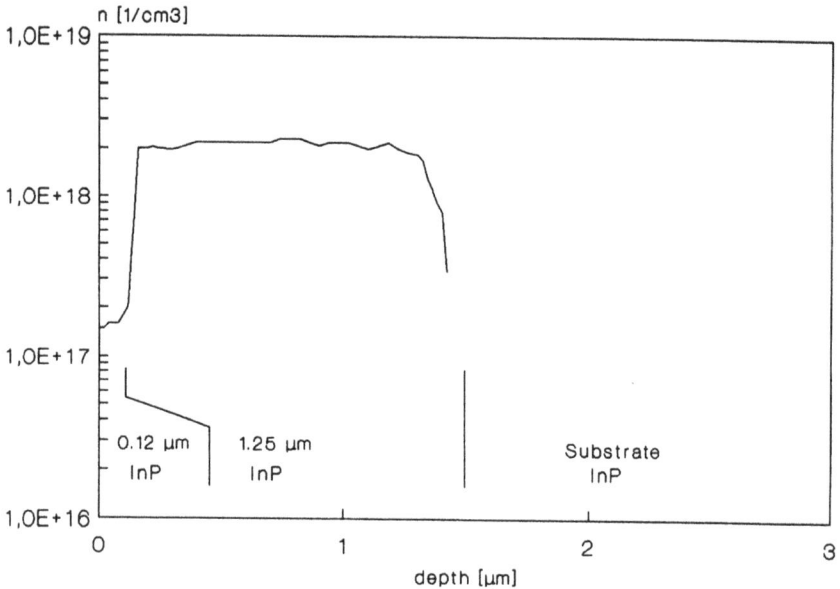

Figure 4.21 Doping profile in a Si-doped InP layer grown by low-pressure MOVPE measured by electrochemical C-V profiling.

In Figure 4.22, the hole concentration in InP layers, determined by the Van der Pauw method, is plotted against Zn precursor partial pressure (diethylzinc). The saturation value can be determined at 2×10^{18} cm^{-3}. Below this value, the hole concentration is linear with the precursor partial pressure.

The activation of acceptors in InP has been a subject of intensive investigations. It has been found that p-doped InP exposed to an atmosphere containing hydrogen radicals, originating from cracked hydrides, shows reduced hole concentration in Van der Pauw measurement during the cooling cycle after growth [26]. This effect also occurs by diffusion of the hydrogen atoms through a GaInAs layer if the InP is covered by a GaInAs layer. Postgrowth heat treatment reactivates the acceptors to their real concentration determined by SIMS.

Glade *et al.* [26] attempts the explanation that the acceptors are considered to be transferred into hydride compounds on interstitial sites in the lattice. The thermal activation process provides group III vacancies because of In atoms diffusing off lattice dislocations. Indications for this model have been found by enhanced activation when slightly mismatched GaInAs top layers have been grown on the InP, inducing dislocations in the bulk InP lattice.

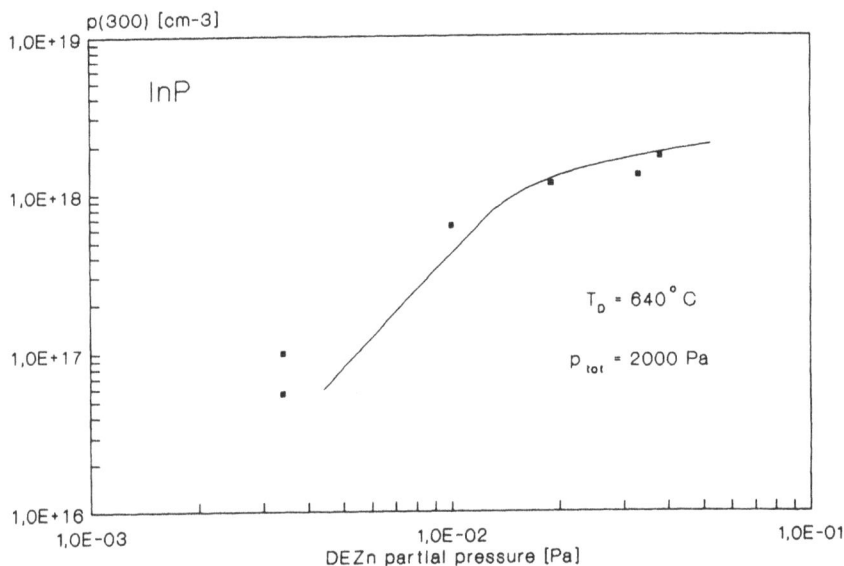

Figure 4.22 Hole concentration in an InP layer doped with Zn measured at room temperature on DEZn partial pressure in the gas phase in the reactor.

In Figure 4.23, the concentration of holes in GaInAs layers, measured by Van der Pauw method, is plotted against the partial pressure of Zn precursors in the gas phase. The relation of net hole concentration is proportional to the precursor concentration with no saturation limit, regardless of whether diethylzinc or dimethylzinc is applied. Both precursors appear to be completely thermalized at the deposition temperature of about 600°C to 650°C, so that only the adsorption of Zn atoms at the substrate surface during the growth process determines the incorporated concentration up to the solubility limit.

The stability and doping abruptness of Zn profiles in GaInAs have been studied by SIMS analysis of doped layer structures with different levels, layer thicknesses and, due to the growth sequence, exposition time to thermal treatment. The evaluation has been made at Heinrich Hertz Institute in Berlin by Grote under ESPRIT contract numbers E 263 and E 927.

In Figure 4.24 such a SIMS profile is plotted. The structure is shown in the insert. Except for the overproportional height of the Zn level in the top layer, all peaks represent very strictly the intended profile by means of layer width and relative level. There is no evidence of smear-out at the doping slope, even in layers exposed to temperature treatment; that is, the first doped layers in this structure. Thus, it can

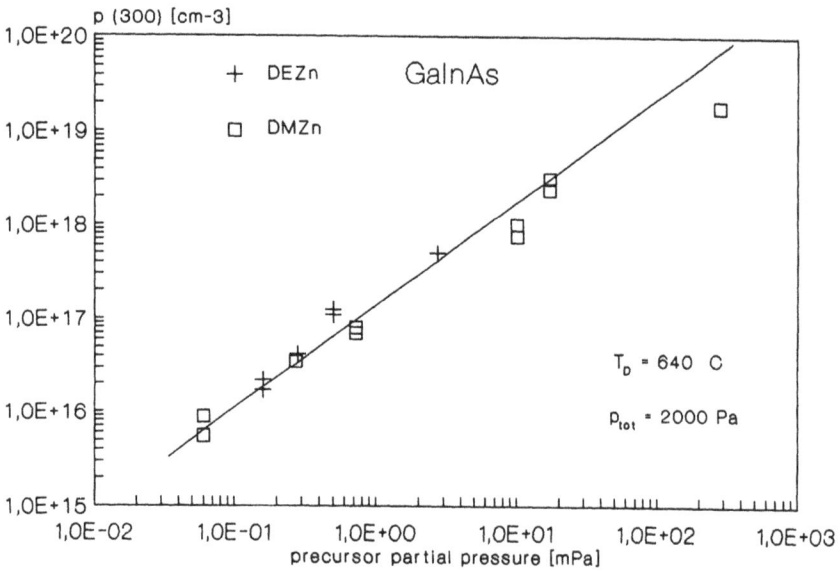

Figure 4.23 Hole concentration in GaInAs layers grown by low-pressure MOVPE dependent on Zn precursor partial pressure.

be concluded that the diffusion of the Zn atoms in low-pressure MOVPE-grown GaInAs layer is very low and definitely acceptable for device structures.

Semi-insulating InP layers through low-pressure MOVPE has gained strong interest because of its importance for integrated circuit structures (OEIC) [27]. To separate high-current device structures, which require substrates with low crystal defect density (i.e., lasers), from surface channel devices (i.e., metal insulator field effect transistor (MISFETs)) for controlling circuits integrated on one chip, high-resistivity separation layers are required.

Two techniques can be used for obtaining semi-insulating materials:

(a) Incorporate a deep-level acceptor to compensate the intrinsic and background n-conductivity of the semiconductor (i.e., Fe from Ferrocene [$Fe(C_5H_5)_2$)]).
(b) Incorporate a deep-level donor and obtain semi-insulating properties by co-doping with a shallow acceptor to compensate the carriers (i.e., Ti as deep-level donor and any shallow acceptor, i.e., Cd as cited in the example).

Iron doping has been extensively studied by various groups in atmospheric pressure MOVPE. However, problems controlling the Fe concentration in InP layers came up. Due to the relatively low Fe concentration required in the gas phase, it

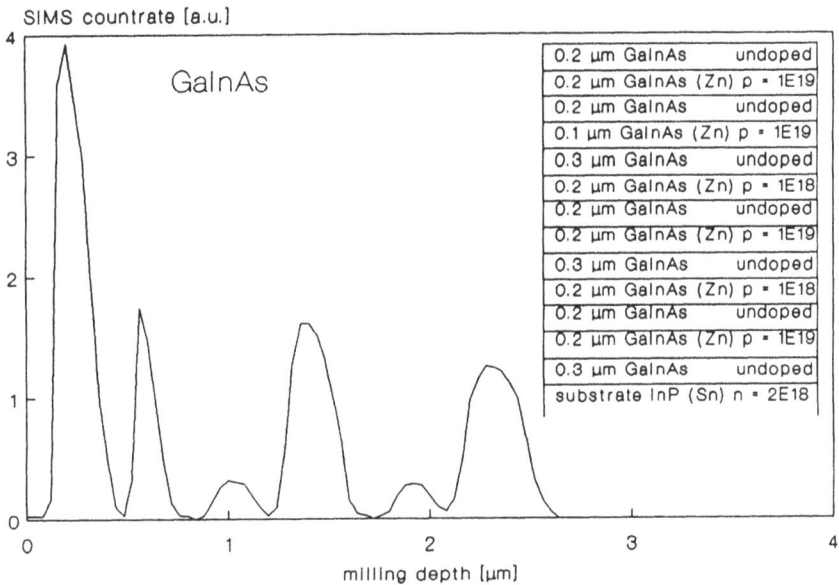

SIMS countrate [a.u.]

GaInAs

0.2 μm GaInAs	undoped
0.2 μm GaInAs (Zn)	p = 1E19
0.2 μm GaInAs	undoped
0.1 μm GaInAs (Zn)	p = 1E19
0.3 μm GaInAs	undoped
0.2 μm GaInAs (Zn)	p = 1E18
0.2 μm GaInAs	undoped
0.2 μm GaInAs (Zn)	p = 1E19
0.3 μm GaInAs	undoped
0.2 μm GaInAs (Zn)	p = 1E18
0.2 μm GaInAs	undoped
0.2 μm GaInAs (Zn)	p = 1E19
0.3 μm GaInAs	undoped
substrate InP (Sn)	n = 2E18

milling depth [μm]

Figure 4.24 SIMS profile of a Zn-doped GaInAs structure for investigation of dopant stability in grown layers. (Measured at Heinrich Hertz Institute in Berlin. The work was supported by the European Community under ESPRIT contract numbers 263 and 927).

has been difficult to adjust the appropriate level of flow rates through the bubbler. The controllability has been improved by applying low-pressure MOVPE. Iron concentration in the range 10^{16} to 10^{17} cm^{-3} could be easily and reproducibly adjusted, yielding reasonable resistivity data for device applications. A resistivity of more than 2×10^{17} Ωcm can be obtained in InP.

The topic of Ti/acceptor codoping is related to the availability of an appropriate Ti precursor. New precursors for this element are presently under development.

4.6 APPLICATION OF LOW-PRESSURE MOVPE FOR DEVICE STRUCTURES

The use of GaInAs/InP layer structures grown by low-pressure MOVPE for a variety of devices has been reported. Some examples will be presented below.

4.6.1 Regrowth and Localized Growth of InP and GaInAs

One application of semiconductor layers through low-pressure MOVPE is the regrowth of device structures such as contact and cladding layers for laser structures.

Usually the materials have to be grown on GaInAsP alloys with different compositions acting as active laser materials or waveguides. The structures defined by dry etching have extremely small dimensions and, as in the case of grids for distributed feedback (DFB) lasers, have to be carefully conserved in shape and size to obtain optimum performance for the devices. Typically, the cladding and contact layers are InP and should not contain As traces; although during heating before start of the InP growth, the structured material has to be conserved using various precautions. Good results in conservation of the surface structure have been obtained by fast heating cycles and reducing the exposure of the bare GaInAsP structures to PH$_3$ [28]. Techniques for localized growth of semiconductor materials such as InP and GaInAs on structured surfaces to obtain spatially separated device structures on the substrate surface have gained similar importance.

It is well known that InP can be grown selectively on InP substrates by covering parts of the substrate surface with SiO$_2$ or Si$_3$N$_4$. The growth by MOVPE will not take place on masked parts of the surface at certain conditions. It is also known that the growth rate is strongly dependent on the orientation of the growth surface [29]. It was found that InP/GaInAs superlattice structures grow under formation of mesa structures limited by (111) and (100) facets. The explanation was that InP shows a reduced growth rate on (111) surfaces while GaInAs does not grow on these crystal planes.

The average growth rate of the materials grown into masks is dependent on the size and the density of the openings in the masked surface. It increases with decreasing lateral dimensions. At the same time the deposition on the masking material decreases. These effects have been explained by the diffusion of constituents via the gas phase from low growth areas to locations with higher material consumption (e.g., the (100) planes forming during growth).

By correctly interpreting these results, it appears that it is possible to grow structures such as buried quantum wire arrays in a single-step epitaxial run.

The influence of the localized growth on the composition of GaInAs on the masked InP substrates or while embedded in InP has been investigated by Finders *et al.* [30] and Kayser *et al.* [31]. It has been found that when using a small lateral structure dimensions, the In incorporation in the alloy tend to increase. On the other hand, decreasing distance between stripes results in decrease of the In content in the grown structures. The variation of the In content that was found in the round spots along the edges could have been reduced by reduction of the reactor pressure.

It can be concluded that the low-pressure MOVPE process is capable of advanced design of laterally resolved device structures.

4.6.2 Photodiodes

Pin-Photodiode structures are an important part of optoelectronic communication technologies. The ability of low-pressure MOVPE processes to obtain high quality device properties has been demonstrated.

An example of an application of advanced high resistivity Fe-doped InP and GaInAs layers grown by low-pressure MOVPE is the so-called *MSM photodiode technology* [27].

Figure 4.25 shows the structure of the photodiode and the photoresponse transient on a laser pulse with a duration of 15 ps. The device characteristic calculated

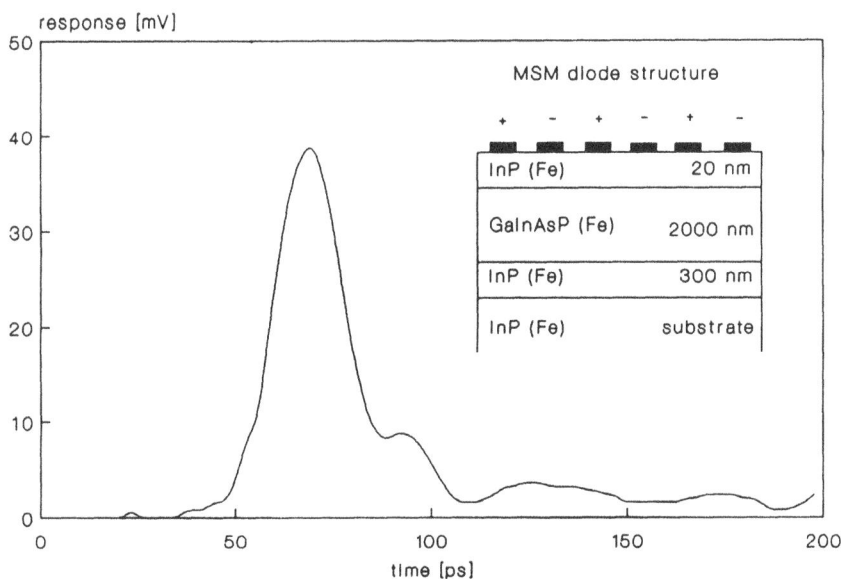

Figure 4.25 Cross section and pulse response of a MSM photodiode structure applying low-pressure MOVPE-grown Fe-doped InP and GaInAs layers. After T. Wolf, A. Krost, F. Reier, P. Harde, D. Kuhl, F. Hieronymi, H. Ullrich, D. Bimberg, and H. Schumann, "Semi-insulating Fe- and Ti-doped InP and InGaAs for Ultrafast Infrared Detectors Grown by LP-MOCVD," *Proceedings of the 6th Conference on Semi-Insulating III-V Materials,* Toronto, May 1990.

from the measured values shows a -3-dB cut-off frequency of 80 GHz. Bias-independent junction capacity gives an indication of the high quality of the semi-insulating InP blocking layer and the GaInAs:Fe active layer. The resistivity of the InP:Fe measured in bulk layers is about $2 \times 10^7 \; \Omega \cdot cm$. GaInAs:Fe with 960 $\Omega \cdot cm$ is also possible.

Commercial GaInAs/InP pin photodetectors as receivers for 1.3- and 1.55-μm wavelengths for fiber-optic applications with bit rates up to 2.4 Gbit/s are routinely produced by low-pressure MOVPE processes [32].

4.6.3 MISFET Device Structures

MISFET devices are another example of the application of low-pressure MOVPE-grown GaInAs bulk layers. Besides the fabrication technology of the devices, such as deposition of the insulating layer and the passivation technique, the quality of the active layer material and the interfaces between substrate, buffer, and active layer are of essential importance.

In Figure 4.26, the structure for a low-pressure MOVPE-grown MISFET is given. MISFETs with transconductances of up to 300 mS/mm at a gate length of 1.5 μm have been realized [33]. The devices, fabricated either with Si_3N_4 [34] or SiO_2 [33,35], show high long-term stability of I_{DS} and could successfully be included in monolithically integrated inverters.

4.6.4 GaInAsP Waveguides

GaInAsP bulk layers grown by low-pressure MOVPE find applications as active layers in double heterostructure (DH) lasers, cladding layers, and waveguides in laser structures.

Responsible for a specific characteristic in waveguide applications are the losses, which are induced by the crystalline quality of the material, interface roughnesses, and, of course, the preparation technique defining the geometry.

The diagram (Figure 4.27) shows loss figures of waveguides produced from low-pressure MOVPE-grown structures, as described in the insert. The waveguides have been defined over lengths of 36 mm and were 3 μm wide. The structures were used as a characterization tool of the grown layers and have been distributed over the entire diameter of a 2-inch wafer. The loss figures were very low, in the majority better than 0.8 dB/cm, and a large number showed less than 0.4 dB/cm, and the best results were below 0.2 dB/cm. This can be considered as proof of the high quality and uniformity of the GaInAsP layer.

GaInAs (SI)	5E18 cm-3	200 nm
GaInAs (SI)	5E17 cm-3	150 nm
InP (Zn)	<5E16 cm-3	200 nm
InP (Fe)	s.i.	substrate

Figure 4.26 MISFET structure used for fabrication of high-transconductance devices.

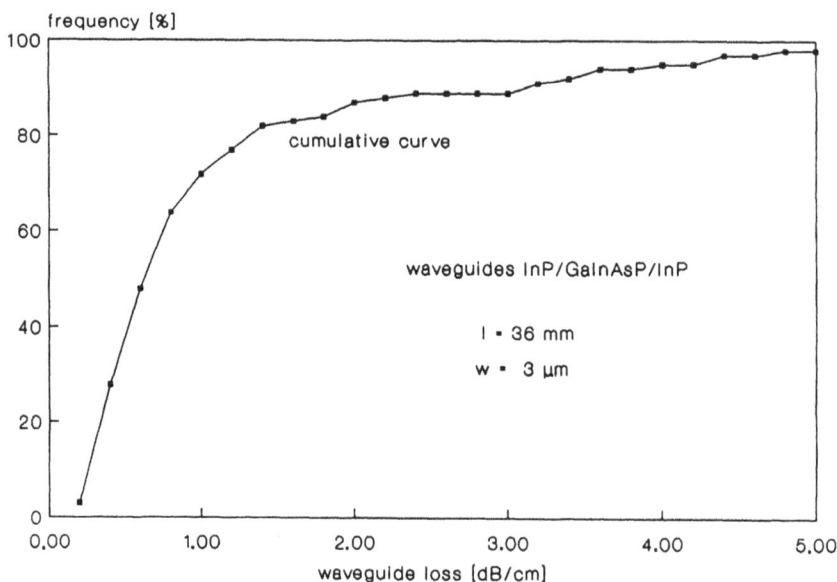

Figure 4.27 Losses of waveguides fabricated on GaInAsP/InP structures grown on 2-inch wafers by low-pressure MOVPE. (Measured at LEP, Avenue Descartes, Limeil Brevannes. The Work was supported by the European Community under ESPRIT contract number 2518).

4.6.5 Quantum Well Structures Grown by Low-Pressure MOVPE

An extensive study on the formation of InP/GaInAs and GaInAsP/GaInAs hetero-interfaces in low-pressure MOVPE growth has been performed by Grützmacher *et al.* [6]. The growth of extended islands at the heterointerfaces, as have been observed before in MBE-grown AlGaAs/GaAs heterostructures [36], were resolved in the photoluminescence (PL) spectrum of GaInAs/InP QWs grown by low-pressure MOVPE. Such features have been observed earlier in low-pressure MOVPE of InP/GaInAsP quantum wells [37].

These extended monolayer steps are larger than the excitonic radius in the low gap material, and they generate doublets and triplets in emission lines of QWs with a thickness of several monolayers. The reason for the formation of these large mono-layer areas was attributed to the manner of atomic rearrangement during the growth of the heterointerfaces [36].

By varying the growth interruption procedures and growth time for the QWs in single QW structures, it was found that the deposition process is predominantly two dimensional. That means monolayer after monolayer is deposited, and the lattice constituents rearrange on the substrate surface after adsorption by lateral diffusion until a nucleus is formed before the next monolayer starts to grow.

The continuous variation of growth time for the QW showed a discontinuous energy shift in the PL line. Figure 4.28 shows a reproduction of the result presented in Grützmacher [6]. It is clearly resolved that the QW grows in large areas of monolayer steps. The energy levels for the thicknesses between six and eight monolayers are indicated in the figure.

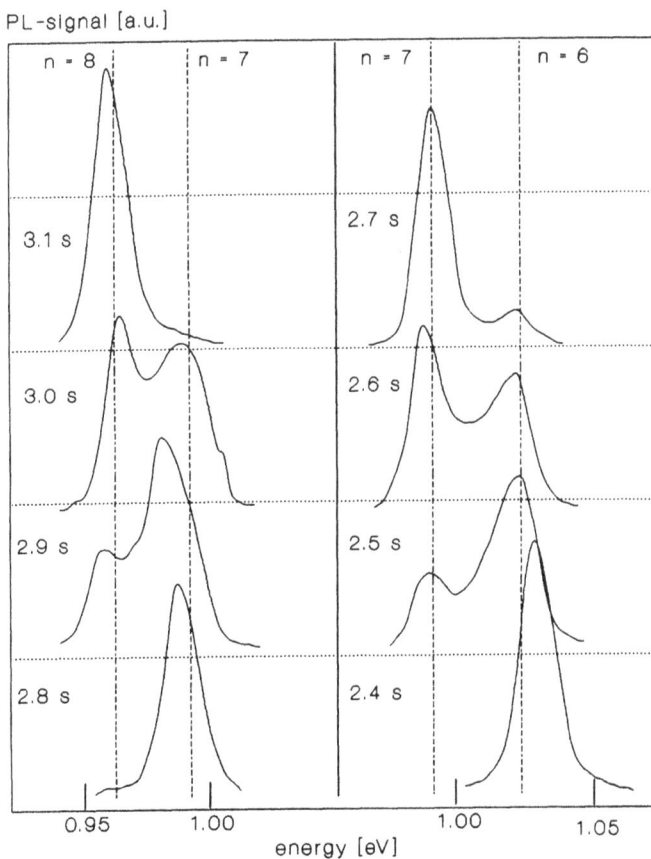

Figure 4.28 Formation of large monolayer areas during growth of QWs observed by discontinuous energy shift in low-temperature PL due to variation of growth time for the QWs. After D. Grützmacher, J. Hergeth, F. Reinhardt, K. Wolter, and P. Balk, "Mode of Growth in LP-MOVPE Deposition of GaInAs/InP Quantum Wells," *J. El. Mat.*, Vol. 19, No. 5, 1990.

With increasing growth time, the number of "islands" grown on a completed QW that are larger than the excitonic radius is increasing. This is indicated by the appearance of a second peak with lower blue shift, demonstrating the capability of the process to control film thickness in monolayer steps.

A result of GaInAsP/InP QW characterized by low-temperature PL measurement is given in the spectrum shown in Figure 4.29. The occurrence of sharp emission lines according to the various QW thicknesses can be observed. Also, in this case doublet and triplet peaks for thin QWs show that a similar theory as that held for GaInAs QW could be held for GaInAsP. Atomically flat interfaces are necessary for the resolution of monolayer growth steps. The result has been presented by Thijs *et al.* [38].

The application of low-pressure MOVPE growth for GaInAs/GaInAsP multiquantum well, broad area lasers; and investigations of the laser performance on the cavity length have been published by Rosenzweig *et al.* [39].

Figure 4.30 shows the threshold current density in broad area lasers with a band structure shown in the insert, plotted against the cavity length of the device.

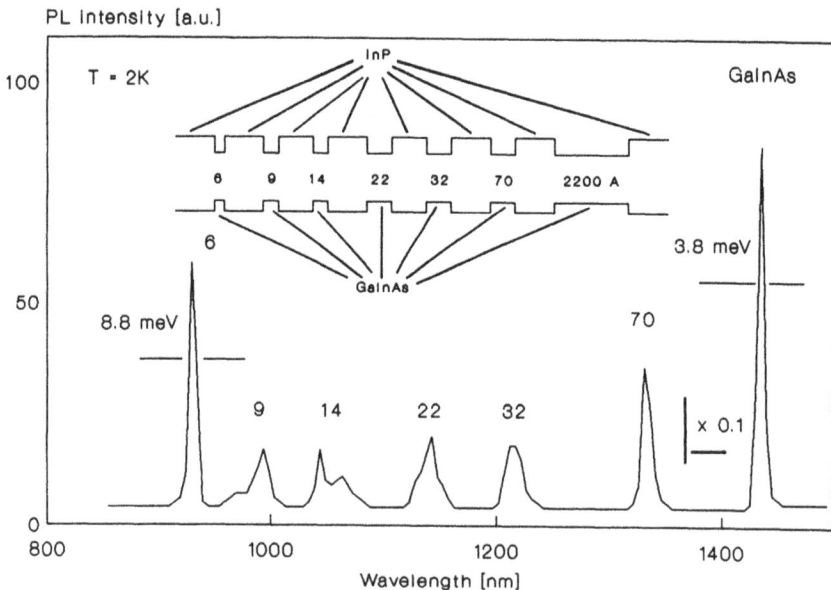

Figure 4.29 Low-temperature PL spectrum of a GaInAsP/InP multi-single-quantum well structure. After P.J.A. Thijs, T.v. Dongen, P.I. Kuindersma, J.J.M. Binsma, L.F. Tiemeyer, J.M. Lagemaat, D. Moroni, and W. Nijman, "High Quality InGaAsP-InP for Multiple Quantum Well Laser Diodes Grown by Low-Pressure OMVPE," *J. Cryst. Growth*, Vol. 93, 1988, pp. 863–869.

The cavity length has been varied by cleaving the device after preparation. The plot shows results for various numbers of QW and, for comparison, the threshold current density of a LPE DH laser. The material for the barriers is GaInAsP ($\lambda = 1.3$ μm). The QW thickness has been chosen to result in an emission wavelength of 1.55 μm.

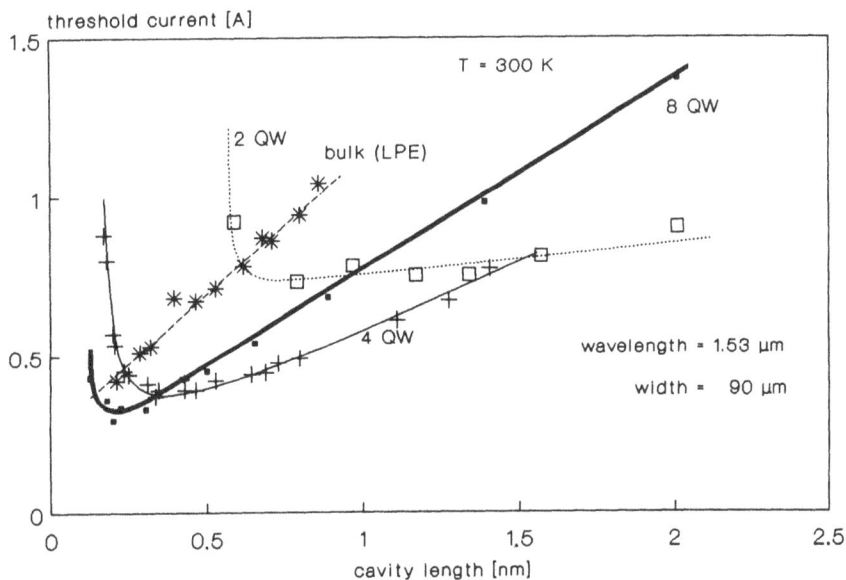

Figure 4.30 Threshold current of broad-area MQW lasers with various numbers of QWs plotted against the cavity length of a laser cell in comparison to a LPE bulk laser. After M. Rosenzweig, W. Ebert, D. Franke, N. Grote, B. Sartorius, and P. Wolfram, "Lasing Characteristics of InGaAs/InGaAsP-MQW-Structures Grown by Low Pressure MOVPE," *Proceedings of 5th International Conference on MOVPE,* Aachen, 1990, to be published in *J. Cryst. Growth.*

The devices prepared were ridge waveguide, 90-μm-wide, broad area lasers. As can be seen, the threshold current densities obtained for the different numbers of QWs extrapolated for infinite cavity length are 290, 450, and 610 A/cm^2 for two, four, and eight QWs, respectively. These values are among the lowest reported for such types of laser devices.

4.6.6 Strained Layer Device Structures

The low-pressure MOVPE process allows the deposition of strained compound semi-conductor layers, even allowing for extreme differences in the lattice constant to the substrate material.

In nonrelaxed layers, the induced strain influences the material band structure. As was predicted by theory, the curvature of the band structures are changed to reduce the effective carrier masses, carrier mobility, and thus the high frequency properties of devices prepared with strained layer superlattices or quantum wells (QWs).

Investigations of microwave devices have been published by Malzahn *et al.* [20]. P-channel MOSFETs with a gate length of 2 μm have been fabricated, and an extrinsic transconductance of 11.5 mS/mm was obtained. An enhancement of sheet carrier concentration and hole mobility in the strained channel due to increased compressive strain (higher In content in the GaInAs) was obtained.

The application of strained GaInAsP layers as MQWs between InP barriers for actual laser devices has been reported by Thijs *et al.* [21]. This publication reports on the results of laser devices with significantly improved properties related to the strained layer technique such as increased efficiency and high output power.

Figure 4.31 shows the low-temperature PL spectrum of a multi-single-quantum well structure consisting of $Ga_{0.2}In_{0.8}As$ wells and InP barriers grown on a nearly exact oriented substrate. Again, multiple emission peaks from one QW indicate the

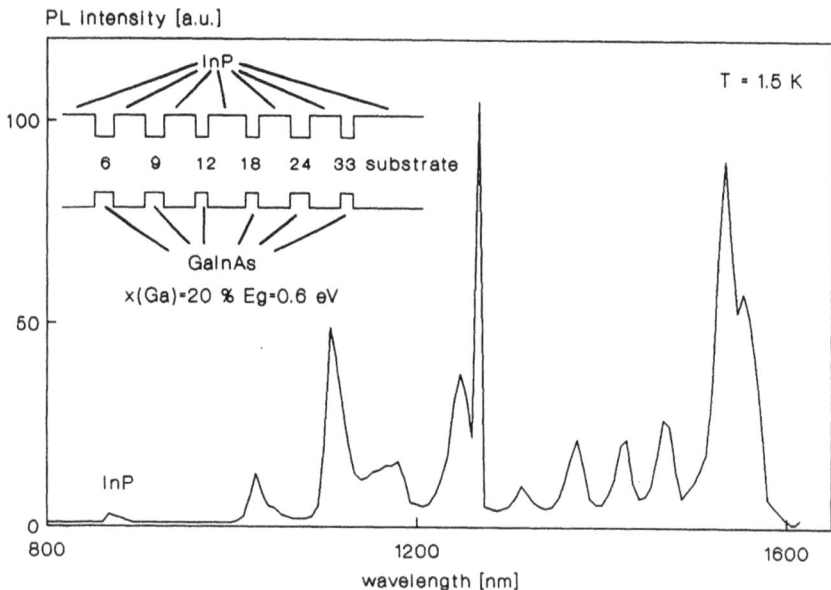

Figure 4.31 Spectrum of an InP/$Ga_{0.2}In_{0.8}As$ strained multi-single-quantum well structure. After P.J.A. Thijs, E.A. Montie, T.v. Dongen, and C.W.T. Bulle-Lieuwman, "Improved 1.5 μm Wavelength Lasers Using High Quality LP-OMVPE Grown Strained-Layer InGaAs Quantum Wells," *J. Cryst. Growth*, special issue ICCBE-2, 1989.

presence of atomically flat areas larger than the excitonic radius. Very narrow line widths indicate very sharp interfaces in the QWs.

The inverse external efficiency for actual laser diodes fabricated from strained layer GaInAs/GaInAsP MQW structures with four QW and Zn modulation-doped barriers and confinement layers of InP-lattice-matched GaInAsP ($\lambda = 1.3\ \mu$m) is plotted against the length of the laser cavity in Figure 4.32. An external efficiency of 82% could be obtained from devices with a cavity length of 265 μm with as-cleaved, uncoated mirrors.

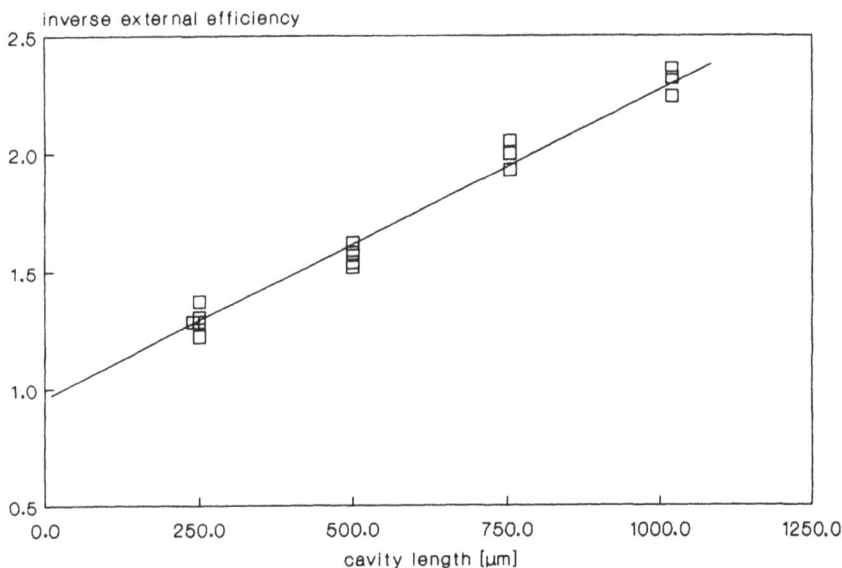

Figure 4.32 External efficiency of strained layer InGaAs/InGaAsP MQW lasers on cavity length. After P.J.A. Thijs, E.A. Montie, T.v. Dongen, and C.W.T. Bulle-Lieuwma, "Improved 1.5 μm Wavelength Lasers Using High Quality LP-OMVPE Grown Strained-Layer InGaAs Quantum Wells," *J. Cryst. Growth,* special issue ICCBE-2, 1989.

4.6.7 Visible Laser Devices

The recent emphasis on and excellent results of visible laser devices produced from the Al-Ga-In-P material system by low-pressure MOVPE make it worth while to note them here. It has been reported by Lee [40] that bulk GaInP and AlGaInP grown by MOVPE shows a room temperature emission wavelength of 674.1 and 628.6 nm.

These materials can be applied to the production of visible wavelength lasers as required for barcode readers and other applications in the consumer electronic field.

Valster *et al.* [41] use GaInP/AlGaInP QW structures to fabricate laser devices emitting at 633 nm (light red), and even recently reported on the success of fabricating yellow as well as 555 nm (green) MQW laser diodes with a differential efficiency of 0.4 W/mA and a CW power of 3 mW.

4.7 CONCLUSION AND FUTURE INVESTIGATIONS

In conclusion, the low-pressure MOVPE process has proven the wide applicability of InP-based compound semiconductors. The crystalline, electrical, and optical properties of the base materials have led to the application of the process technique for production of advanced device structures.

From intense investigations of the properties of such structures, it can be concluded that the process is capable of defining layer structures with atomically sharp interfaces and outstanding uniformity of physical properties. The production of QW and 2-DEG device structures using the outstanding capabilities of MOVPE is state of the art. The same features could be obtained in different material systems, which demonstrates the flexibility and advantages of the versatile MOVPE technique.

Devices produced on layer structures grown by low-pressure MOVPE have gained state-of-the-art properties and are setting production standards in industry. Various industrial laboratories have exploited this technique for mass production of high-quality optoelectronic devices.

In the field of reactor development, considerable progress has been made toward the application of multiwafer reactors. The planetary gas foil rotation technique has already generated production systems for a variety of applications.

The application of new precursors (e.g., organic compounds) as replacements for the hydride gases AsH_3 and PH_3 is presently under development and may in the future even simplify the technique.

ACKNOWLEDGMENTS

The author would like to thank Dr. H. Jürgensen and Dr. M. Heyen for fruitful discussions and final editing of this work. I gratefully acknowledge the support of Joachim Hergeth for supplying data collections and preparing a great deal of the artwork.

REFERENCES

1.   K.-H. Bachem and M. Heyen, "Chemical Vapour Phase Deposition of GaAs and AlGaAs," *Proceedings of the Colloque Matériaux et Technologie pour le Microelectronique,* Montpellier, France, 1976.

2. D. Grützmacher, "Großflächiges Wachstum von GaInAs/InP Heterostrukturen mit Niedrigdruck MOVPE," Master thesis at Institute of Semiconductor Electronics University of Aachen, 1987.

3. N. Pütz, "Zum Wachstum von GaAs aus Metall-Organischen Ga-Verbindungen und Arsin," Ph.D. thesis at Institute of Semiconductor Electronics University of Aachen.

4. F.C. Evershteym, I.W.P. Severin, and C.H.J. Brehel, *J. Electrochem. Soc.*, Vol. 117, 1970, pp. 925–931.

5. R. Lückerath, P. Tommack, A. Hertling, H.J. Koss, P. Balk, K.F. Jensen, and W. Richter, "CARS In Situ Diagnostics in MOVPE Thermal Decomposition of AsH$_3$ and PH$_3$," *J. Cryst. Growth*, Vol. 93, 1988, pp. 151–158.

6. D. Grützmacher, J. Hergeth, F. Reinhardt, K. Wolter, and P. Balk, "Mode of Growth in LP-MOVPE Deposition of GaInAs/InP Quantum Wells," *J. Electron. Mat.*, Vol. 19, No. 5, 1990.

7. M. Heyen, M. Heuken, G. Strauch, D. Schmitz, H. Jürgensen, and K. Heime, "Low Pressure Growth of GaAs/AlGaAs Layers on 2″ and 3″ Substrates in a Multiwafer Reactor," *Proc. of the "MRS Spring Meeting,"* San Diego, CA, April 24–29, 1989, p. 245.

8. A. Mircea, A. Ougazzaden, P. Dasté, Y. Gao, C. Kazmierski, J.-C. Bouley, and A. Carenco, "Extremely Uniform, Reproducible Growth of Device Quality InGaAsP:InP Heterostructures in the T-Shaped Reactor at Atmospheric Pressure," *J. Cryst. Growth*, Vol. 93, 1988, pp. 235–241.

9. P.J.M. Frijlink, "A New, Versatile, Large Size MOVPE Reactor," *J. Cryst. Growth*, Vol. 93, 1988, p. 207.

10. D. Schmitz, G. Strauch, M. Heyen, and H. Jürgensen, "The Application of the Gas Foil Rotation for Growth of InP Based III-V Compound Semiconductors in a Horizontal LP-MOVPE Reactor," *Proc. of the "GaAs 90,"* Rome, April 19–20, 1990, p. 32.

11. P.J.M. Frijlink, J.L. Nicolas, and P. Suchet, "Layer Uniformity in a Multiwafer MOVPE Reactor for III-V Compounds," *Proceedings of 5th International Conference on MOVPE*, Aachen, 1990, to be published in *J. Cryst. Growth*.

12. E. Luthardt, H. Jürgensen, W. Fabian, H. Roehle, and P. Wolfram, "Exhaust Gas Scrubbing for Semiconductor Processes," *Proc. of Hazmacon*, 1986, p. 419.

13. M. Razeghi, M. Defour, F. Omnes, P. Maurel, E. Bigan, J. Nagle, F. Brillouet, and C. Portal, "MOCVD Challenge for III-V Semiconductor Materials for Photonic and Electronic Devices on Alternative Substrates," *J. Cryst. Growth*, Vol. 93, 1988, pp. 776–781.

14. P. Speier, U. Koerner, A. Nowitzki, F. Grotjahn, F.J. Tegude, and K. Wünstel, "MOVPE Studies for a Monolithically Integrated DH Laser/HBT Laser Driver," *J. Cryst. Growth*, Vol. 93, 1988, pp. 885–891.

15. J. Knauf, D. Schmitz, G. Strauch, H. Jürgensen, and M. Heyen, "Comparison of Ethyldimethylindium (EDMIN) and Trimethylindium (TMIn) for GaInAs and InP Growth by LP-MOVPE," *J. Cryst. Growth*, Vol. 93, 1988, p. 34.

16. G. Strauch, D. Schmitz, M. Heyen, and H. Jürgensen, "Production of High Resistivity GaAs Buffer Layers by LP-MOCVD," *Proceedings of the 6th Conference on Semi-Insulating III-V Materials*, Toronto, May 1990.

17. W. Walukiewicz, J. Lagowski, L. Jastrzebski, P. Rava, M. Lichtensteiger, C.H. Gatos, and H.C. Gatos, "Electron Mobility and Free-Carrier Absorption in InP; Determination of the Compensation Ratio," *J. Appl. Phys.*, Vol. 51(5), May 1980, pp. 2659–2668.

18. M. Razeghi, Vol. 1, "A Survey of GaInAsP-InP for Photonic and Electronic Applications," Vol. 1 of *The MOCVD Challenge*, pp. 50ff.

19. P.J.A. Thijs, P.I. Kuindersma, W. Dijksterhuis, T.v. Dongen, L. Tiemeyer, J.J.M. Binsma, G.L. v.d. Hofstad, and W. Nigman, "DCCPBH Laser Diodes with an MOVPE Grown Separate Confinement (SC) Multiple Quantum Well Active Region, emitting at $\lambda = 1.5 \mu$m," Presented at "13th European Conference on Optical Communication," Helsinki, Sept. 13–17, 1987.

20. E. Malzahn, M. Heuken, D. Grützmacher, M. Stollenwerk, and K. Heime, "High Mobility and High Sheet Density p-Type Pseudomorphic MODFET," to be published in the *Proceedings of 17th Int'l Symposium on GaAs and Related Compounds*, St. Helier/Jersey, September 1990.

21. P.J.A. Thijs, E.A. Montie, T.v. Dongen, and C.W.T. Bulle-Lieuwma, "Improved 1.5 μm Wavelength Lasers Using High Quality LP-OMVPE Grown Strained-Layer InGaAs Quantum Wells," *J. Cryst. Growth* (special issue ICCBE-2), 1989.

22. H. Hardtdegen, KFA Jülich, private communication.

23. G.H. Olsen and T.J. Zamerowski, "Vapor-Phase Growth of (In,Ga)(As,P) Quaternary Alloys," *IEEE J. Quant. Electr.*, Vol. QE-17, No. 2, February 1981, pp. 128–138.

24. P. Wiedemann, M. Klenk, W. Koerner, R. Weinmann, E. Zielinski, and P. Speier, "MOVPE of In(GaAs)P/InGaAs MQW Structures," *Proceedings of 5th International Conference on MOVPE*, Aachen, 1990, Aachen, to be published in *J. Cryst. Growth*.

25. A.W. Nelson and L.D. Westbrook, "A Study of *p*-Type Dopants for InP Grown by Adduct MOVPE," *J. Cryst. Growth*, Vol. 68, 1984, pp. 102–110.

26. M. Glade, D. Grützmacher, R. Meyer, E.G. Woelk, and P. Balk, "Activation of Zn and Cd Acceptors in InP Grown by Metalorganic Vapor Phase Epitaxy," *Appl. Phys. Lett.*, Vol. 54 (24), June 12, 1989, pp. 2411–2413.

27. T. Wolf, A. Krost, F. Reier, P. Harde, D. Kuhl, F. Hieronymi, H. Ullrich, D. Bimberg, and H. Schumann, "Semi-Insulating Fe- and Ti-Doped InP and InGaAs for Ultrafast Infrared Detectors Grown by LP-MOCVD," *Proceedings of the 6th Conference on Semi-Insulating III-V Materials*, Toronto, May 1990.

28. P. Dasté, Y. Miyake, M. Cao, Y. Miyamoto, S. Arai, Y. Suematsu, and K. Furuya, "Fabrication Technique for GaInAsP/InP Quantum Wire Structure by LP-MOVPE," *J. Cryst. Growth*, Vol. 93, 1988, pp. 365–369.

29. Y.D. Galeuchet, P. Roentgen, and V. Graf, "GaInAs/InP Selective Area Metalorganic Vapor Phase Epitaxy for One-Step-Grown Buried Low-Dimensional Structures," *J. Appl. Phys.*, Vol. 68, No. 2, July 15, 1990, pp. 560–568.

30. J. Finders, J. Geurts, A. Kohl, M. Weyers, B. Opitz, O. Kayser, and P. Balk, "Composition of Selectively Grown $In_xGa_{1-x}As$ Structures from Locally Resolved Raman Spectroscopy," *Proceedings of 5th International Conference on MOVPE*, Aachen, 1990, to be published in *J. Cryst. Growth*.

31. O. Kayser, B. Opitz, R. Westphalen, U. Niggebrügge, K. Schneider, and P. Balk, "Selective Embedded Growth of GaInAs by Low Pressure MOVPE," *Proceedings of 5th International Conference on MOVPE*, Aachen, 1990, to be published in *J. Cryst. Growth*.

32. Euro III–V Review, "Update," Vol. 3, No. 3, 1990, p. 8.

33. J. Splettstößer and H. Beneking, "$Ga_{0.47}In_{0.53}As$ Enhancement- and Depletion-Mode MISFETs with Very High Transconductance," submitted for publication to IEEE.

34. M. Renaud, P. Boher, J. Schneider, J. Barrier, M. Heyen, and D. Schmitz, *Electronics Letters*, June 9, 1988, Vol. 24, No. 12, pp. 750–752.

35. F. Schulte, J. Splettstößer, A. Trasser, D. Schmitz, and H. Beneking, "Microwave Properties of $Ga_{0.47}In_{0.53}As$ MISFETs with SiO_2 Insulator."

36. D. Bimberg, J. Christen, T. Fukunaga, H. Nakashima, D.E. Mars, and J.N. Miller, "Cathodoluminescence Atomic Scale Images of Monolayer Islands at GaAs/GaAlAs Interfaces," *J. Vac. Sci. Technol.*, Vol. B5, 1987, p. 1191.

37. P.J.A. Thijs, T.v. Dongen, P.I. Kuindersman, J.J.M. Binsma, L.F. Tiemeyer, J.M. Lagemaat, D. Moroni, and W. Nijman, "High Quality InGaAsP-InP for Multiple Quantum Well Laser Diodes Grown by Low-Pressure OMVPE," *J. Cryst. Growth*, Vol. 93, 1988, pp. 863–869.

38. P.J.A. Thijs, L.F. Tiemeyer, T.v. Dongen, P.I. Kuindersma, J.J.M. Binsma, G.L. van de Hofstad, and W. Nijman, "High Performance InGaAsP-InP 1.5 μm DCPBH Laser Diodes with LP-MOCVD Grown Bulk and MQW Active Layers," *Proceedings of European Conference on Optical Communication*, Brighton, September 1988.

39. M. Rosenzweig, W. Ebert, D. Franke, N. Grote, B. Sartorius, and P. Wolfram, "Lasing Characteristics of InGaAs/InGaAsP-MQW-Structures Grown by Low Pressure MOVPE," *Proceedings of 5th International Conference on MOVPE*, Aachen, 1990, to be published in *J. Cryst. Growth*.

40. Dr. Lee, Materials Research Laboratories, Hsinchu, Taiwan, private communication.

41. A. Valster, C.T.H.F. Liedenbaum, M.N. Finke, A.L.G. Severens, M.J.B. Boermans, D.E.W. Vandenhout, and C.W.T. Bulle-Lieuwma, "High Quality $Al_xGa_{1-x-y}In_yP$ Alloys Grown by MOVPE on (311)B GaAs Substrates," *Proceedings of 5th International Conference on MOVPE*, Aachen, 1990, to be published in *J. Cryst. Growth*.

42. F. Scholz, private communication.

Chapter 5
Doping of InP as Grown by Metalorganic Chemical Vapor Deposition

E.K. Byrne
AT&T Bell Laboratories

5.1 INTRODUCTION

Control of the impurity properties of InP and its alloys is central to the performance of optoelectronic devices fabricated from these materials. While it is important to understand the mechanisms of incorporation of residual impurities in these compounds in order to minimize background doping and maximize carrier mobility, it is equally important to understand how intentional dopants are incorporated, because they are essential for virtually every device design. The ability to insert a dopant atom into a growing host lattice with monolayer control (and without diffusion) is a worthy goal for any crystal grower. To achieve this goal, it is necessary to understand the variables involved in the growth of intrinsic material, and then factor in the additional complexity of the dopant chemistry.

The growth technique to be discussed in this paper is metalorganic chemical vapor deposition (MOCVD), also known as organometallic vapor phase epitaxy (OMVPE), as well as several other terms of similar meaning [1]. These names all refer to epitaxial growth in a cold-wall reactor using (for III/V compounds) metal-organic precursors for the group III elements (and some dopants and group V elements) and, usually, hydride sources, such as PH_3 and AsH_3, for the group V elements. The reactor pressure is typically 1 to 760 torr. MOCVD has gained popularity in recent years because of its versatility in producing many different III/V alloys, and because of its potential scalability for large-area, multiwafer production. It is now widely used on a research scale for fabrication of state-of-the-art lightwave devices. In the InP-based system of alloys, reports of structures containing MOCVD-grown quantum wells with layer thicknesses on the order of tens of angstroms have recently appeared [2]. As the technology for producing these alloys improves, so must the doping technology improve for the production of the devices made from them. This chapter presents the status of doping of InP as of late 1990.

Progress in the field of epitaxial crystal growth is frequently driven by the need for improved processes for device fabrication. Indeed, device performance is often used as a scale for comparing material quality. InP and its alloys with Ga and As are used most commonly for the production of lightwave devices for optical fiber communication. This use is due to the loss and dispersion minima of glass fiber at wavelengths comparable to the bandgap emission wavelength of different alloy compositions of InGaAsP [3]. The design of diode lasers, LEDs, and their attendant detectors all require precise control of p-n junction placement in or near the narrow bandgap active material ($In_xGa_{1-x}As_yP_{1-y}$), as well as control of the electrical properties of the adjoining InP cladding layers. Many of the authors cited herein will favor the choice of one dopant over another for properties relevant to a particular device application. This review is therefore organized by grouping doping elements by their electrical activity in InP: shallow donors, shallow acceptors, deep donors and acceptors, and rare earth elements used to generate photoluminescence at relevant wavelengths (such as 1.55 μm). The substrates employed in these studies are all cut within 3° of exact (100) InP, except where noted.

5.2 SHALLOW DONORS

5.2.1 Device Applications

Doping of InP with shallow donors to generate n-type material is easily accomplished by MOCVD. Controlled doping over several orders of magnitude (10^{16} to 10^{19} atoms/cm^3) has been reported using several different elements. Typical device applications for *n*-InP include cladding layers for InGaAsP injection lasers and InGaAs light-emitting diodes (LED), multiplication region for avalanche photodiodes (APDs), high-speed electronic devices such as high electron mobility transistors (HEMT) or heterojunction bipolar transistors (HBT), and solar cells.

5.2.2 Minimizing Background Doping

The impurities associated with nominally undoped InP are typically shallow donors. Hsu *et al.* [4] carried out a study that measured background carrier concentration as a function of V/III ratio. They found that the optimum ratio was about 60 for a growth temperature of 650°C, and that variations in the V/III ratio by a factor of three in either direction caused less than a factor of two increase in background doping. These authors, as well as others in the mid-1980s [5], concluded that levels of background impurities in InP were highly dependent on the source of the In compound used as a precursor. As vendors improved the purity of the various In sources, the purity of the resulting InP improved as well. Briggs and Butler [5b] carried out a 1987 study of the growth parameters necessary to achieve optimum mobility in InP grown by MOCVD. Their In source was $(CH_3)_3In$ purified by the diphos-adduct route [6]. Growth temperature was the first variable examined. In a single run, the temperature was stepped between 580°C and 720°C, and the optimum temperature for pure InP was determined to be 580°C. Carrier concentrations ranged from 5×10^{14} cm^{-3} at 580°C to 2×10^{16} cm^{-3} at 720°C. SIMS profiles of these layers showed that concentrations of Si closely resembled the stepped increases in carrier concentration as measured by a polaron electrochemical profiler. In this particular sample, no Se was detected by SIMS, and the S level was invariant with growth temperature and was lower in the epilayer than in the substrate. Thus, they concluded that Si is the major impurity in the growth of undoped InP. The growth rate was then varied at constant V/III ratio and growth temperature, with the lowest background doping found at the highest growth rate. Since higher growth rates at equal V/III ratios require higher flow rates of $(CH_3)_3In$ (TMIn) and PH_3, the growth rate dependence on background doping indicates the impurities did not arise from the In or P sources in this case. Briggs and Butler also confirmed the result of Hsu *et al.* [4] and others [5a, b], showing no dependence of background doping on the V/III ratio. They proposed a source for Si incorporation which fits all the data: Si comes from the reduction of quartz comprising the reactor chamber during heating. They, therefore, recommended growth at the lowest possible temperature and highest possible growth rate for maximum mobility.

Finally, two recent reports of high mobilities at room temperature and 77K indicate the state of the art in growth of pure InP. In 1987, Thrush *et al.* [7] found a low-temperature mobility of 264,000 cm^2 V^{-1} s^{-1} for InP grown at 570°C and 150 torr using "diphos"-purified TMIn. Razeghi *et al.* [8] reported a 50K mobility of 200,000 cm^2 V^{-1} s^{-1} in 1988, corresponding to a background doping of 3×10^{13} cm^{-3}. They also achieved their results using low-pressure MOCVD (100 mbar) at relatively low growth temperatures (550°C).

5.2.3 Silicon

Silicon is a group IV element, and as such it is expected to be amphoteric when incorporated into InP. If Si substitutes for In, it will be a donor, while on a P site it would act as an acceptor. Although Si is closer in atomic radius to P than to In, in fact this element behaves almost exclusively as a donor in InP. The most common dopant precursors for Si in MOCVD are SiH_4 and Si_2H_6. Both may be obtained in high purity and both provide sharp doping interfaces because of the involatility of Si and a small diffusion coefficient.

Several authors have confirmed the linear relationship between molar flow of Si precursor and the resulting carrier concentration, using both SiH_4 and Si_2H_6. Table 5.1 summarizes the growth conditions and doping levels they examined.

In addition, many of these studies also included the dependence of carrier concentration on growth temperature for InP:Si. Qualitative agreement among three authors [13–15] using Si_2H_6 as the dopant source showed two doping regimes: the first was highly temperature dependent between ~500°C and ~600°C; above ~600°C, Si incorporation became nearly temperature invariant. Clawson, Vu, and Elder [11] found a similar strong temperature dependence of Si incorporation below 550°C using SiH_4, but they also noted that carrier concentration continued to rise approximately a factor of five between 600°C and 700°C. They also explored several other growth parameters for their effects on Si doping of InP. Their studies showed that incorporation of Si increases linearly with reactor pressure, presumably because of residence time effects; it is generally agreed that the temperature behavior of Si incorporation is caused by the kinetic limitation of SiH_4 decomposition. Clawson *et al.* [11] also demonstrated the linear dependence of carrier concentration on inverse growth rate (at 650°C). This is the result expected for a low vapor pressure dopant substituting on the group III site in mass-transport-limited growth [16].

Differences between Si incorporation from SiH_4 and Si_2H_6 arise from the differences in decomposition rates of the two compounds at a given growth temperature. Blaauw *et al.* [14] examined the pressure dependence of Si incorporation from Si_2H_6 at two temperatures. Noting that the decomposition rate of Si_2H_6 (to produce SiH_2 and SiH_4) is faster than that of SiH_4 (which results in SiH_2 and H_2), they found no pressure dependence of Si concentration of 625°C. Under these conditions, doping is limited by transport of Si to the surface because Si_2H_6 is largely dissociated. In contrast, they found a similar pressure dependence to that shown by Clawson *et al.* [11] for SiH_4 when Si_2H_6 is used at a growth temperature of 575°C. Thus, Rose *et al.* [13] concluded that Si_2H_6 is a superior doping source to SiH_4, citing better doping

Table 5.1

Si precursor	Growth T	Growth P	Range (N_D–N_A)	Reference
SiH_4	575°C	100 mb	10^{15}–10^{18}	[9] (1985)
SiH_4	655°C		10^{17}–4×10^{18}	[10] (1985)
SiH_4	600°C	1 atm	6×10^{16}–4×10^{17}	[4] (1986)
SiH_4	650°C	9 torr	10^{17}–6×10^{18}	[11] (1987)
Si_2H_6	600°C	1 atm	5×10^{16}–4×10^{18}	[12] (1987)
Si_2H_6	650°C	1 atm	2×10^{16}–10^{18}	[13] (1989)
Si_2H_6	610°C	75 torr	10^{17}–2×10^{19}	[14] (1990)

uniformity over 50-mm wafers. Their results were obtained at 650°C, so it is likely that temperature uniformity played a role in their study. These authors also found that carrier concentration was proportional to the square root of the PH_3 flow rate when Si_2H_6 was used.

Where it was discussed, all of these authors agreed that Si provides abrupt doping interfaces, and was not subject to *memory effects*, or evidence for Si contamination in subsequent runs when it was not intentionally introduced. Indeed, di Forte-Poisson *et al.* [17] used SiH_4 to achieve *delta doping* of InP by low-pressure MOCVD. They attempted to completely substitute a monolayer of In sites with Si atoms by interrupting growth (turning off the flow $(C_2H_5)_3In$) and admitting SiH_4 to the system under PH_3. A (100) surface of InP normally contains 6×10^{14} In atoms/ cm^2. The authors reported a SIMS half-width at the half-maximum Si peak width of 26Å, which yielded an integrated Si density of 4×10^{12} cm^{-2}.

5.2.4 Tin

Like Si, Sn is a group IV element that acts as a shallow donor when incorporated on an In site in InP. Since the covalent radius of Sn is quite similar to that of In, atomic size considerations strongly favor In substitution for this element.

Pinzone *et al.* [18,19] have studied the incorporation properties of Sn (derived from $(C_2H_5)_4Sn$, TESn) in InP. In growth at atmospheric pressure and 660°C, they showed that *n*-doping between 1.5×10^{17} cm^{-3} and 3.3×10^{19} cm^{-3} could be achieved with good morphology. Linear doping with Sn mole fraction was observed. SIMS measurements of atomic Sn concentration were in close agreement with net electron concentrations obtained from Hall measurements, indicating that, within experimental error, all of the Sn is incorporated as electrically active. The authors were able to grow undoped InP with background doping of 6.7×10^{14} cm^{-3} immediately after growing heavily Sn-doped layers, establishing the lack of a memory effect for Sn from TESn. A single run comprising growth of InP:Sn at three different growth temperatures (600°C, 650°C, and 700°C, with other growth parameters remaining constant), revealed a temperature dependence of Sn incorporation corresponding to an activation energy of 1.9 eV for a TESn mole fraction of 2.4×10^{-7} for InP.

5.2.5 Sulfur

As a group VI element, S is expected to substitute as a donor on the P site in InP. Atomic size also favors P substitution. In contrast to Si and Sn, however, S cannot behave as an amphoteric dopant in III/V semiconductors. Sulfur (derived from H_2S) has occasionally been used in InP-based device structures of recent years [20]; however, there has appeared only one fundamental study of its doping characteristics to date [21]. In this study, Logan *et al.* found carrier concentration to be linear with dopant mole fraction for doping levels in the region from mid 10^{17} cm^{-3} to low 10^{18} cm^{-3}. At doping levels below 1×10^{17} cm^{-3}, dopant incorporation was observed with a much higher slope relative to H_2S mole fraction. Layers were grown at atmospheric pressure and 625°C, and characterized by van der Pauw Hall-effect measurements and electrochemical capacitance voltage (C-V) profiling. The authors found good agreement between the Hall-effect measurements of carrier concentration and the C-V measurements of net donor concentration, $|N_D\text{-}N_A|$, and thus concluded that S behaves as a shallow donor, which is essentially fully ionized at room temperature. A series of S-doped layers were then grown at different temperatures from 525°C to 625°C, with other parameters remaining constant. The measured carrier concentrations for these layers followed a linear dependence on inverse growth temperature, up to a maximum electron concentration of nearly 10^{20} cm^{-3} at $T_g = 525$°C. The increase in doping level of nearly two orders of magnitude between 625°C and 525°C was attributed to the desorption of S from the growth interface at the higher growth temperatures. This behavior is as expected qualitatively for a high vapor pressure dopant, and indeed it was shown by Stringfellow [16] that the inverse growth temperature dependence of the carrier concentration of S:GaAs closely parallels the inverse temperature dependence of the reciprocal vapor pressure of S.

5.2.6 Selenium

Selenium (from H_2Se) and tellurium (from Et_2Te, DETe) behave solely as donors in InP, as expected by analogy to their lighter homologue, S. Based on Stringfellow's theory [16], one might predict a smaller growth temperature dependence of dopant incorporation due to the lower vapor pressures of these heavier elements.

Perhaps because of an early report of memory effects with the use of H_2Se for n-doping of GaAs [22] by MOCVD, no detailed studies of Se doping of InP have appeared. However, a few papers have been published that refer to the use of Se as an *n*-dopant in particular device applications. Specifically, Sogou *et al.* [23] used Se to generate *n*-InP for laser structures. They obtained *n*-doping of 10^{18}–10^{19} cm^{-3} for growth at 76 torr. Also, Mircea and coworkers [24] demonstrated early growth of InGaAsP by atmospheric pressure MOCVD. These authors reported using Se as an *n*-dopant, achieving a doping level of 6×10^{17} cm^{-3}. Finally, Ban *et al.* [25] fabricated InP/InGaAsP HBTs using low-pressure (76 torr) MOCVD, employing Se as an *n*-dopant.

5.2.7 Tellurium

Clawson, Vu, and Elder [11] investigated Te incorporation in comparison with Si doping. They found the usual linear dependence of carrier concentration on DETe molar flow for doping levels between 10^{17} and 5×10^{18} cm^{-3}. Also as expected, carrier concentration increased linearly with inverse growth temperature between 500°C and 700°C, presumably due to increased evaporation from the surface at higher temperatures. In addition, they noted no dependence of Te doping with changes in reactor pressure between 9 and 150 torr (most of their studies were conducted at 9 torr).

An earlier paper by Hsu *et al.* [4] investigated Te doping of InP at atmospheric pressure. In contrast to the results of Clawson *et al.* these authors noted a superlinear dependence of carrier concentration on mole fraction of DETe for carrier concentrations between 10^{17} and 10^{19} cm^{-3}. They could not explain the cause of this superlinearity.

Since Te substitutes for P in the InP lattice, one might predict a dependence of Te incorporation on inverse PH$_3$ mole fraction similar to the dependence of Si doping on the inverse mole fraction of TMIn (and, therefore, also inverse growth rate) or substitution on the group III site. Clawson *et al.* [11] found that, indeed, carrier concentration was proportional to inverse PH$_3$ mole fraction, for values of PH$_3$ mole fraction between 0.04 and 0.01 and carrier concentrations between 4 and 9×10^{16} cm^{-3}. However, contrary to expectations, it was also found that carrier concentration varied linearly with inverse TMIn mole fraction between 5×10^{16} and 3×10^{17} cm^{-3}. Clawson *et al.* observed that this behavior was in contrast to incorporation of S from H$_2$S and Se from H$_2$Se in GaAs grown at atmospheric pressure, as reported by other authors [26,27]. They proposed that this unexpected behavior might be explained by differences in decompositon/incorporation chemistry between the alkyl tellurium source and the hydrides; that is, that incorporation of Te from a metalorganic source is limited by mass transport of Te to the surface, and, therefore, will exhibit an inverse growth rate dependence based on dilution of a fixed dopant concentration in a more rapidly growing layer.

These authors also noted that the incorporation of Te from DETe was not uniform throughout a 0.5-μm-thick layer, based on Polaron electrochemical carrier concentration *versus* depth profiles. They observed a continuous increase in carrier concentration as the growth progressed. This behavior has also been observed by Bhat *et al.* [28] for DETe mole fractions above 3×10^{-9} (for a doping level above 7×10^{16} cm^{-3}). Since this phenomenon was not observed in Te:GaAs, Te:InGaAs, or Te:AlGaAS, they concluded that the large size of the TE ion compared with that of P hindered substitutional incorporation and caused accumulation of Te at the surface as the growth proceeded.

5.3 SHALLOW ACCEPTORS

5.3.1 Device Applications

The controlled doping of InP with shallow acceptors to generate *p*-type material has been a challenge to crystal growers since the first *p-n* junction was fabricated from this material several years ago. Most doping studies of acceptor incorporation have concluded that there is an electrical saturation limit for p-InP of somewhere between 3×10^{18} and 8×10^{18} cm^{-3}, depending on the particular dopant and crystal growth technique. Tokumitsu [29] has published a theory that predicts a relatively low electrical saturation for p-InP based on Fermi-level pinning energies for various III/V materials.

The most important devices fabricated in the InP materials system also require the most precise control of p-doping in order to generate uniform p-n junctions. These devices are diode lasers and LEDs, which use InP cladding layers for the low-bandgap ternary or quaternary material. Typical limitations to achieving high hole concentrations and sharp doping interfaces include high diffusivity of the dopant atom in the lattice, low incorporation efficiency, and also extreme toxicity of some of the potential dopants and their precursors. These properties will be highlighted throughout this section.

5.3.2 Zinc

5.3.2.1 Basic Doping Studies

Group II elements are used as acceptor dopants in InP. Zn, Cd, and Hg are all members of the group IIB elements. These elements have one fewer valence electron than In and will, therefore, exhibit acceptor activity when incorporated on the group III site in the InP lattice. Zn is typically derived from $(C_2H_5)_2Zn$ (DEZn) or $(CH_3)_2Zn$ (DMZn). The vapor pressure of DEZn is much lower than that of DMZn, and, therefore, DEZn is more useful for applications requiring low p-doping.

The discussion of Zn incorporation into InP begins with the usual studies measuring the dependence of doping on various relevant growth parameters. Nelson and Westbrook [30] first examined this relationship and found that the net acceptor density was proportional to the (partial pressure of Zn vapor)$^{1/2}$ for atmospheric pressure growth at 650°C using the adduct TMIn-PEt$_3$. They also found that above a dopant partial pressure of 3×10^{-6} atm, carrier density saturated at about 3×10^{18} cm^{-3}. Many other authors have confirmed this electrical saturation limit for InP:Zn. Interestingly, Nelson and Westbrook also characterized these same samples by SIMS and, while quantitative data was not available, the measured ion intensities paralleled the carrier concentrations, and the total amount of Zn incorporated into InP also seemed to reach a saturation value at the same input flux of Zn vapor. In contrast, Hsu *et al.* [4] reported a superlinear dependence of hole concentration on DMZn flow rate for doping levels between 10^{17} and mid 10^{18} cm^{-3}. Their growth conditions were 650°C and 1 atmosphere. A recent study of Zn doping of InP by Molassioti *et al.* [31] found a linear dependence of sheet carrier concentration on Zn precursor flow for molar flows between 0.01 and 0.1 μmoles/min (corresponding to carrier concentrations up to about 10^{17} cm^{-3}). These authors also employed the adduct TMIn-Pet$_3$ and used a growth temperature of 580°C. The three examples of sublinear, linear, and superlinear dependence of Zn doping level on molar flow indicate a lack of complete understanding of the mechanism of Zn incorporation into InP.

Perhaps because of the varying results obtained in plots of carrier concentration *versus* Zn molar flow, no studies have been published showing the dependence of Zn incorporation on inverse growth temperature. As a high vapor pressure dopant, one would predict Zn concentration to decrease significantly with increasing growth temperature. This is the case for Zn:GaAs, and, using data published by Glew [32], Stringfellow has shown that the slope of the line from the distribution coefficient *versus* inverse growth temperature plot parallels the inverse vapor pressure dependence of Zn [16].

There also appears to be no definitive study of Zn incorporation *versus* inverse growth rate (for InP). If Zn were a low vapor pressure dopant, one would predict a linear relationship between Zn concentration and (growth rate)$^{-1}$, since it substitutes on the group III sublattice. However, due to the volatility of Zn, one would instead predict linear doping behavior with PH$_3$ partial pressure [16]. Although this study has not been done, Bass and Oliver have confirmed this group V dependence for Zn:GaAs [26].

Yang *et al.* [33] studied Zn incorporation as a function of substrate orientation. They performed their experiments using atmospheric pressure growth at a temperature of 730°C, using DEZn as the dopant precursor. The measured hole concentrations were 6.5×10^{17} cm^{-3} for the (100) substrate orientation and 1.9×10^{17} cm^{-3} for a (111B) substrate. However, since the background doping of these substrates for nominally undoped InP was higher (and *n*-type) for the (111B) substrate, the authors were hesitant to attribute the Zn incorporation dependence on substrate effects only. They do cite earlier work by Manasevit *et al.* [34] who found that Zn incorporation for a given input concentration of DEZn was most effective for growth on (111A) substrates, least effective for (111B), and intermediate for (100). This result is consistent with doping efficiency being related to the number of available substitutional sites at the substrate surface.

5.3.2.2 Diffusion

One of the biggest challenges associated with the use of Zn as a dopant in InP is control of diffusion. As stated previously, precise control of *p-n* junction placement is critical to achieving optimum performance for diode lasers, and diffusion processes complicate this placement. Many studies have been made to determine the mechanism of Zn diffusion in InP [35], however there is still not widespread agreement on a mechanism that can explain all experimental results to date.

The recent study by Blaauw, Shepherd, and Eger [35] cites some of the papers that report the various results obtained on the determination of the mechanism of Zn diffusion in InP. Their findings indicate perhaps one reason for the lack of agreement on this mechanism; evidence is presented that the doping level of the layers into which Zn is to diffuse plays a critical role in the diffusion profile. Blaauw *et al.* grew several *n-p-n-p-n* InP structures at 75 torr and 625°C, using Si as an *n*-dopant and DEZn as the Zn source. They characterized these samples by SIMS and polaron electrochemical C-V profiling, each layer being about 0.5 μm thick. When the spacer layers were undoped instead of *n*-type, the authors found little diffusion at [Zn] = 4×10^{17} cm^{-3} and essentially complete undoped -> *p*-type conversion for [Zn] > 3×10^{18} cm^{-3}. As progressively more Si is added to the spacer layers, from 2×10^{16} cm^{-3} to 5×10^{17} cm^{-3}, and at a constant [Zn] of 2×10^{18} cm^{-3}, the Zn level found in the Si-doped layers increases. In each of these cases, the polaron profile shows the entire sample to be *p*-type. When more Si is added to the spacer layers than the 2×10^{18} cm^{-3} found in the Zn-doped layers, the electrochemical profile shows *n-p-n-p-n* as expected, and the SIMS profiles of Zn show progressively sharper and narrower spikes as the Si concentration approaches 3×10^{19} cm^{-3}. Blaauw *et al.* propose a model to explain Zn diffusion both in the presence and the absence of Si donors in the spacer layers. They first present the popular view that Zn diffusion is dominated by the fraction of total Zn present as mobile interstitials, Zn_i. Substitutional Zn acceptors, Zn_{In}, are not as mobile and probably exist in equilibrium with Zn_i:

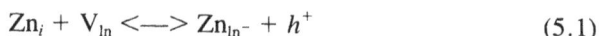

$$Zn_i + V_{In} <\longrightarrow> Zn_{In^-} + h^+ \qquad (5.1)$$

V_{In} represents an In site vacancy. The authors then include the contribution of the donor species, suggesting that a donor-acceptor complex forms, which immobilizes the diffusing Zn:

$$Zn_i + V_{In} + Si_{In} + <\longrightarrow> (Zn_{In} - Si_{In^+}) + h^+ \qquad (5.2)$$

This model explains the Zn pileup at Si-doped interfaces for highly Si-doped spacer layers, and also accounts for the extensive Zn diffusion into the low-doped Si spacer layers. Finally, the authors integrated the amount of Zn that diffused into the various Si-doped layers and plotted it *versus* (diffusion time)$^{1/2}$ (i.e., the time during which the grown Si-Zn-doped interface remained at the growth temperature) and found linear relationships for the different Si concentrations consistent with a diffusion-controlled process.

5.3.2.3 Deactivation of Acceptors

A curious phenomenon has appeared recently in the literature which contributes to the difficulty in controlling acceptor concentrations in MOCVD-grown InP. Reports first appeared in 1988 from Cole *et al.* [36,37] and Antell *et al.* [38], which showed that the carrier concentration found in p-InP layers grown as part of a structure containing *p*-InP and p^+-InGaAS layers was affected by the cooling ambient used for the run. Specifically, Cole *et al.* [36] reported that the measured carrier concentration in a sample cooled in AsH_3/H_2 was about an order of magnitude less than when an identically grown sample was cooled in PH_3/H_2, and that this level in turn was a factor of two lower than that for a sample cooled in H_2 alone. SIMS measurements of total Zn concentration indicate no loss of dopant, and it was also found that the doping level of the sample cooled in AsH_3 could be restored by a heat treatment in PH_3 or H_2 for a short time. The authors therefore concluded that the lower doping levels must result from deactivation or compensation of the Zn dopant. They proposed that this deactivation is caused by atomic H present in the as-grown crystal, and they showed SIMS evidence for increased [H] in the AsH_3-cooled samples as compared with the PH_3-grown ones. Hydrogen atoms would form electrically inactive complexes with the dopant atom. The H is thought to have come from the pyrolysis of the hydrides, and the authors cite a paper that discusses such pyrolysis by semiconductor surfaces. AsH_3-cooled samples would be expected to exhibit greater deactivation because of the lower thermal stability of AsH_3 relative to PH_3 and H_2. In a separate publication [37], Cole *et al.* extended their work to include another acceptor (Cd). They also showed that the cool-down ambient does not affect *n*-doping levels when S or Sn is used as donors. In addition, when a 1:1 mixture of AsH_3:PH_3 in H_2 was used as the cooling ambient for a *p*-InP sample, no thermal degradation of the surface was observed. The authors then determined that the capping InGaAs layer played no role in the dopant deactivation effect observed during the cool-down phase.

Antell *et al.* [38] examined this effect with respect to *n*-type spacer layers. The test structure used was 1 μm *p*-InP (Zn), 1 μm *n*-InP (Si), 1 μm *p*-InP (Zn), and finally 0.5 μm p^+-InGaAS (Zn). The carrier concentration profile of this structure showed that the acceptor level in the second-grown *p*-layer was about 30% lower than that of the first *p*-InP layer. The flow of DMZn was the same for each *p*-InP layer, and the Zn concentration was confirmed to be the same for these layers by SIMS. At the same time, [H] was examined by SIMS and found to be present in only the *p*-InP layers, with the level in the first-grown layer somewhat lower. This H was removed by an anneal at 800°C for 5 sec. These authors showed that the deactivation effect could be avoided by the use of an *n*-InGaAs capping layer, because *n*-type material can impede the entry of H into buried p-InP layers. Their experiments supported a model which maintains that the solubility of H in InP is not high at typical growth temperatures of 650°C, but that it becomes bound to a P atom during cooling at a temperature where the P-H bond in InP becomes stable. This model of a P-H bond formation contrasts with the model of Cole *et al.* [36], who proposed a Zn-H complex. Antell *et al.* maintain that the H compensation found in *p*-InP is due to the fact that since the As-H bond is weaker than P-H, pyrolysis of

AsH$_3$ continues to form H at cooling temperatures where the P-H bond is stable, and thus incorporates P-H in the InP lattice. Conversely, compensation of *p*-InGaAs does not occur because arsine pyrolysis has ceased at the temperatures where As-H bonds become stable in the lattice. This model explains the lack of deactivation of *p*-InGaAs and predicts no deactivation for p-InP cooled in PH$_3$ only.

Glade *et al.* [39] investigated the reactivation process of annealing the hydrogenated *p*-InP. They found that the time dependence of the carrier concentration during activation for different annealing temperatures was consistent with a diffusion-controlled process. It was also noted that the carrier concentration profile was flat for the *p*-InP layers for all the times studied at an annealing temperature of 400°C. The flat profiles and the fact that the presence of InGaAs capping layers increased the rates of activation led the authors to believe that H effusion was not the rate-determining step in the activation process. Furthermore, the activation energy for this process was independent of the specific *p*-dopant (Zn or Cd). The authors proposed a model for the rate-determining step of the activation to be diffusion of intrinsic defects, such as In vacancies, which are uniformly distributed throughout the layers.

A 1990 paper from Cole *et al.* [40] showed deactivation effects from cooling a *p*-InP sample in DMZn and Ph$_3$/H$_2$. They found that the carrier concentration decreased as the DMZn was switched off at progressively lower cool-down temperatures. Suspecting that the DMZn may be catalyzing the pyrolysis of PH$_3$ to yield H atoms, SIMS profiles of H were performed on the various samples. These profiles showed that [H] decreased as the temperature at which the DMZn was switched off decreased, a result inconsistent with a hydrogen passivation mechanism. In fact, the authors cooled a *p*-InGaAs-capped sample in DMZn/H$_2$ alone and found the carrier concentration was the same as that of a sample cooled in DMZn/PH$_3$/H$_2$: PH$_3$ and its pyrolysis played no role in the acceptor deactivation. A model for the deactivation was proposed which invoked increased Zn diffusion when the DMZn was left on longer during cool-down. This increased diffusion of Zn caused incorporation as interstitials, since the existing substitutional sites would be occupied. The interstitials would be present as donors and would then compensate the substitutional acceptors and lower the *p*-doping level.

5.3.3 Cadmium

Cadmium, like Zn, behaves as a shallow acceptor in InP by substituting for In and having one fewer valence electron. Me$_2$Cd (DMCd) is the only Cd precursor reported in the literature to date available for use as an InP dopant. This source is quite toxic. In most cases, the toxicity and incorporation properties of Cd in InP make it less attractive than Zn as a *p*-dopant for InP-based device structures.

Nelson and Westbrook, in their 1984 study of InP *p*-dopants, explored the incorporation properties of Cd using atmospheric pressure, TMIn-PEt$_3$ adduct MOCVD

[41]. They reported achieving doping levels of 5×10^{15} to 1×10^{18} cm^{-3}. The doping level varied linearly with DMCd vapor flow, as did the concentration of Cd in the grown layers as measured by SIMS. The authors noted that the doping efficiency, defined as the ratio of the p-doping level to the dopant precursor vapor flow, was much lower for Cd (from DMCd) than for Zn (from DMZn). In fact, a molar flow of DMCd almost three orders of magnitude greater than DMZn was required to achieve a doping level of 1×10^{18} cm^{-3}. The growth temperature dependence of Cd doping was also examined and, in accordance with the results of Zn doping, Nelson and Westbrook found increasing carrier concentration with decreasing growth temperature at the same dopant molar flow rate. For example, using DMCd flow of 2×10^{-5} moles/min, carrier concentrations ranged from 1×10^{17} cm^{-3} at 650°C to 1.4×10^{18} cm^{-3} at 550°C. The higher doping levels at lower temperatures were explained by decreased desorption from the surface. Increased background n-doping at higher growth temperatures may also have compensated the Cd and reduced the carrier concentration. Finally, diffusion coefficients were measured using SIMS profiles. While the absolute values of the measured diffusion coefficients differed by two to three orders of magnitude from previously reported studies, the relative diffusivities are in general agreement with other literature reports: Zn diffuses one to two orders of magnitude faster in InP than does Cd.

In a more recent paper, Blaauw, Emmerstorfer, and Springthorpe [42] carried out a study of Cd doping at various reactor pressures using TMIn as the In source instead of the adduct sources used by Nelson and Westbrook. Blaauw *et al.* confirmed the linear dependence of carrier concentration on DMCd molar flow rate at atmospheric pressure, and also found that the linear dependence held for growth at 75 torr. Interestingly, the DMCd flow rates required to achieve a certain doping level were ten times higher at 75 torr than at atmospheric pressure. In addition, layer morphology began to deteriorate at flow rates greater than 2×10^{-4} moles/min for growth at 75 torr, corresponding to a doping level of about 7×10^{16} cm^{-3}. To explore further the effect of ambient pressure on Cd incorporation, the authors varied the growth pressure from 75 torr to 225 torr to 675 torr in a single run, using a Cd molar flow rate of 2×10^{-4} moles/min. The SIMS profile for this run showed the atomic CD concentration varied in the ratio of 9.7:2.7:1, similar to the ratio of pressures used for the individual layers (9:3:1). The SIMS data also confirmed that the growth rates were independent of reactor pressure. Blaauw *et al.* explained the linear dependence of Cd incorporation with reactor pressure using the thermodynamic model of Stringfellow [16]. Since DMCd is expected to be completely dissociated at normal growth temperatures (it is reportedly highly dissociated at 200°C–350°C [43]), the partial pressure of Cd at the growing layer surface is expected to be the same as the partial pressure of DMCd at the reactor inlet. For high vapor pressure dopants having a low incorporation efficiency, Stringfellow predicted that the concentration of the dopant in the solid would be proportional to the partial pressure of the dopant at the layer surface, with the proportionality constant being the thermodynamic distribution coefficient. A plot presented by Blaauw *et al.* [42] shows that the carrier concentration varies linearly with Cd partial pressure for a number of growth techniques, including VPE, LP-MOCVD, AP-MOCVD, and adduct AP-MOCVD.

5.3.4 Magnesium

Magnesium is a third element that has been explored as a *p*-type dopant in MOCVD-grown InP. It, too, has two valence electrons and becomes an acceptor when substituting on the group III lattice site. The most common sources for Mg in MOCVD applications are bis(cyclopentadienyl)magnesium, Cp_2Mg, and bismethylcyclopentadienyl)magnesium, $(MCp)_2Mg$.

Nelson and Westbrook included Mg in their 1984 study of *p*-dopants [41]. They used Cp_2Mg to study Mg incorporation by adduct-MOCVD, and found they could achieve *p*-type doping in the range 3×10^{16} to 2×10^{18} cm^{-3}. However, the doping level varied as the square of the dopant vapor flow over this range, making precise carrier concentrations more difficult to obtain. SIMS measurements confirmed the square law dependence of the atomic Mg concentration on the input Cp_2Mg molar flow.

Several other authors have noted the superlinear increase of carrier concentration with dopant flow. Bacher *et al.* [44,45] examined Mg incorporation using Cp_2Mg at atmospheric pressure and found that the slope of the log-log plot of N_A-N_D *versus* $[Cp_2Mg]/[TMIn]$ had a slope of 1.82, almost precisely the same slope that Nelson and Westbrook [41] obtained. (If their Cp_2Mg flows are converted to $[Cp_2Mg]/[TMIn]$ values at a typical TMIn molar flow, the comparable slope is 1.85.) Bacher *et al.* were able to obtain carrier concentrations up to nearly 1×10^{19} cm^{-3}. Blaauw *et al.* [46] studied Mg incorporation at reduced growth pressures (75 torr) using $(MCp)_2Mg$. These authors plotted [Mg] (from SIMS) *versus* $(MCp)_2Mg$ flow rate for [Mg] from about 1×10^{18} cm^{-3} to 3.2×10^{19} cm^{-3} and found a superlinear incorporation behavior. This behavior was interpreted as linear above a minimum flow required to participate in parasitic gas phase reactions with residual O_2 or H_2O; the minimum flow corresponded to a Mg concentration of about 3×10^{18} cm^{-3}. Blaauw *et al.* also noted that for Mg concentrations up to about 2×10^{18} cm^{-3}, a linear relationship existed between carrier concentrations by C-V profiling and [Mg] by SIMS. However, above [Mg] = 2×10^{18} cm^{-3}, the carrier concentrations decreased to a value of 8×10^{16} cm^{-3} for [Mg] = 3×10^{19} cm^{-3}. Finally, Veuhoff *et al.* [47] also used $(MCp)_2Mg$ at low pressures (100 mbar) to examine Mg incorporation into InP by MOCVD. They found a cubic increase in carrier concentration (obtained from Hall measurements) with increasing $(MCp)_2Mg$ flow rate, over a hole concentration range of 8×10^{16} cm^{-3} to 2×10^{18} cm^{-3}. Above a maximum hole concentration of 2×10^{18} cm^{-3}, Mg continued to incorporate, presumably both as a substitutional acceptor and as an interstitial donor. The hole concentration begins to decrease, and the authors present photoluminescence and mobility data consistent with the formation of donor-acceptor pairs at high [Mg]. Surface morphology did not degrade for [Mg] less than 9×10^{19} cm^{-3}.

Some of these same authors have studied the diffusion behavior of Mg in InP. Blaauw *et al.* [46] noted significant diffusion of Mg into the S- and Fe-doped substrates used in their work. The Mg diffusion depths depended on the S- or Fe-doping level, with maxima of 0.1 μm at 10^{19} cm^{-3} for a 10^{19} cm^{-3} S-doped substrate, and 32 μm at 10^{17} cm^{-3} for a 10^{17} cm^{-3} Fe-doped substrate. The formation of a complex between Mg and the substrate dopant, which immobilizes the diffusing Mg, was proposed to explain these results. Veuhoff *et al.* [47,48] also looked at Mg diffusion. They reported strong diffusion of Mg into sandwiched undoped layers during growth for Mg concentrations above 10^{18} cm^{-3}, and concluded that the diffusing species was interstitial Mg. This species becomes important in their incorporation mechanism for [Mg] above 10^{18} cm^{-3}. The diffusion could be reduced, however, by doping the interlayers with Si to render them *n*-type.

5.4 DEEP DONORS AND ACCEPTORS

5.4.1 Device Applications

The use of deep donors and acceptors as traps for shallow dopants has increased in recent years with the need for current-confining, semi-insulating layers to minimize current leakage in index-guided diode lasers fabricated using the InP/InGaAsP materials system [49]. Such semi-insulating layers eliminate the need to use reverse-biased *p-n* junctions for current confinement, and thus avoid the problem of parasitic capacitance, which limits high-speed operation. Since the background doping of InP is normally *n*-type, deep acceptors are often used to compensate the residual shallow donors to achieve semi-insulating material with typical resistivities of 10^7–10^8 $\Omega \cdot$ cm. These semi-insulating layers may also be used for device isolation in optoelectronic integrated circuits fabricated on InP substrates.

5.4.2 Iron

Iron is by far the most widely employed deep acceptor for InP, and the preferred epitaxial technique for generating InP:Fe has been MOCVD. Substituting for In in the InP lattice, the Fe level lies 0.65 eV below the conduction band [50]. It is an acceptor level by virtue of the Fe^{+3}/Fe^{+2} transition. Iron doping of InP by MOCVD was first demonstrated by Long, Riggs, and Johnston in 1984 [51]. They grew InP at atmospheric pressure using the adduct TMIn-PEt$_3$ as the In source. The Fe precursors were iron pentacarbonyl, $Fe(CO)_5$, and ferrocene, $(C_5H_5)_2Fe$, abbreviated Cp$_2$Fe. Long *et al.* preferred Cp$_2$Fe as the FE source because its vapor pressure is four orders of magnitude lower than that of $Fe(CO)_5$, and is therefore easier to control the small flows required for doping. They obtained resistivities of up to 2×10^8 $\Omega \cdot$ cm using Cp_2Fe.

Long *et al.* [52] extended their work on Fe doping of InP to include two other Fe precursors, butadiene iron tricarbonyl, $(C_4H_6)Fe(CO)_3$, and cyclooctatetraene iron tricarbonyl, $(C_8H_8)Fe(CO)_3$. Both compounds were used as solids. Although high-resistivity layers of InP:Fe were grown using both Fe sources, $(C_4H_6)Fe(CO)_3$ was significantly more volatile and yielded layers with an iron concentration high enough to degrade the surface morphology. SIMS analysis of such a layer showed the FE concentration was 3×10^{19} cm^{-3}, while the resistivity was concomitantly reduced to 7×10^5 $\Omega \cdot$ cm. The authors stated that the semi-insulating properties are affected only at Fe concentrations above 8×10^{18} cm^{-3}. This may indicate that a level of FeP precipitates is reached which give rise to conduction paths through the epilayers. Such FeP precipitates were observed at Fe concentrations as low as 10^{17} cm^{-3}, and have been further characterized by Chu [53], Nakahara [54], and coworkers. When ferrocene was used as the FE source, an increase in the growth temperature from 650°C to 700°C resulted in almost an order of magnitude greater concentration in the resulting epilayer, as measured by SIMS. This result was interpreted as an increase in ferrocene decomposition at the higher growth temperature, making more Fe available for incorporation.

Following the initial reports from Long *et al.* a number of other papers appeared reporting the growth of InP:Fe by MOCVD. Speier, Schemmel, and Kuebart [55] also employed ferrocene as the FE source at atmospheric pressure. They achieved resistivities as high as 1.5×10^9 $\Omega \cdot$ cm at 294K. Huang and Wessels [56] grew Fe-doped InP layers using Cp_2Fe at atmospheric pressure and studied resistivity as a function of Fe mole fraction. Resistivities ranged from 0.14 $\Omega \cdot$ cm to 7×10^4 $\Omega \cdot$ cm for Fe mole fractions from 2.1×10^{-14} to 2.8×10^{-8}. Nakai *et al.* [57] found the maximum amount of Fe that could be incorporated as electrically active was 7×10^{16} cm^{-3}. They also showed through SIMS measurements that the FE concentration depended linearly on the Cp_2Fe flow rate. Hess *et al.* [58] demonstrated FE doping using low-pressure (70 torr) MOCVD and $Fe(CO)_5$ as an Fe source. These authors achieved a maximum resistivity of 5×10^7 $\Omega \cdot$ cm, and also found that growth at 700°C, instead of their usual 640°C, resulted in a nonplanar, nonepitaxial surface morphology characterized by island growth. Like the results of Long *et al.* [52], this roughened surface likely arose from an increased Fe incorporation at the higher growth temperature. Analysis of the islands by x-ray diffraction and EDAX spectra was consistent with nonstoichiometric In-Fe-P solid solution phases. The resistivity of the 700°C-grown film was reduced to 1×10^5 $\Omega \cdot$ cm.

Two papers have recently appeared that discuss the diffusion behavior of Fe in InP. Franke *et al.* [59] studied Fe incorporation at atmospheric pressure using Cp_2Fe as an Fe source. They demonstrated a linear increase in doping with an increase in ferrocene mole fraction for [Fe] between 1×10^{16} cm^{-3} and 3×10^{17} cm^{-3}

as measured by SIMS. Higher Fe concentrations resulted in inhomogeneous distributions of Fe because of the aforementioned FeP precipitates. The diffusivity of Fe was studied by growing five layers of successively increasing [Fe] sandwiched between undoped InP layers. The highest [Fe] for this experiment was 7×10^{16} cm^{-3}, just below the solubility limit. SIMS measurements indicated well-defined doping spikes, with the Fe signal returning to background in each of the undoped regions. There was virtually no change in the SIMS Fe profile upon annealing this sample for 4 hr at 650°C. A similar study using FE-doped layers of concentrations from 8×10^{16} to 6×10^{17} cm^{-3} revealed the presence of Fe in the undoped regions at a level of about 1 to 2×10^{16} cm^{-3}, while heat treatment increased this value slightly. Young and Fontijn [60] found that when an Fe-doped layer was grown adjacent to a Zn-doped layer ([Zn] = 1×10^{18} cm^{-3}), not only does Zn diffuse into the InP:Fe, but almost all the Fe diffuses into the InP:Zn layer. The Fe concentration in the intentionally Fe-doped layer was intended to be about 2×10^{17} cm^{-3}; however, it was reduced to 2×10^{16} cm^{-3} for the 0.1 μm closest to the Fe-/Zn-doped interface. In addition, the Fe concentration was determined to be almost 1×10^{17} cm^{-3} by SIMS in the intentionally Zn-doped layer. This behavior was explained by the authors as competition of Fe and Zn for interstitial lattice sites. Interstitial Zn kicks out substitutional Fe, which then becomes interstitial and mobile.

5.4.3 Other Transition Elements

Several other transition elements have been used as deep traps to generate semi-insulating InP. Among these is Co, which has been studied by Hess and coworkers [61] at reduced growth pressure (70 torr). They used cobaltnitrosyltricarbonyl, Co(NO)(CO)$_3$, or CNT, as the Co source. Colbalt is also a deep acceptor (Co^{+3}/Co^{+2} transition), with a trap level at 0.32 eV above the valence band in InP. Hess *et al.* were able to incorporate Co into InP at a level of about 3×10^{16} cm^{-3}, as determined by SIMS. However, the SIMS profile revealed an extremely variable [Co] with depth through the 6-μm-thick layer. Surface morphologies were smooth and reflective, and resistivities were of the order of 10^5 Ω·cm. Thus despite the varying [Co] with depth, the authors were able to fabricate semi-insulating buried crescent lasers having thresholds and output powers comparable to those containing InP:Fe blocking layers [61,62].

Titanium has also been employed as a dopant to generate semi-insulating InP. Dentai, Joyner, and Weidman [63] have grown InP:Ti by atmospheric pressure MOCVD using tetrakis(diethylamino)titanium, (Et$_2$N)$_4$Ti, as well as other Ti sources; they have achieved total Ti concentrations up to 10^{20} cm^{-3}. Unlike Fe and Co, Ti is a deep donor in InP (transition Ti^{+3}/Ti^{+4}), with a trap level lying 0.63 eV below the conduction band. As a deep donor, Ti would be expected to compensate shallow acceptors in InP up to the maximum electrically active Ti concentration. Dentai *et al.* showed they could produce semi-insulating behavior by growing InP:Ti on *p*-type substrates with N_A-N_D up to 9×10^{17} cm^{-3}. The resistivity of such a layer was greater than 10^7 $\Omega \cdot$ cm, and had a total [Ti] of 4×10^{18} cm^{-3}. Ti-doped layers grown on more highly doped substrates or on *n*-type substrates were conductive. However, the authors demonstrated semi-insulating behavior (resistivities greater than 10^7 $\Omega \cdot$ cm) for epilayers co-doped with Fe and Ti grown on both *n*- and lightly doped (N_A-N_D = 5×10^{17} cm^{-3}) *p*-type substrates.

5.5 RARE EARTH IONS

5.5.1 Device Applications

Rare earth elements such as Yb and Er exhibit sharp luminescence spectra in the infrared arising from intra-4f-shell transitions. These transitions are nearly independent of host material and show only a weak temperature dependence [64]. This luminescence may be optically or electrically excited in semiconductor hosts and is therefore potentially useful for various optoelectronic devices [65].

5.5.2 Ytterbium

Uwai *et al.* [66] first reported the growth of InP:Yb by MOCVD in 1986. They deposited their layers at 0.1 atm growth pressure using tris-cyclopentadienyl ytterbium, Cp$_3$Yb. Surface morphologies varied with [Yb], as measured by SIMS. Specular surfaces were obtained for [Yb] up to 10^{15} cm^{-3}. The surfaces gained a wavy structure for [Yb] between 10^{15} and 5×10^{16} cm^{-3}. Increasing the [Yb] to 1×10^{18} cm^{-3} resulted in an increasingly rough surface. All of the Yb-doped layers showed *n*-type conductivity. The photoluminescence spectra recorded for these samples exhibited the expected sharp emission lines near 1.23 eV (1.00 μm) at a temperature of 4.2 K. However, the PL intensity diminishes rapidly above 80 K and is barely observable above 200 K. In a separate paper [67], these same authors presented SIMS results that showed a flat Yb incorporation profile for a total Yb concentration of about 6×10^{18} cm^{-3}. The SIMS spectra also revealed significant amounts (sometimes greater than 10^{16} cm^{-3}) of Fe and Mn, elements which were not present in undoped layers and which probably derive from the Cp$_3$Yb source.

Several authors have studied Yb incorporation into InP using other Yb precursors. Weber *et al.* [68] synthesized tris-methylcyclopentadienyl ytterbium, (MeCp)$_3$Yb, and used it at a source temperature of 70° C to 90° C to grow InP:Yb by adduct MOCVD. They achieved total [Yb] up to 5×10^{19} cm^{-3}, and also found that all the InP:Yb samples showed *n*-type conduction (carrier concentrations were between 9×10^{14} and 2×10^{16} cm^{-3}). These authors also mentioned a memory effect of Yb^{+3} luminescence observed in the PL of nominally undoped layers grown after Yb doping experiments. In a separate paper [69], Weber and coworkers stated that the doping level was difficult to control using $(MeCp)_3Yb$ due to evaporation rate fluctuations in the solid. They then synthesized tris-isopropylcyclopentadienyl ytterbium, $(IpCp)_3Yb$, which was used as a liquid above its melting point of 47°C. Again, all layers exhibited *n*-type conduction, and the authors claimed better surface morphologies, reproducible incorporation, and no memory effects using the new $(IpCp)_3Yb$ precursor. Samples grown at Yb concentrations of about 3×10^{17} cm^{-3} with $(IpCp)_3Yb$ exhibited higher Yb^{+3} emission intensities at 100 times lower total [Yb] than samples grown using $(MeCp)_3Yb$ as the Yb source. Layers co-doped with S and Yb with carrier concentrations greater than 10^{17} cm^{-3} exhibited a drastic reduction in PL intensity, while intensity was maintained in alternating InP:Yb/InP:S multilayer structures. A direct interaction between Yb and S was proposed to explain this decrease. Finally, Williams and Wessels [70] used commercially available, solid tris-heptafluorodimethyloctanedionate ytterbium, $Yb(fod)_3$, in atmospheric pressure MOCVD of InP:Yb. The source temperature of the $Yb(fod)_3$ was 120° C to 160° C. Again, all samples were *n*-type, and some evidence for impurities such as Fe and Mn was found in the photoluminescence spectra. These impurities did not appear in spectra of undoped samples.

5.5.3 Erbium

Uwai *et al.* [71] also made the first report of the growth of InP:Er by MOCVD in 1987. They used tris-cyclopentadienyl erbium, Cp_3Er, to incorporate Er into InP. Uniform doping profiles were observed through SIMS profiles for [Er] as high as 1.5×10^{19} cm^{-3} and could be obtained reproducibly. Again, the main unintentional impurities were Fe and Mn. Photoluminescence spectra exhibited peaks around the expected value of 1.54 μm, both at 77 K and 300 K. The same authors extended their work [72] to include $(MeCp)_3Er$ as the Er source. This source is to be preferred to the Cp_3Er source because the former must be used at temperatures approaching 200° C, while the higher vapor-pressure $(MeCp)_3Er$ may be used at 100° C to obtain Er concentrations above 10^{18} cm^{-3}. Er doping (as measured by SIMS) was linearly dependent on Er precursor flow rate for both Cp_3Er and $(MeCp)_3Er$. The Er concentration was invariant for changes in growth temperature between 600° C and 700° C. Weber *et al.* [73] also used alkylcyclopentadienyl derivatives to improve ER incorporation. They employed $(MeCp)_3Er$ and the isopropyl derivative, $(IpCp)_3Er$, as Er sources in adduct MOCVD of InP. Like its Yb counterpart, $(IpCp)_3Er$ was used as a liquid above its melting point of 47° C. The highest ER concentrations achieved (measured by SIMS) were 2×10^{19} cm^{-3} for $(MeCp)_3Er$, and 2×10^{18} cm^{-3} for $(IpCp)_3Er$. Co-doping the InP:Er layers with Zn or S gave ER^{+3} luminescence intensity the same or slightly higher than the intensity observed for InP:Er alone. This result contrasted with their earlier result, which showed a decrease in Yb^{+3} intensity upon co-doping with S. Van der Pauw-Hall experiments indicated that layers with [Er] greater than 10^{19} cm^{-3} were semi-insulating.

5.6 SUMMARY

In this chapter we reviewed studies that have probed the incorporation properties of various dopant elements for InP as grown by MOCVD. We discussed specifically the effects of growth rate, growth temperature and pressure, V/III ratio, and dopant molar flow on incorporation.

ACKNOWLEDGMENTS

I thank R.F. Karlicek, V.R. McCrary, and J.L. Zilko for their comments on the manuscript.

REFERENCES

1. G.B. Stringfellow, *Organometallic Vapor Phase Epitaxy: Theory and Practice*, Boston: Academic Press, 1989.
2. (a) V.R. McCrary, J.W. Lee, S.N.G. Chu, S.E.G. Slusky, M.A. Brelvi, G. Livescu, P.M. Thomas, L.J. Ketelsen, and J.L. Zilko, submitted to *J. Appl. Phys.* (b) D. Gruetzmacher, K. Wolter, M. Zachau, H. Juergensen, H. Kurz, and P. Balk, *Inst. Phys. Conf. Ser., GaAs Rel. Cmpds.*, Vol. 91, 1987, p. 613.
3. G.P. Agrawal and N.K. Dutta, *Long-Wavelength Semiconductor Lasers*, New York: Van Nostrand Reinhold, 1986, pp. 15–18.
4. C.C. Hsu, J.S. Yuan, R.M. Cohen, and G.B. Stringfellow, *J. Cryst. Growth*, Vol. 74, 1986, p. 535.
5. (a) S.J. Bass, C. Pickering, and M.L. Young, *J. Cryst. Growth*, Vol. 64, 1983, p. 68. (b) A.T.R. Briggs, B.R. Butler, *J. Cryst. Growth*, Vol. 85, 1987, p. 535. (c) M.A. di Forte-Poisson, C. Brylinski, and J.P. Duchemin, *App. Phys. Lett.* Vol. 46, 1985, p. 476.
6. A.H. Moore, M.D. Scott, J.I. Davies, D.C. Bradley, M.M. Faktor, and H. Chudzynska, *J. Cryst. Growth*, Vol. 77, 1986, p. 19.
7. E.J. Thrush, C.G. Cureton, J.M. Trigg, J.P. Stagg, and B. R. Butler, *Chemtronics*, Vol. 2, 1987, p. 62.
8. M. Razeghi, Ph. Maurel, M. Defour, F. Omnes, G. Neu, and A. Kozacki, *Appl. Phys. Lett.*, Vol. 52, 1988, p. 117.
9. M.A. diForte-Poisson, C. Brylinski, and J.P. Duchemin, *Appl. Phys. Lett.*, Vol. 46, 1985, p. 476.
10. M. Oishi, S. Nojima, and H. Asahi, *Jpn. J. Appl. Phys.*, Vol. 24, 1985, p. L380.
11. A.R. Clawson, T.T. Vu, and D.I. Elder, *J. Cryst. Growth*, Vol. 83, 1987, p. 211.
12. E. Woelk and H. Beneking, *Inst. Phys. Conf. Ser., GaAs Rel. Cmpds.*, Vol. 91, 1987, p. 497.
13. B. Rose, C. Kazmierski, D. Robein, and Y. Gao, *J. Cryst. Growth*, Vol. 94, 1989, p. 762.
14. C. Blaauw, F. R. Shepherd, C.J. Miner, and A.J. Springthorpe, *J. Elect. Mat.*, Vol. 19, 1990, p. 1.

15. E. Woelk and H. Beneking, *J. Appl. Phys.*, Vol. 63, 1988, p. 2874.
16. G.B. Stringfellow, *J. Cryst. Growth*, Vol. 75, 1986, p. 91.
17. M.A. di Forte-Poisson, C. Brylinski, E. Blondeau, D. Lavielle, and J.C. Portal, *J. Appl. Phys.*, Vol. 66, 1989, p. 867.
18. C.J. Pinzone, N.D. Gerrard, R.D. Dupuis, N.T. Ha, and H.S. Luftman, *Elect. Lett.*, Vol. 25, 1989, p. 1315.
19. C.J. Pinzone, N.D. Gerrard, R.D. Dupuis, N.T. Ha, and H.S. Luftman, *J. Appl. Phys.*, Vol. 67, 1990, p. 6823.
20. (a) R.D. Dupuis, H. Temkin, and L.C. Hopkins, *Elect. Lett.*, Vol. 21, 1985, p. 60. (b) A. Molassioti, F. Scholz, A. Forchel, and Y. Gao, *J. Elect. Mat.*, Vol. 19, 1990, p. 851.
21. R.A. Logan, T. Tanbun-Ek, and A.M. Sergent, *J. Appl. Phys.*, Vol. 65, 1989, p. 3723.
22. C.R. Lewis, M.J. Ludowise, and W.T. Dietze, *J. Elect. Mat.*, Vol. 13, 1984, p. 447.
23. S. Sogou, A. Kameyama, Y. Miyamoto, K. Furuya, and Y. Suematsu, *Jpn. J. Appl. Phys.*, *Part 1*, Vol. 23, 1984, p. 1182.
24. A. Mircea, R. Azoulay, L. Dugrand, R. Mellet, K. Rao, and M. Sacilotti, *J. Elect. Mat.*, Vol. 13, 1984, p. 603.
25. Y. Ban, S. Kimura, M. Morisaki, M. Ogura, and J. Shibata, *J. Cryst. Growth*, Vol. 93, 1988, p. 924.
26. S.J. Bass and P.E. Oliver, *Inst. Phys. Conf. Ser.*, *GaAs Rel. Cmpds.*, Vol. 33b, 1977, p. 1.
27. H. Asai and H. Sugiura, *Jpn. J. Appl. Phys.*, *Part 2*, Vol. 24, 1985, p. L815.
28. R. Bhat, J.R. Hayes, H. Schumacher, M.A. Koza, D.M. Hwang, and M.H. Meynadier, *J. Cryst. Growth*, Vol. 93, 1988, p. 919.
29. E. Tokumitsu, *Jpn. J. Appl. Phys.*, Vol. 29, 1990, p. L698.
30. A.W. Nelson and L.D. Westbrook, *J. Appl. Phys.*, Vol. 55, 1984, p. 3103.
31. A. Molassioti, F. Scholz, and Y. Gao, *J. Cryst. Growth*, Vol. 102, 1990, p. 974.
32. R.W. Glew, *J. Cryst. Growth*, Vol. 68, 1984, p. 44.
33. J.J. Yang, R.P. Ruth, and H.M. Manasevit, *J. Appl. Phys.*, Vol. 52, 1981, p. 6729.
34. H.M. Manasevit, K.L. Hess, P.D. Dapkus, R.P. Ruth, J.J. Yang, A.G. Campbell, R.E. Johnson, L.A. Moudy, R.H. Bube, L.B. Fabick, A.L. Bahrenbruch, and M.-J. Tsai, *Conf. Rec. 13th IEEE Photovoltaic Spec. Conf.*, New York: IEEE, 1978, p. 165.
35. C. Blaauw, F.R. Shepherd, and D. Eger, *J. Appl. Phys.*, Vol. 66, 1989, p. 605.
36. S. Cole, J.S. Evans, M.J. Harlow, A.W. Nelson, and S. Wong, *Elect. Lett.*, Vol. 24, 1988, p. 929.
37. S. Cole, J.S. Evans, M.J. Harlow, A.W. Nelson, and S. Wong, *J. Cryst. Growth*, Vol. 93, 1988, p. 607.
38. G.R. Antell, A.T.R. Briggs, B.R. Butler, S.A. Kitching, J.P. Stagg, A. Chew, and D.E. Sykes, *Appl. Phys. Lett.*, Vol. 53, 1988, p. 758.
39. M. Glade, D. Gruetzmacher, R. Meyer, E.G. Woelk, and P. Balk, *Appl. Phys. Lett.*, Vol. 54, 1989, p. 2411.
40. S. Cole, W.J. Duncan, E.M. Marsh, P.J. Skevington, and G.D.T. Spiller, *Elect. Lett.*, Vol. 26, 1990, p. 391.
41. A.W. Nelson and L.D. Westbrook, *J. Cryst. Growth*, Vol. 68, 1984, p. 102.
42. C. Blaauw, B. Emmerstorfer, and A.J. Springthorpe, *J. Cryst. Growth*, Vol. 84, 1987, p. 431.
43. M.R. Czerniak and B.C. Easton, *J. Cryst. Growth*, Vol. 68, 1984, p. 128.
44. F.R. Bacher and W.B. Leigh, *J. Cryst. Growth*, Vol. 80, 1987, p. 456.
45. F.R. Bacher, H. Cholan, and W.B. Leigh, *Mat. Res. Symp. Proc.*, Vol. 90, 1987, p. 5.
46. C. Blaauw, R.A. Bruce, C.J. Miner, A.J. Howard, B. Emmerstorfer, and A.J. Springthorpe, *J. Elect. Mat.*, Vol. 18, 1989, p. 567.
47. E. Veuhoff, H. Baumeister, O. Brandt, and R. Treichler, *J. Cryst. Growth*, Vol. 105, 1990, p. 353.

48. E. Veuhoff, H. Baumeister, R. Treichler, and O. Brandt, *Appl. Phys. Lett.*, Vol. 55, 1989, p. 1017.

49. (a) D.P. Wilt, J.A. Long, W.C. Dautremont-Smith, M.W. Focht, T.M. Shen, and R.L. Hartman, *Elect. Lett.*, Vol. 22, 1986, p. 869. (b) J.L. Zilko, L.J.-P. Ketelsen, Y. Twu, D.P. Wilt, S.G. Napholtz, J.P. Blaha, K.E. Strege, V.G. Riggs, D.L. Van Haren, S.Y. Leung, P.M. Nitzsche, J.A. Long, G. Przybylek, J. Lopata, M.W. Focht, and L.A. Koszi, *IEEE J. Quant. Elect.*, Vol. 25, 1989, p. 2091.

50. G.W. Iseler, *Inst. Phys. Conf. Ser., GaAs Rel. Cmpds.*, Vol. 45, 1979, p. 144.

51. J.A. Long, V.G. Riggs, and W.D. Johnston, Jr., *J. Cryst. Growth*, Vol. 69, 1984, p. 10.

52. J.A. Long, V.G. Riggs, A.T. Macrander, and W.D. Johnston, Jr., *J. Cryst. Growth*, Vol. 77, 1986, p. 42.

53. S.N.G. Chu, S. Nakahara, J.A. Long, V.G. Riggs, and W.D. Johnston, Jr., *J. Electrochem. Soc.*, Vol. 132, 1985, p. 2795.

54. S. Nakahara, S.N.G. Chu, J.A. Long, V.G. Riggs, and W.D. Johnston, Jr., *J. Cryst. Growth*, Vol. 72, 1985, p. 693.

55. P. Speier, G. Schemmel, and W. Kuebart, *Elect. Lett.*, Vol. 22, 1986, p. 1216.

56. K. Huang and B.W. Wessels, *J. Appl. Phys.*, Vol. 60, 1986, p. 4342.

57. K. Nakai, O. Ueda, T. Odagawa, T. Takanohashi, and S. Yamakoshi, *Inst. Phys. Conf. Ser., GaAs Rel. Cmpds.*, Vol. 91, 1987, p. 199.

58. K.L. Hess, S.W. Zehr, W.H. Cheng, and D. Perrachione, *J. Elect. Mat.*, Vol. 16, 1987, p. 127.

59. D. Franke, P. Harde, P. Wolfram, and N. Grote, *J. Crystal Growth*, Vol. 100, 1990, p. 309.

60. E.W.A. Young and G.M. Fontijn, *Appl. Phys. Lett.*, Vol. 56, 1990, p. 146.

61. K.L. Hess, S.W. Zehr, W.H. Cheng, J. Pooladdej, K.D. Buehring, and D.L. Wolf, *J. Cryst. Growth*, Vol. 93, 1988, p. 576.

62. W.H. Chang, J. Pooladdej, S.Y. Huang, K.D. Buehring, A. Appelbaum, D. Wolf, D. Renner, K.L. Hess, and S.W. Zehr, *Appl. Phys. Lett.*, Vol. 53, 1988, p. 1257.

63. A.G. Dentai, C.H. Joyner, and T.W. Weidman, *Inst. Phys. Conf. Ser., GaAs Rel. Cmpds.*, Vol. 91, 1987, p. 283.

64. H. Ennen and J. Schneider, *Proc. 13th Int. Conf. Defects in Semiconductors*, New York: AIME, 1985, pp. 115–127.

65. (a) W.T. Tsang and R.A. Logan, *Appl. Phys. Lett.*, Vol. 49, 1986, p. 1686. (b) W. Koerber, J. Weber, A. Hangleiter, and K.W. Benz, *J. Cryst. Growth*, Vol. 79, 1986, p. 741. (c) P.S. Whitney, K. Uwai, H. Nakagome, and K. Takahei, *Elect. Lett.*, Vol. 24, 1988, p. 740.

66. K. Uwai, H. Nakagome, and K. Takahei, *Inst. Phys. Conf. Ser., GaAs Rel. Cmpds.*, Vol. 83, 1986, p. 87.

67. K. Uwai, H. Nakagome, and K. Takahei, *Appl. Phys. Lett.*, Vol. 50, 1987, p. 977.

68. J. Weber, A. Molassioti, M. Moser, A. Stapor, F. Scholz, G. Hoercher, A. Forchel, A. Hammel, G. Laube, and J. Weidlein, *Appl. Phys. Lett.*, Vol. 53, 1988, p. 2525.

69. J. Weber, M. Moser, A. Stapor, F. Scholz, G. Hoercher, A. Forchel, G. Bohnert, A. Hangleiter, A. Hammel, and J. Weidlein, *J. Cryst. Growth*, Vol. 100, 1990, p. 467.

70. D.M. Williams and B.W. Wessels, *Appl. Phys. Lett.*, Vol. 56, 1990, p. 566.

71. K. Uwai, H. Nakagome, and K. Takahei, *Appl. Phys. Lett.*, Vol. 51, 1987, p. 1010.

72. K. Uwai, H. Nakagome, and K. Takahei, *J. Cryst. Growth*, Vol. 93, 1988, p. 583.

73. J. Weber, M. Moser, A. Stapor, F. Scholz, G. Bohnert, A. Hangleiter, A. Hammel, D. Weidmann, and J. Weidlein, *J. Cryst. Growth*, Vol. 104, 1990, p. 815.

Chapter 6
Growth of InP and Related Compounds by GSMBE and MOMBE

C.R. Abernathy
AT&T Bell Laboratories

6.1 INTRODUCTION

The use of ultrahigh vacuum (UHV) growth techniques for deposition of III-V semiconductors offers several desirable features. The absence of gas phase interactions allows for precise control of both thickness and composition beyond what can be achieved by other growth methods. In addition, the use of substrate rotation produces excellent uniformity of $\pm 1.5\%$ across substrates up to 3 inches in diameter. These advantages have made molecular beam epitaxy (MBE) the dominant method for growth of GaAs/AlGaAs structures, which require abrupt interfaces and a high degree of compositional control, such as the high electron mobility transistor (HEMT) [1]. While MBE is ideally suited for the growth of As-containing compounds, it has been less successful for fabrication of structures that require P. The allotropic nature of P makes the attainment of a stable and reproducible flux from elemental P difficult. This has led to the introduction of gaseous sources to the MBE growth environment. The use of the group V hydrides AsH_3 and PH_3 in conjunction with the elemental sources of conventional MBE has come to be known as gas-source MBE (GSMBE) [2]. This method allows the advantages of MBE to be retained, while the capability of growing P-containing materials is added.

Just as GSMBE possesses the advantages of MBE, it also retains the disadvantages. Among these are the surface defects generated from the use of group III effusion ovens, and the need to open the UHV growth system to air when source materials are depleted. In addition, the potential for scaleup is limited because of the configuration limit imposed by the use of liquid metal sources.

In response to these problems, the replacement of elemental group III sources with gaseous sources, primarily metalorganic (MO) compounds, was initiated. This technique came to be known as metalorganic MBE (MOMBE) [3] or, alternatively, chemical beam epitaxy (CBE) [4]. While this approach does appear to address all of the problems discussed above, it introduces much greater complexity in terms of equipment design and growth kinetics. Thus, though MOMBE will most likely become the preferred method in the future, GSMBE is at present the easier of the two methods to implement. This chapter will provide a discussion of the important aspects of both methods, and will review the current state of the art for their use in deposition of InP and materials that can be lattice matched to InP substrates.

6.2 EQUIPMENT DESIGN

6.2.1 GSMBE

6.2.1.1 Gas Handling Design

Since group V hydrides are the only gaseous sources used in this technique, the gas handling equipment can be kept rather simple. Mass flow controllers (MFC) are the only active control elements required. Due to the interaction of hydrides with Viton and the need to maintain a gas manifold whose leak integrity matches that of the UHV growth system, the use of all metal-sealed MFCs is greatly preferred.

In order to obtain abrupt interfaces between layers, it is necessary to switch rapidly from one group V source to another during growth. This is usually accomplished by establishing the hydride flow into a vent line before introduction to the chamber. This vent line may be evacuated by any pump suitable for use with corrosive gases. Switching of the gas is accomplished through the simultaneous opening of the valve on the growth chamber and closing of the valve on the vent line. This approach has proven adequate for rapid switching of the group V beams. The question of interface abruptness will be discussed in greater detail in later sections.

Since the group V hydrides will not sufficiently decompose on the substrate during growth, it is necessary to crack these compounds prior to their arrival at the wafer surface [2]. This is necessary both to allow for growth and to prevent a dangerous buildup of uncracked hydrides on the cryopanels. While various methods and materials have been employed to decompose the hydrides [2,5–8], the most common method at present is the catalytic, or low pressure, approach. A typical low pressure cracker design is shown in Figure 6.1. The gas is introduced at the back of the cell and flows around a catalytic material, typically Mo or Ta. When heated to temperatures of ~950°C or higher, this catalyst allows for the efficient decomposition of the hydrides to group V dimers or monomers [6,7], even though the pressure inside the cracker cell is only ~10^{-3} torr. While both AsH_3 and PH_3 may be cracked simultaneously, there may be some residual memory in the cracker that will produce

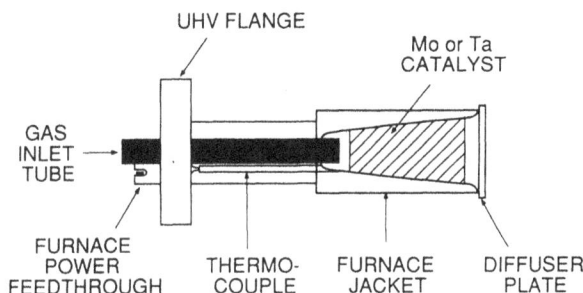

Figure 6.1 Schematic of low-pressure cracker with Mo or Ta catalyst. (Courtesy of INTEVAC)

incorporation of one of the species even after the gas flow has been switched off. For this reason it is probably desirable to use separate crackers for AsH_3 and PH_3. These cracker cells are approximately the same dimensions as the standard MBE effusion ovens, and, therefore, may be used as direct replacements for other cells in the MBE system.

6.2.1.2 Growth Chamber

For both GSMBE and MOMBE, the growth apparatus is based on conventional MBE equipment (Fig. 6.2). This system consists of a UHV growth chamber containing liquid-nitrogen-cooled cryopanels around the substrate heater assembly and the source flange. This chamber is usually isolated from the load lock by a buffer chamber in order to minimize contamination of the growth environment, and to allow for outgassing of wafers prior to growth. In conventional MBE, these chambers are pumped with either ion pumps or cryopumps. In GSMBE, large quantities of hydrogen are generated from the decomposition of the hydrides; therefore, ion pumps are not generally used on the growth chamber.

As in conventional MBE, effusion cells that contain elemental charges are used to supply the various group III and dopant sources. These cells, shown in Figure 6.3, consist of pyrolytic boron nitride (PBN) or carbon crucibles housed inside a resistively heated cylinder. Thermocouples placed near the base of the crucible are used to monitor the temperature of the cell. The flux of the group III source is directly related to its vapor pressure. Therefore, the flux may be varied by altering the temperature of the cell, as shown in Figure 6.4. Switching of the beam is accomplished by moving a shutter over the mouth of the oven. Great progress has been made in the design of effusion ovens so that defect densities on the layer surface generated from the group III ovens have been reduced in many cases to <100 cm^{-2}

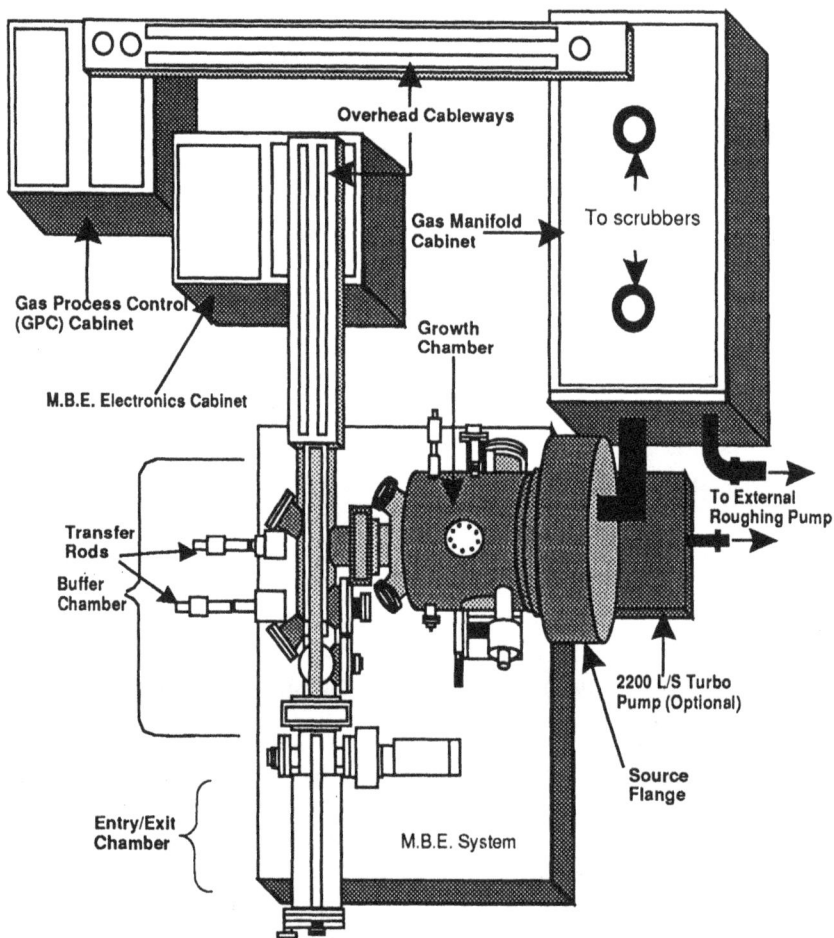

Figure 6.2 Schematic of GSMBE or MOMBE growth apparatus. (Courtesy of INTEVAC)

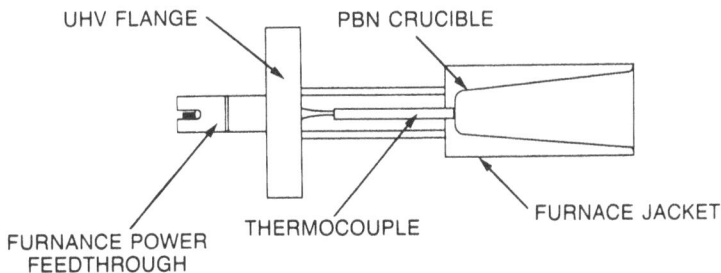

UHV FLANGE

PBN CRUCIBLE

FURNANCE POWER
FEEDTHROUGH

THERMOCOUPLE

FURNACE JACKET

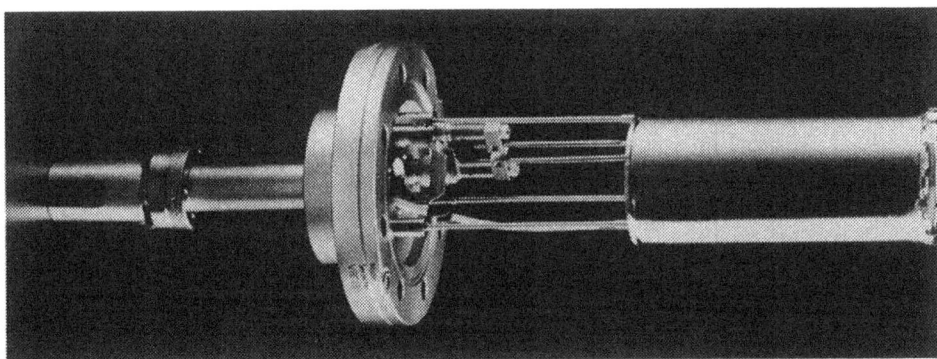

Figure 6.3 Standard 60-cc. group III effusion cell shown in schematic and actual form. (Courtesy of INTEVAC)

[9]. Further advances have made the cells quite thermally stable so that the composition and deposition rate in the grown layer are highly uniform in the growth direction.

Though the fluxes are generally stable during growth, reproducibility of flux from run to run can be difficult. As the group III charge is depleted, the relationship between the surface temperature of the melt and the temperature measured by the thermocouple may change. Thus, for a given set point on the temperature controller the actual flux will change with time. This effect necessitates the use of in-situ calibration methods for growth rate. The two most common methods at present are beam flux measurement and reflection high energy electron diffraction (RHEED). Beam flux measurement simply requires placing an ion gauge in front of the beam so that the gauge can measure the partial pressure. Correction for differences in ionization behavior of the various elements must then be accounted for if the flux is to be accurately determined.

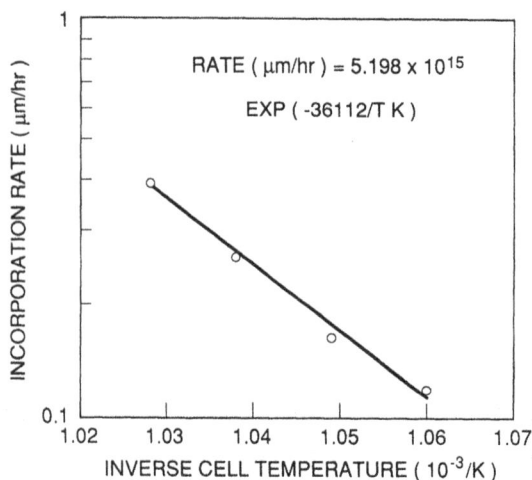

Figure 6.4 Variation of In deposition rate with effusion cell temperature. (Courtesy of INTEVAC)

Though more complex, RHEED measurement is more commonly used as a calibration method because of its high degree of accuracy [10]. This technique monitors the actual growth rate by observing the intensity of an electron beam scattered from the growth surface. Specular scattering is at a minimum for a surface coverage of approximately one-half monolayer. The intensity returns to a maximum when the layer is completed. Thus, the frequency of the oscillation is exactly equal to the monolayer deposition rate. This technique not only allows for calibration of the binary growth rate, it may also be used for calibration of the group III composition in ternary and quaternary layers, since the total group III incorporation rate is simply the sum of the individual group III rates.

6.2.2 MOMBE

All of the considerations discussed for GSMBE apply to MOMBE as well. However, the introduction of gaseous group III or dopant sources adds unique complications to the growth chamber and the gas handling manifold. Modifications beyond those required for GSMBE will be discussed in the following sections.

6.2.2.1 Gas Handling Design

There are two methods of controlling gas flow into the growth chamber. One uses mass flow controllers to regulate the flow of a carrier gas through the MO source bubbler while the other relies on accurate control of the pressure behind an orifice. The carrier gas approach is very similar to that used in most metalorganic chemical vapor deposition (MOCVD) systems with slight modifications to allow for differences in the MOMBE process. A schematic diagram of the gas manifold is shown in Figure 6.5. Alkyls are introduced through the use of a H or He carrier gas. The carrier gas flow rate is regulated through a mass flow controller into the alkyl bubbler. The output of the bubbler is then fed past a capacitance manometer and through a proportional valve. The proportional valve is used to maintain a constant pressure above the bubbler, and thus allows one to prevent back diffusion of reactants by balancing the pressure in the various lines. It also allows more freedom in the setting of the hydrogen and alkyl flows, since the pickup rate of the MO is determined by the total pressure according to the following relation:

$$F_{MO} = \frac{F_{CG} P_{MO}}{P_T - P_{MO}} \tag{6.1}$$

where P_{MO} is the vapor pressure of the MO, P_T is the bubbler pressure, and F_{CG} and F_{MO} are the flow rates of the carrier gas and MO, respectively. The major advantage of this design is that most of the sources can be maintained at or below room temperature.

The alternative design is based on the premise of controlling the pressure behind an orifice, and does not involve the use of mass flow controllers or carrier gases. A schematic of this manifold is also shown in Figure 6.5. The flux of the

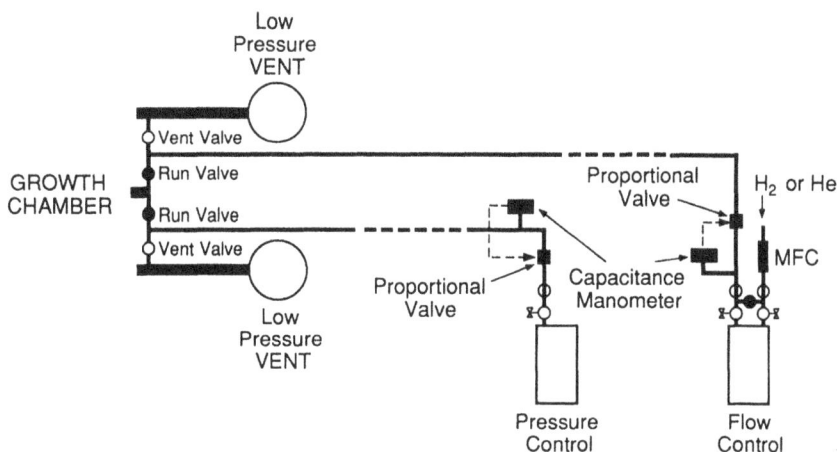

Figure 6.5 Schematic of gas manifold for handling gaseous group III or dopant sources. (Courtesy of INTEVAC)

alkyl is regulated by controlling the temperature of the bubbler, and thus the line pressure, and by adjusting the orifice size in a variable leak valve or proportional valve. The advantages of this approach are reduced pumping loads and less complicated flow dynamics. The major disadvantage is the need to heat the sources so that the vapor pressure will be high enough to provide an adequate flux. This can be particularly troublesome when low-stability sources are employed.

For both types of manifold, the MOs are introduced through a cell heated sufficiently to prevent condensation on the cell walls. The gases are switched in a manner similar to that of the group V sources discussed earlier. As seen in Figure 6.6, the switching speed is quite sufficient to provide the same sort of abruptness as in GSMBE.

INJECTOR SWITCHING TRANSIENTS

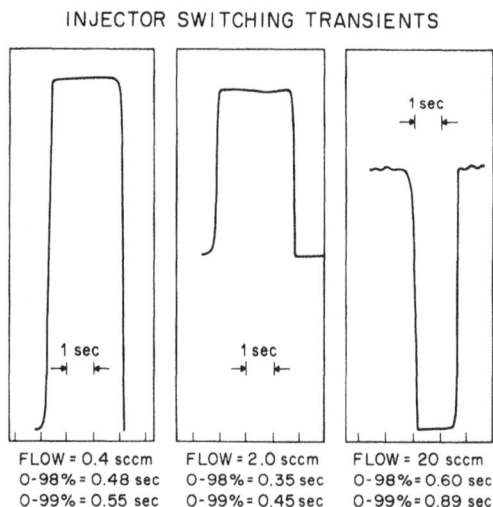

FLOW = 0.4 sccm	FLOW = 2.0 sccm	FLOW = 20 sccm
0-98% = 0.48 sec	0-98% = 0.35 sec	0-98% = 0.60 sec
0-99% = 0.55 sec	0-99% = 0.45 sec	0-99% = 0.89 sec

Figure 6.6 Switching speed of gas manifold as measured by beam flux ion gauge. (Courtesy of INTEVAC)

6.2.2.2 Growth Chamber

The addition of MO compounds to the growth chamber raises a serious issue regarding long-term degradation of various components in the chamber due to carbon contamination. For example, ion gauge filaments exposed to MOs exhibit a marked reduction in lifetime relative to what is normally observed in GSMBE. While this problem can be overcome by maintaining the filaments at ~100°C when not in use [11], it does serve to illustrate the care which must be taken in design and operation of MOMBE systems.

Perhaps the greatest concern is damage to the pumping system on the growth chamber. The long-term effects of MOs on cryopumps, for example, is not well documented. Therefore, the general approach has been to use turbo-molecular or diffusion pumps as the main pumping units. Both of these systems have been demonstrated to provide reliable service without degradation in performance.

In addition to the main pumping unit, which operates continuously, an additional pump may be needed to assist in removal of gaseous species during growth. This is particularly important if the MOs are transported with a carrier gas, because the additional load of the carrier gas may overwhelm the main pumping unit. For example, it has been found for growth of AlGaAs that the growth rate is dependent upon the pumping capacity of the system [12] (Fig. 6.7). This effect is believed to be caused by screening by the background carrier gas, which limits the pumping efficiency of the liquid nitrogen cryoshroud for MOs [13]. When the background pressure is reduced through the use of an additional pump, the growth rate approaches the level one would expect based on incorporation at lower flux rates. Since the booster pump is used only during growth, when the cryopanels are cold, it should not encounter large quantities of MOs. Therefore, any pump that is UHV compatible and capable of pumping hydrogen should work satisfactorily, including cryopumps.

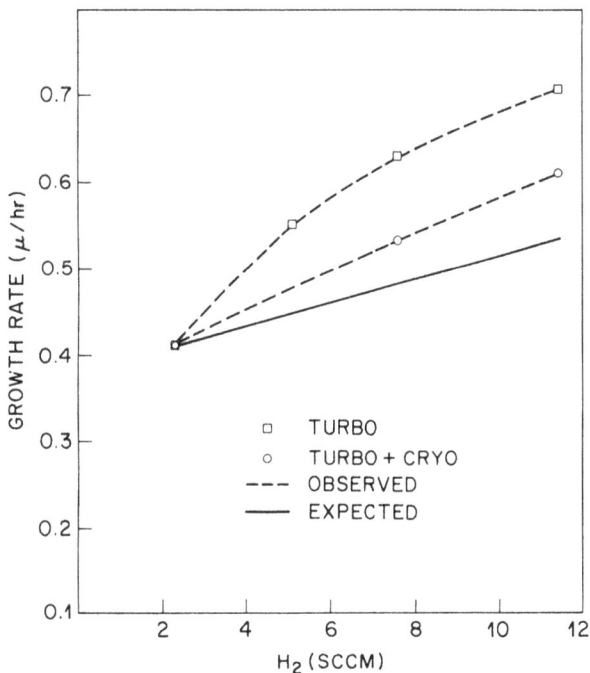

Figure 6.7 Variation in AlGaAs growth rate with carrier gas flow through the triethylaluminum (TEA) bubbler for different pumping capacities. Growth temperature was 600°C. Triethylgallium (TEG) and As_4 were the other sources employed [12].

In-situ calibration techniques may also be affected by the use of gaseous group III sources. Measurement of the group III flux with beam flux gauges is possible only if the pressure control method of MO introduction is used. If the carrier gas approach is employed, the output of the beam flux gauge will be primarily a measure of the carrier gas flux, since the MO flux is only a small part of the beam. Even when using the pressure control method, reproducible flux measurement is difficult because of degradation of the ion gauge from exposure to MOs.

The other major in-situ measurement technique, RHEED, may be used with either alkyl delivery system. The utility of RHEED has been demonstrated for MOMBE growth of binary compounds [14,15]; however, its usefulness for calibration of ternary or quaternary composition is currently doubtful. In contrast to elemental group III sources, gaseous group III compounds show significant interaction on the growth surface. Thus, the ternary composition is no longer a simple addition of the binary fluxes as shown in Figure 6.8. The InGaAs compositon predicted by RHEED measurement of the binary growth rates is generally lower in In than ternary material grown from the same fluxes [14]. If RHEED is to be used for in-situ calibration of ternary composition, software algorithms must be developed to compensate for the interaction of the MOs on the substrate.

As will be discussed in a later section, this surface interaction of MOs is strongly temperature dependent. Therefore, the growth temperature must be reproducible and uniform over the entire substrate if the ternary composition is to be both reproducible and uniform. Fortunately the latter requirement is not difficult to achieve given the present state of commercially available equipment. Attaining reproducibility of growth temperature from run to run, however, is much more difficult. Various methods can be employed to provide in-situ calibration of the growth temperature. Among the most common are observation of the InSb melting point (525°C) and the use of an

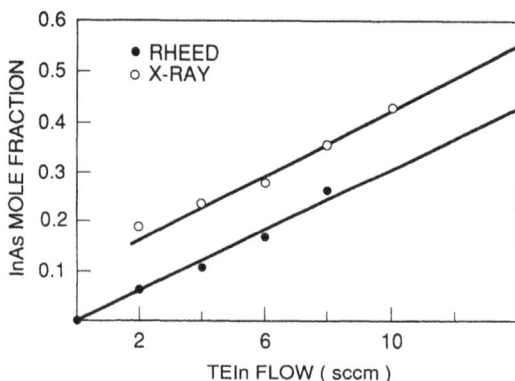

Figure 6.8 Comparison of RHEED and x-ray measurements of InAs mole fraction *versus* TEI flux [after 14].

infrared pyrometer. The former suffers from the disadvantage of requiring In mounting, while the latter is difficult to keep operating because of internal coating of the viewport in front of the pyrometer.

6.3 GROWTH KINETICS

6.3.1 GSMBE

As in MBE, the incorporation kinetics of group III sources in GSMBE are roughly independent of one another and of the group V flux. The group III incorporation rate is therefore only a function of the group III flux and the sticking coefficients of the various group III elements. Below ~540°C, the sticking coefficient of In is essentially unity, and the In incorporation rate is independent of temperature. Above this temperature, the growth rate becomes strongly dependent on temperature because of In re-evaporation. Ga evaporation does not occur until ~710°C and is also unity below this temperature. Thus, most In-containing compounds are grown at temperatures of less than 540°C to avoid losing In from the surface. Surface mobility places a lower limit on the growth temperature of InP-based materials. As the growth temperature is reduced, the atoms become increasingly sluggish and thus are less likely to migrate to lattice sites. While growth temperatures as low as 366°C [16] have been employed, standard growth temperatures range from 450°C to 530°C.

Unlike the group III species, group V sources show an incorporation efficiency strongly influenced by the group III surface coverage. Generally, group V species are incorporated in a one-to-one ratio with group III elements, with the excess group V material evaporating from the surface. For this reason, growth in GSMBE is usually conducted with the minimum group V flux needed to either maintain a stable surface or to match the group III incorporation rate.

For materials containing only one group V element, the incorporation behavior is thus quite straightforward. However, for materials which contain both As and P (i.e., $In_xGa_{1-x}As_{1-y}P_y$) the incorporation of the group V species is slightly more complicated, as shown in Figure 6.9. For alloys rich in As, $y < 0.2$, the relative amounts of As and P incorporated in the grown layer are equivalent to the As/P ratio in the beam, regardless of growth temperature. For higher P compositions, the As/P ratio in the grown layer is actually higher than the As/P ratio in the beam [17,18]. This disparity becomes even more pronounced for growth temperatures below 528°C [19]. The cause of this effect has been variously ascribed to a greater surface residence time for As than for P [20], and to the need for thermal decomposition of P_x species on the surface where x may be two or higher [19]. As yet, it can only be stated conclusively that there exists a kinetic barrier to the incorporation of P in quaternary alloys containing both As and P.

Little information is available regarding the growth of AlInAs by GSMBE. Preliminary results suggest that lattice-matched material with good morphology can

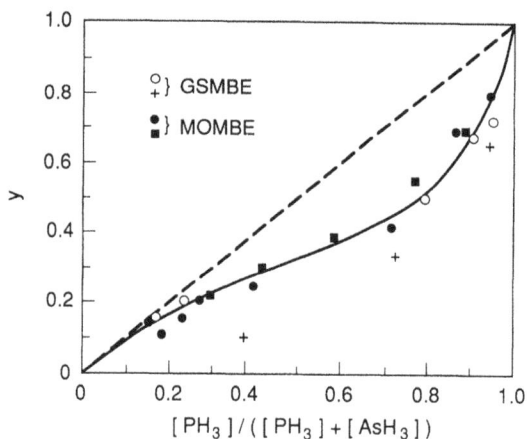

Figure 6.9 Variation of phosphorous content, y, in $In_xGa_{1-x}As_{1-y}P_y$ *versus* PH_3 fraction of hydride beam [after 41].

be obtained quite readily [21]. In principle, the electrical and optical quality of AlInAs grown by GSMBE should be comparable to that obtained by MBE.

6.3.2 MOMBE

As in GSMBE, MOMBE uses P and As beams generated by the catalytic decomposition of the hydrides. Thus, the desorption behavior of the group V elements is the same as in the previous discussion. In contrast, the incorporation kinetics of the group III precursors used in MOMBE are significantly different than for the elemental sources used in GSMBE. These differences arise from the need to remove the elimination. This mechanism involves the transfer of a hydrogen atom from an attached ethyl side group to the group III atom. The ethyl species then forms ethylene and breaks away from the group III atom. Because of the weaker In-ethyl bond, TEI yields a constant growth rate over a lower temperature range of 350°C to 400°C [25] (Fig. 6.10). There does appear to be some variation between 400°C and 525°C. In principle, the weaker bond should lead to reduced carbon incorporation. Similar comparisons in the GaAs system show several orders of magnitude of variation in carbon level when trimethylgallium (TMG) is replaced with triethylgallium (TEG) [26,27]. As will be discussed in later sections, this advantage does not appear to be necessary for the growth of high-quality InP, since the In-methyl bond cleaves much more readily than the GA-methyl bond.

In addition to temperature, the group V species also plays a role in the decomposition of the alkyl molecule. Below ~550°C, simple pyrolysis is not sufficient to

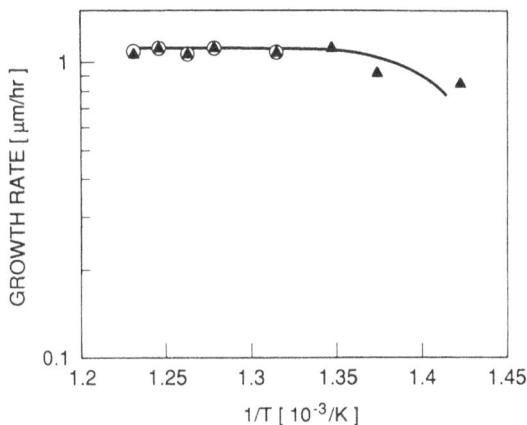

Figure 6.10 InP growth rate from TEI *versus* inverse growth temperature for two PH₃ fluxes: 10 sccm (O) and 5 sccm (O) [after 25].

allow efficient decomposition of the MO sources. The presence of a group V species is also needed in order to prevent desorption of the alkyl group III compounds. This effect can be exploited to achieve highly selective deposition [28,29]. Since the MOs do not readily decompose on surfaces that are group V deficient, mask materials such as SiN remain free of growth during deposition on the unmasked region. On the group V stabilized surface, however, the InP growth rate from trimethylindium (TMI) is independent of the V/III ratio [24], as shown in Figure 6.11. In contrast, the InP growth rate from triethylindium (TEI) shows a slight dependence on V/III ratio for temperatures below 500°C [25] (Fig. 6.11). This effect has also been observed in GaAs growth by MOMBE [27,30], and is believed to be caused by site blocking by group V atoms which inhibit adsorption of alkyl group III species [23]. While one would expect similar V/III dependences with TMI and TEI, it is quite possible that adsorbed P interacts hydrocarbon chains from the group III atoms before growth can occur. The mechanisms by which these side groups are removed thus play a large role in determining the growth kinetics from these sources.

It is generally believed that the first hydrocarbon side group can be removed quite easily, with the rate-limiting step being the removal of the second hydrocarbon molecule [22,23]. At low temperatures, there is insufficient energy to remove the second carbon group and the alkyl metal desorbs from the surface without incorporating into the growth front. Thus, unlike the elemental sources in GSMBE, the alkyl sources require a minimum substrate temperature in order for growth to occur.

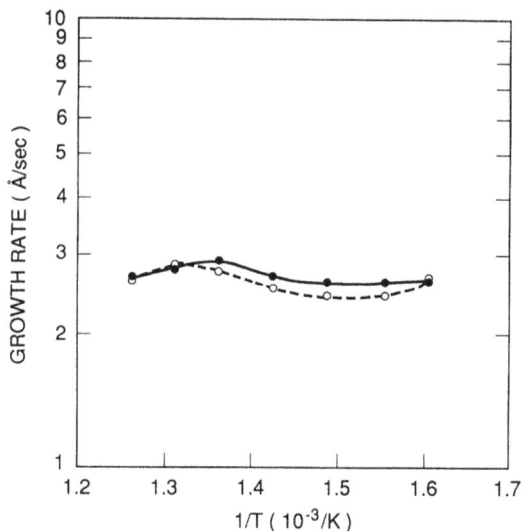

Figure 6.11 InP growth rate from TMI *versus* inverse growth temperature for a PH_3 flux of 5×10^{-3} Pa (Δ) or 9.7×10^{-3} Pa (O) [after 24].

6.3.2.1 InP

The most commonly used alkyl-In source is TMI. This compound, like TMG, appears to decompose on the growth surface through pyrolytic cleavage of the metal-methyl bond. Unlike the Ga-CH_3 bond, the In-CH_3 bond is quite weak. This allows the TMI to decompose quite efficiently on the growth surface. Because of this, the InP growth rate from TMI is roughly independent of temperature over the range of ~470°C to 540°C [24] (Fig. 6.11). Below 470°C, pyrolysis of the alkyl-In species becomes increasingly less efficient; while above 540°C, In desorption reduces the growth rate. Desorption of P at temperatures above 540°C begins to adversely affect the growth as well [24].

TEI has also been explored as a potential In source [25]. Due to the longer hydrocarbon chain, the In-ethyl bond is weaker than the In-methyl bond. In addition, the ethyl side group may be removed by a process known as β-hydride with CH_3 radicals to form volatile PCH_3 species. Such an interaction would effectively lower the V/III ratio at the growth surface and prevent site blocking by the P. In contrast, C_2H_4 formed from the decomposition of TEI is less reactive and forms a less volatile P hydrocarbon species, and would thus be less likely to reduce the surface V/III ratio.

6.3.2.2 InGaAs

Though no dependence of InGaAs composition on V/III ratio has been observed [31,32], there is a strong variation with temperature [32–35], as shown in Figure 6.12. It is this dependence which necessitates the temperature calibration methods described in Section 6.2.2. Below 500°C, Ga incorporation becomes increasingly less efficient due to incomplete pyrolysis of the Ga-ethyl bonds. This causes the material to become increasingly rich in In as the growth temperature is reduced.

The sensitivity to temperature above 500°C arises from the interaction of the group III precursors on the wafer surface. In particular, the Ga incorporation rate from TEG is reduced by the presence of In. This interaction is enhanced as the growth temperature is increased; thus, the composition becomes increasingly In rich. As shown in Figure 6.13, this effect occurs regardless of whether TMI or TEI is used. Similar results have been obtained with elemental In [37]; therefore, this effect cannot be ascribed to exhange reactions of the alkyls between the adsorbed Ga and In species. Mass spectrometric studies of the desorption behavior of triethylgallium show that the presence of TEI enhances the desorption of alkyl gallium from the growth surface [38]. It has been suggested that this desorption may be caused by In segregation at the surface, as has been observed in MBE [16,37,39]. If this is indeed the cause, then a less stable Ga source, which would decompose more readily and therefore be less likely to desorb, may be required if the sensitivity of composition

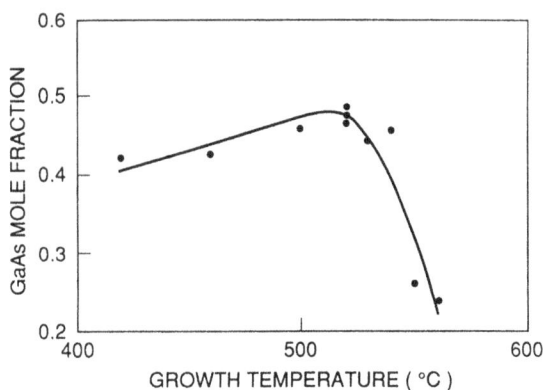

Figure 6.12 Variation InGaAs composition with growth temperature for fixed TMI and TEG fluxes [after 35].

to temperature is to be overcome. Though several gaseous Ga sources appear promising, none have yet been demonstrated for the growth of InGaAs. Alternatively, precracking of the TEG may also reduce the magnitude of the temperature sensitivity by reducing the volatility of the adsorbed Ga species. This approach also remains to be demonstrated.

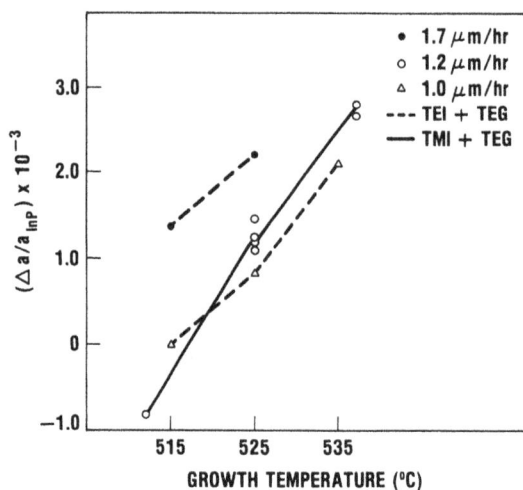

Figure 6.13 Effect of growth temperature on InGaAs/InP lattice mismatch for (TEG + TEI) and (TEG + TMI) [36].

6.3.2.3 InGaAsP

Due to the precracking of the group V beams, the As/P ratio can be controlled in much the same way as in GSMBE. As shown in Figure 6.9, the incorporation efficiencies of the group V species are similar to those obtained when using elemental sources [40,41]. This suggests that, to first order, the group V incorporation kinetics are independent of the group III interactions on the growth surface. In contrast, the group III precursors undergo the same type of surface interactions discussed for the growth of InGaAs. Thus, as shown in Figure 6.14, the In/Ga ratio of InGaAsP displays the same type of sensitivity to growth temperature as that observed for InGaAs.

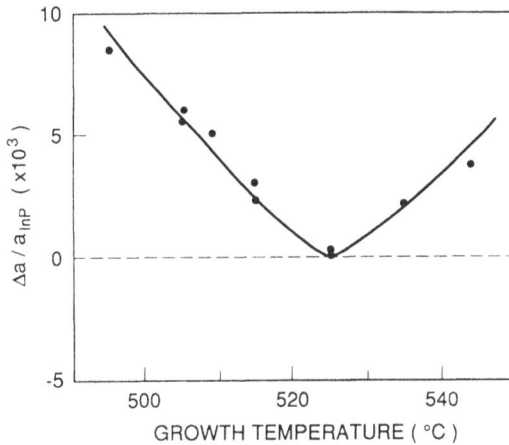

Figure 6.14 Variation of $In_xG_{1-x}As_{1-y}P_y$ lattice constant with growth temperature [after 41].

6.3.2.4 AlInAs

While AlInAs has been used extensively in structures grown by MBE, virtually no work has been reported on the growth of this compound by MOMBE. This is most likely due to the difficulty involved in growth of high-purity Al containing compounds using conventional alkyl Al sources. The most commonly used Al source, triethylaluminum (TEA), has been shown to produce high levels of both carbon and oxygen in AlGaAs grown with TEG and AsH_3 [42,43]. While the carbon is due primarily to the Al-C bond contained in the precursor molecule itself, the oxygen has been shown to arise from oxygen contamination of the TEA source which forms volatile Al-O species [43,44]. Thus, similar contamination of AlInAs films grown from TEA would be expected, as is seen in AlGaAs grown from the same Al source.

In order to overcome the limitations of the alkyl Al sources, a new compound, trimethylamine alane (TMAA), has been introduced as an alternative [43,45,46]. Since this source has no direct Al-C bond the amount of carbon incorporated into the epitaxial layer can be greatly reduced. Most important, this compound does not form volatile Al-O-containing species, and thus does not readily transport oxygen to the growth surface. Use of this source for growth of AlGaAs has resulted in dramatic reductions of both carbon and oxygen. Though no information is yet available on the utility of TMAA for growth of AlInAs, it is reasonable to expect that this source will spur the investigation of AlInAs growth by MOMBE because of the promising results obtained for AlGaAs.

6.4 MATERIAL QUALITY

6.4.1 GSMBE

Continued advances in In and PH_3 source purity have led to significant improvement in the quality of InP and InGaAs grown by GSMBE. Using In of 7N purity and a growth temperature of 500°C, Lambert *et al.* [47] have reproducibly achieved InP with 77 K electron concentrations $<10^{15}$ cm^{-3}, and 77 K mobilities of 70,000 to 100,000 cm^2/V s, with a maximum of 112,000 cm^2/V s. Such mobilities suggest an absence of compensation due to shallow acceptors, further confirmed by the absence of acceptor-related peaks in the 4.2K PL spectra [19,47]. High-purity InGaAs and InGaAsP have also been demonstrated with 77 K mobilities $\geq$40,000 cm^2/V s and $n_{77} \sim 2 \times 10^{15}$ cm^{-3} for InGaAs [48], and $\mu_{77} = 25,000$ cm^2/V s and $n_{77} < 2 \times 10^{15}$ cm^{-3} for InGaAsP ($\lambda = 1.55$ μm) [18].

As expected, excellent interface abruptness has also been demonstrated. Quantum well structures containing InGaAs or InGaAsP show intense luminescence, and from the spectral shift appear to be abrupt to at least two monolayers [20].

6.4.2 MOMBE

In spite of the presence of hydrocarbons at the growth surface, very-high-purity InP and InGaAs have been demonstrated by MOMBE. InP layers with 77 K mobilities $>$110,000 cm^2/V s, and 77 K carrier concentrations $\leq$2 $\times$ 10^{14} cm^{-3} have been demonstrated from both TEI [25] and TMI [24]. The purity of the material is further confirmed by photoluminescence spectra that show excitonic features comparable to those obtained in the best material grown by other techniques. The comparable quality of material grown with TMI and TEI suggests that the weaker In-C bond eliminates the need to use the ethyl-bonded In precursors, which are somewhat unstable [49], in order to obtain high-purity material.

Both In source compounds show a marked increase in free-electron concentration with decreasing substrate temperature [24,25] (Fig. 6.15). This background doping is partially caused by increased carbon incorporation at low temperatures; however, carbon alone cannot account for all of the carriers observed in the material; thus, other impurities or defects are also responsible for the variation in carrier concentration with growth temperature. The nature and source of these donors as yet remain unclear.

Like temperature, V/III ratio must be optimized in order to obtain the highest quality material. In contrast to GaAs growth, where increasing the V/III ratio produces a dramatic reduction in the background impurity concentration [27,42], InP layers show a steady increase in carrier concentration with increasing V/III ratio [24,25] (Fig. 6.16). Since this increase is accompanied by a reduction in mobility

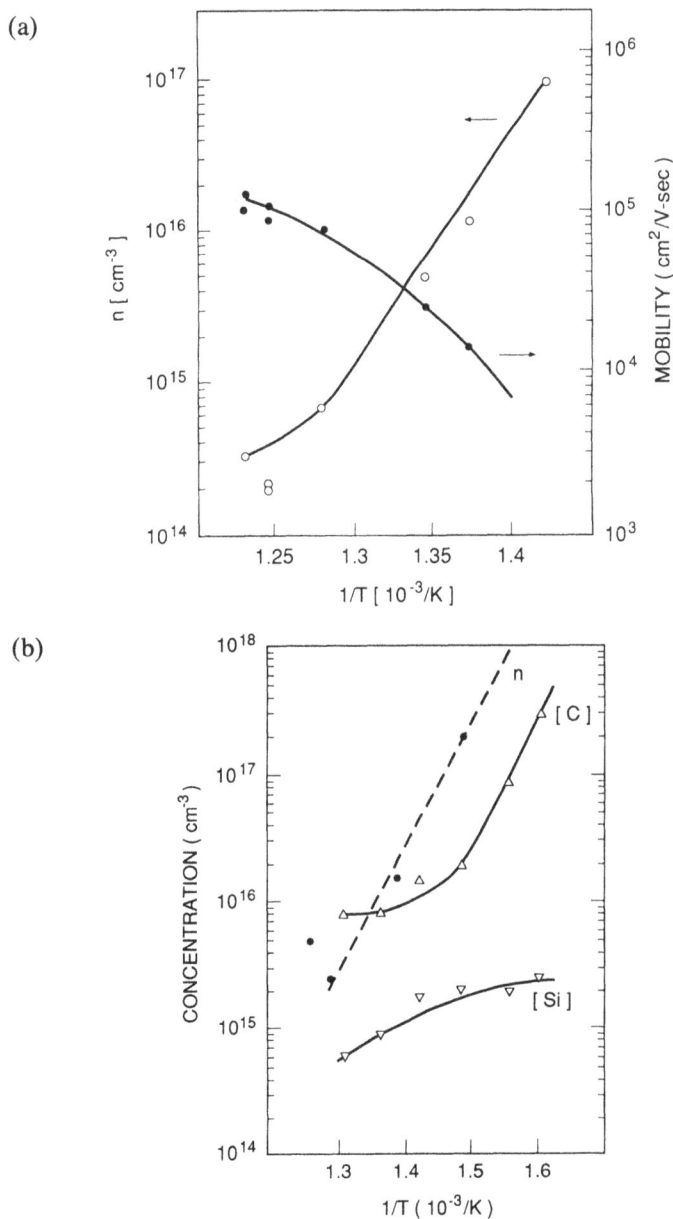

Figure 6.15 Variation of InP electron concentration with inverse growth temperature using a) TMI and b) TEI [after 24,25].

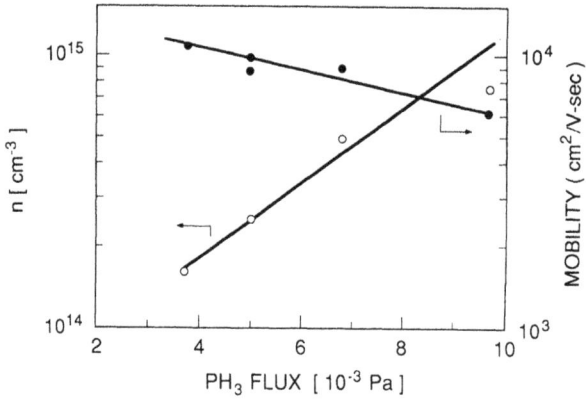

Figure 6.16 Carrier concentration and mobility *versus* PH_3 flux for InP grown from TMI [after 24].

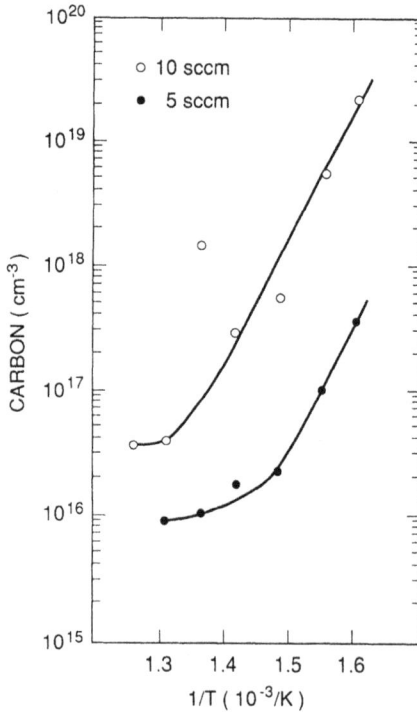

Figure 6.17 Effect of PH_3 flow on carbon concentration for InP grown from TEI at various growth temperatures [after 25].

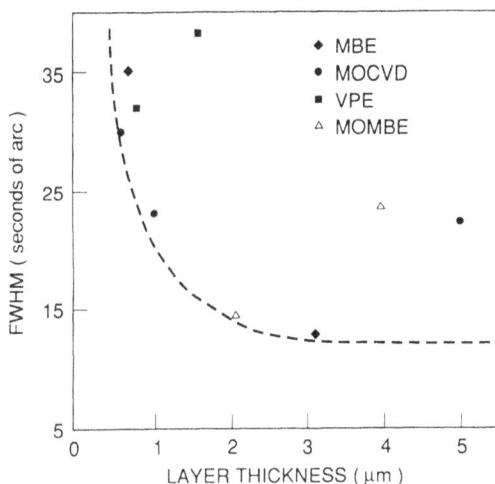

Figure 6.18 Full-width-at-half-maximum intensity (FWHM) *versus* layer thickness for InGaAs grown by various methods. Dashed line represents theoretical limit [after 51].

[24], it is clear that the total impurity density is actually increasing as well. It is quite possible that the PH_3 is a source of *n*-type impurities, thus increasing the donor concentration as the PH_3 flow is increased. However, work by Benchimol *et al.* [25] has revealed the presence of a surprising increase in carbon concentration with increasing V/III ratio, as shown in Figure 6.17. The exact mechanism giving rise to this effect is unclear. It may involve a complex interaction between adsorbed alkyl group III species and P_2, or it may suggest that the PH_3 source is contaminated with carbon-containing species, such as CH_4, which may decompose in the catalytic cracker.

Like InP, InGaAs grown by MOMBE has been demonstrated to be of excellent purity. Using TMI, TEG, and AsH_3, 77 K mobilities up to 67,000 cm^2/V s, and free carrier concentrations in the range of 5×10^{14} cm^{-3} to 5×10^{15} cm^{-3} have been achieved [50]. Further evidence of the material's purity can be seen from 2 K PL spectra, which show almost no donor-to-acceptor recombination. The observation of a very narrow, 1.2-meV, exciton peak also indicates that the material is quite compositionally uniform in the growth direction [50]. This uniformity has also been confirmed by x-ray analysis, which shows (004) Bragg reflection peaks with full-width-at-half-maximum (FWHM) of only 15 to 20 arcsec for 0.2 μm layers [31,50,51]. As shown in Figure 6.18, these are among the best reported for any technique.

As for InP, InGaAs shows a decrease in both carbon and free-electron concentration as the substate temperature is increased to 500°C (Fig. 6.19). Above this temperature these levels remain fairly constant. The effect of V/III ratio on background impurity concentration remains somewhat unclear. While the free carrier concentration increases with increasing V/III ratio [52], the electron mobility increases

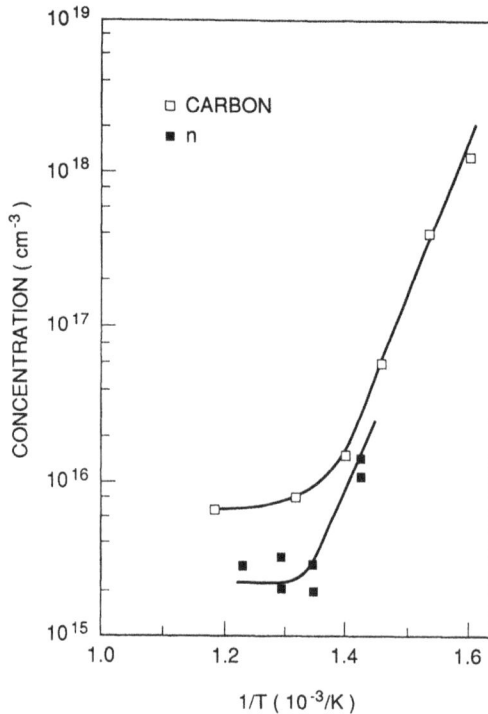

Figure 6.19 Variation in free electron concentration, n, and carbon concentrations with substrate temperature for InGaAs grown from TEG, TEI, and As$_4$ [after 52].

as well [53], suggesting that increasing the V/III ratio reduces the concentration of shallow acceptors. The origin of these shallow acceptors is unknown, but could be related to the presence of carbon, since carbon is an amphoteric impurity in InGaAs. Reduced carbon acceptor concentrations at high V/III ratio have been demonstrated for GaAs and AlGaAs grown from TEG [27,42]; therefore, it is reasonable to expect InGaAs grown from TEG to show some dependence of carbon content on V/III ratio also.

As discussed previously, switching speed does not appear to be a problem in MOMBE even though gaseous sources are used. InGaAs/InP quantum wells grown by MOMBE have been reported to yield very sharp intense luminescence peaks for well thicknesses down to 6Å [54]. In fact MOMBE material has produced some of the narrowest PL linewidths obtained to date, as shown in Figure 6.20.

High-purity In$_x$Ga$_{1-x}$As$_{1-y}$P$_y$ has also been demonstrated by MOMBE with 77 K mobilities of 2.2 to 6.7 $\times$ 10^4 cm^2/V s and $n = 10^{15} - 10^{16}$ cm^{-3} depending upon the composition [40]. As with InGaAs, low-temperature PL spectra are free of peaks

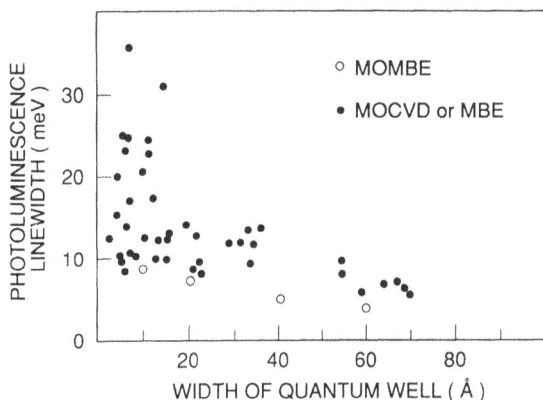

Figure 6.20 4.2 K PL peak linewidth *versus* quantum well width for InGaAs/InP quantum wells grown by various methods [after 78].

related to donor-to-acceptor pair recombination, and they exhibit linewidths of 3 to 5 meV, which are comparable to those predicted solely for alloy broadening, ~2 to 3 meV [40]. Little information is available regarding the influence of growth parameters on layer quality. The best results reported thus far have been achieved at the relatively high growth temperature of ~555°C [40]; therefore, it is quite probable that quaternary material quality improves with increasing substrate temperature as has been seen for InGaAs and InP.

6.5 DOPING

6.5.1 Elemental Sources: GSMBE

Elemental source dopants behave similarly in GSMBE as in MBE. Dopant elements are therefore chosen with regard both to their behavior in the lattice (i.e., activation energy) and to their incorporation behavior. The most common *n*-type dopants are Si and Sn, due to their small activation energies and to their high sticking coefficients. Si is particularly attractive because it shows almost no tendency to diffuse or segregate at normal growth temperatures. Si doping is limited, however, because of its amphoteric behavior to concentrations in the mid 10^{19} cm^{-3} range.

In contrast to Si, Sn offers the promise of higher carrier concentrations, up to 10^{20} cm^{-3}, due to its smaller tendency toward amphoteric behavior. However, in InP, elemental Sn does not incorporate in a straightforward manner. Before the steady-state doping level for any given flux is attained, ~0.5–1 monolayer of Sn must accumulate at the growth surface [55]. Once this coverage exists, the incorporated dopant concentration becomes linearly dependent on the Sn flux arriving at the surface. This not only makes the turn-on of the dopant profile diffuse, it also leads to problems in confining the dopant once the beam has been turned off. If InP growth is continued after switching off the Sn source, the Sn profile will not drop abruptly because of the Sn still left on the surface. Rather, the profile will decrease gradually until all of the segregated Sn is incorporated. If the subsequently grown material is one that does not exhibit severe segregation, however, all of the Sn at the growth surface will incorporate readily, producing a spike in the Sn profile. Neither of these cases is desirable for device structures where abrupt doping profiles are quite often essential. Thus, Sn does not appear to be a suitable dopant for structures in which buried n-type InP layers are required.

In contrast, Sn doping of InGaAs works quite well for growth temperatures in the 450°C to 500°C range, as shown in Figure 6.21. Even at Sn concentrations of $\sim 10^{21}$ cm^{-3}, only 10% of the arriving Sn is rejected at the growth surface [56]. As the Sn flux is reduced below this level, the percentage of rejected Sn decreases as well. While Sn concentrations of 10^{21} cm^{-3} can be achieved, self-compensation limits the useful doping limit to $\sim 10^{20}$ cm^{-3}.

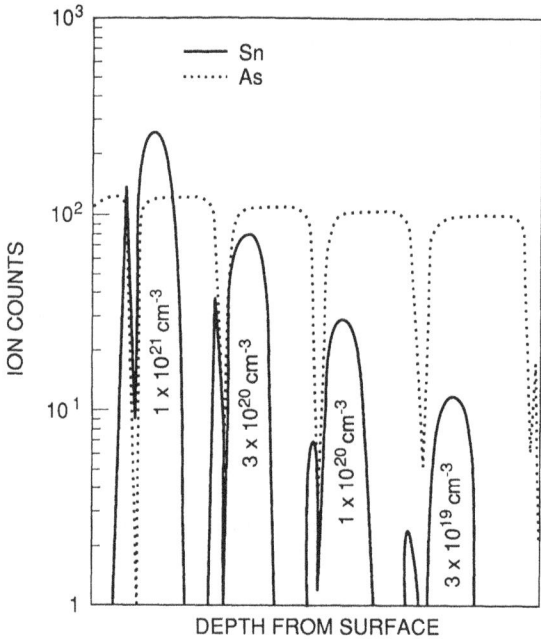

Figure 6.21 SIMS profile of Sn doping spikes in InGaAS grown at a temperature of 480°C. InGaAs layers are separated by 500Å of InP [after 56].

The most common *p*-type dopant in GSMBE is Be because of its near-unity sticking coefficient. In InP, Be concentrations up to $\sim 4 \times 10^{18}$ cm^{-3} have been achieved; however, dramatic redistribution of Be during growth occurs for concentrations above $\sim 2 \times 10^{18}$ cm^{-3}, even for growth temperatures as low as $\sim 450°C$ [57]. Similar behavior is observed for InGaAs, though the onset of diffusion occurs at a much higher Be concentration, typically ~ 20 times higher than for InP. The maximum Be concentration that can be incorporated in InGaAs without incurring enhanced diffusion is a strong function of temperature [17], as shown in Figure 6.22. As the growth temperature is reduced, the maximum hole concentration that can be achieved without catastrophic diffusion increases. Similar behavior is observed for Be [58] and Zn [59] doping in GaAs as well. The cause of this dramatic change in diffusion behavior with Be concentration is believed to be the formation of Be interstitials [60]. As the Be concentration increases, the growing layer becomes saturated with substitutional Be acceptors, leading to both rejection of Be at the growth front, and to incorporation of Be interstitials. The Be interstitials diffuse away from the doped layer toward the substate until the Be interstitial population reaches equilibrium with substitutional Be (Fig. 6.23). This causes a broadening of the profile in the direction of the substrate. Once the Be beam is turned off, the Be atoms at the surface then incorporate substitutionally in subsequently grown material. This causes the profile to broaden in the direction of the growth front. Given this mechanism it is not surprising that the onset of redistribution is quite temperature sensitive, since one would expect the diffusion coefficient and the equilibrium between

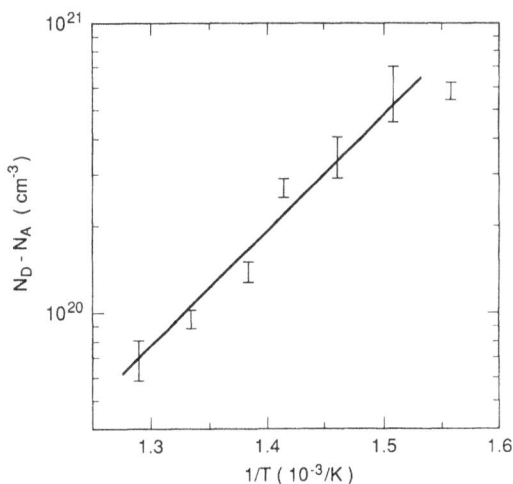

Figure 6.22 Maximum hole concentration obtained from Be doping of InGaAs as a function of growth temperature [after 17].

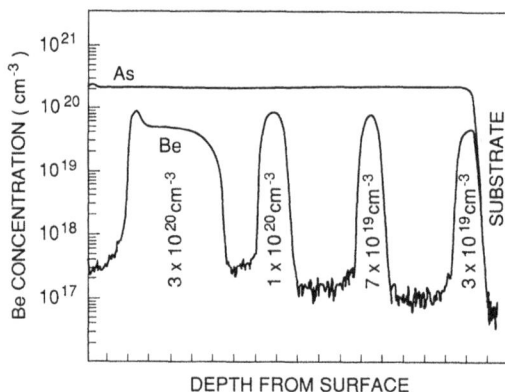

Figure 6.23 SIMS profile of Be doping spikes in InGaAs grown by GSMBE at a growth temperature of 450°C [after 47].

substitutional and interstitial populations to be strongly dependent on temperature as well.

In addition to shallow dopants, impurities that produce deep levels can also be introduced via elemental sources. In InP, Fe behaves as a deep acceptor and can be used to produce semi-insulating buffer or blocking layers. Resistivities as high as 10^9 Ω-cm have been obtained for Fe concentrations up to 10^{17} cm^{-3} [61]. As the Fe concentration is increased beyond 10^{17} cm^{-3}, the morphology is degraded because of the formation of FeP precipitates.

6.5.2 Elemental Sources: MOMBE

The incorporation of elemental dopants in MOMBE is similar to the behavior observed in GSMBE, suggesting little interaction between the elemental sources and adsorbed hydrocarbon species on the growth surface. Si incorporation in InP at concentrations up to 5×10^{19} cm^{-3} has been reported, with essentially all of the Si electrically active as donors [25]. In InGaAs, free carrier concentrations up to 1.5×10^{18} cm^{-3} have been obtained [62], and it is likely that even higher concentrations can be achieved if necessary. For both materials, the Si dopant profiles are quite abrupt and show no evidence of surface segregation or diffusion.

While the incorporation behavior at the growth surface does not appear to be influenced by hydrocarbons, the Si charge in the effusion cell does appear to react over time to form SiC [63]. This results in a reduction of Si flux for any given cell temperature. This degradation in flux can be forestalled somewhat by idling the cell at 100°C when not in use and by heating the cell to very high temperatures (~1550°C) under hydrogen prior to operation [64]. However, this procedure itself can shorten the lifetime of the heating elements and the thermocouple inside the effusion cell.

As in GSMBE, Sn doping of InP from elemental Sn is problematic because of surface segregation during growth [65]. This segregation occurs at growth temperatures as low as 480°C, resulting in smeared dopant profiles. While doping levels up to 7.5×10^{18} cm^{-3} can be achieved, the difficulty in controlling the dopant profile severely limits the use of this source for InP doping. As with Si, Sn effusion cells show some interaction with MO species in the growth chamber [63]. Unintentional Sn incorporation at concentrations up to 10^{18} cm^{-3} has been observed in InP grown in a system in which the Sn cell is heated, but kept shuttered. This anomalous doping behavior is believed to result from the information of alkyl Sn species on the Sn cell shutter. Thus, as with Si, great care must be taken if elemental Sn is to be used in a MOMBE environment.

As discussed for GSMBE, Be is the most commonly used p-type dopant in MOMBE growth of InP and InGaAs. In InP, Be concentrations up to 2×10^{19} cm^{-3} have been demonstrated, however, only 5 to 6×10^{18} cm^{-3} is electrically active [66]. In addition, severe segregation occurs at concentrations above $\sim 7 \times 10^{18}$ cm^{-3} [65]. Therefore, the 5 to 7×10^{18} cm^{-3} range appears to be a practical limit for Be doping of InP.

Much higher levels of electrical activation have been achieved for Be-doped InGaAs with 2×10^{20} cm^{-3} being the highest level reported to date [67]. It is not yet clear whether such high levels are well confined. Concentrations up to 4×10^{19} cm^{-3} appear to incorporate without inducing diffusion or segregation [62]. While degradation of Be effusion ovens in the MOMBE environment is not as well documented as for Si and Sn, anecdotal reports suggest that there may be some degradation of Be flux with time, as observed for Si. It is not yet clear whether such degradation may prove to be a limiting factor in p-type doping of InP compounds in MOMBE.

As in GSMBE, deep-level incorporation through the introduction of elemental Fe can be used to produce InP buffer layers with resistivities as high as 9×10^{7} $\Omega \cdot$cm for Fe concentrations of $\sim 9 \times 10^{18}$ cm^{-3} [68]. As the Fe concentration is increased beyond $\sim 2 \times 10^{19}$ cm^{-3}, the resistivity begins to decrease from precipitation of FeP.

6.5.3 Gaseous Sources: MOMBE

Due to the problems with dopant ovens, discussed in the previous section, the use of gaseous dopants in MOMBE is highly desirable. While some issues remain unchanged by the nature of the dopant source, such as maximum doping level allowed by the solid solubility limit at any given temperature, other concerns arise when gaseous sources are used in place of elemental dopants. In particular, the increased volatility of gaseous species often reduces the incorporation efficiency of dopants. Poor efficiency necessitates the use of high dopant fluxes which can degrade the growth surface or, in exterme cases, lead to the type of memory effects that have been observed for some dopants in MOCVD. Thus, the transition to gaseous dopants may also entail a transition to different dopant species from those commonly used in MBE. The following sections will discuss the relative tradeoffs of each of the commonly used gaseous dopants, keeping in mind the stringent requirements for growth of device quality structures.

6.5.3.1 n-type Dopants

While both SiH_4 [69] and Si_2H_6 [71] have been explored extensively for use in the GaAs/AlGaAs system, little has been reported for InP/InGaAs materials. Since InP is typically grown at lower temperatures than GaAs, one might expect poorer incorporation efficiency in this material. In fact, it has been reported that Si_2H_6 yields an incorporated donor concentration in InP approximately 10^{-5} or less than the amount of Si_2H_6 introduced into the growth chamber [71]. This suggests that Si_2H_6 is too stable to be used as a convenient dopant source for InP unless recracking, such as through the use of an ECR plasma, can be employed. Fortunately, other n-type sources, while not 100% efficient, do appear to decompose at a rate sufficient for growth of high-quality layers. Tetraethyltin (TESn) is particularly attractive because of the high uncompensated doping levels that can be obtained with Sn. Though the incorporation efficiency of this source is somewhat low [71], $\sim 10^{-4}$, it has been shown that the carrier concentration in either InP or InGaAs can be reproducibly varied over a wide range by varying the TESn flow rate [72]. The dopant concentration increases linearly with increasing TESn flux, which suggests that the incorporation efficiency is independent of Sn surface concentration. This linear variation combined with the absence of a strong temperature dependence [71] makes dopant control relatively easy using this source. Moreover, materials grown using TESn as the dopant source do not exhibit the diffuse dopant profiles normally observed with doping from elemental Sn [72–74]. The absence of dopant segregation allows for the fabrication of the abrupt junctions needed for most device applications. Thus, TESn appears to be the most promising candidate for gaseous n-type doping of InP materials.

6.5.3.2 p-type Dopants

Though Be is the most commonly used p-type dopant in GSMBE, its use in an all gas source process is problematic due to the extreme toxicity of gaseous Be compounds. Diethylberylium (DEBe) has been explored for doping of InP, however, the mobilities achieved were lower than can be obtained with elemental Be, possibly because of oxygen contamination [71]. Given the safety concerns associated with this compound, it is doubtful that DEBe will become a commonly used source.

Zn doping from diethylzinc (DEZn) has also been studied [75], though Zn typically exhibits a high diffusion coefficient that can lead to degradation of device performance during postgrowth processing [76]. The effects of diffusion can also be seen during growth when high Zn doping levels are employed. As shown in Figure 6.24, even at growth temperatures below ~500°C, Zn concentrations above ~1.5 × 10^{18} cm^{-3} lead to smearing of the dopant profile [75]. High concentrations of Zn at the growth surface also appear to reduce the InP incorporation rate, which leads to a decrease in the InP growth rate and reduction in the In content in InGaAs and InGaAsP. The utility of Zn doping in MOMBE, therefore, appears to be limited to levels ≤10^{18} cm^{-3}. While this level is suitable for some applications, device structures that require heavy p-type doping will necessitate the use of other dopant sources.

Perhaps the most promising dopant from the point of view of toxicity and dopant confinement is Mg. Biscyclopentadienylmagnesium (Cp$_2$Mg) has been successfully used in MOCVD but initial reports suggest that the electrical activation of Mg acceptors derived from this source is quite low [71]. When used in MOMBE growth of InP, this effect is probably due to the interaction of Mg with oxygen, which is transported to the surface via the gaseous sources. Thus, if Mg is to be used successfully as a p-type dopant in MOMBE, efforts will be required to reduce the surface oxygen concentration like those carried out for growth of AlGaAs by MOMBE. If the interaction of Mg with oxygen can be controlled, Cp$_2$Mg may prove to be a useful p-type dopant source. At the present time, however, none of the gaseous p-type dopant sources are capable of providing the same range of performance as elemental Be. Thus, elemental Be seems to be the most viable source for now.

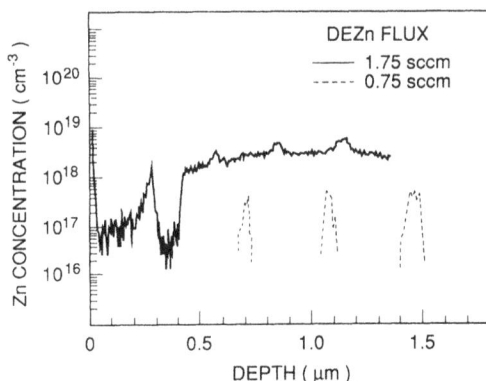

Figure 6.24 SIMS profiles of Zn delta-doped InP grown by MOMBE at a growth temperature of 490°C using two DEZn fluxes [after 75].

6.5.3.3 Deep-Level Dopants

To date, Fe doping of InP through the use of gaseous Fe sources has not been reported. Given the success of FE doping in MOCVD [76], however, it is quite likely that compounds such as ferrocene will eventually replace the elemental Fe source, discussed previously, in order to maintain an all-gas-source approach.

6.6 DEVICES

6.6.1 GSMBE

The ability to grow high-quality ternary and quaternary layers with abrupt interfaces has allowed the fabrication of a number of photonic devices from structures grown by GSMBE. Optical amplifiers [48] and lasers (MQW [48], DFB-MQW [48], DH [18], BH [18,48], and SCH [5,18,20]), in which the active regions are grown by GSMBE, all show performance comparable to the best achieved by MOCVD.

While most of the device efforts in GSMBE have focussed on optical devices, great promise has been demonstrated for electronic devices as well. InGaAs/InP heterojunction bipolar transistors with $f_t = 165$ GHz and $f_{max} = 100$ GHz have been reported [77]. The integration of high performance lasers with high-speed electronics is an area which will no doubt receive a great deal of attention in future GSMBE work.

6.6.2 MOMBE

A wide range of photonic devices have been fabricated by MOMBE, including various types of double heterostructure and quantum well lasers, as well as InP/InGaAs detectors [78]. The wide range of devices attests to the versatility of this technique. Generally, the performance of these devices is comparable to that which can be achieved by other techniques. For example, as shown in Figure 6.25, 1.55-μm InGaAsP/InP double heterostructure lasers exhibit threshold currents quite similar to those obtained with LPE or MOCVD [78]. Unlike the latter two techniques, MOMBE has been reported to produce excellent uniformity and wafer-to-wafer reproducibility.

The use of MOMBE for growth of electronic devices has not been explored as extensively as for photonic applications. The attainment of an InGaAs/InP two-dimensional electron gas (2DEG), with a 4.2 K mobility of 1.3×10^5 cm^2/V s [79], however, suggests that the material purity and interface quality are quite sufficient for fabrication of high-speed electronics.

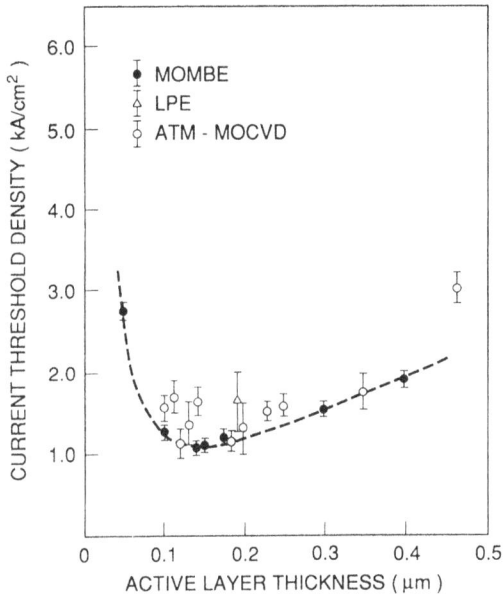

Figure 6.25 Variation of current threshold density *versus* active layer thickness for 1.55-μm DH lasers grown by various methods. Dashed line represents best results reported for lasers grown by LPE [after 78].

REFERENCES

1. See for example: S.J. Pearton and N.J. Shah in *High Speed Semiconductor Devices*, ed. S.M. Sze, John Wiley & Sons, 1990, p. 283.
2. M.B. Panish, *J. Electrochem. Soc.*, Vol. 127, 1980, p. 2730.
3. E. Veuhoff, W. Pletschen, P. Balk, and H. Lüth, *J. Cryst. Growth*, Vol. 55, 1981, p. 30.
4. W.T.Tsang, *Appl. Phys. Lett.*, Vol. 45, 1984, p. 1234.
5. M.B. Panish, H. Temkin, and S. Sumski, *J. Vac. Sci. Technol.*, Vol. B3, 1985, p. 657.
6. A.R. Calawa, *Appl. Phys. Lett.*, Vol. 38, 1981, p. 701.
7. M.B. Panish and S. Sumski, *J. Appl. Phys.*, Vol. 55, 1984, p. 3571.
8. D. Huet, M. Lambert, D. Bonnevie, and D. Defresne, *J. Vac. Sci. Tech.*, Vol. B3, 1985, p. 823.
9. D.G. Schlom, W.S. Lee, T. Ma, and J.S. Harris, Jr., *Proc. of the MBE Workshop*, 1988.
10. See for example: B.A. Joyce, P.J. Dobson, J.H. Neave, and J. Zhang in *Two-Dimensional Systems: Physics and New Devices*, Springer-Verlag, 1986.
11. J. Czerny and C.R. Abernathy, unpublished.
12. C.R. Abernathy, unpublished.
13. A. Robertson, Jr., unpublished.
14. T.H. Chiu, W.T. Tsang, J.E. Cunningham, and A. Robertson, Jr., *J. Appl. Phys.*, Vol. 63, 1987, p. 2302.

15. Y. Morishita, S. Maruno, M. Gofoda, Y. Nomura, and H. Ogata, *Appl. Phys. Lett.*, Vol. 53, 1988, p. 42.
16. A. Bensaoula, H. Hansen, H.C. Chen, J.T. Zborowski, K.D. Jamison, A. Ignatiev, and Y.K. Tao, *J. Cryst. Growth*, Vol. 105, 1990, p. 227.
17. R.A. Hamm, M.B. Panish, R.N. Nottenburg, Y.K. Chen, and D.A. Humphrey, *Appl. Phys. Lett.*, Vol. 54, 1989, p. 2586.
18. L. Goldstein, M. Lambert, B. Fernier, D. Bonnivie, C. Starck, and M. Boulou, *Inst. Phys. Conf. Ser. No. 91*, 1988, p. 211.
19. H. Asonen, K. Rakennus, K. Tappura, M. Hovinen, and M. Pessa, *J. Cryst. Growth*, Vol. 105, 1990, p. 101.
20. M.B. Panish, *Prog. in Cryst. Growth and Charact.*, Vol. 12, 1986, p. 1.
21. J.M. Kuo, unpublished.
22. A. Robertson, T.H. Chiu, W.T. Tsang, and J.E. Cunningham, *J. Appl. Phys.*, Vol. 64, 1988, p. 877.
23. A.J. Murrell, A.T.S. Wee, D.H. Fairbrother, N.K. Singh, J.S. Foord, G.J. Davies, and D.A. Andrews, *J. Cryst. Growth*, Vol. 105, 1990, p. 199.
24. H. Heinecke, B. Baur, R. Höger, and A. Miklis, *J. Cryst. Growth*, Vol. 105, 1990, p. 143.
25. J.L. Benchimol, F. Alaoui, Y. Gao, G. Le Roux, E.V.K. Rao, and F. Alexandre, *J. Cryst. Growth*, Vol. 105, 1990, p. 135.
26. N. Pütz, H. Heinecke, M. Heyen, P. Balk, M. Weyers, and H. Lüth, *J. Cryst. Growth*, Vol. 74, 1986, p. 292.
27. C.R. Abernathy, S.J. Pearton, F. Ren, W.S. Hobson, T.R. Fullowan, A. Katz, A.S. Jordan, and J. Kovalchick, *J. Cryst. Growth*, Vol. 105, 1990, p. 375.
28. O. Kayer, *J. Cryst. Growth*, vol. 107, 1991, p. 989.
29. D.A. Andrews, M.A.Z. Rejman-Greene, B. Wakefield, and G.J. Davies, *Appl. Phys. Lett.*, Vol. 53, 1988, p. 97.
30. T.H. Chiu, J.E. Cunningham, and A. Robertson, Jr., *J. Cryst. Growth*, Vol. 95, 1989, p. 136.
31. H. Heinecke, B. Baur, R. Höger, A. Miklis, and B. Jobst, *J. Cryst. Growth*, Vol. 107, 1991, p. 1062.
32. G.J. Davies and D.A. Andrews, *Chemtronics*, Vol. 3, 1988, p. 3.
33. W.T. Tsang, *J. Electron. Mat.*, Vol. 15, 1985, p. 235.
34. Y. Kawaguchi, H. Asahi, and H. Nagai, *Conf. on Solid State Devices*, Tokyo, 1986, p. 619.
35. F. Genova, G. Morello, S. Autore, and L. Gastaldi, *J. Cryst. Growth*, Vol. 107, 1991, p. 1065.
36. C.R. Abernathy, M.B. Panish, and R.A. Hamm, unpublished.
37. N. Kobayashi, J.L. Benchimol, F. Alexandre, and Y. Gao, *App. Phys. Lett.*, Vol. 51, 1987, p. 1907.
38. T.R. Martin and C.R. Whitehouse, *J. Cryst. Growth*, Vol. 105, 1990, p. 57.
39. F. Houzay, C. Guille, F. Barthe, and M. Van Rompay, *J. Cryst. Growth*, Vol. 95, 1989, p. 35.
40. W.T. Tsang, E.F. Schubert, T.H. Chiu, J.E. Cunningham, E.G. Burkhardt, J.A. Ditzenberger, and E. Agyekum, *Appl. Phys. Lett.*, Vol. 51, 1987, p. 761.
41. J.L. Benchimol, G. Le Roux, H. Thibierge, C. Daguet, F. Alexandre, and F. Brillouet, *J. Cryst. Growth*, Vol. 107, 1991, p. 978.
42. B.J. Lee, Y.M. Houng, J.N. Miller, and J.E. Turner, *J. Cryst. Growth*, Vol. 105, 1990, p. 168.
43. C.R. Abernathy, S.J. Pearton, F. Baiocchi, A.S. Jordan, D.A. Bohling, G.T. Muhr, and T. Ambrose, *J. Cryst. Growth*, in press.
44. V. Frese, G.K. Regel, H. Hardtdegen, A. Brauers, P. Balk, M. Hostalek, M. Lokai, L. Pohl, A. Miklis, and K. Werner, *J. Electron Matt.*, Vol. 19, 1990, p. 305.
45. C.R. Abernathy, A.S. Jordan, S.J. Pearton, W.S. Hobson, D.A. Bohling, and G.T. Muhr, *Appl. Phys. Lett.*, Vol. 56, 1990, p. 2654.
46. F. Ren, C.R. Abernathy, S.J. Pearton, T.R. Fullowan, J. Lothian, and A.S. Jordan, *Electron. Lett.*, Vol. 26, 1990, p. 724.

47. M. Lambert, A. Péralès, R. Vergnaud, and C. Starck, *J. Cryst. Growth*, Vol. 105, 1990, p. 97.

48. L. Goldstein, *J. Cryst. Growth*, Vol. 105, 1990, p. 93.

49. D.A. Bohling, private communication.

50. W.T. Tsang, A.H. Dayen, T.H. Chiu, J.E. Cunningham, E.F. Schubert, J.A. Ditzenberger, and J. Shah, *Appl. Phys. Lett.*, Vol. 49, 1986, p. 170.

51. J. F. Carlin, A. Rudra, R. Houdré, J.L. Stachli, and M. Ilegems, *J. Cryst. Growth*, Vol. 107, 1991, p. 1057.

52. J.L. Benchimol, F. Alexandre, Y. Gao, and F. Alaoui, *J. Cryst. Growth*, Vol. 95, 1989, p. 150.

53. C.R. Abernathy, M.B. Panish, and R.A. Hamm, unpublished results.

54. W.T. Tsang and E.F. Schubert, *Appl. Phys. Lett.*, Vol. 49, 1986, p. 220.

55. M.B. Panish, R.A. Hamm, and L.C. Hopkins, *Appl. Phys. Lett.*, Vol. 56, 1990, p.2301.

56. M.B. Panish, R.A. Hamm, L.C. Hopkins, and S.N.G. Chu, *Appl. Phys. Lett.*, Vol. 56, 1990, p. 1137.

57. M.B. Panish, R.A. Hamm, D. Ritter, H.S. Luftman, and C. Cotell, *J. Cryst. Growth*, in press.

58. P. Enquist, G.W. Wicks, L.F. Eastman, and C. Hitzman, *J. Appl. Phys.*, Vol. 58, 1985, p. 4130.

59. W.S. Hobson, S.J. Pearton, and A.S. Jordan, *Appl. Phys. Lett.*, Vol. 56, 1990, p. 1251.

60. See for example: H.C. Casey, *Atomic Diffusion in Semiconductors*, ed. Don Shaw, Plenum Press, 1973.

61. M. Lambert, L. Goldstein, F. Gaborit, A. Perales, and M. Boulou, "Sixth Int. Conf. on MBE San Diego, 1990," to be published in *J. Cryst. Growth*.

62. T. Uchida, T.K. Uchida, K. Mise, N. Yokouchi, F. Koyama, and K. Iga, *Jpn. J. Appl. Phys.*, Vol. 29, 1990, p. 1771.

63. P.J. Skevington, D.A. Andrews, and G.J. Davies, *J. Cryst. Growth*, Vol. 105, 1990, p. 371.

64. C.R. Abernathy, unpublished.

65. W.T. Tsang, B. Tell, J.A. Ditzenberger, and A.H. Dayem, *J. Appl. Phys.*, Vol. 60, 1986, p. 4182.

66. Y. Gao, S. Godefroy, J.L. Benchimol, F. Alaoui, F. Alexandre, and K. Rao, *Surf. and Interface Anal.*, Vol. 16, 1990, p. 36.

67. T.K. Uchida, T. Uchida, N. Yokouchi, F. Koyama, and K. Iga, *Jpn. J. Appl. Phys.*, Vol. 29, 1990, p.2146.

68. W.T. Tsang, A.S. Sudbo, L. Yang, R. Camarda, and R.E. Leibenguth, *Appl. Phys. Lett.*, Vol. 54, 1991, p. 2336.

69. H. Heinecke, K. Weiner, M. Weyers, H. Lüth and P. Balk, *J. Cryst. Growth*, Vol. 81, 1987, p. 270.

70. T. Fujii, A. Sandhu, H. Ando, Y. Kataoka, and H. Ishikawa, *Jpn. J. Appl. Phys.*, Vol. 29, 1990, p. 2386.

71. M. Meyers, J. Musolf, D. Marx, A. Kohl, and P. Balk, *J. Cryst. Growth*, Vol. 105, 1990, p. 383.

72. Y. Kawaguchi and K. Nakashima, *J. Cryst. Growth*, Vol. 95, 1989, p. 181.

73. J. Musolf, D. Marx, A. Kohl, M. Weyers, and P. Balk, *J. Cryst. Growth*, Vol. 107, 1991, p. 1043.

74. C.R. Abernathy, S.J. Pearton, F. Ren, and J. Song, *J. Cryst. Growth*, in press.

75. W.T. Tsang, F.S. Choa, N.T. Ha, *J. Elect. Mat.*, in press.

76. J. Long, V.G. Riggs, and W.D. Johnston, Jr., *J. Cryst. Growth*, Vol. 69, 1984, p. 10.

77. Y.K. Chen, A.F.J. Levi, R.N. Nottenburg, P.H. Beton, and M.B. Panish, *Appl. Phys. Lett.*, Vol. 55, 1989, p. 1789.

78. For a review see: W.T. Tsang, *J. Cryst. Growth*, Vol. 105, 1990, p. 1.

79. W.T. Tsang, *Appl. Phys. Lett.*, Vol. 49, 1986, p. 960.

Chapter 7
Ion Beam Processing of InP
and Related Materials

S.J. Pearton and U.K. Chakrabarti
AT&T Bell Laboratories

7.1 INTRODUCTION

The sophistication of the device processing of InP and related materials is at a somewhat earlier stage than GaAs, simply because there has been less work done on the former, and also because InP is a more difficult material to work with in terms of fragility, thermal stability, and sensitivity to ion-induced damage.

The binary nature of the InP lattice makes ion implantation damage removal and dopant activation steps more complex than for Si. The situation is even more complicated for ternary (InGaAs, AlInAs) and quaternary (InGaAsP) materials. Amorphization of the In-based III-V semiconductors should be avoided in order to achieve the best electrical activation [1,2], and these materials show poor recrystallization characteristics. This is basically due to the creation of regions with local deviations from stoichiometry resulting from the different displacement properties of the lattice constituents, which have unequal masses [3]. Incongruent evaporation of P or As from the sample surface upon high temperature annealing is also a severe problem for InP and related materials.

The dry etching of In-based semiconductors is attracting a great deal of interest because of the increasing importance of electronic and photonic devices based on these materials. Building these devices requires the development of controlled methods for fabricating mesa-type structures. For small-area devices, wet chemical etching is simply unable to achieve the kind of anisotropic removal of material necessary to maintain the target dimensions.

In this chapter we review recent work on carbon implantation and activation in InP, InGaAs and AlInAs, the rapid thermal annealing of these materials in a graphite susceptor that provides a reproducible group V partial pressure. The dry etching of In-based materials using CH_4/H_2 mixtures in either the conventional reactive ion etching (RIE) or electron cyclotron resonance (ECR) plasma etching modes is also discussed. Finally, the ion milling rates and damage depths in InP using low-energy (200–800 eV) Ar^+ ions are also reported.

7.2 ION IMPLANTATION FOR DOPING

In this section we will review the use of ion implantation for creating work-doped regions in In-based III-V materials. For all III-V compounds, implant activation is observed only when essentially all of the gross damage and most of the point defect damage has been removed by high temperature annealing. In other words, damage removal and activation are sequential processes.

7.2.1 InP

Implant damage in InP can consist of either amorphous layers or extended crystalline defects such as dislocations and stacking faults, depending on the ion, the dose, the dose rate, and the implant temperature [4,5]. In InP, amorphous layers recrystallize epitaxially during annealing at 150°C to 200°C, but the recrystallized layer is invariably highly defective, consisting of twins, stacking faults, and other defects. Local deviations in stoichiometry are created because of unequal recoil of In and P, and the diffusion lengths of the atoms are not long enough to prevent twinning at the amorphous-crystalline interface. Annealing at 400°C to 500°C leaves a high density of dislocation loops, and these are gradually reduced in concentration for higher temperature annealing. For a 300-keV Se^+ implant at room temperature, the critical dose for amorphization is $\sim 2 \times 10^{13}$ cm^{-3} [4], corresponding to a nuclear stopping energy density of 3×10^{20} keV cm^{-3}. This is on the order of 10 eV per molecule to cause amorphization, compared to a requirement of ~ 100 eV per molecule for Si. In III-Vs, damage accumulation and possible eventual amorphization are modelled using either a heterogeneous mechanism, in which individual damage clusters are considered to be amorphous and overlapping of these regions results in complete amorphization (heavy ions), or a homogeneous mechanism, in which the crystal becomes unstable and collapses to an amorphous state when the defect concentration reaches a critical value (light ions).

As the annealing temperature for implanted InP is increased, dopant atoms make a short-range diffusion to a substitutional site around 550°C, but optimum electrical activation is not achieved until ~ 750°C for donors and ~ 700°C for acceptors. For InP, n-type dopants give higher carrier concentrations (up to $\sim 2 \times 10^{14}$ cm^{-3}) compared to acceptors (up to $\sim 2 \times 10^{18}$ cm^{-3}) [1]. Above ~ 500°C, the InP surface must be protected against dissociation by P loss. Implantation at elevated temperatures ($\sim$150°C–180°C) of heavier ions (S, Se, Zn) prevents the formation of an amorphous layer, resulting in improvement in activation over the case where amorphization occurred [1]. The acceptors in InP display very rapid diffusion at high concentration and annealing temperatures. This diffusion can be reduced significantly by using a co-implant of P or As with the acceptor ion [6]. The purpose is twofold: to force the acceptor to occupy a substituted Ga site, and to create enough vacancies to ensure a high degree of substitutional over interstitial dopants. For room temperature implantation, amorphization of InP occurs at much lower doses than for GaAs or InAs. In GaAs for LN_2 temperature implantation temperature, the defect concentration is determined by the nuclear stopping energy deposition, and amorphization occurs by the heterogeneous mechanism. With increasing implantation, temperature-homogeneous nucleation becomes dominant. By contrast, in InP it appears that even at room temperature, heterogeneous defect nucleation plays a role, and less energy deposition is required to damage the material [7].

In crystalline semiconductors, unwanted channeling of implanted ions can be reduced or eliminated by preamorphizing the near-surface region. This is not feasible in III-Vs, and, in practice, to minimize both axial and planar channeling, InP wafers are oriented with an appropriate azimuthal or twist direction (the angle between the wafer flat and the direction of beam tilt) in addition to being tilted with respect to the beam direction. This is especially necessary for low-dose implants where control of the ion profile is critical for reproducible device performance. Channeling is generally minimized for tilt angles of $\geq$7°C and rotation angles of $\geq$30°. Increasing the ion dose creates damage in the uppermost InP region, which randomizes the beam and reduces subsequent channeling. Increasing the implant temperature can either increase or decrease channeling through two mechanisms: dynamic annealing, which reduces scattering out of channels, and increased lattice vibrations, which increases randomization of the beam. Implantation through a thin layer deposited on the sample surface can also reduce channeling.

A summary of the activation characteristics of the most common dopants implanted into InP (and also InGaAs and AlInAs) is given in Table 7.1. For Se implanted into InP and annealed at 750°C for 20 min we have measured substitutional fractions of 100% and 54%, respectively, for 200°C or room-temperature (RT) implantation at a dose of 2×10^{15} cm^{-2}. For a higher dose implant of 5×10^{15} cm^{-2} at 200°C, annealing at 750°C for 20 min produced a substitutional Se percentage of 86% [8]. The peak Se concentrations in these cases are well above 10^{20} cm^{-3}, whereas the electrically active donor concentration is an order of magnitude less. In other words, substitutionality of an implanted dopant in InP is a necessary but not sufficient condition for electrical activity. The likely reasons for the fact that the dopant can be substitutional but not electrically active is the complexing of the dopant with a native defect which actually causes the resulting center to be an acceptor, compensating the simple substitutional donors. A similar explanation is required to rationalize the low doping efficiency of acceptors in InP. In that case, native donor

Table 7.1

Summary of Implantation Characteristics in InP, InGaAs, and AlInAs.

	InP	InGaAs	AlInAs
Common Implant Species	Si(RT)	Si(RT)	Si(RT)
	Se(180°C)	Be(RT)	Be(RT)
	Be(RT)		
	Zn(180°C)		
Activation Temperature			
Donor	750°C	700°C	900°C
Acceptor	700°C	700°C	900°C
Activation Percentage			
Donors			
—low dose	80% 90%	50%	
—high dose	$N_m = 2 \times 10^{19}$ cm^{-3}	$N_m = 2 \times 10^{19}$ cm^{-3}	$N_m = 1 \times 10^{18}$ cm^3
Acceptors			
—low dose	60%	60%	80%
—high dose	$P_m = 2 \times 10^{18}$ cm^{-3}	$P_m = 2 \times 10^{18}$ cm^{-3}	$P_m = 2 \times 10^{18}$ cm^{-3}

defects act to compensate the acceptors. This behavior is observed in all of the III-V compounds for doping either by crystal growth, diffusion, or ion implantation.

Recently carbon has attracted a great deal of interest as an acceptor dopant in GaAs and AlGaAs, both because of its low diffusion coefficient relative to other acceptors and because of the recent demonstration of carbon doping levels above 10^{20} cm^{-3} during metalorganic molecular beam epitaxy (MOMBE) growth [9,10]. Surprisingly, very little is known about carbon in InP, InGaAs, and AlInAs. Undoped melt-grown InP is usually n-type because of S or Si contamination, and a similar situation holds for vapor phase growth techniques. The major residual acceptor in the latter case is usually Zn, and in no situation does carbon appear to be the dominant residual impurity in InP.

We have found carbon to act predominantly as a donor in all implanting and annealing conditions in InP [11]. Implantation of C + P at 200°C, followed by annealing at 700°C for 10 sec led to the best n-type doping levels (up to 2×10^{19} cm^{-3}). The effect of P co-implantation was to increase the net donor activity, which is consistent with increased occupation of In sites by carbon. In contrast, the effect of group III (Ga, B, or Al) co-implants was to decrease the net donor activity. Figure 7.1 shows the relative activation efficiency of C as a function of ion dose for 700°C, 10-sec anneals. The activation percentages are still quite low (~10% at a dose of 10^{14} cm^{-2}) and lower than for the more conventional n-type dopants Si and S. The

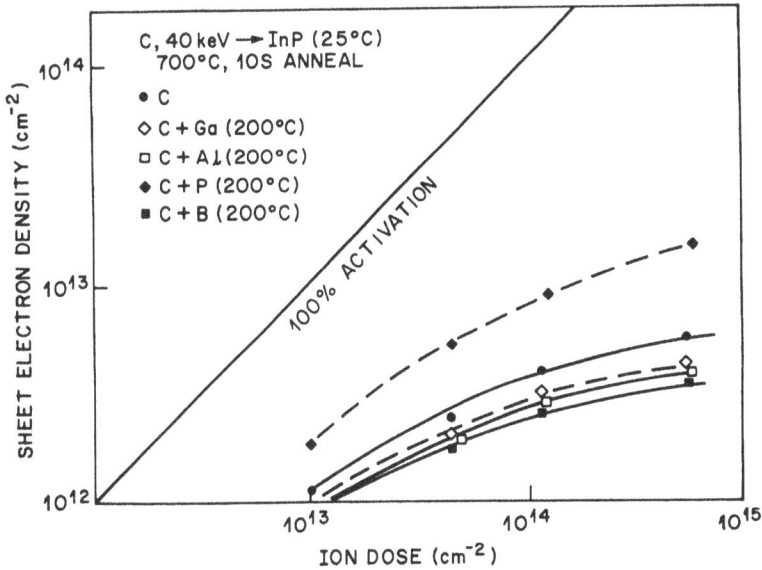

Figure 7.1 Sheet electron density obtained for C, C + P, C + Ga, C + B, or C + Al implantation into InP, as a function of ion dose for an annealing temperature of 700°C (10 sec). The dose of C and the group III or group V ions were the same in each case.

increase in sheet electron density with increasing dose is sublinear and shows saturation due to self-compensation. An upper limit for the diffusivity of C at 800°C of $\leq 3 \times 10^{14}$ cm^2 s^{-1} was established from SIMS measurements. While the modes of incorporation are obviously different, our experience with all other amphoteric dopants in III-Vs is that implantation leads to the same conductivity type as doping during growth. We expect, therefore, that carbon will behave predominantly as a donor in intentionally doped InP.

The temperatures needed for optimum implant activation in In-based materials are well above the temperature at which incongruent evaporation of the group V species occurs, and, therefore, some method of minimizing this loss of P or As is required. Each of the possible solutions have problems associated with them. For example, the commonly used dielectric encapsulants (SiN, SiO$_2$, AlN, PSG) either have adhesion problems at the high annealing temperatures involved, or else they induce considerable stress into the underlying wafer [12]. The use of an AsH$_3$ or PH$_3$ ambient presents considerable safety concerns, although in principle this is the ideal way to minimize surface degradation. A less effective, but more common method of reducing loss of the group V element is the so-called proximity annealing method, in which the wafer of interest is placed face-to-face with another wafer of the same type [13]. As the wafers are heated up, each begins to lose a small amount of the group V species, but an overpressure is created between the wafers which prevents further dissociation. This is ultimately an unsatisfying solution because it relies on a loss of P or As. Moreover, the edges of the wafers often show pitting of the surface, and movement of the wafers relative to each other leads to microscratches, which reduce device yield.

An enclosed graphite cavity provides a uniform heating environment for the wafer because of the high emissivity of the graphite, and reduces problems with slip formation. This type of cavity eliminates the need for a guard ring around the circumference of the wafer to prevent temperature nonuniformities, and it also provides protection of the surface at the same time. A more detailed discussion of the characteristics of annealing InP in these susceptors is given later in the chapter.

7.2.2 InGaAs

The activation characteristics of Si (donor) and Be (acceptor), the two most common implanted ions used for doping InGaAs, are also given in Table 7.1. In general, the behavior of implanted species in InGaAs is similar to that in InP. Much higher *n*-type doping can be achieved relative to *p*-type doping. Once again this is similar to the situation for doping during crystal growth. There has been little work to date on the regrowth and damage removal characteristics in implanted InGaAs, but all indications are that these are even more complex than in the binary case (InP or GaAs).

It would be particularly useful if carbon could be made to act predominantly as an acceptor in InGaAs so that it could be used as the base dopant in InGaAs / InP heterojunction bipolar transistors (HBT). This would obviate the need for trying to achieve high acceptor concentrations using low temperature growth of Be-doped layers.

For carbon-only implants into undoped InGaAs or AlInAs, we did not observe any activation under any annealing condition [14]. With co-implantation of Ga or As into InGaAs, or Al or As into AlInAs, we observed relatively high acceptor sheet densities for annealing at 700°C, 10 sec (InGaAs), or at 900°C, 10 sec (AlInAs). Higher hole concentrations were obtained for the group III species co-implant with C, as expected if its role is to enhance As-site occupation by the C. In both materials annealing above the optimum temperatures led to decreases in carrier density and mobility, possibly due to carbon self-compensation by site switching. The results can be understood if the C^+ ions are too light to create a high enough vacancy concentration for a significant density of the C to occupy substitutional sites where it is electrically active. The role of the heavier co-implant species is to create enough collisional mixing-induced displacements for the carbon to occupy the resultant vacancies upon annealing. When the co-implant species is Ga or Al, the additional effect is to enhance As site occupation by the C, whereas As co-implantation leads to more compensated material by forcing a relatively greater number of carbon ions to become substitutional in group III sites.

Figure 7.2 shows the relative activation efficiency of C co-implanted with Ga, As, Al, or Ar as a function of the ion dose for each species at annealing temperatures of 700°C (InGaAs) or 900°C (AlInAs). In each case a co-implant of carbon with the group III species leads to the best activation, while C + As gives the lowest net acceptor sheet densities. We also examined the diffusion behavior of C in both materials by SIMS measurements, and found an upper limit of 3.3×10^{-14} cm^2 s^{-1} at 800°C for substitutional carbon in InGaAs and AlInAs. These results give hope that high carbon doping of these materials can be achieved during crystal growth.

For Be implantation in InGaAs, of particular interest for many device applications, it is often observed that rapid thermal annealing leads to better activation than furnace annealing because of a reduction in the amount of Be out-diffusion [15]. As with InP, the degree of redistribution of the Be can be significantly reduced by co-implantation of a heavier species, preferably As.

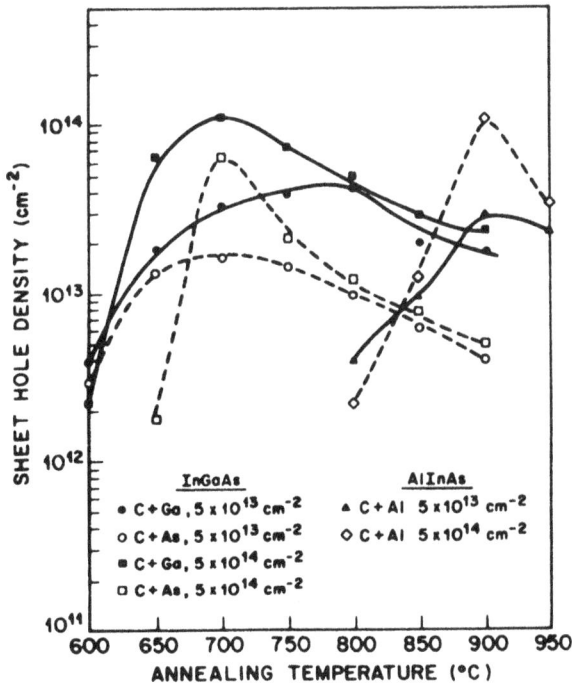

Figure 7.2 Sheet hole density obtained for C + Ga or C + As implantation into InGaAs or for C + Al implantation into AlInAs as a function of ion dose and annealing temperature (10-sec anneals).

7.2.3 AlInAs

The relatively wide bandgap and ternary nature of AlInAs makes it difficult to achieve high doping levels for either donor or acceptor implantation. The activation efficiency of implanted Be falls with increasing dose, from ~75% for 2×10^{13} cm^{-2} to <50% at 2×10^{14} cm^{-2}. Peak carrier concentrations are generally limited to ≤2 $\times 10^{18}$ cm^{-3} for both donor and acceptor implants. Once again, the acceptors can diffuse rapidly during annealing, and rapid thermal processing offers the advantage of less loss of the dopant to the surface.

7.2.4 InGaAsP

The electrical activation of implanted dopant impurities in InGaAsP follows similar trends to those in implanted epitaxial InP, with good activation of donor impurities and more moderate activation of implanted acceptors.

The most common implant species for the creation of n-type layers in InGaAsP alloys are Si and S, and for p-type layers, Be and Zn are the species of choice. The redistribution of the acceptors can be severe during the activation anneal. This diffusion can be reduced significantly by using a co-implant of P or As with the acceptor. Most information is available on implant activation in In$_{0.53}$Ga$_{0.47}$As [16–20]. For donor implants at low doses (<10^{13} cm^{-2}) the activation percentage is ~90%, while the maximum carrier concentration that can be achieved is ~2×10^{19} cm^{-3}. The electron mobility in these conditions is on the order of 2500 cm^2 V^{-1} s^{-1}, leading to a resistivity of 1.25×10^{-4} $\Omega \cdot$ cm for 1-μm thick layers. For acceptor implants at low doses (<10^{13} cm^{-2}), the activation percentage is ~60%, and the maximum hole concentration that can be achieved is ~4×10^{18} cm^{-3}. For this carrier density the hole mobility is typically around 280 cm^2 V^{-1} s^{-1}, yielding a resistivity of 5.5 $\times 10^{-2}$ $\Omega \cdot$ cm for 1-μm thick layers.

From limited data, we have observed similar results for the quaternaries with the 1.3-μm and 1.55-μm compositions. Donor activation is always more effective than acceptor activation, and similar maximum carrier concentrations to those obtained in In$_{0.53}$Ga$_{0.47}$As were observed.

The minimum resistivities obtainable in In$_{0.53}$Ga$_{0.47}$As epitaxial layers by donor implantation are of the order of 1.25×10^{-4} $\Omega \cdot$ cm for 1-μm thick layers, and 5.5 $\times 10^{-2}$ $\Omega \cdot$ cm for acceptor implantation into similar layers. Although there is much less data available for the quaternaries, similar results to these are obtained with both the 1.3-μm and 1.55-μm compositions.

7.2.5 InSb

InSb is a narrow-gap (0.18 eV) semiconductor with very high electron mobility, up to 10^5 cm^2 V^{-1} s^{-1} at 300K and 10^6 cm^2 V^{-1} s^{-1} at 77K [21]. In recent years it has attracted interest as a material for long-wavelength (5.3 μm) infrared detectors and emitters. It would be convenient to fabricate such detectors, and also high-speed electronic devices operating at cryogenic temperatures, in order to take advantage of the extremely high electron mobility by selective area implantation to define ohmic contacts. There have been early reports of *p-n* junction formation in InSb from the implantation of S, Zn, and Be, [22,23] and also of the production of *n*-type layers resulting from proton irradiation [24]. In this case the radiation-induced damage was assumed to have donor levels associated with it that caused the *n*-type doping. Solid-phase epitaxial regrowth of amorphized InSb has also been reported [25], with a damaging energy density of $(2-3) \times 10^{20}$ keV cm^{-3} necessary to create an initial amorphous layer. The regrowth of InSb is similar to that of GaAs in the sense that thick amorphous layers cannot be completely recrystallized because of the condensation of microcrystallites in regions of stoichiometrical excess or deficiency resulting from the initial unequal recoil of the In and Sb atoms [25].

One of the other widely studied aspects of implantation into InSb is the swelling of the implanted region for high dose, heavy ion implantation [26–28]. Step heights of up to 4200Å have been reported for room-temperature Cd implants at a dose of 2×10^{14} cm^{-2}. The surface in this case was found to be strongly oxidized and deficient in In [29]. Marked differences have been observed in damage removal rates between room-temperature implanted samples, and those implanted at elevated temperatures [21,30]. Amorphous layers did not form for Ar or S implantation at a substrate temperature of 200°C or greater, and surface swelling was not observed for ion masses less than C [26,30].

It is clear from past experience with high dose implants in InSb that amorphization should be avoided, simply because, as with all III-Vs, recrystallization is much less successful than the single-stage, solid-phase epitaxial regrowth of amorphized Si [31,5]. Figure 7.3 shows random and aligned Rutherford back scattering (RBS) spectra from InSb implanted with 100-keV Si^{+} ions at a dose of 10^{14} cm^{-2}. The minimum backscattering yield χ_{min} rose from 4% in the unimplanted sample to ~39% in the implanted InSb. Upon annealing at 300°C for 20 sec, the χ_{min} decreases to 11.5%, consistent with substantial removal of lattice disorder. Implantation of Be at the same dose created approximately half the disorder introduced by Si.

There is essentially no redistribution of implanted Be in InSb for rapid thermal anneal (RTA) at 400°C, 20 sec. At 450°C, some loss of Be to the surface is observed. For a dose of 10^{13} cm^{-2} Be^{+}, an activation percentage of 65% is obtained, corresponding to a peak hole density of ~10^{18} cm^{-3}. For implantation of Si^{+} we observed substantial diffusion of the Si during annealing at 300°C to 400°C for 20 sec. This diffusion was reduced for LN$_2$-implanted material subsequently annealed at 400°C [32].

We observed a number of similarities in both the activation and damage removal processes in implanted InSb to those in other III-V semiconductors. First, there is a considerable amount of damage after annealing, mainly in the form of

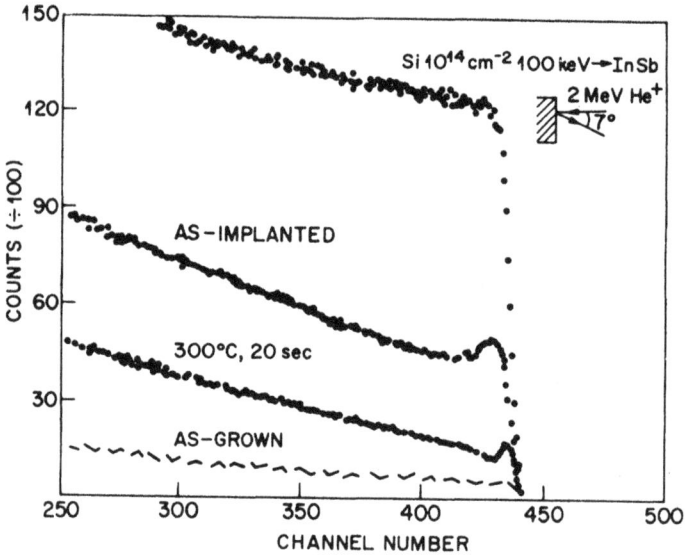

Figure 7.3 RBS spectra from InSb samples implanted nominally at room temperature with Si (100 keV, 10^{14} cm^{-2}) and subsequentially annealed at 300°C for 20 sec.

small dislocation loops extending to the end of the implanted ion range. In the case where amorphization is prevented, there is no detectable oxygen uptake into the implanted region. The second similarity occurs with the redistribution behavior of implanted Be, where there is significant loss of Be to the surface, with essentially no redistribution on the substrate side of the implanted profile. This appears to be typical of acceptor species behavior in III-V materials during capless RTA. The motion of Si in InSb appears to be unique among the materials we have investigated. In most III-V semiconductors (GaAs, InP, GaP, InAs), Si shows basically minimum redistribution during annealing. In the case of GaSb, there is some loss of the dopant in capless annealing, but in InSb there appears to be extensive in-diffusion during the implant step itself and during subsequent annealing. This can be reduced by lowering the substrate temperature during implantation.

7.2.6 InAs

InAs can be used to form nonalloyed contacts with other III-V materials, providing lower contact resistance than the usual alloyed contacts on the semiconductor. We were unable to amorphize InAs by implantation at temperatures above ~80°C, observing only small ($\leq$20%) increases in the backscattering yield from samples receiving even doses of 1×10^{15} cm^{-2} of either Si or Mg. The exact temperature above which the material cannot be made amorphous by the implantation of medium mass ions was not determined and, in any case, will be a function of the ion species, and 80°C is an upper limit for Si implants. This contrasts with values of ~150°C for InP and ~180°C for GaAs [1]. To try to examine the amorphous-to-crystalline transition in InAs, a sample was implanted with Si ions at two energies, 70 and 160 keV, each to a dose of 6×10^{14} cm^{-2}, with the sample held at 77K during the implantation. Figure 7.4 shows the RBS spectrum of this sample both before and after a 500°C, 10-sec anneal. The height of the aligned spectrum of the implanted section does not reach that of the random spectrum, indicating that, although it is highly disordered, the near-surface region was not made amorphous under our conditions. The annealing treatment produces only a modest improvement in crystalline quality, and the backscattering yield is still far above that of the as-grown sample.

For the introduction of a large amount of damage, but not full amorphization, annealing of InAs causes regrowth of the damaged region with a remnant buried dislocation band near the end of the projected range of the ion. There is no significant diffusion of implanted Si in InAs, but Mg and Be show substantial redistribution after annealing at 500°C and above, caused primarily by a loss of the acceptors to the surface, rather than an in-diffusion [31].

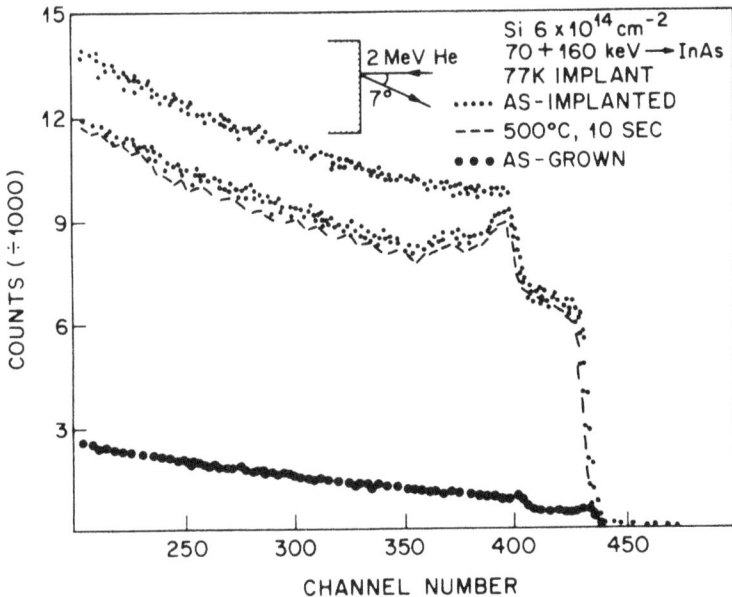

Figure 7.4 RBS spectra from InAs samples implanted at 77K with Si at two energies, 70 keV (6×10^{14} cm^{-1}) and 160 keV (6×10^{14} cm^{-2}) and subsequentially annealed at 500°C for 10 secs.

7.3 ION IMPLANTATION FOR ISOLATION

Bombardment with energetic ions can be used to create high-resistivity regions selectively in InP-based structures that already contain doped layers. This procedure is generally not as effective in In-containing materials as it is for GaAs and AlGaAs. There are two basic methods for creating high-resistivity layers [33]:

(a) Creating midgap, damage-related levels to compensate the free carriers in the material. This type of compensation is stable only to the temperature at which the damage-related levels are annealed out.

(b) Creating chemically-induced deep levels by implanting a species like Fe that has an electronic level in the middle of the bandgap. This type of compensation is thermally stable.

The deep level centers created in process (a) are due to elastic nuclear collisions of the implanted ion with nuclei or whole atoms of the solid, so-called nuclear stopping. The simplest defects created are Frenkel pairs, consisting of a vacancy and the displaced atom. More complex defects such as divacancies, bivacancies, *et cetera*, can be created, along with clusters of vacancies or interstitials with impurity atoms. Line defects, such as dislocations caused by an accumulation of point defects, are common in implanted material.

Since ions require a certain threshold energy for the production of damage, the maximum of the damage distribution is always closer to the surface than that of the ion profile [21]. The damage profile is generally obtained from Monte Carlo calculations of energy deposition given up in atomic stopping processes, or, equivalently, of vacancy production per incoming ion. To create a thick, high-resistance layer by ion bombardment, it is necessary to use multiple energy implants that overlap to form a uniformly damaged region.

Dielectric layers such as SiN_x or SiO_2, photoresist, or metal layers have all been successfully used as masks for implant isolation. The thickness of the mask should obviously be larger than the range, R, of the implanted ion to ensure that none penetrate into the semiconductor. The fraction stopping in the mask can be generally approximated from [34]

$$\frac{\phi_{mask}}{\phi} = \frac{1}{2}\{1 + erf[(\ell - R_p)/\sqrt{2}\,\Delta R_p]\} \tag{7.1}$$

where ℓ is the thickness of the mask and R_p and ΔR_p are the projected range and straggle of the ion in the masking material. In general for a thick mask, the fraction of the dose deposited beyond a depth, d, is [34]

$$\frac{\phi_d}{\phi} = \frac{1}{2}\,erfc[(d - R_p)/\sqrt{2}\,\Delta R_p] \tag{7.2}$$

For a masking effectiveness of 99.99%, the mask thickness should be R_p + 3.72 ΔR_p. One of the major potential problems with implant isolation is the lateral straggle ΔR_{pL}, which is always on the same order as the longitudinal straggle ΔR_p.

7.3.1 InP

The defects created in InP tend to pin the Fermi level in the upper half of the band-gap, and, therefore, the resistivity of *n*-type material can only be increased to the 10^3 to 10^4 $\Omega \cdot$ cm range in general. If beginning with *p*-type InP, a judicious choice of implant dose will move the Fermi level to midgap, but higher doses may cause a type conversion from *p*-type to *n*-type conductivity [1]. The resistivity in this case is again in the 10^3 to 10^4 $\Omega \cdot$ cm range [1,35–37].

Most of the early work on isolation of InP was done with protons, but it was found by Focht and Schwartz [35] that deuterium gives high resistivity material over a wider dose range than protons. From I-V measurements they were able to obtain the average resistivity of the implanted region of the InP samples [35]. Figure 7.5 shows the average resistivity of p^+-InP implanted with 200-keV deuterons as a function of the ion dose. Full compensation is obtained in this case for a dose of ~3 ×

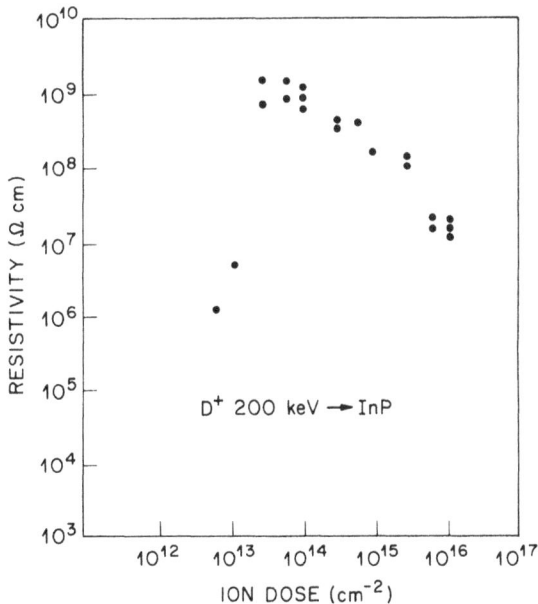

Figure 7.5 Average resistivity of initially p^+ InP implanted with 200 keV deuterons, as a function of ion dose [after 35].

10^{13} cm^{-2}. Figure 7.6 shows the measured average resistivity of He$^+$-ion-bombarded n- and p-type InP as a function of the ion dose. In n-type material, resistivities of only ~10^3 $\Omega \cdot$ cm were obtained, whereas in p-type material semi-insulating values could be achieved. Implantation with ^{3}He$^+$ appeared to give a broader range of doses over which highly resistive layers could be obtained. Some idea of the energy levels responsible for the isolation can be obtained by measuring the temperature dependence of the sample resistivity. In the case of H$^+$ or O$^+$ bombardment of n^+-InP, the activation energies are in the range 0.2 to 0.3 eV. The fact that these are not on the order of 0.7 eV (midgap of InP) explains the relatively low values of resistivity obtainable by ion bombardment of n-type InP [38].

The resistivity of InP may actually be lowered in some cases by ion bombardment [38,39]. It has been demonstrated in n-type InP that ion bombardment creates carriers in certain regions of the implanted ions profile, while in other regions there

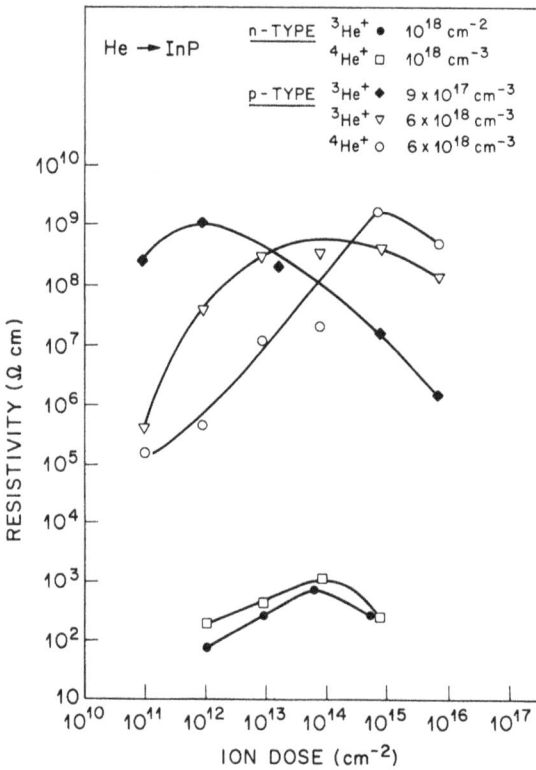

Figure 7.6 Average resistivity of n^+ and p^+ InP implanted with 200 to 275-keV ^{3}He$^+$ or ^{4}He$^+$ as a function of ion dose [after 35].

is strong compensation of the initial doping level of the material [40]. Similar results are obtained in InGaAs [39]. This phenomenon can be observed by implanting ions into semi-insulating InP (Fe) and then performing differential Hall measurements to obtain the "created" carrier profiles [38]. Figure 7.7 shows the carrier profiles in InP (Fe) after the implantation of 100-keV O^+ ions at a dose of 10^{14} cm^{-2} and annealing at 300°C, where the conductivity does not change significantly with dose. The profile was essentially identical at 77K, indicating that the ion-induced carriers are relatively shallow donor levels. Indeed, Kamiya et al. measured an activation energy of 15 meV for these centers [39]. The shape and depth of the profile coincide with the region of excess In created by the unequal recoil of In and P atoms under bombardment from O. It has previously been reported that P vacancies behave as shallow donors because of the degenerately doped behavior of ohmic contacts formed on InP from which P has been preferentially lost [41]. Since the ion-induced carriers are also created in InGaAs, it may be more likely that In interstitial-type defects behave as shallow donors in In-based materials. This is consistent with the fact that the ion-induced carrier profiles in InGaAs are relatively shallower than in InP, corresponding to the closer masses of the lattice constituents, and, hence, the less-pronounced regions of stoichiometric excess created by ion bombardment.

The bombardment-induced carrier profiles are a strong function of the ion dose. Figure 7.8 shows the carrier profile in InP (Fe) implanted with 100-keV O^+ ions to a dose of 5×10^{14} cm^{-2}. In this case, the carriers, after annealing at 300°C, occur near the back of the oxygen distribution. Hall measurements again indicated n-type conductivity. It has been postulated that the end-of-range disorder created by the O^+ ions basically consists of a slight excess of donor-type defects over acceptors, and that at the higher doses used here this effect becomes more significant [38]. By contrast, in the region of In excess there may be competition between the creation of In-related donors and implant-induced deep levels that compensate these donors. At higher doses there appears to be a balance between carrier creation and deep-level compensation up to the maximum of the nuclear damage profile, while there is a buildup of damage-related conductivity at the end of range. This concept is consistent with the result that the 0.4 level of Macrander and Schwartz was not observed at the end of range [42]. Quantitatively similar results are obtained with B^+ implantation, and therefore the carrier creation is not related to O itself. One of the key points to emphasize from this spatial variation in carrier removal and production in InP is that overlapping multiple-energy implants should be used to create high-resistance regions rather than relying on a single-energy bombardment. This approach will distribute the deep-level compensation centers associated with the end-of-range damage throughout the layer being implanted.

In summary, carriers may actually be created in InP by implantation damage. This reaction is dose dependent, and the regions in which conductivity is induced correspond for low doses to the excess In caused by nonequal recoil of the lattice

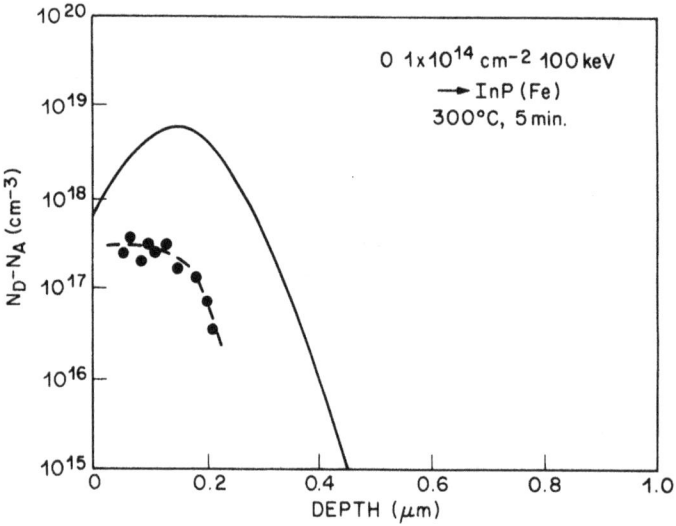

Figure 7.7 Carrier profile in initially semi-insulating InP after implantation with 100-keV O$^+$ ions (10^{14} cm^{-2}) and annealing at 300°C. The solid line is the ion profile while the dots represent the carrier profile.

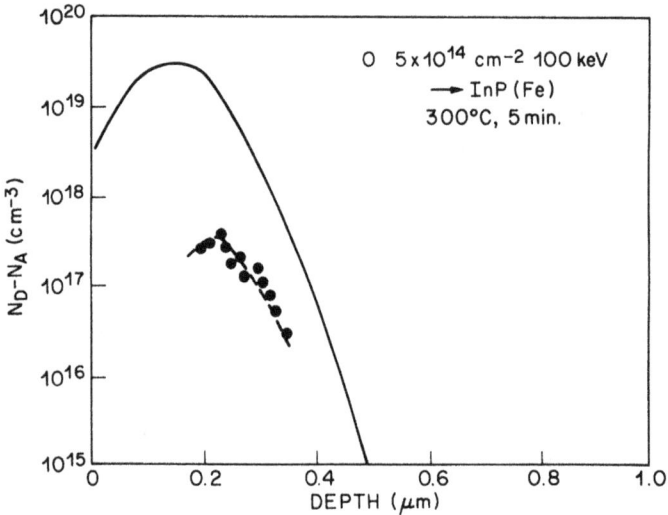

Figure 7.8 Carrier profile in initially semi-insulating InP after implantation with 100-keV O$^+$ ions (5×10^{14} cm^{-2}) and annealing at 300°C. The solid line is the ion profile, while the dots represent the carrier profile.

constituents. For higher doses, the induced carriers appear to be associated with end-of-range disorder. The maximum resistivities achievable in In-based materials by implant isolation are given in Table 7.2.

Table 7.2
Maximum Resistivities Typically Achieved by Ion Bombardment of InP, InGaAs, and AlInAs.

Material	Resistivity ($\Omega \cdot cm$)
p-InP	$>10^7$
n-InP	$10^3 - 10^4$
p-InGaAs	500
n-InGaAs	50
p-AlInAs	$>10^7$
n-AlInAs	$>10^7$

7.3.2 InGaAs

Since the bandgap of InGaAs is only 0.74 eV, the intrinsic resistivity of this material is small, approximately 10^3 $\Omega \cdot cm$ [43–45]. It is therefore to be expected that ion bombardment will not produce adequate isolation for InGaAs-based electronic devices. For some optical device applications, however, resistivities of this order are acceptable. In n-type InGaAs, maximum resistivities of only a few ohm-centimeters were reported by a number of authors after ion bombardment. In p-type material, a maximum resistivity of 580 $\Omega \cdot cm$ has been reported after B implantation at a dose of 10^{13} cm^{-2} [43,45].

For conventional ion doses of O$^+$ and H$^+$, even lower resistivities are obtained. Figure 7.9 shows the sheet resistance of 0.5-μm-thick n^+-InGaAs layers grown on semi-insulating InP substrates, as a function of the H$^+$ or O$^+$ ion dose and the subsequent annealing temperature [38]. In the case of O$^+$ implantation, the maximum sheet resistance occurs after annealing the higher-dose implants at ~300°C, with complete removal of the effect by 500°C. In the case of H$^+$ implantation, there was actually a reduction in sheet resistance from the unimplanted value (~100 $\Omega/\square$) for anneals above 200°C. The material retained n-type conductivity for all of the implant conditions, although the increase in sheet resistance in the O-implanted samples was predominantly accounted for by a decrease in the carrier density rather than a decrease in carrier mobility. Tell and Brown-Goebeler [47] reported similar results for B$^+$ implantation into undoped InGaAs on InP (Fe) substrates. The maximum measured reduction in B-implanted samples was only a factor of four at a dose of 10^{13}

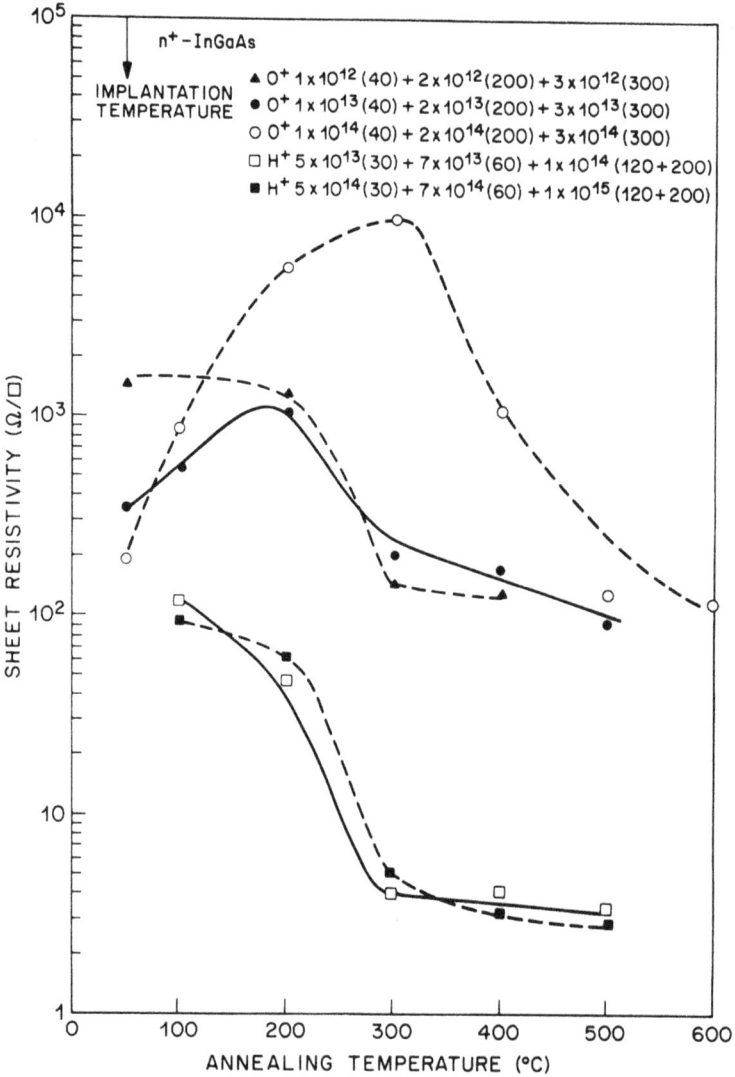

Figure 7.9 Sheet resistance of n^+ InGaAs after H^+ or O^+ implantation at various doses, as a function of postimplant annealing temperature. The ion's dose is given first, followed by the ion energy.

cm^{-2}, while the mobility monotonically decreased by about three orders of magnitude over the dose range of 10^{11} to 10^{15} cm^{-2}. For protons implanted at 10^{13} to 10^{14} cm^{-2}, there was actually an increase in sheet carrier concentration, together with a decrease in mobility, leading to an overall decrease in sheet resistance. The temperature dependence of the sheet resistance of bombarded InGaAs, showed activation energies in the range of 0.20 to 0.27 eV for O$^+$, or B$^+$ ion doses of 10^{13} to 10^{14} cm^{-2}.

Qualitatively similar results are obtained for ion-bombarded p^+-InGaAs [38]. For material doped with Be to a density of 10^{18} cm^{-3}, oxygen implantation at multiple energies and doses of $\sim 10^{14}$ cm^{-2} produced a sheet resistance of $10^3 \Omega/\square$, increased from an as-grown value of 200 $\Omega/\square$. Annealing at 400°C led to a maximum value of $\sim 10^5 \Omega/\square$, corresponding to a resistivity of 5 $\Omega \cdot$ cm. The activation energy for this resistance was 0.28 eV. Rao *et al.* [43] investigated the use of very low dose $(10^{10}$–10^{12} cm$^{-2})$ H$^+$, He$^+$, and B$^+$ implants into p-type ($p = 5 \times 10^{16}$ cm^{-3}) InGaAs. Figure 7.10 shows the variation of the resistivity of the material with ion concentration (in cm^{-3}) for these three species. The resistivity increases appear to be mainly due to the introduction of compensating donor levels associated with the implant damage. For all of these species, type conversion (to n-type conductivity) was observed as the ion dose was increased. The resistivity then decreased with a large increase in mobility (from typically 80 to 100 cm^2 V^{-1} s^{-1} to 2000 to 4000 cm^2 V^{-1} s^{-1}). An example for proton implants is shown in Figure 7.11 [43]. For

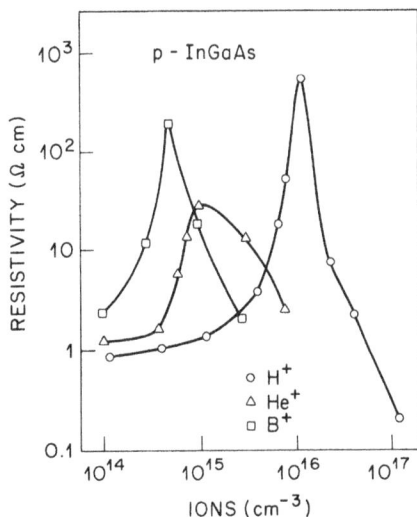

Figure 7.10 Resistivity of p-type InGaAs (5×10^{16} cm^{-3}) as a function of H$^+$, He$^+$, or B$^+$ ion concentration (in cm^{-3}) implanted into the material [after 43].

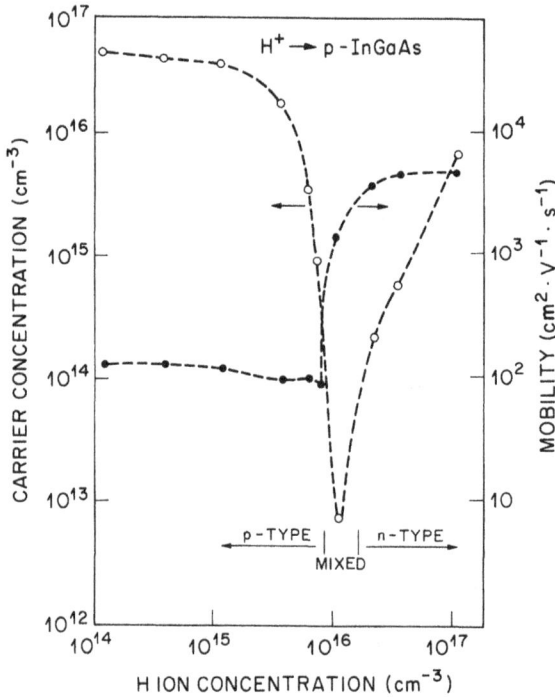

Figure 7.11 Proton concentration dependence of carrier concentration in *p*-InGaAs (5×10^{16} cm^{-3}) [after 43].

proton implants, the increased resistivity was stable only to 200°C while for boron implants the resistivity decreased gradually over the temperature range 225 to 400°C. From photoluminescence results it was concluded that the bombardment-induced donors had energy levels at 43 to 50 meV below the conduction band [43]. Recombination at damage sites reduces the minority carrier lifetime from 2 to 4 ps in as-grown InGaAs to less than 1 ps [48]. Tell and Brown-Goebeler [47] observed the introduction of significant concentrations of bombardment-induced donors in undoped *n*-type InGaAs. These caused a reduction in mobility, and an increase in carrier concentration. Figure 7.12 shows the ion-induced carrier profiles after 400- or 800-eV Be implantation at 10^{12} cm^{-2} [47]. These samples were not annealed, so the carriers are due to damage-related centers only. Note that these profiles are in close correlation with the profiles of energy deposition by nuclear stopping. Similar results were obtained for implantation into semi-insulating InP.

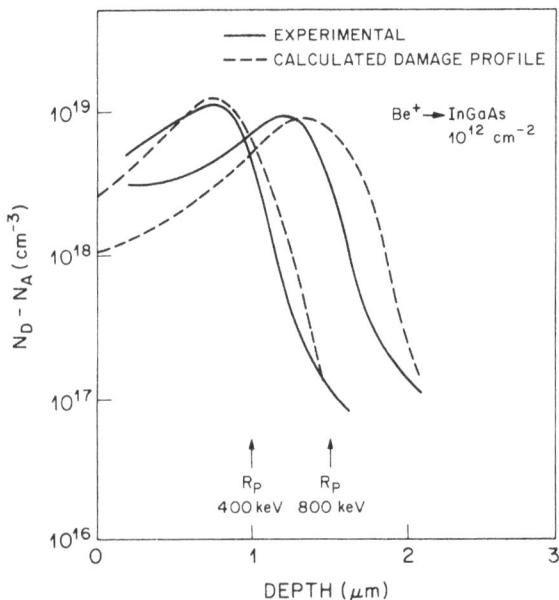

Figure 7.12 Net donor concentration profiles in InGaAs implanted with 400- or 800-keV Be$^+$ ions at a dose of 10^{12} cm^{-2} prior to annealing. Also shown are the nuclear energy loss profiles [after 47].

7.3.3 AlInAs

The wide bandgap (1.47 eV) material, AlInAs, is gaining increasing interest for use as the emitter in heterojunction bipolar transistors (HBT), and the quaternary system AlInGaAs can be grown lattice matched to InP [49,50]. Tell *et al.* [51] have investigated the sheet resistance, Hall mobility, and sheet carrier density as a function of the He ion dose for In$_{0.52}$Al$_{0.48}$As, In$_{0.52}$Ga$_{0.13}$Al$_{0.35}$As, In$_{0.53}$Ga$_{0.26}$Al$_{0.21}$As, and In$_{0.53}$Ga$_{0.47}$As. The Si-doped ($n \approx 10^{18}$ cm^{-3}) layers were grown on InP (Fe) substrates, and implanted with 400-keV He$^+$ ions to doses between 5×10^{12} and 2.5×10^{14} cm^{-2}. Figure 7.13 shows the sheet resistance of the material as a function of ion dose. The Al-containing compounds reach high resistances ($>10^7$ $\Omega/\square$), while the InGaAs is limited to 2000 $\Omega/\square$ (0.2 $\Omega \cdot$ cm). At the highest doses, the sheet resistance in the Al-based materials is stable to annealing at 425°C [51]. These changes in resistance were caused by large decreases in the sheet carrier densities, and smaller decreases in mobility.

In electronic device applications it is necessary to achieve very good isolation of the conducting layers in the layer structure. AlInAs-InGaAs has a number of advantages over the more established AlGaAs-GaAs heterostructure, including the

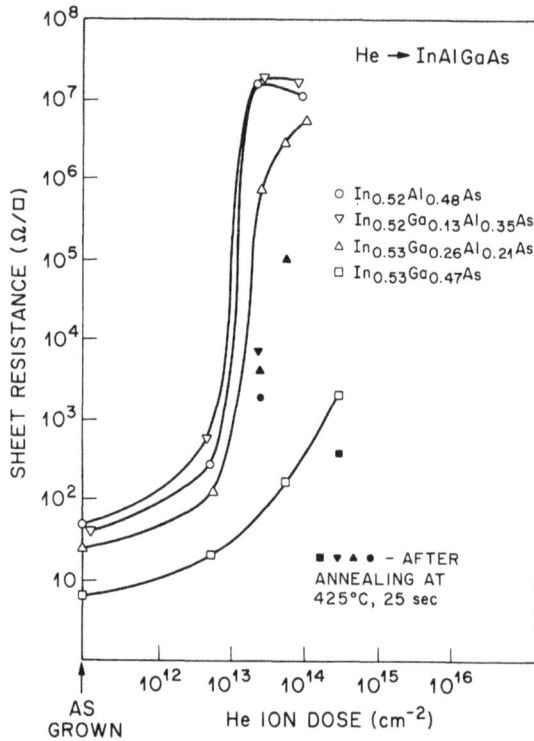

Figure 7.13 Sheet resistance of InGaAlAs samples of different composition as a function of 400-keV H^+ ion dose [after 51].

higher electron mobility and velocity in InGaAs compared with GaAs, the lower surface recombination velocities for In-based materials, and the exceptionally large conduction band discontinuity between AlInAs and InGaAs ($\sim$0.5 eV). These advantages promise HBTs with performance superior to comparable GaAs-AlGaAs devices, and, recently, AlInAs/InGaAs HBTs demonstrating near-ideal emitter scaling characteristics have been reported [52]. In addition, the AlGaInAs system is seen as a possible replacement for InGaAsP alloys in lasers and photodetectors because the concentrations of only two of its constituents (Al and Ga) need to be adjusted to vary the bandgap while retaining lattice matching, whereas in the InGaAsP system the ratio of all four of its constituents must be altered [53].

It might be expected that, analogous to the situation in AlGaAs, oxygen incorporation into n-type AlInAs would create compensated, high-resistivity material. This type of compensation in AlGaAs is thermally stable. However, the implantation of O^+ into n^+-AlInAs layers, followed by annealing at various temperatures, showed that oxygen does not produce chemical compensation in n^+-AlInAs [54]. Figure 7.14 shows the evolution of the sheet resistance of O-implanted material as a function of

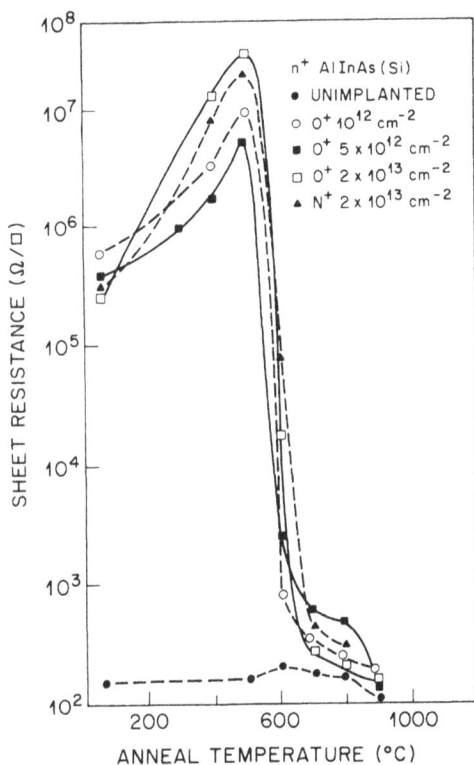

Figure 7.14 Sheet resistance of n^+ AlInAs layer implanted with O at various doses, as a function of postimplant annealing temperature.

both ion dose and annealing temperature. The results are very similar to those obtained for oxygen-implanted GaAs, and it is reasonable to assume the explanations forwarded for GaAs can also be applied to AlInAs. For the doses used in device applications, generally, no line defects are introduced. Specifically relating to implant isolation, there appears to be a high concentration of damage sites introduced by irradiation with any ion. Each implanted ion is capable of removing many electrons from the conduction band (or holes from the valence band in p-type material) of the AlInAs by trapping them at defects created along the ion path. The as-implanted resistivity of the material is orders of magnitude higher than its unimplanted value because of this carrier removal. When the implant dose is high enough, essentially all of the electrons from the dopant are trapped, but they can hop from damage site to damage site, leading to an intermediate value of sheet resistivity (10^5–10^6 $\Omega/\square$). As the material is annealed at progressively higher temperatures, the trap density is of roughly the same order as the carrier density. As the annealing temperature is increased still further, the point-defect concentration is reduced below the carrier concentration, and so electrons (or holes in p-type material) are returned to the conduction (valence) band and the resistivity of the material gradually returns to its initial, unimplanted value.

The activation energies obtained from the temperature dependence of the sheet resistance are 75 meV (hopping conduction) after 400°C annealing and 0.68 eV for 500°C annealing (midgap, damage-related level).

7.3.4 InGaAsP

The bandgap of $In_xGa_xAs_yP_{1-y}$ varies from 0.35 eV (InAs) to 2.26 eV (GaP), and this quaternary is used to produce lasers operating in the wavelength range of 1.3 to 1.6 μm. Lasers emitting at 1.3 μm and 1.55 μm are of great technological interest because of their application in optical fiber communications [55]. For 1.3-μm lasers, $x = 0.28$ and $y = 0.6$, while at 1.6 μm the Ga fraction x, is 0.22 and the As fraction is 0.7. These layers are grown lattice matched to InP substrates. For electronic devices, the most important form is the ternary $In_{0.53}Ga_{0.47}As$, which is also lattice matched to InP. Because InGaAs has a rather small bandgap (0.74 eV at 300K), its intrinsic resistivity is also rather small, approximately 10^3 $\Omega \cdot$cm. Therefore, it is to be expected that ion bombardment would not produce adequate isolation for InGaAs-based electronic devices. For some optical device applications, however, resistivities of this order are acceptable.

Very little information is available on the sheet resistivities obtainable by ion bombardment of the 1.3- and 1.55-μm compositions of $In_{1-x}Ga_xAs_yP_{1-y}$. Our own limited data shows that for both compositions and for both n- and p-type material, the maximum resistivities are around 500 to 800 $\Omega \cdot$cm. This would be expected because of the relatively small bandgap (0.95 eV for 1.33 μm and 0.8 eV for 1.55 μm). Even if all of the extrinsic carriers are removed by trapping at damage-induced deep levels, then the intrinsic carrier generation at room temperature is still significant. Similarly, pure InAs suffers from the same problem because of its small bandgap. In n-type InGaAs, maximum resistivities of only a few ohm-centimeters (up to 5 $\Omega \cdot$cm) have been reported after ion bombardment. In p-type material a maximum resistivity of 580 $\Omega \cdot$cm has been reported after B^+ implantation at a dose of 10^{11} cm^{-2} [43]. In contrast, in the wide bandgap GaP, both n- and p-type material display large specific resistivities with ion bombardment. Values of 10^8 $\Omega \cdot$cm for proton doses of 10^{14} cm^{-2}, and 10^{14} $\Omega \cdot$cm at a dose of 4×10^{14} cm^{-2} have been reported [56]. Above this dose the resistivity decreases, due presumably to hopping conduction. This compensation was stable to at least 400°C, where a resistivity of $>10^{10}$ $\Omega \cdot$cm was obtained after proton implantation at 10^{14} cm^{-2}. These results are expected because midgap levels are created by the ion implantation and because of the large bandgap of GaP. This leads to its large intrinsic resistivity.

In all cases of ion bombardment into InGaAsP alloys, the compensation mechanism for the increase in resistivity is the creation of damage-related deep levels which trap the free carriers in the material. This type of compensation is stable only to the temperature at which these damage-related levels are annealed out. It is possible to create chemically-induced deep levels by implanting a species that has an electronic level in the middle of the bandgap. This type of compensation usually requires the implanted species to be substitutional, and, therefore, annealing is required to promote the ion onto a substitutional site. In the absence of out-diffusion or precipitation of this species, the compensation effect is thermally stable. Examples of such species are Fe and Ti, but these do not overcome the low intrinsic resistivity of InGaAsP alloys. Another significant problem with trying to create highly resistive layers in InGaAs is the creation of shallow donor levels associated with the implant damage [38,39]. Because of these levels, a conversion of lightly *p*-type InGaAs to *n*-type conductivity is often observed upon ion bombardment. From photoluminescence measurements it was concluded that the bombardment-induced donors had energy levels at 43 to 50 meV below the conduction band [43]. The profile of these shallow donors correlates closely with the profiles of energy deposition from nuclear stopping of the implanted ions.

In summary, in terms of creating high-resistivity layers in $In_{1-x}Ga_xAs_yP_{1-y}$ alloys by ion bombardment, the relatively small bandgaps for low Ga contents effectively preclude achieving the kind of resistivities obtained in GaAs, AlGaAs, and GaP. A further problem is the creation of shallow donor levels as a result of the ion bombardment.

7.3.5 InSb

Since the bandgap of InSb is only 0.18 eV, the intrinsic resistivity of this material is too low for the achievement of adequate device isolation. There have also been reports of donors being created by proton implantation, as with InP and InGaAs.

7.3.6 InAs

In a similar fashion to that of InSb, the small bandgap of InAs (0.36 eV) precludes the achievement of high resistivities through ion bombardment.

7.3.7 Thermally Stable Isolation Schemes

Apart from the creation of midgap, damage-related levels that compensate the free carriers in a bombarded region, it is also possible to achieve high resistivity behavior by creating chemically-induced deep levels by implantation of a species that has an electronic level in the middle of the bandgap. This type of compensation usually requires the implanted species to be substitutional, and, therefore, annealing is required to promote the ion onto a substitutional site. In the absence of out-diffusion or precipitation of this species, the compensation effect is thermally stable.

In this section we detail some specific examples of chemically-induced isolation in In-based semiconductors. As described above, this requires implantation of species that create chemical deep levels in the particular material. In many respects this mechanism is complementary to the damage-induced method, because the latter is thermally stable only up to the temperature at which the damage is annealed out, which in most III-V materials is also the temperature at which chemically-active species become substitutional. This is not simply a coincidence, because the temperature at which damage is removed, or equivalently there is short-range motion of vacancies and interstitials, will also be the temperature at which interstitial impurities can move onto a nearly substitutional site. The annealing-temperature dependence of the sheet resistivity in material implanted with a deep-level impurity is similar to that displayed by material implanted with an inert species for temperatures up to the point at which the damage begins to anneal out. For higher temperatures, the conductivity in the damaged-only material returns toward its unimplanted value. It remains low in the chemically active implanted material. This effect is shown schematically in Figure 7.15.

It was first shown by Donnelly and Hurwitz [37] that implantation of Fe into n-type InP led to the formation of highly resistive material after annealing to promote the Fe onto a substitutional site. Iron has been used to produce semi-insulating regions for isolation in a variety of lightwave devices. For example, Chevy et al. [57]

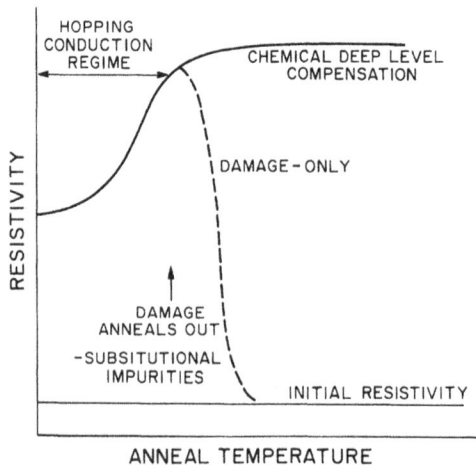

Figure 7.15 Schematic of the annealing temperature dependence of sheet resistance in a III-V material implanted with a species that creates a chemical deep level.

have fabricated semiconductor-insulator-semiconductor n^+-SI-n^+-InP structures using Fe implantation into n^+-InP substrates. These structures are used for current confinement in channeled substrate buried heterostructure (CSBH) lasers. Chevy *et al.* [57] have observed apparent Fe solubilities of $>7 \times 10^{17}$ cm^{-3} in n^+-InP for implant temperatures of $\sim 200°C$. This value exceeds the reported solubility limit of Fe in undoped InP at 700°C, which is found to be $\sim 10^{17}$ cm^{-3}. The Fe acceptor was found to be at a level of $E_C - (0.60–0.65)$ eV in the implanted and annealed material [57].

Following an activation anneal at 700°C, room-temperature implants of Fe develop two peaks in their atomic profiles, located at 0.8 R_p and $R_p + \Delta R_p$ [58,59]. The Fe appears to be gettered to the region of maximum implant disorder (0.8 R_p) and to the region of In deficiency created by unequal recoil during the implant. This gettering is particularly evident for doses at which the InP is amorphized. However, the Fe pileup does not occur if the implant step is performed at elevated temperatures ($\sim 200°C$) to reduce the damage accumulation [58]. InP does not become amorphous for implant temperatures above 180°C. There is usually a significant amount of Fe diffusion during elevated-temperature implantation in InP.

Figure 7.16 shows the sheet resistance of n^+-InP layers grown epitaxially on semi-insulating substrates as a function of Fe dose and postimplant annealing temperature [38]. For the higher dose where the Fe concentration is in excess of the dopant concentration, the InP shows a thermally stable resistance of $\sim 2 \times 10^6$ $\Omega/\square$. Assuming that the epitaxial layer and the substrate, both of whose electrical properties are controlled by the Fe midgap level, are of similar resistivity, then this corresponds to a resistivity of $\sim 10^5$ $\Omega \cdot$ cm. For doses below the threshold value for chemical compensation, there is an evolution of the resistivity with annealing temperature caused by damage-related compensation. The activation energy of the resistivity obtained for the high-dose implants, followed by annealing at 550°C to promote the Fe onto a substitutional site, is 0.67 eV, consistent with the acceptor level of substitutional Fe [57].

Similar experiments were carried out on n^+-In$_{0.53}$Ga$_{0.47}$As epilayers on InP substrates. Fe also appears to create a deep acceptor in n-type InGaAs, with a level around 0.45 eV from the conduction band. The sheet resistivity of the Fe-implanted epitaxial layer was still only on the order of 5 $\Omega \cdot$ cm because of the small bandgap of InGaAs.

There are a number of applications for implant isolation of InP-based devices. Proton implantation has been used in the fabrication of stripe geometry InP/InGaAsP lasers in which the top p^+-InP is converted to n^- material, effectively confining the current to the unbombarded region [60]. CSBH InP/InGaAsP lasers have also been demonstrated using a hybrid technique of Fe implantation to form a high-resistivity current confinement barrier layer, followed by epitaxial regrowth [61].

One of the limiting parasitics in an HBT is the extrinsic base-collector capacitance. A reduction in this capacitance can be achieved by using a proton implant at doses of $\geq 5 \times 10^{12}$ cm^{-2} and energy of ~ 140 keV [62], or oxygen at 380 keV

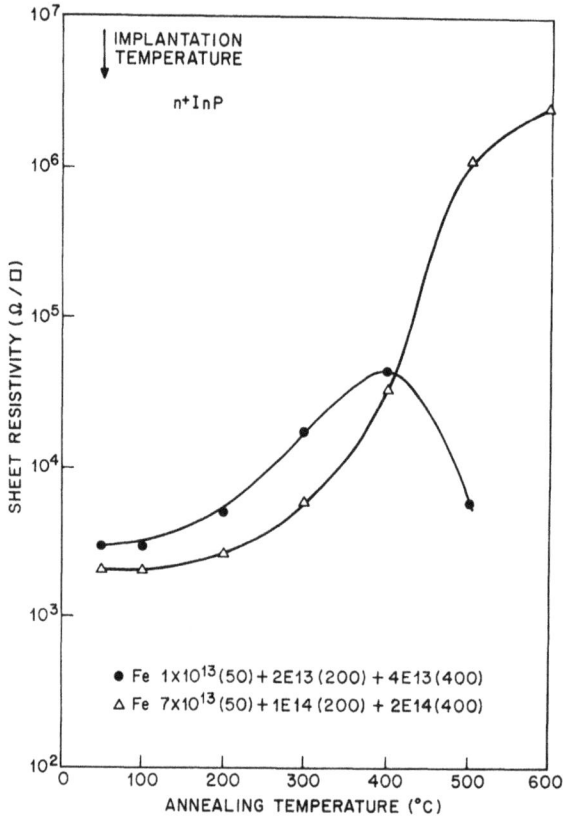

Figure 7.16 Sheet resistance of n^+ InP epitaxial layers as a function of Fe^+ ion dose and postimplant annealing temperature. The implantation temperature was ~50°C.

and doses of ~10^{14} cm^{-2}, which penetrate through the highly doped p^+-GaAs base and compensate the doping in the n-type GaAs collector region [63]. The advantage of using oxygen is that the collector is usually doped in the (3 to 5) × 10^{16} cm^{-3} range, and so thermally stable compensation can be achieved. This enables the selective Be implants for base contacts and the O implants for reduction of the collector capacitance to be performed sequentially, followed by an anneal to activate the Be and promote the O onto a substitutional site. The O or proton implant dose is low enough that the doping in the p^+-base layer is not affected. The ideality factor of the forward current in the base-collector junction is usually increased substantially as a result of the implantation, due to an increased density of deep levels in the collector [62–64]. The speed of the HBTs is improved by 10% to 30% as a result of the implant into the collector to reduce its capacitance [62].

Similar results are obtained in HBTs fabricated [65] in the InGaAs/InAlAs material system. Lee and Fonstad [65] have used O implants (400 keV at 10^{14} cm^{-2} dose) to create a compensated layer beneath the extrinsic base-emitter junction of an inverted (emitter-down) HBT. This reduces parasitic current injection into the base region. Provided that the device was not heated above 750°C, the implanted HBTs displayed higher current gain than the unimplanted samples.

7.4 RAPID THERMAL ANNEALING

RTA remains of great interest for activating implanted dopants in both GaAs and InP electronic and photonic devices. It offers a number of advantages over conventional furnace annealing, in particular, for limiting the redistribution of *p*-type dopants in both materials. In addition, RTA furnaces can be used for more reproducible contact alloying in comparison to graphite strip heaters, and the smaller amount of time spent annealing leads to reduced degradation of selectively doped heterostructures during high-temperature annealing steps. The fact that implant activation in III-V materials requires temperatures above the incongruent evaporation point means that some form of surface protection must be employed during the anneal [66].

The most commonly used method of surface protection during RTA is the proximity technique, in which the wafer of interest is placed face-to-face with another uncapped wafer of the same type. The obvious problem with this method is the presence of microscratches and baked-on contamination on the wafer surface. These are clearly more important as device packing densities are increased. RTA furnaces allowing the use of arsine of phosphine (or equivalent) atmospheres are the ideal solution, but present considerable safety concerns [67].

Another solution would be the use of a stable encapsulating layer deposited at low temperatures with low O content, which prevents erosion of the III-V surface, does not allow in-diffusion of impurities either from the cap or external to it, does not induce strain in the near-surface region, and is easily removed by benign means. This cap does not exist, at least to our knowledge. Numerous deposition methods including chemical vapor deposition, sputtering, e-beam evaporation, and spinning have produced films of SiO_2, SiN_x, AIN, PSG, SiO_xN_4, and combinations thereof, all of which have produced satisfactory results on small samples or small batches of wafers, but reproducible results are very difficult to obtain. We have found that RTA is a more severe test of encapsulating layers than furnace annealing because of the higher temperature and heating rates involved. Because of the sensitivity of GaAs and InP surfaces to subsequent processing steps, the idea of encapsulating the wafer at the start of the fabrication procedure and implanting through the cap is gaining increasing interest. This has the added advantage of providing the highest doping density at the surface, provided the cap thickness and implant energy are selected correctly. The requirement, of course, for spatially uniform doping means that the capping layer must be extremely uniform over the full wafer area.

The proximity annealing method is an inherently unsatisfying solution because it relies on a loss of As or P to create an overpressure between the two wafers to prevent further dissociation. A variation of this scheme is to coat the proximity wafer with Sn to increase the As_2 or P_2 partial pressure, but such an approach is not easily amenable to the sequential annealing of large numbers of wafers.

There is a relatively simple solution to problems associated with annealing of InP and associated materials. The use of SiC-coated graphite susceptors, into which is placed the wafer to be annealed, eliminates slip formation during implant activation treatments and provides much better protection against surface degradation at the edges of wafers compared to the more conventional proximity method.

Two types of SiC-coated graphite susceptors were examined, as shown in Figure 7.17. In the standard type, the wafer is placed within the depression milled into the bottom plate and covered with a graphite lid. To provide an adequate overpressure between the lid and the wafer, it is necessary to heat up a sacrificial GaAs or InP wafer substrate prior to annealing the wafer of interest. This coats the inside surfaces of the susceptor with the group V element; when subsequent anneals are performed, this As or P is vaporized to provide a partial pressure over the wafer. The obvious shortcoming of this approach is that the susceptor has to be recharged frequently to ensure an adequate partial pressure of the group V species for each new wafer that is annealed. The second type of susceptor has four small reservoirs that can be filled with material that provides a group V element partial pressure, eliminating the need for continual recharging of the susceptor. Since the reservoir material is at the same temperature as the wafer being annealed, pure P or As cannot be used because of their excessively high vapor pressures at the temperatures needed for implant activation (900°C for GaAs, 750°C for InP).

The thermal mass of the susceptors is obviously greater than that of the conventional proximity arrangement and thus the heating and cooling rates of the wafer

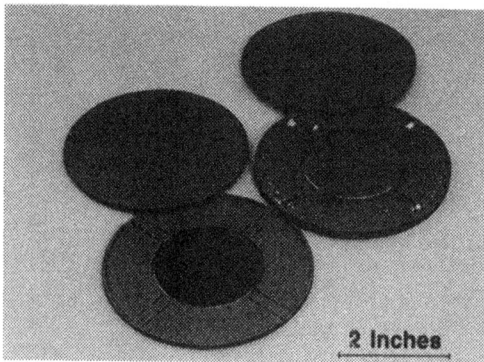

Figure 7.17 Photograph of the conventional graphite susceptor containing a 2-inch diameter wafer (left) and the susceptor with four GaAs-filled reservoirs.

in the susceptor are much slower than those obtained with the proximity geometry. Figure 7.18 shows the time-temperature profiles for 2-inch-diameter InP wafers annealed at 900°C for 10 seconds in either type of susceptor and in the proximity geometry. The slower heating and cooling rates in the susceptors are advantageous in reducing slip generation. To more clearly delineate slip lines in the annealed wafers, they were etched in a selective etchant. Optical micrographs of 2-in diameter InP wafers annealed at either 800°C or 850°C for 10 sec in a susceptor or in the proximity geometry are shown in Figure 7.19. In this case the wafers annealed in the proximity arrangement show severe degradation around their edges with In droplets visible. At the highest annealing temperature used, this actually causes the wafers to bond together, and they break if there is any attempt to separate them. By contrast the wafer annealed in the simpler susceptor shows no apparent degradation. Slip lines oriented in $\langle 1\bar{1}0 \rangle$ and $\{111\}$ directions are visible in the samples annealed in the proximity geometry. These slip lines are heaviest around the periphery of the wafer.

Optical micrographs of InP and GaAs wafers annealed sequentially in the simple susceptor following an initial charging procedure (850°C for 2 min with an InP wafer, or 950°C for 5 min with a GaAs wafer) showed that under these conditions effective surface protection is provided for up to five sequential high-temperature (900°C, 10 sec) anneals of GaAs, or only one anneal (750°C, 10 sec) of InP, before recharging is necessary.

Clearly, for capless rapid annealing of large numbers of InP and GaAs wafers, it is necessary to have a continuous reservoir available to provide an adequate group V partial pressure for more than just a few annealing runs. We investigated the use

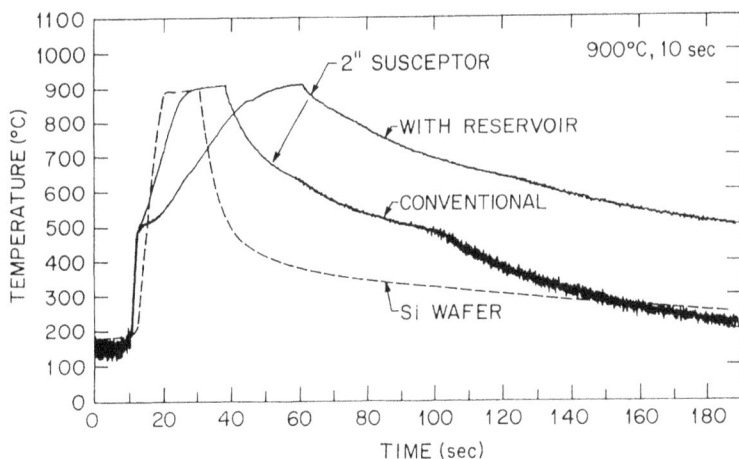

Figure 7.18 Temperature response as a function of time for the two types of susceptors and a conventional proximity annealing arrangement.

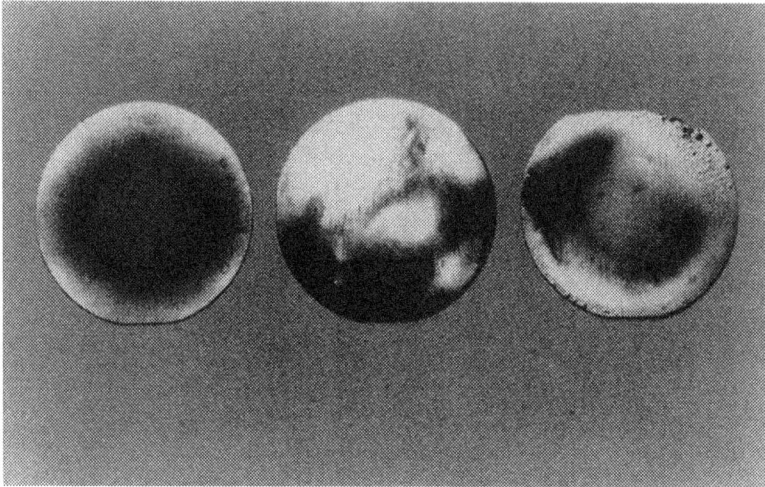

Figure 7.19 Optical micrographs of 2-inch diameter InP wafers annealed at 850°C, 10 sec in a susceptor (at left) or at 800°C, 10 sec (center) or 850°C, 10 sec (at right) in the proximity geometry.

of various reservoir materials in conjunction with that type of susceptor. We tried filling the reservoirs with many small pieces (each on the order of ~1 mm^3) of either InP, InAs, or GaAs. Basically, all of these materials provided good surface protection for either GaAs or InP wafers, and, hence, either P or As overpressure can be used for either material. In practice, however, the InAs melts below 570°C and forms balls within the reservoirs. These balls have a smaller effective surface area than the original slivers and, therefore, provide a lower As partial pressure.

Auger electron spectroscopy (AES) survey spectra and depth profiles of InP control samples and of samples annealed in a susceptor containing InP in its reservoirs showed that there was no difference in the stoichiometry of the two samples, and no loss of P from the surface appears to have occurred. Moreover, there was no additional oxidation of the annealed InP relative to the control material, indicating that all of the air initially trapped within the susceptor when it was loaded was effectively removed by the 45-sec flush with N$_2$:H$_2$ prior to commencing the heating cycle within the RTA system. Finally, there was no additional C on the surface of the annealed sample relative to the control piece, indicating that there were no problems involving contamination from the susceptor itself. Similar results were obtained for GaAs annealed with either GaAs or InAs in the reservoirs.

Using the susceptor method, it is possible to anneal patterned metallized wafers with getting pitting around the periphery of the metal contacts. The use of a SiC-coated graphite susceptor containing reservoirs for providing a group V element partial pressure allows high-temperature annealing of both InP and GaAs with no discernible surface degradation or slip formation. The ability to provide this partial pressure without the need for continual recharging of the susceptor means that it is possible, by our estimates, to sequentially anneal dozens of wafers at high temperatures without inducing surface deterioration.

7.5 DRY ETCHING

The III-V semiconductors, particularly InP and related materials, are particularly sensitive to ion-induced damage during device processing. In this section we will discuss some of the characteristics of these materials and their use in a variety of different etching techniques, ranging from pure physical removal (ion milling) to combined physical plus chemical removal (reactive ion etching). A more detailed discussion of reactive ion etching of InP is given in another chapter in this book [68].

7.5.1 Ion Milling

A considerable body of information exists on the characteristics of ion beam etching of GaAs, due mainly to the extensive use of this technique for mesa isolation of GaAs-based electronic and photonic devices [69–72]. Briefly, it is usually observed that the etch rate goes through a maximum at an incidence angle of the ion beam of about 60°, with the sputtering yield being proportional to the ion energy [73]. These energetic ions create deep-level traps leading to significant carrier compensation at depths of up to ~2000Å from the surface. At shallower depths, the apparent net electron concentration in n-type GaAs can actually increase when ion energies above 400 eV are used [74]. In these samples, low-temperature (400°C) annealing can actually lead to further degradation of the electrical quality of the GaAs. In general, such low-temperature annealing leads to improvements in diode ideality factors (n), but the Schottky barrier height (ϕ_B) remains below its unetched value. Pang *et al.* showed that the ion damage depth is shallower with low energy and heavy ion species, as expected, and that the introduction of an adsorbed gas on the sample surface can act as a protective layer to further reduce ion damage effects [75].

By sharp contrast to the situation for GaAs, very little is known about ion damage to InP surfaces. There has been one report of the introduction of deep-level centers in InP during ion beam etching with 0.5- to 2-keV Ar^+ or Cl_2^+ ions [76]. These levels were found to anneal out approximately 300°C and to be more prevalent in the case of Ar ion beam etching than with Cl_2 etching under the same conditions. Electrical and structural changes in the near-surface (~1000Å) of InP reactively ion etched in C_2H_6/H_2 or CCl_2F_2/O_2 discharges have also been reported [77].

The average mill rate, normalized to the Ar^+ ion beam current for InP and GaAs as a function of Ar^+ ion energy at 45° incidence angle and a temperature of 10°C, is shown in Figure 7.20. Within experimental error (±10%), in each case the

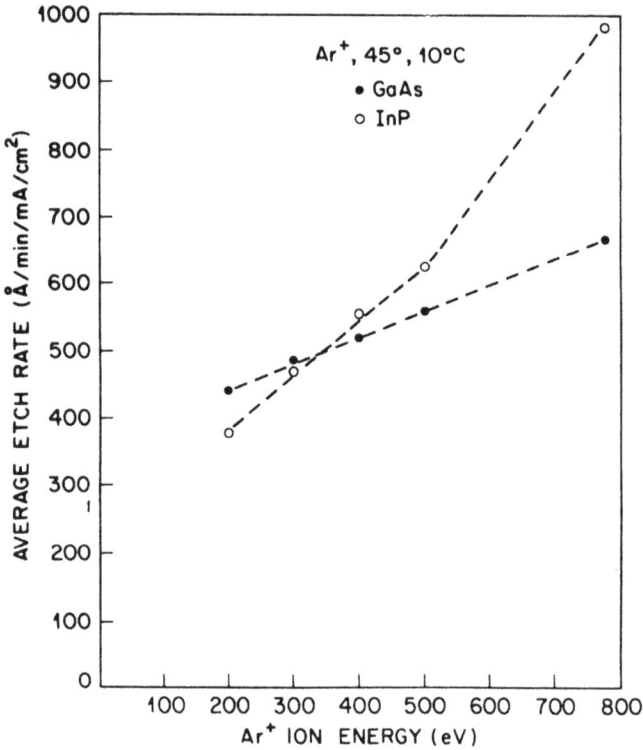

Figure 7.20 Average etch rate normalized to ion current for Ar$^+$ ion milling of InP and GaAs, as a function of ion energy. The beam incidence angle was 45°, and the sample temperature 10°C.

etch rate depends linearly on the Ar$^+$ ion energy, with slopes of ~0.4 (Å × min^{-1} × mA^{-1} cm^{-2} × eV^{-1}) and 0.8, respectively, for GaAs and InP. For GaAs this appears to be consistent with some previously reported data [69,78] in that a doubling of the ion energy leads to an increase in mill rate of ~20%. For InP, the mill rate increases more rapidly than for GaAs. There is very little data available for the sputtering yield of compound semiconductor targets [72], and for the case of InP, preferential removal of P relative to the much heavier In atoms might be expected. Our measured mill rates correspond to sputtering yields for InP of 0.30 atoms per ion at 200 eV Ar$^+$ ion energy, 0.42 atoms per ion at 500 eV, and 0.80 at 800 keV when the Ar ion currents are factored in. For GaAs, the sputtering yield is 0.35 atoms per ion at 200 eV, and 0.54 atoms per ion at 800 eV Ar$^+$ ion energy.

The experimentally observed mill-rate dependence on the angle of incidence of the Ar ion beam for both materials at 500 eV Ar$^+$ ion energy is shown in Figure

7.21. The mill rate increases with increasing incidence angle up to ~60°, where the mill rate is ~40% to 50% higher than at normal incidence. The mill rate then decreases for high beam incidence angles, as expected from linear cascade theory [72]. Our data for GaAs are consistent with the previously reported angular dependence [79]. For InP the sputtering yields were 0.36, 0.45, 0.49, and 0.26 atoms per ion, respectively, at 0°, 45°, 60°, and 75° angles of incidence. For GaAs the corresponding values were 0.31, 0.33, 0.39, 0.44, and 0.23 atoms per ion at 0°, 15°, 45°, 60°, and 75° incidence angles. For GaAs the corresponding values were 0.31, 0.33, 0.39, 0.44, and 0.23 atoms per ion at 0°, 15°, 45°, 60°, and 75° incidence angles. Both materials show a similar angular dependence of the ion milling rates, with the InP removal rate being faster at all incidence angles for this Ar^+ ion energy of 500 eV [80].

The surface morphologies of the ion-milled materials were examined on patterned samples by SEM. Figure 7.22 shows results for InP samples, milled at 45°

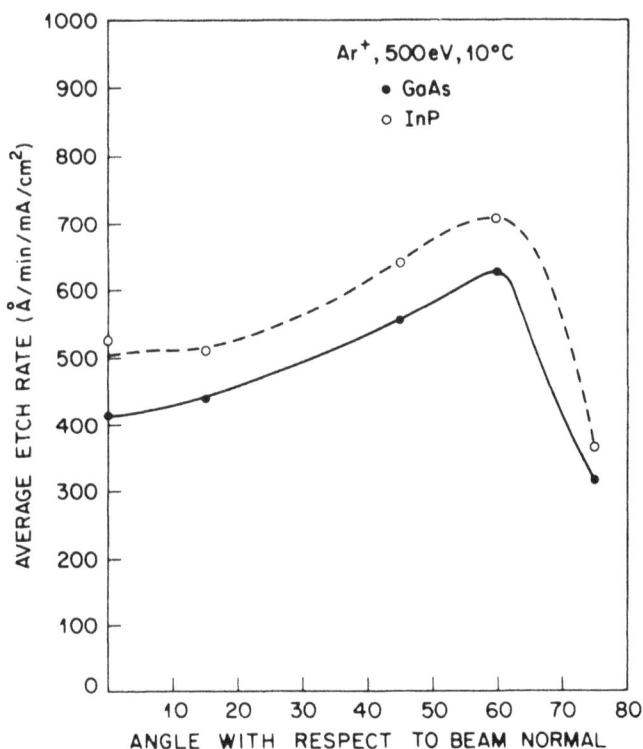

Figure 7.21 Average etch rate normalized to ion current for Ar^+ ion milling of InP and GaAs, as a function of beam incidence angle measured from the normal.

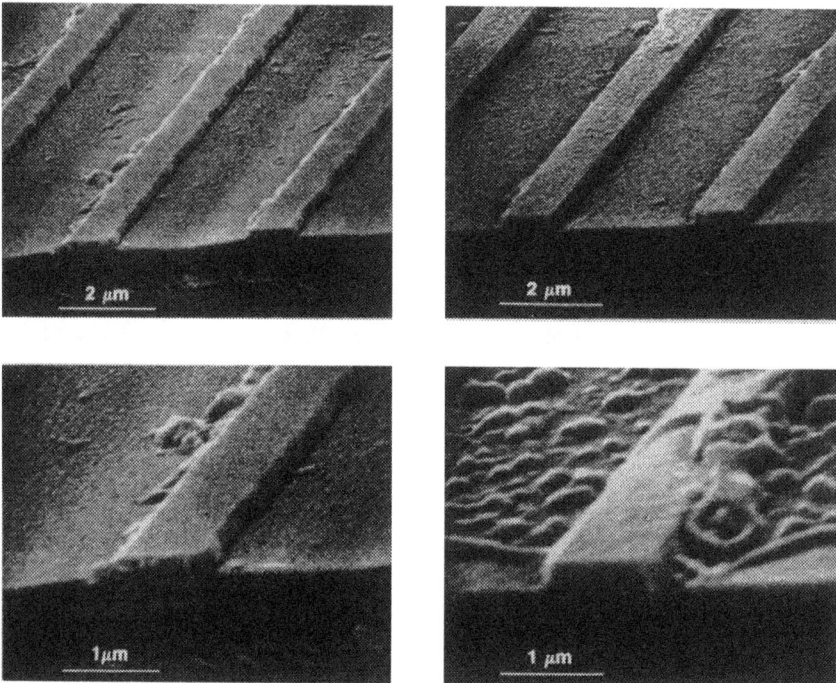

Figure 7.22 SEM micrographs of features ion milled into InP at 45°C incidence angle of the beam with Ar⁺ ion energies of 200 eV (top left and bottom left), 500 eV (top right-hand side) and 800 eV (bottom right-hand side).

incidence angle at Ar^+ ion energies of 200 eV (top left and bottom left), 500 eV (top right), and 800 eV (bottom right). The surfaces are reasonably smooth except in the latter case where preferential sputtering of P during the ion milling treatment leads to In droplets remaining on the surface. In some cases, the In spills over onto the unetched regions.

A useful monitor of the introduction of near-surface damage into III-V semiconductors is the measurement of the total photoluminescent (PL) intensity from a sample before and after a particular treatment. If deep-level, nonradiative defects associated with lattice damage are introduced as a result of a processing step, then the band-to-band PL intensity from the sample will be reduced. Figure 7.23 shows the 300K PL spectra from Fe-doped InP samples both before and after ion milling at different Ar ion energies. The beam was at 45° incidence angle for each treatment. The PL intensity is reduced by approximately an order of magnitude after even the lowest ion energy (200 eV) milling treatment, and is further reduced for the higher ion energies (500 and 800 eV). These decreases are significantly greater than we

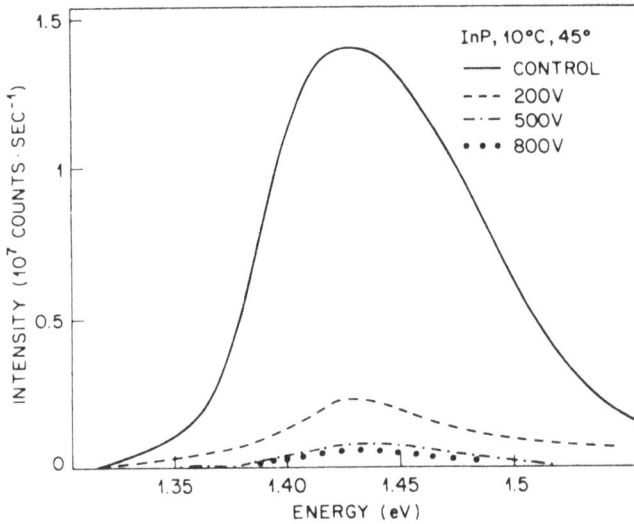

Figure 7.23 PL spectra recorded at 300K of InP (Fe) samples ion milled using 200-, 500-, or 800-eV Ar$^+$ ions.

have observed for reactive ion etching (RIE) of InP, where the maximum ion energies are comparable to those used for ion milling [81]. This is presumably because the etch rate is slower for ion milling relative to reactive ion etching in which there are both physical and chemical components to the etching. For ion milling, therefore, there is more accumulation of lattice disorder because this disordered material is not being removed as quickly as in RIE. The associated reduction in PL intensity is greater in ion milling as a result of the slower etch rates, compared to plasma etching techniques. It is interesting to note that annealing at 700°C for 30 sec produced only moderate increases (~20%) in the PL intensity.

Ion milling of InP led to surfaces to which we could not fabricate rectifying contacts. Both Hg and evaporated Au metallizations on milled InP samples showed ohmic behavior, as evidenced by the I-V characteristics of Figure 7.24. Similar data were observed for all ion energies (200 to 800 eV) and angles of incidence (0 to 75° from the vertical). The damaged, near-surface region was progressively removed by wet chemical etching in 1HCl:5H$_2$O and the etch depth directly measured by Dektak stylus profilometry after masking a section of the sample with apiezon wax. Evaporated Au-Schottky contacts were then deposited onto the etched surface. Figure 7.24 also shows I-V characteristics from samples from which 390Å or 945Å were removed prior to deposition of the Schottky contact. Removal of 390Å of material was not sufficient to restore the I-V characteristics, while after taking off 945Å, the I-V data was identical to that of a control sample. It was found that for 500-eV Ar$^+$

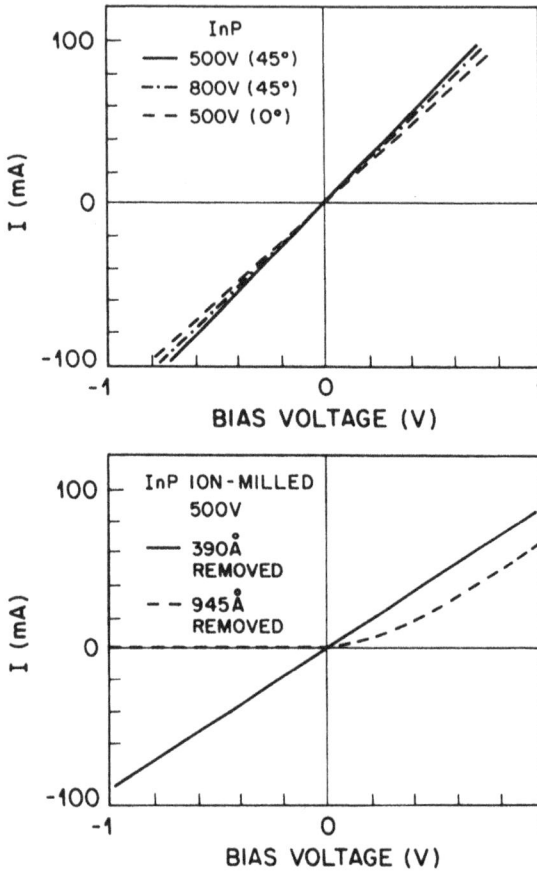

Figure 7.24 I-V data from Au-InP Schottky diodes fabricated on ion milled, *n*-type InP (top) and after removal of 390Å or 945Å of material prior to evaporation of the Au contact (bottom).

ion milling, at least 600Å of InP had to be removed by wet chemical etching to restore the initial electrical characteristics of the material as evidenced by the I-V data.

It has commonly been assumed that ion milling of III-V materials leads to the formation of a shallow amorphous layer in the near-surface region [69,74]. Surprisingly, this is stated without supporting data. A close examination of publications dealing with crystalline quality after ion beam etching actually show no evidence for amorphous layer creation [75,77]. We examined a variety of InP and GaAs samples after ion milling and after subsequent annealing. Figure 7.25 shows ion channeling spectra from InP ion milled for 4 min at either 200 or 800 eV (45° incidence angle),

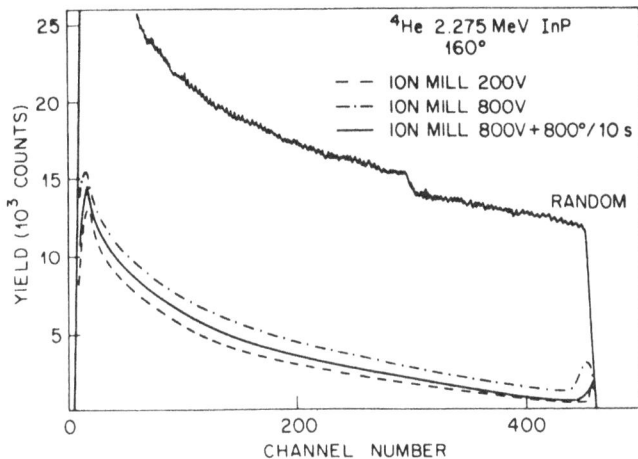

Figure 7.25 Ion channeling spectra recorded at 160° detection angle from ion milled InP before and after annealing.

and also, from the latter sample, after an 800°C, 10-sec anneal. These spectra were recorded by the 160° detector. We used both the minimum backscattering yield (χ_{min}) and the areal density calculated from the surface peaks in the 105° spectra as indicators of the crystalline quality after ion milling. The χ_{min} is defined as the height in the channeled spectrum divided by the height in the rotating random spectrum at the same channel number. In this case, the height was measured at the point where the channelled spectrum was at a minimum. Due to the differences in depth resolution, the χ_{min} values obtained at 105° and 160° should not be compared directly to each other. The areal densities give an idea of the number of displaced surface atoms. For InP, only the In backscattering peak was measured, while for GaAs both the Ga and As peaks were measured because of the close proximity of the masses of Ga and As. The InP sample ion milled at 200 eV had a χ_{min} of 2.9% at 105° and 3.1% at 160°, similar to that of an unetched control sample. The sample etched with an 800-eV beam showed a significant level of damage in the first 200Å, with a χ_{min} of 9.1% behind this damaged region. The crystalline quality of this sample actually became worse after a 500°C/10-sec anneal with a χ_{min} of 13.4% measured in the 105° spectrum. This is consistent with other reports using electrical characterization techniques [24,82]. Annealing at 800°C does produce a substantial reduction in gross lattice disorder, with the χ_{min} at 105° returning to 5.3%. The In surface peak returned to 1.9×10^{16} cm^{-2}, compared to a value of 1.1×10^{16} cm^{-2} for an unetched control sample. The sample annealed at 800°C for 10 sec has less crystal damage in the upper 100Å than does the sample ion milled at 200 eV, but the disorder peak after annealing extends to ~800Å. The damaged region of the 500-eV milled sample after

500°C/10-sec annealing also extends to greater depths than before annealing, indicating a migration of the disorder.

A similar picture of ion milling damage emerges, as was postulated for dry etch damage in InP [77]. For 500-eV ion milling, gross damage is detectable by ion channelling to depths of ~200Å. Beyond this is a region containing point defects that alter the carrier concentration to depths of ~1000Å. Migration of near-surface (~200Å) defects occurs during annealing, and ion channelling shows a broader disordered region. In the as-milled condition, removal of 500Å to 600Å by wet chemical etching is enough to essentially restore the electrical characteristics of the material, because this takes off the grossly disordered region and a section of the electrically compensated layer. The remaining region where there is carrier compensation is then totally within the zero bias depletion depth and has little effect on the I-V characteristics of simple diodes.

7.5.2 Plasma Etching Techniques and Chemistries

The dry etching of InP and related compounds is experiencing a resurgence of interest, largely due to the need to achieve high-resolution, anisotropic etching in device applications. This etching has a variety of requirements, such as fast etch rate for creation of deep (≥ 1 μm) trenches, high selectivity for one material over another (e.g., InGaAs over AlInAs) or, conversely, equi-rate etching for these materials.

There are two basic classes of gas mixtures used for the etching of III-V materials [83,84]. The first is based on Cl or Br, particularly the former because Ga and other group III chlorides are volatile at relatively low temperatures, as opposed to the stable gallium fluorides. It is common, then, to use Cl_2-based etching for III-V materials, in contrast to the F-based etching prevalent for Si. Most of the Cl-containing gases also contain C, and often problems are encountered with the deposition of polymer films during etching. Some of the gas mixtures used include Cl_2, CCl_4, BCl_3, $SiCl_4/Cl_2$, $SiCl_4/SF_6$, $CHCl_3$, $COCl_2$ and CCl_2F_2 (with O, He, or Ar). The advantages of using Freon 12 (CCl_2F_2) are that it is a nontoxic, noncorrosive gas which contains both Cl for etching and F to provide an etch stop upon reaching an underlying AlGaAs or AlInAs layer [85]. The etch stop mechanism involves the formation of an involatile AlF_3 layer, allowing several hundred selections of GaAs-to-AlGaAs and InGaAs-to-AlInAs.

The second general class of gas mixture is based on methane or ethane and H [86]. This has attracted considerable recent attention for etching both Ga- and In-based semiconductors. This nonchlorinated mixture shows controlled, smooth, highly anisotropic etching of all III-V materials. The etch products are thought to be AsH_3 or PH_3 for the group V element, and most likely some form of methyl adduct (e.g., $(CH_3)_nGa$) for the group III species.

It is common to use the normal boiling points or vapor pressures of the possible etch products as an indication of the suitability of a particular discharge for the dry etching of a III-V semiconductor. Strictly speaking, it is necessary to know the volatility under ion bombardment of these etch products, and in some cases the actual etch product is not known. An example is the group III product for CH_4/H_2 RIE of III-V materials.

The Cl-based etching is usually faster for III-V materials than CH_4/H_2. Comparisons between the etch rates of InP, InGaAs, and InAlAs in CCl_2F_2/O_2 and C_2H_6/H_2 are shown in Figure 7.26, where it is seen that the etch rates with the former mixture are a factor of three to five times faster. In many applications, notably mesa etching of heterostructures, a slower etch rate is actually an advantage because it gives better control of the etch depth. In other cases, such as the etching of via holes right through a thinned-down substrate, a high etch rate is obviously more important. Similar data for RIE of InAs, GaSb, and InSb with the same mixtures are shown in Figure 7.27. From these results it appears that gallium and antimony chlorides are more volatile under RIE conditions than their indium and arsenic counterparts. This

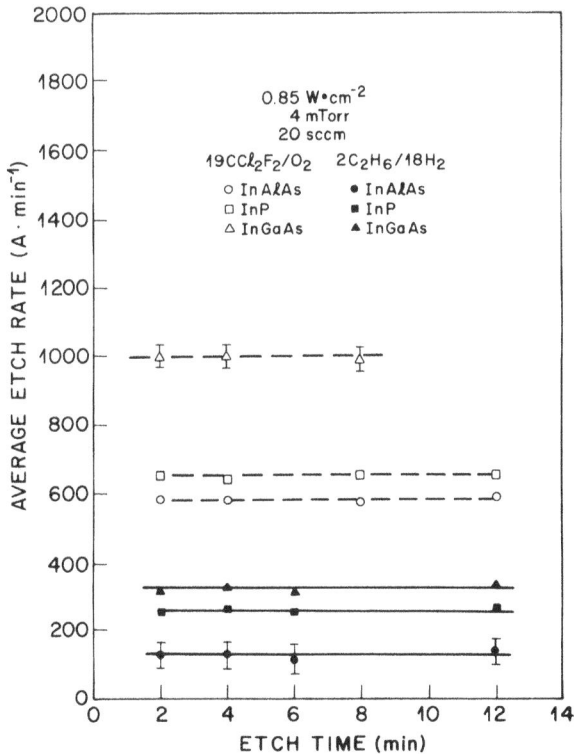

Figure 7.26 Average etch rates of InP, InGaAs, and InAlAs as a function of time in either 2 C_2H_6 : 18H_2 or 19 CCl_2F_2 : 1O_2 discharges (4 mtorr, 0.56W cm^{-2}).

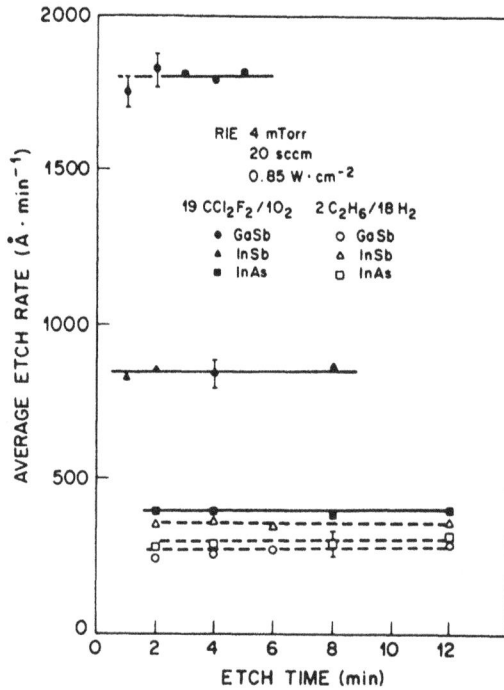

Figure 7.27 Average etch rate of GaSb, InSb and InAs as a function of time for 4 mtorr, 0.85W cm^{-2} discharges.

result would not be expected from a consideration of the boiling point data. Due to the possible restrictions on the production and use of chlorofluorocarbons, we have investigated the etching characteristics of the hydrogenated chlorofluorocarbons (HCFCs), notably CHCl$_2$F (Freon 21), and CHClF$_2$ (Freon 22). Etch depth as a function of time for five In-based materials with these two gases is shown in Figure 7.28. Based on such data we see that the etch rates with CHCl$_2$F/O$_2$ are ~20% slower for all materials relative to those with CCl$_2$F$_2$/O$_2$ under the same conditions. There was no evidence of an incubation time required before the onset of etching. All of the III-V materials, including GaAs, AlGaAs, and GaSb exhibit smooth surface morphologies over a wide range of RIE parameters. Thin (20Å–30Å) residue layers containing 3 to 9 at. % Cl and 1 to 3 at. % F (24 at. % for AlInAs) are present after dry etching with the HCFCs, although this contamination can be moved by solvent cleaning. The formation of a high concentration of AlF$_3$ on AlInAs provides a natural etch stop for removal of InGaAs layers in HBTs based on these materials.

Etch rates for the In-based materials with Cl$_2$ and SiCl$_4$ are considerably slower than for GaAs if the sample temperature is kept low during the plasma exposure.

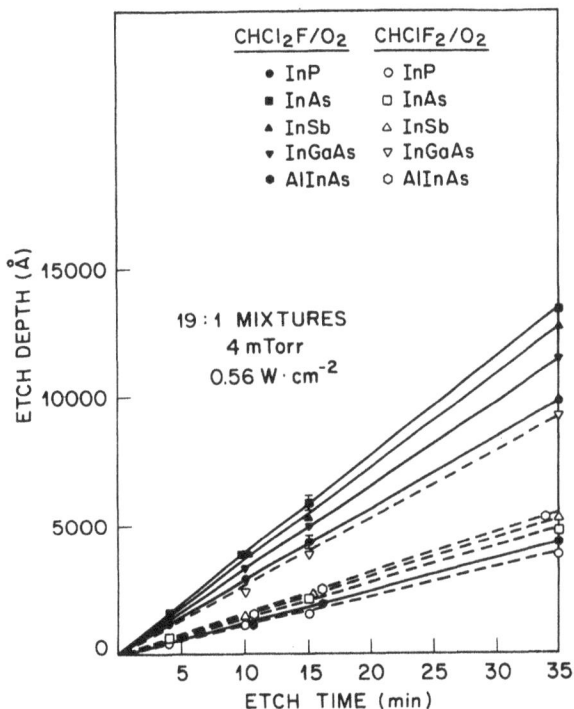

Figure 7.28 Etched depth in InP, InAs, InSb, InGaAs, or AlInAs as a function of exposure time to either 19:1 CHCl$_2$FO$_2$ or CHClF$_2$:O$_2$ discharges (4 mtorr, 0.56W cm^{-2}).

Much faster and smoother etching of InP and related materials are obtained at elevated ($\geq$100°C) temperatures, where desorption of the group III chlorides becomes easier. Extremely fast etch rates of InP are possible using Cl$_2$ or SiCl$_4$ at these temperatures, although in general this is impractical in many device processing sequences. A list of the comparative etch rates of GaAs, Al$_{0.3}$Ga$_{0.7}$As, InP, and InGaAs in the various gas mixtures at low pressure (4 m torr), and power densities (0.56 W cm^{-2}, self-biases of ~300V) are shown in Table 7.3.

While there has been extensive characterization of the damage caused to the near-surface region of GaAs and AlGaAs layers as a result of the ion bombardment occurring during RIE [87–89], there has been little work on modifications to the electrical and structural properties of InP. This is an important issue not only from the viewpoint that InP-based lasers and other photonic devices are used extensively in lightwave communication systems, but also because the surfaces of In-based materials have much lower carrier recombination velocities than GaAs and AlGaAs, and are not subject to Fermi level pinning. These surfaces may, therefore, be more disrupted by dry-etching-induced changes than those of GaAs and AlGaAs.

Table 7.3

Typical Etch Rates of GaAs, InP, $Al_{0.3}Ga_{0.7}As$, and InGaAs in Different Gas Mixtures as 4 mtorr and $0.56 \ W \ cm^{-2}$.

	Etch Rate ($\mathring{A} \ min^{-1}$)			
Mixture	GaAs	$Al_{0.3}Ga_{0.7}As$	InP	InGaAs
Cl_2/Ar	20,000	5,000	150	150
$SiCl_4/Ar$	5,000	3,000	130	320
CCl_2F_2/O_2	750	500	650	1,000
CH_3Br	600	400	220	300
CH_4/H_2	190	140	220	280
C_2H_6/H_2	220	160	250	320
$CHCl_2F/O_2$	600	300	300	320
$CHClF_2/O_2$	400	250	110	130

Figure 7.29 shows near-surface carrier profiles in etched InP as a function of post-RIE annealing temperature for 30-sec anneals in the Heatpulse 410 system. For the case of C_2H_6/H_2 etching, the reduction in the net carrier concentration extends to ~1000Å, and the initial profile is recovered by 500°C annealing. This reduction in the doping density is assumed to be due to the creation of deep acceptors in the InP by bombarding ions from the discharge, which trap the conduction electrons in the material and are not thermally ionized at room temperature. This effect has been well documented by Pang and others [87–89] for dry etching damage in GaAs. We rule out H passivation of the shallow donor impurities in the InP as contributing to the carrier loss, since this appears to be significant only for shallow acceptor dopants (i.e, it is apparent only in p-type material) [90]. The depth of the carrier reduction is also considerably greater than the projected range of ions crossing the plasma sheath, and this may be a result of axial and planar channeling of these ions once they enter the lattice. For example, the projected range of 380-eV H^+ ions is estimated to be ~38Å with a straggle of ~98Å from a Monte Carlo simulation [91], and this is much less than the distance over which we see carrier removal. If we take the relation that the final ion distribution has fallen to 10^{-3} of its peak value at a distance of 3.72 ΔR_p [21] where ΔR_p is the longitudinal straggle, then the H should be present to a depth of only ~365Å. Channeling, however, can increase this distance by factors of up to eight, and damage created by the channeling ions alone could account for the depth of the carrier removal without invoking recombination-enhanced motion of the H.

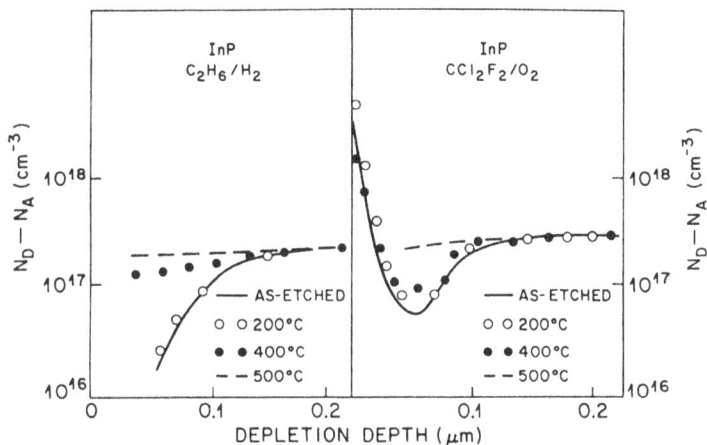

Figure 7.29 Carrier profiles in uniformity doped, n-type (1.5×10^{17} cm^{-3}) InP etched in either a $1C_2H_6:10H_2$ or $19CCl_2F_2:10_2$ discharge for 4 min, as a function of post-RIE annealing temperature (30-sec anneals).

In the case of CCl_2F_2/O_2 RIE of InP, the profiles appear more complicated. The as-etched samples and those annealed up to 400°C show an apparent increase in carrier concentration right at the surface. It is not clear if this is indeed a real effect due to preferential loss of P which we see from auger spectroscopy, or is an artifact of the difficulty in making a good Schottky contact to the etched surface. It is known that P-poor surfaces created by ion bombardment are n-type, and the removal of this high near-surface doping upon annealing at 500°C may either be ascribed to out-diffusion of P from the bulk to restore the stoichiometry in this region, or annealing of disorder, which allows better Schottky contact to the material. Stripping Hall measurements also appear to detect an increase in conductivity right at the surface of the dry-etched InP, which supports the idea of P vacancy-related donors, and this is currently under further study. We observed no evidence of Cl- or F-related residues on the surface after the RIE treatment. The damage-induced compensation of the doping profile further into the material appears to be slightly more resistant to annealing than for C_2H_6/H_2 etching, and this may be related to the heavier ions involved in CCl_2F_2/O_2 chemistry.

Reverse voltage-current characteristics from InP ($n = 6 \times 10^{15}$ cm^{-3}) etched in C_2H_6/H_2 or CCl_2F_2/O_2 are shown in Figure 7.30. Upon RIE, Au contacts no longer showed rectifying behavior for InP etched in C_2H_6/H_2, but were essentially ohmic for either polarity of bias applied to the contact. Once again this is presumably

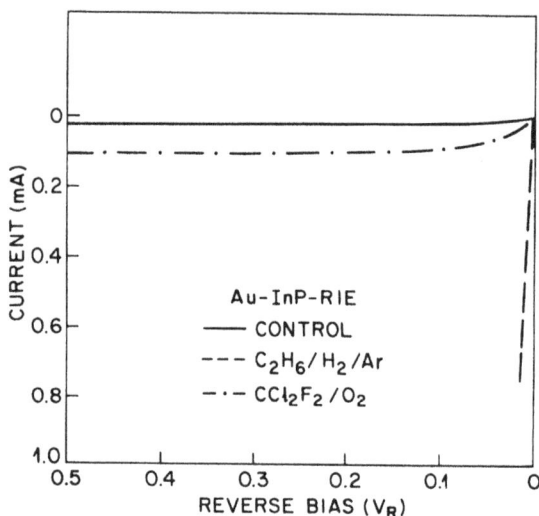

Figure 7.30 Reverse bias I-V characteristics from Au contacted n-type InP ($n = 6 \times 10^{15}$ cm^{-3}) after etching in $1C_2H_6/10H_2$ or $19CCl_2F_2/O_2$ discharges prior to Au deposition. The reverse breakdown voltage on the control sample was ~ 2.5V.

due to the creation of a nonstoichiometric near-surface region. Chemical analysis of the surface showed a deficiency of P to a depth of ~ 150Å. By sharp contrast, RIE in a CCl_2F_2/O_2 discharge leads to an approximately fivefold increase in reverse bias current, but the I-V characteristic is still rectifying. Chemical analysis of the near-surface region of InP etched in this type of discharge showed only a slight deficiency of P within 20Å of the surface. The difference between the two gas mixtures is clearly the high concentration of atomic hydrogen which can preferentially remove P from InP in the form of PH_3. This appears to be a fundamental problem with the C_2H_6/H_2 mixture, since the use of lower H concentrations leads to increasing polymer deposition.

Ion channeling spectra from InP samples exposed for 4 min to a C_2H_6/H_2 discharge are shown in Figure 7.31. The upper part of the figure compares the aligned spectra with a random InP spectrum, while the bottom part shows the aligned spectra plotted on an expanded scale. The introduction of lattice disorder to a depth of ~ 400Å is readily seen for this etching condition, with significant restoration of the crystallinity upon 400°C, 30-sec annealing. We note that care was taken to check that these results were indeed caused by the introduction of a thin disordered region during RIE and not simply due to enhanced oxidation of the surface. The dry-etched samples were transferred immediately to the backscattering chamber upon completion of the RIE treatment, and Auger spectroscopy of companion samples showed similar native oxide film thicknesses for both the control and dry-etched material, and the absence

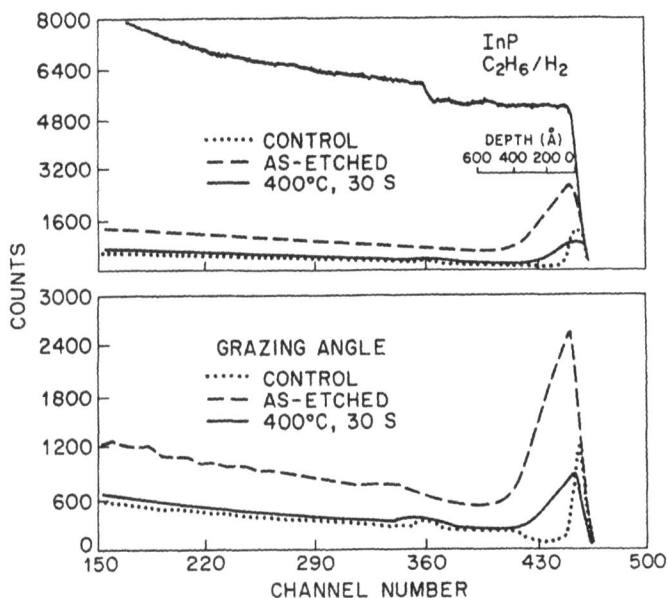

Figure 7.31 Ion channeling spectra from InP reactivity ion etched in a $1C_2H_6:10H_2$ discharge for 4 min, and after annealing at 400°C for 30 sec. All spectra were measured in glancing angle geometry to improve the depth resolution. The bottom figure shows the same spectra plotted with an expanded vertical scale.

of any additional surface residues on the RIE samples. The depth of the lattice disorder is significantly less than that determined from the C-V measurements, and indicates that point defects inaccessible to ion channeling are predominantly responsible for the carrier removal. This viewpoint corresponds to the current thinking for the mechanism of carrier compensation in ion-implanted compound semiconductors [33]. In samples etched in the CCl_2F_2/O_2 mixture, there was no detectable disorder, which implies that its depth was $<200Å$.

Based on these results we can postulate the following picture of the near-surface region of InP, after RIE, in the gas mixtures used in this work. The immediate surface ($\leq150Å$) for C_2H_6/H_2 etching is highly disordered and nonstoichiometric, being In rich. This degradation in the electrical and structural properties is severe enough to cause the complete absence of rectifying behavior in Au contacts deposited on the surface. Below this region there is another, ranging in depth from 150Å to 400Å, in which there is significant lattice disorder, both defect aggregates detectable by ion channeling and point defects which compensate the shallow dopant impurities in the region. This layer appears to be essentially stoichiometric in composition. Below this region there appears to be point-defect introduction to a depth of $\sim1000Å$.

These defects produce carrier compensation, but there is no detectable disruption to the crystalline order in the material, at least for ion channeling analysis. Annealing at 500°C appears to remove both the point defects and the more extended defects, leaving the material with similar electrical properties to that of virgin InP. For CCl_2F_2/O_2 etching, the immediate surface region shows less disorder and is more nearly stoichiometric due to the absence of H in the discharge. Point-defect introduction is again apparent, however, leading to carrier compensation at shallower depths to that in the C_2H_6/H_2 etching.

Smooth etching of all of the III-V materials is obtained with CH_4/H_2 or C_2H_6/H_2 mixtures, provided that the CH_4-to-H_2 ratio is kept between approximately $1:1$ and $1:3$. For high CH_4 concentrations, polymer deposition on the sample leads to micromasking and rough surfaces, whereas at high H_2 concentrations there is a preferential loss of P or As relative to In or Ga, and this also leads to rough surface morphologies.

The etching of all of the III-V materials in the Freon-based mixtures is relatively smooth under RIE or ECR conditions when Ar, He, or O is used as the dilutent. As mentioned earlier, these mixtures usually leave a surface residue 20Å to 50Å thick, although solvent cleaning removes essentially all of this film.

The composition and chemical bonding in the near-surface region after RIE in both types of mixture CCl_2F_2/O_2 or C_2H_6/H_2 was examined by auger electron spectroscopy (AES) and small-area x-ray photoelectron spectroscopy (XPS). Figure 7.32 shows AES surface scans of an InP control sample, and samples etched for 4 min in either $19:1$ CCl_2F_2/O_2 or $2:18$ C_2H_6/H_2. There are several notable differences between the samples. First, after C_2H_6/H_2, the InP shows a deficiency in P in the near-surface region relative to the control sample or the one etched in CCl_2F_2/O_2. Second, the C_2H_6/H_2 chemistry appears to leave a much cleaner surface than CCl_2F_2/O_2. In the latter case, Cl is detected, as well as a much higher C signal. By contrast, the sample etched in C_2H_6/H_2 appears to have even less C on the surface than the control. This is consistent with previous results showing that H-containing plasmas are very effective in removing C from semiconductor surfaces [92]. The P deficiency after this type of etching is also expected because of the high probabilities for formation of PH_3, and has been demonstrated by a number of authors [93–96]. Figure 7.33 shows AES depth profiles of P, In C, O, and Cl in these samples. The results confirm that there is less C on the surface of the C_2H_6/H_2 etched sample and also that there is Cl present to a depth of $\sim$30Å in the InP etched in CCl_2F_2/O_2. The P deficiency in the former sample appears to persist only to a depth of $\leq$30Å, and there is more oxide on the surface after C_2H_6/H_2 etching, as we discussed earlier in connection with the ion channeling data.

Based on this type of data, the average elemental composition in the top 100Å of reactively ion-etched InP, InGaAs, and AlInAs was estimated and is reported in Table 7.4. The surface C concentration for the control samples or those etched in

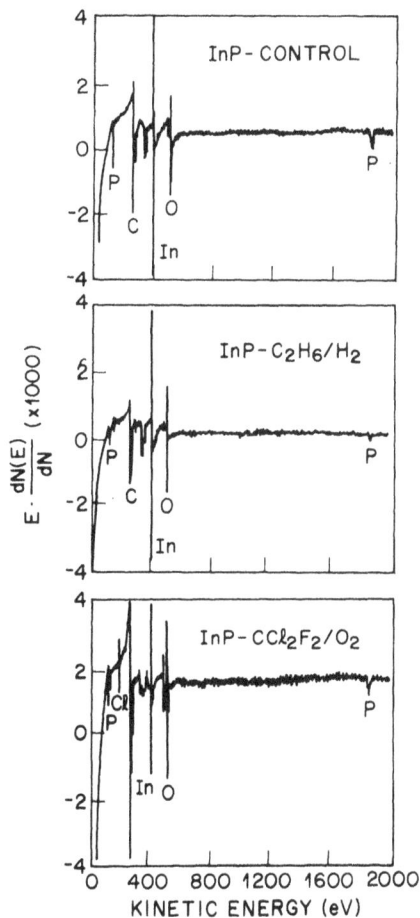

Figue 7.32 AES survey spectra from control and RIE InP samples etch either in $2C_2H_6/18H_2$ (0.85W cm^{-2}, 4 mtorr) or $19CCl_2F_2/1O_2$ (0.85W cm^{-2}, 4 mtorr).

CCl_2F_2/O_2 varied from 40 to 45 at. %, which is within the range expected for atmospheric contamination. The oxygen concentration in the near-surface region is clearly higher on the C_2H_6/H_2 etched samples, and this may be related to a higher chemical reactivity of these surfaces after the particular type of ion bombardment they incurred during the etch. Chlorine, and in some cases fluorine, contamination is evident for the material exposed to the CCl_2F_2/O_2 discharges.

A comparison of the etching characteristics of C_2H_6/H_2 and CCl_2F_2/O_2 for InP, InGaAs, and InAlAs yields the following main conclusions:

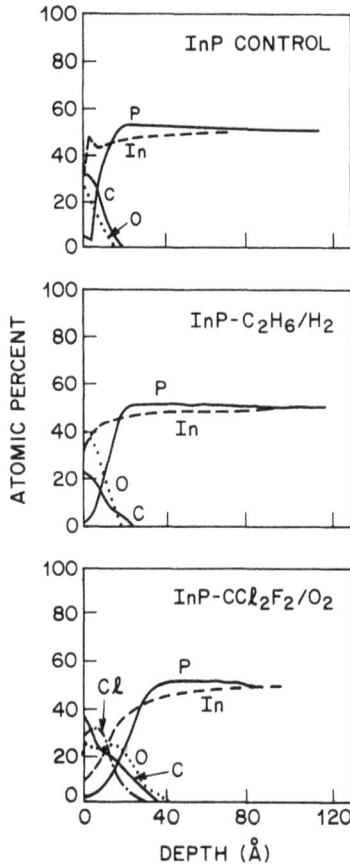

Figure 7.33 AES depth profiles of elemental composition in the near-surface region of control or RIE InP samples etched under the conditions of Figure 7.32.

(a) The etch rates of these materials under standard conditions are a factor of three to five higher for CCl_2F_2/O_2 relative to C_2H_6/H_2. There is no delay in initiation of the etching upon ignition of the plasma for either mixture, and the maximum etch rates for the latter chemistry occurs at a concentration of 25% by volume in the discharge. Some degree of ion bombardment appears to be necessary for efficient desorption of the etch products for C_2H_6/H_2 RIE.

(b) The etched surface morphology is smooth for InP, InGaAs, and InAlAs for C_2H_6 compositions of $\leq 40\%$ in the C_2H_6/H_2 chemistry, while for CCl_2F_2/O_2 etching, all of the materials show generally rough surfaces over the whole

Table 7.4

XPS Elemental Composition Data Measured from the Top 100Å of Each Sample and Expressed in Atomic Percent Units for the Units for the Elements Detected

Sample	*C*	*O*	*F*	*Al*	*P*	*Cl*	*Ga*	*As*	*In*
InP control	42.7	24.0	—	—	17.3	—	—	—	16.0
InP-C$_2$H$_6$/H$_2$	30.9	38.0	—	—	4.8	—	—	—	27.3
InP-CCl$_2$F$_2$/O$_2$	45.0	24.0	—	—	16.4	4.3	—	—	10.3
InGaAs-control	45.0	27.0	—	—	—	—	6.7	8.9	12.4
InGaAs-C$_2$H$_6$/H$_2$	34.0	37.0	—	—	—	—	6.4	13.0	9.6
InGaAs-CCl$_2$F$_2$/O$_2$	44.0	26.0	2.0	—	—	2.1	5.6	8.6	11.7
InAlAs-control	40.0	26.0	—	6.0	—	—	—	16.4	11.6
InAlAs-C$_2$H$_6$/H$_2$	20.5	49.5	—	17.7	—	—	—	6.9	5.4
InAlAs-CCl$_2$F$_2$/O$_2$	40.0	24.2	9.0	8.1	—	3.0	—	8.9	6.8

Etch Conditions :2 C$_2$H$_6$/18H$_2$ 0.85 W cm^{-2}, 4 mtorr.
:19 CCl$_2$F$_2$/O$_2$ 0.85 W cm^{-2} 4 mtorr.

range of plasma parameters. No lattice disorder is detectable by ion channeling for CCl$_2$F$_2$/O$_2$ RIE, even for high power density (1.3 W·cm^{-2}) discharges, but for C$_2$H$_6$/H$_2$ etching, damage is detectable to ~300Å in InP under these conditions. InAlAs appears to be more damage-resistant than InP.

(c) The C$_2$H$_6$/H$_2$ chemistry appears to leave an inherently cleaner surface on all three materials than CCl$_2$F$_2$/O$_2$. There is less C remaining on the InP, InGaAs, and InAlAs after C$_2$H$_6$/H$_2$ RIE than on the control samples, but more oxide is present. Chlorine-containing residues are detectable on all three materials after CCl$_2$F$_2$/O$_2$ RIE.

7.5.3 Electron Cyclotron Resonance Plasma Etching

Various methods have been developed for reducing ion energies in the discharge while trying to maintain anisotropic etching. These include the so-called triode reactor, in which a second plasma-generating electrode is included within the process chamber, or the addition of magnetic fields configured to reduce electron loss from the discharge and thus reduce the potential between it and the sample. This form of magnetically enhanced etching is generally divided into two types: magnetron RIE or electron cyclotron resonance (ECR) plasma etching. In ECR discharges, free electrons in the plasma are forced to orbit about magnetic field lines while absorbing microwave energy. At the cyclotron resonance condition outer shell, electrons from gas molecules in the discharge may also be liberated, leading to a very high degree of ionization in the plasma [97]. Since the motion of the electrons is constrained by the external magnetic field, fewer are lost by collisions with the reactor walls than in a conventional radio frequency (RF) plasma, and therefore the plasma potential relative to ground is much lower. The resultant energies of an ion reaching the sample to be etched are typically ≤ 15 eV. Since this is less than the displacement threshold for damage in most semiconductors, ECR etching should lead to much lower levels of damage than conventional RIE processes. A schematic of the ECR reactor we use is shown in Figure 7.34 (Plasma-Therm SL720). The sample is manually loaded into the load lock and then transferred into the etch chamber on a robotic arm. The system uses a 2.45-GHz microwave excitation source with additional RF bias superimposed at the wafer position (13.56 MHz). Pumping is accomplished through a very high conductance pump manifold linking the process chamber to a $1000\text{-}\ell s^{-1}$ turbomolecular pump.

Electron cyclotron resonance is provided by a plasma source of the multipolar, tuned cavity design. The microwave cavity plasma source has been described extensively elsewhere [97], and only a brief description is given here. The resonant cavity is a 17.8-cm in diameter brass cylinder terminated at the top by an adjustable short. A variable length launching probe enters the resonant cavity at the side, impressing microwave energy from the magnetron/waveguide assembly to an evacuated 100-mm diameter quartz "cup" where the plasma resides. The brass resonant

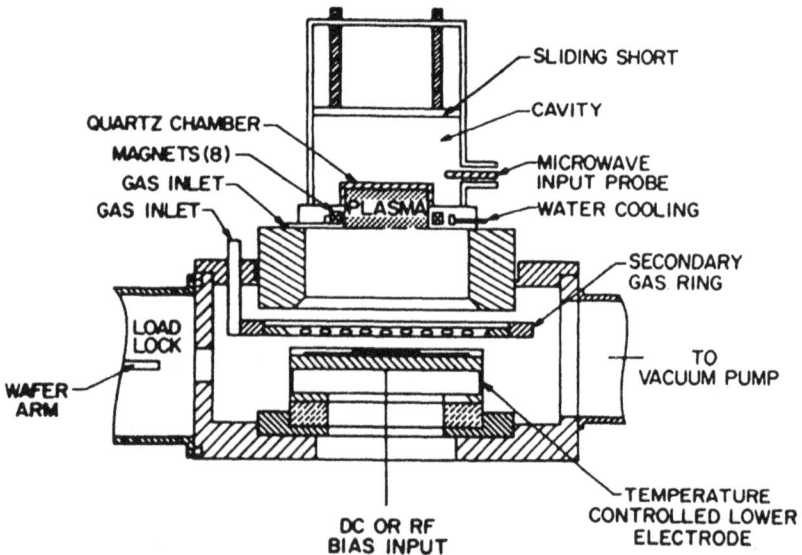

Figure 7.34 Schematic diagram of multipolar ECR plasma-etching system.

cavity is at atmosphere. Within the baseplate are eight high-strength rare-earth magnets which produce the B field level of 0.0875 T necessary for resonance.

Cyclotron resonance occurs on an ECR surface within the quartz cup. In as much as large solenoidal magnets are not used, there is no degrading B field to Coulombically accelerate ions to the wafer position. Therefore, 0V RF bias at the wafer will indeed correspond to zero additional acceleration energy on the etching species; arrival mechanisms for 0V RF are dominated by ambipolar and free-fall diffusion. The ions will still, however, be accelerated through a sheath potential even with zero RF biasing. We have demonstrated low-damage dry etching of III-V materials using $CH_4/H_2/Ar$ and CCl_2F_2/O_2 ECR discharges. In essence, the provision of microwave power provides faster etch rates at the same bias relative to conventional RIE, or, equivalently, with ECR discharges, it is possible to use lower self-biases and still obtain useable etch rates [98].

It is particularly important to minimize ion-induced damage to InP surfaces, which have much lower recombination velocities than GaAs and related compounds, and are also subject to a reduced degree of Fermi-level pinning. At this point, there have been few systematic studies of the plasma etching of InP using ECR discharges.

Lower levels of ion-induced damage are present when microwave ECR discharges are used because of the lower ion energies relative to conventional RF plasmas. An example of the relatively benign nature of ECR etching with regard to InP is given by the forward current-voltage (I-V) measurements shown in Figure 7.35.

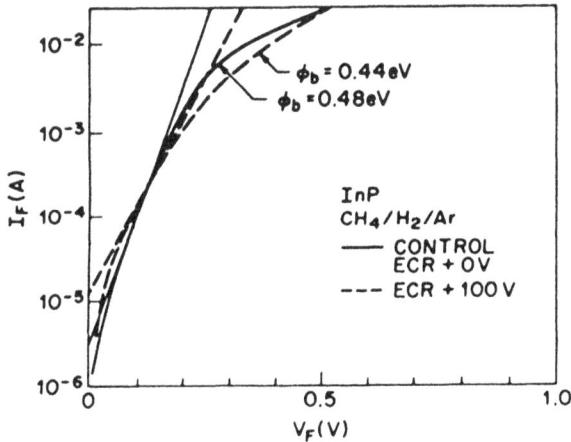

Figure 7.35 Forward current-voltage characteristics from Au-InP Schottky diodes etched in ECR 5 $CH_4/15H_2/7$ Ar discharges (0 or 100V substrate self-bias) prior to deposition of the Au contacts. The straight lines in each case are used to give the intercept and slope of the characteristic. The ECR + 0V sample had forward I-V curves very close to those of an unetched control sample.

Samples etched under ECR conditions with no additional biasing gave I-V characteristics very close to those of an unetched control sample, with a Schottky barrier height (ϕ_B) of 0.48 eV and diode ideality factor (n) of 1.1, both derived from the forward I-V plots, assuming thermionic emission. Once again our past experience with RIE on InP using this type of gas chemistry has been that Au deposition onto the RIE surface results in ohmic behavior, and a rectifying characteristic is not observed until at least 100Å is removed from the sample by wet etching prior to the Au deposition. With RIE we also observe substantial In enrichment of the near-surface region, but to much greater depths than with ECR. Even with the addition of 100V substrate bias during the ECR etching, we observe only a relatively small reduction of the Schottky barrier height to 0.44 eV, while the ideality factor shows a greater degradation, to a value of 1.6. This is a convincing demonstration of the much lower degree of disruption to the semiconductor surface using ECR discharges compared to conventional RIE [81].

The dependence of etching rate of InP, InAs, InSb, InGaAs, and AlInAs on applied RF power for 5 CH_4/17 H_2/8 Ar, 10 mtorr discharges (total flow rate = 30 sccm) is shown in Figures 7.36 and 7.37. With the exception of AlInAs, the etching rates are linearly dependent on the RF power as we have observed previously under RIE conditions using C_2H_6/H_2 discharges [99]. Since the DC bias on the samples increases linearly with RF power [99], it is plausible that the increased etching rates are largely due to more efficient sputter-induced desorption of the etch products at the higher powers. It has previously been established that CH_4/H_2 etching requires some ion bombardment in order to achieve measurable rates under RIE conditions [96]. The other features obvious from Figures 7.36 and 7.37 are that the presence of the microwave excitation, in this case 150W, increases the etching rates by about a factor of two over the RIE configuration. This is presumably due to the increased density of active species in the discharge available for etching. Once again the exception is AlInAs, where the microwave power produces no increase in etching rate. The CH_4/H_2/Ar gas mixture has very slow rates for Al-containing compounds [86,88,100], and high biases are required to obtain practical etch rates.

Provided the microwave power in our reactor is kept below ~150W, we obtain smooth etching of all In-based III-V materials. For microwave powers greater than 150W, we observed progressively rougher surface morphologies for all of the In-based semiconductors. Chemical analysis of these surfaces showed them to be very In-rich. The most likely explanation, therefore, for this rough etching is that at high microwave powers there is an imbalance of active hydrogen species over methyl species, leading to a faster removal of the group V element relative to the group III element. This leaves In droplets on the surface which act as micromasks and lead to rough morphologies. Limiting the microwave power to 150W or less at our gas mixture composition was effective in preventing this surface roughening. We ruled out local heating of the substrate due to the increased power above 150W, since fluoroptic probe measurements showed that the wafer temperature was below 80°C

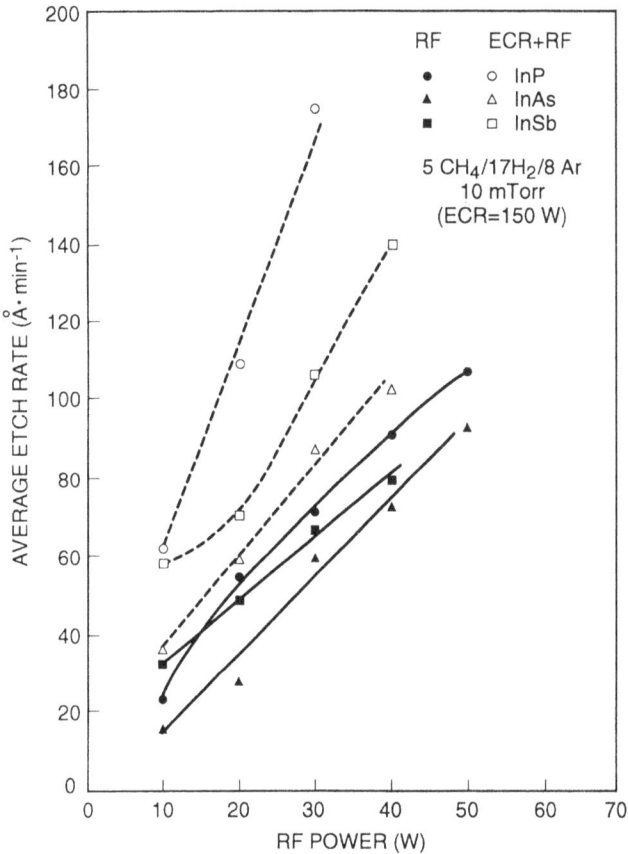

Figure 7.36 Average etch rate over a 30 minute period of InP, InAs, and InSb in 5 CH_4/17H_2/8 Ar, 10 mtorr, or 150W (microwave) discharges as a function of applied RF power.

under all conditions. Another obvious possibility is to increase the CH_4/H_2 ratio in order to keep the removal rates of group III and group V species nearly equivalent at high microwave powers. This can be a problem, however, since excessive polymer deposition can occur at CH_4 to H_2 ratios above approximately 1:1 [86].

The surface degradation of InP by pure H_2 plasmas is a well-documented phenomenon [93], and intentional hydrogenation of this material requires the presence of some form of surface protection [90]. This protection can be achieved either with a thin encapsulating layer such as SiN_x, which is permeable to the diffusing hydrogen, or by supplying the hydrogen via the dissociation of PH_3 or AsH_3 so that there is an overpressure of the group V species available to prevent surface dissociation [101].

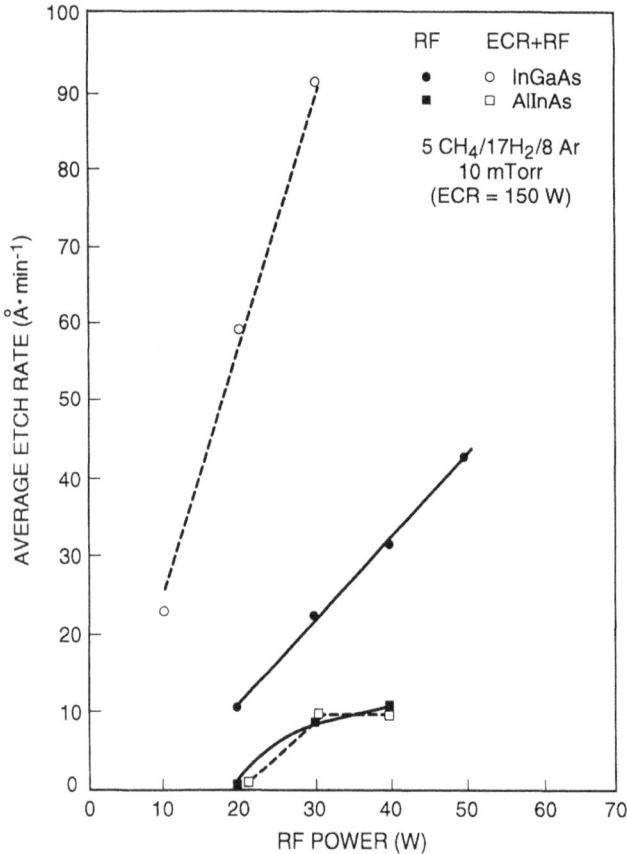

Figure 7.37 Average etch rate over a 30-min period of InGaAs, and AlInAs in 5 CH_4/$17H_2$/8 Ar, 10 mtorr, 0 or 150W (microwave) discharges as a function of applied RF power.

The effect of the CH_4-to-H_2 ratio in the gas mixture on the optical properties of InP has also been investigated. For sample biases that are very low (25V, top) or moderate (100V, bottom), the pure H_2 plasmas lead to the biggest reduction in PL intensity. The addition of CH_4 to the mixture reduces this PL degradation, and our standard composition of CH_4/H_2/Ar leads to the lowest reduction. Low-temperature annealing produced no change in PL intensity for anneals below 500°C; above this temperature the PL intensity decreased further due to surface degradation. For 25V bias, the standard discharge composition produces virtually no change in the PL intensity; but at 100V bias, there is a significant reduction. This is most likely due to the introduction of near-surface deep levels associated with ion bombardment damage. These deep levels act as nonradiative recombination centers to reduce the optical emission from the InP [101].

The addition of PCl$_3$ to the plasma is also effective in reducing surface degradation by providing a source of P. I-V characteristics from n-type InP samples etched in CH$_4$/H$_2$/Ar discharges with 100V bias as a function of PCl$_3$ addition show that the reverse currents decrease with increasing PCl$_3$ flow rate and change from pure ohmic behavior with zero PCl$_3$ to rectifying behavior as PCl$_3$ is added. These results are consistent with the reduced degradation of InP substrates in H$_2$ plasmas when either P vapor [101] or PH$_3$ [102] is added to the discharge.

The etching rates of the In-based semiconductors for low DC bias conditions are of particular interest from the viewpoint of dry etching device structures. Since it is usually necessary only to remove 1000Å to 8000Å of material in each step requiring dry etching in HBTs, even a relatively small etch rate will be acceptable, provided the ion energies are also low. Figures 7.38 and 7.39 show the etching rates

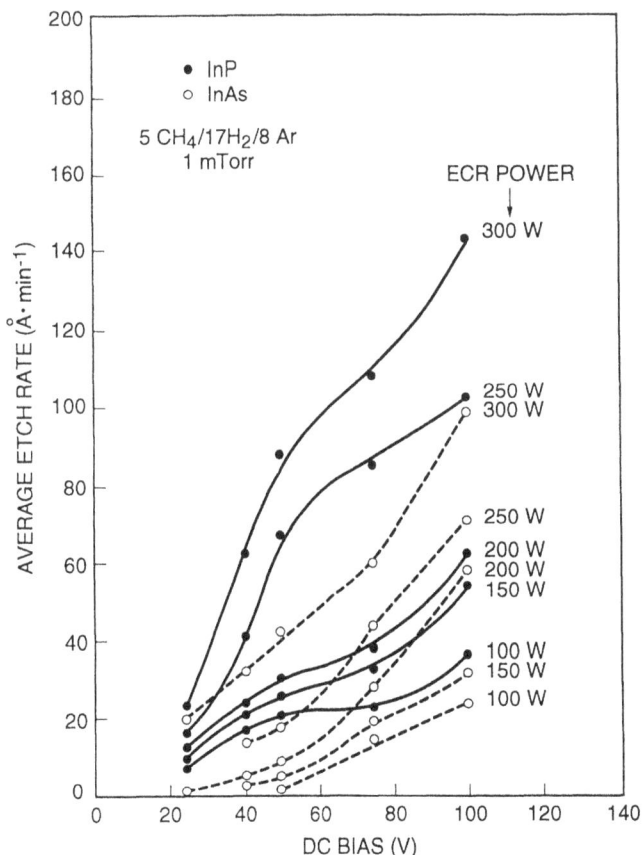

Figure 7.38 Average etch over a 30-min period of InP and InAs in 5 CH$_4$/17 H$_2$/8 Ar, 1 mtorr discharges as a function of microwave power and DC bias.

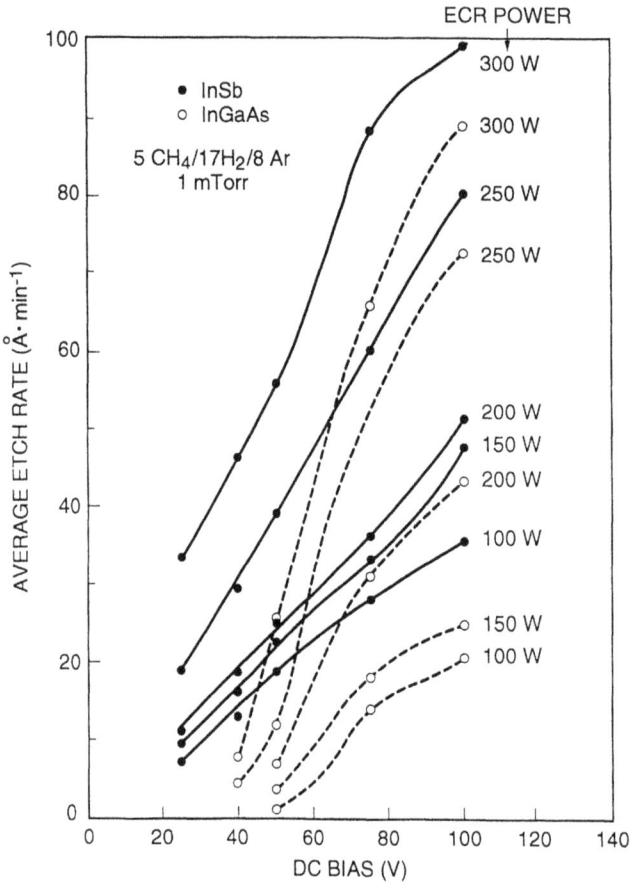

Figure 7.39 Average etch rate over a 30-min period of InSb, and InGaAs in 5 CH$_4$/17 H$_2$/8 Ar, 1 mtorr discharges as a function of microwave power and DC bias.

of InP, InAs, InSb, and InGaAs with 5 CH$_4$/17 H$_2$/8 Ar, 1-mtorr discharges as a function of DC bias (25–100V) on the samples, and as a function of the applied microwave power (100W–300W). The rates of all of these materials are greater than 20Å min^{-1} for biases of 80V or less and microwave powers of 150W or less. The most important dry etch step in the fabrication of an InGaAs/AlInAs HBT is the etching of ~1500Å of InGaAs (the contact layer) before stopping at the underlying AlInAs emitter. This is achievable with complete selectivity in ~75 min using a microwave power of 150W and a DC bias on only 75V. Even shorter biases are necessary with higher gas flow rates. A further etching step, in which 6000Å to 8000Å of InGaAs (the collector and sub-collector) is removed down to the InP substrate in order to isolate the device, is less critical since the etching is not stopping on an active layer and, therefore, higher biases can be used without affecting the device performance. The advantage of the additional microwave excitation of the discharge in enhancing the etching rate at low biases is also apparent from the data in Figures 7.38 and 7.39.

The maximum etching rates for the In-based materials under ECR conditions are also of interest for a noncritical device application like mesa formation. Figure 7.40 shows the etching rates of InP, InGaAs, and AlInAs with and without the applications of 300W of microwave power for DC bias voltages of 400V to 1000V. Under these conditions, the maximum etching rate of InP is ~1000Å min^{-1}, and that of InGaAs is ~430Å min^{-1}. Once again the AlInAs displays the slowest etching rates, in this case ~250Å min^{-1} at 1000V bias and 300W microwave power. It is clear that these rates are much too slow for applications requiring very large etch

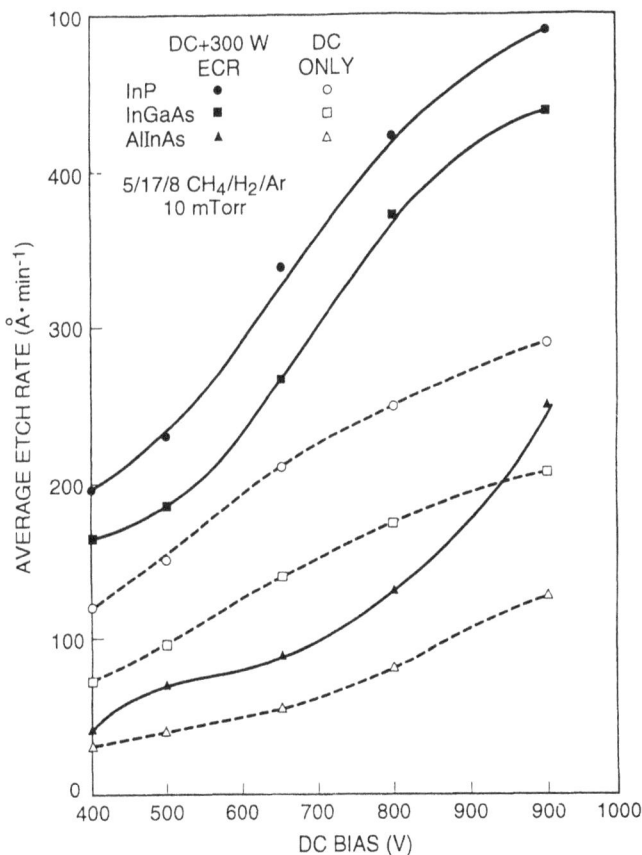

Figure 7.40 Average etch rate over a 30-min period of InP, InGaAs, and AlInAs in 5 CH$_4$/17 H$_2$/8 Ar, 1 mtorr discharges as a function of microwave power (0 or 300W) and DC bias.

depths (>20 μm). For example, for via hole fabrication, where at least 75 μm of InP must be removed, it will still be necessary to use elevated-temperature, Cl-based etching [84,103].

Under high bias conditions, the surface morphologies can still be quite reasonable, even for the highest microwave powers. Since the ion-enhanced component of the etching is increased, it is easier to remove the group III species at higher, rather than at lower biases, and so there are fewer In or Ga droplets remaining to act as micromasks.

The issue of which mask material to use for ECR etching with CH_4/H_2 mixtures has obvious practical implications because the mask should have two important characteristics:

(a) It should not erode during the dry etching.
(b) It should be easily removable at the completion of the etching, either by wet chemical or dry processing methods.

We found that a brown polymer was deposited on each of the three masking materials we investigated, namely AZ 1350J photoresist, CVD SiO_2, and sputtered W. The SEM micrographs in Figure 7.41 show some of the characteristics we found. First, W is a stable masking material, provided the self-bias during the etching treatment is less than ~125V. Moderate amounts of polymer are deposited on the W during such runs and these are readily removed by O_2 plasma barrel etching. For bias voltages above 125V, sputtering of the W onto the surface during etching occurs, leading to micromasking and the appearance of "grass" on this surface. Second, the greatest amount of polymer deposition is observed when using photoresist masks. The SEM micrograph in the center of Figure 7.41 shows an InP sample with the photoresist mask still in place. The mask is covered with a thick polymer film. If the etching depth on the InP is greater than ~4 μm, it can become essentially impossible to remove the photoresist mask because it appears that the etch products (most likely the In product) is incorporated into the polymer, forming a mechanically and chemically resistant film. For moderate etch depths, the photoresist is easily removed by wet stripping in acetone or barrel etching with C_2F_6/O_2. The use of photoresist on top of W obviously prevents the sputtering problem we mentioned above, and similar amounts of polymer are deposited as for a single layer of photoresist without the underlying W. The least amount of polymer deposition was found to occur on SiO_2 masking layers. Similar conclusions were reached by Adesida et al. [104] for RIE of InP in methane-based plasmas. The SEM micrograph at the bottom of Figure 7.41 shows a SiO_2-masked InP substrate after ECR etching at the same conditions as for the photoresist masked sample in the center of the figure. Only a thin polymer film was found on the SiO_2 and this was easily removed by O_2 barrel etching. The polymer obviously becomes thicker for extended dry etching treatments in CH_4/H_2, but the SiO_2 mask can always be removed in HF solution.

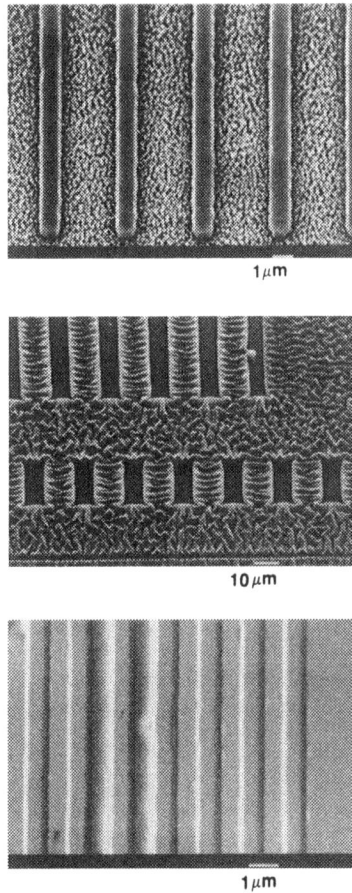

Figure 7.41 SEM micrographs from InP etched in 5 CH$_4$/17 H$_2$/8 Ar 1 mtorr, 150V, 150W (microwave) discharges with either a W (top), photoresist (center) or SiO$_2$ (bottom) mask. The masks are still in place.

For large etch depths, therefore, SiO$_2$ is the preferred masking material of the three we investigated.

The results of a systematic investigation of ECR + RF CH$_4$/H$_2$/Ar plasma etching show a number of significant features:

(a) The etching rates of InP, InAs, InSb, and InGaAs are increased by approximately a factor of two for additional microwave excitation of the discharge relative to RF only.

(b) The etched morphologies are smooth for all of the In-based materials, provided the microwave power is kept below ~150W. Above this value preferential loss of the group V species occurs.

(c) The addition of PCl_3 to the $CH_4/H_2/Ar$ discharge prevents some of the degradation of electrical characteristics of InP resulting from dry etching.

(d) Low bias voltage (<100V) etching at usable rates is possible for InP, InAs, InSb, and InGaAs. This has particular importance in device applications.

(e) The use of methyl chloride, CH_3Cl, has no advantage over CH_4 in terms of enhancing the etch rates of the In-based materials.

(f) For the etching of relatively deep features (≥ 2 μm), SiO_2 is the preferred masking material over photoresist or W.

(g) For 2-in diameter wafers the uniformity of etching is $\pm 3\%$ of the average etch depth in our particular ECR system.

REFERENCES

1. J.P. Donnelly, *Nucl. Inst. Meth.*, Vol. 182/183, 1981, p. 553.
2. C.W. Farley and B.G. Streetman, *J. Electron. Mat.*, Vol. 13, 1984, p. 401.
3. L.A. Christel and J.F. Gibbons, *J. Appl. Phys.*, Vol. 52, 1981, p. 5050.
4. D.K. Sadana, *Nucl. Inst. Meth.* in *Phys. Rev.*, Vol. B7/8, 1985, p. 375.
5. J.S. Williams, Chapter 10 of *Laser Annealing of Semiconductor*, ed. J.M. Poate, Sydney: Academic Press, 1984.
6. K.W. Wang, *Appl. Phys. Lett.*, Vol. 51, 1987, p. 2127.
7. W. Wesch, E. Wendler, G. Gotz, and N.D. Kehelidse, *J. Appl. Phys.*, Vol. 65, 1989, p. 519.
8. Q.F. Xiao, S. Hasimoto, W.M. Gibson, and S.J. Pearton, *Nucl. Inst. Meth.* in *Phys. Rev.*, Vol. B45, 1990, p. 464.
9. C.R. Abernathy, S.J. Pearton, R. Caruso, F. Ren, and J. Kovalibick, *Appl. Phys. Lett.*, Vol. 55, 1989, p. 1750.
10. K. Saito, E. Tokumitsu, T. Akusuku, M. Miyauchi, T. Yamada, M. Konagai, and T. Takukashi, *J. Appl. Phys.*, Vol. 64, 1988, p. 9475.
11. S.J. Pearton, U.K. Chakrabarti, C.R. Abernathy, and W.S. Hobson, *Appl. Phys. Lett.*, Vol. 55, 1989, p. 2014.
12. G.J. Valco and V.J. Kapoon, *J. Elect. Soc.*, Vol. 134, 1987, p. 564.
13. B. Molnar, *Appl. Phys. Lett.*, Vol. 36, 1980, p. 927.
14. S.J. Pearton, W.S. Hobson, A.P. Kinsella, J. Kovalchick, U.K. Chakrabarti, and C.R. Abernathy, *Appl. Phys. Lett.*, Vol. 56, 1990, p. 1263.
15. M.V. Rao, S.M. Gulwadi, P.E. Thompson, A. Fathimulla, and O.A. Aina, *J. Electron. Mat.*, Vol. 18, 1980, p. 131.
16. J.H. Wilkie, G.D.T. Spiller, I.D. Henning, and B.J. Sealy, *J. Cryst. Growth*, Vol. 3, 1987, p. 433.
17. E.U.K. Rao, M. Djamei, and P. Krauz, *Jap. J. Appl. Phys.*, Vol. 25, 1986, p. L458.
18. M. Maier and J. Selders, *J. Appl. Phys.*, Vol. 60, 1986, p. 2783.
19. K.W. Wang, J. Long, and D. Mitsham, *J. Appl. Phys.*, Vol. 63, 1988, p. 4455.
20. B. Tell, R.F. Leheny, A.S.H. Liao, T.J. Bridges, E.G. Buckbardt, T.Y. Chang, and E.D. Beebe, *Appl. Phys. Lett.*, Vol. 44, 1984, p. 438.
21. H. Ryssel and I. Ruge, *Ion Implantation*, New York: John Wiley & Sons, 1986, p. 576.
22. P.J. MacNally, *Rad. Eff.*, Vol. 6, 1970, p. 149.

23. C.E. Hurwitz and J.P. Donnelly, *Solid State Elec.,* Vol. 18, 1975, p. 753.

24. A.G. Foyt, W.T. Lindley, and J.P. Donnelly, *Appl. Phys. Lett.,* Vol. 16, 1970, p. 335.

25. W.G. Opyd and J.F. Gibbons, *Mat. Res. Soc. Symp. Proc.,* Vol. 45, 1985, p. 273.

26. M. Shanaan, R. Kalish, and V. Richter, *Nucl. Inst. Meth. Phys. Res.,* Vol. B7/8, 1985, p. 443.

27. D. Kleitman and H.J. Yearian, *Phys. Rev.,* Vol. 108, 1957, p. 901.

28. G.L. Destefanis and J.P. Gailliard, *Appl. Phys. Lett.,* Vol. 36, 1980, p. 40.

29. H.W. Alberts, *Mat. Res. Soc. Symp. Proc.,* Vol. 27, 1984, p. 335.

30. H.W. Alberts and R. Cilliers, *Solid State Phenom.,* Vol. 1/2, 1988, p. 371.

31. S.J. Pearton, A.R. VonNeida, J.M. Brown, K.T. Short, L.J. Oster, and U.K. Chakrabarti, *J. Appl. Phys.,* Vol. 64, 1988, p. 629.

32. S.J. Pearton, S. Nahakara, A.R. Von Neida, K.T. Short, and L.J. Oster, *J. Appl. Phys.,* Vol. 66, 1989, p. 1942.

33. S.J. Pearton, *Mat. Sci. Rep.,* Vol. 4, 1990, p. 313.

34. S.K. Ghandi, Chapter 6 of *VLSI Fabrication Principles—Si and GaAs,* New York: John Wiley & Sons, 1983.

35. M.W. Focht and B. Schwartz, *Appl. Phys. Lett.,* Vol. 42, 1983, p. 970.

36. P.E. Thompson, S.C. Binari, and H.B. Dietrich, *Solid State Electron.,* Vol. 26, 1983, p. 805.

37. J.P. Donnelly and C.E. Hurwitz, *Solid State Electron.,* Vol. 30, 1980, p. 727.

38. S.J. Pearton, C.R. Abernathy, M.B. Panish, R.A. Hamm, and L.M. Lunardi, *J. Appl. Phys.,* Vol. 66, 1989, p. 656.

39. Y. Kamiya, K. Shinomura, and T. Itoh, *J. Electrochem. Soc.,* Vol. 133, 1986, p. 780.

40. A.T. Macrander, B. Schwartz, and M. Focht, *J. Appl. Phys.,* Vol. 55, 1984, p. 3595.

41. W.C. Dautremont-Smith, P.A. Burnes, and J.W. Stayt, Jr., *J. Vac. Sci. Technol.,* Vol. B2, 1984, p. 620.

42. A.T. Macrander and B. Schwartz, *Mat. Res. Soc. Symp. Proc.,* Vol. 35, 1985, p. 293.

43. M.V. Rao, R.S. Baber, H.B. Dietrich, and P.E. Thompson, *J. Appl. Phys.,* Vol. 64, 1988, p. 4755.

44. B. Tell, K.F. Brown-Goebeler, T.J. Bridges, and E.G. Burkhardt, *J. Appl. Phys.,* Vol. 60, 1986, p. 665.

45. H. Krautle, O. Lintjorn, and H. Beneking, *Solid State Electron.,* Vol. 26, 1983, p. 1033.

46. J. Barnard, C.E.C. Wood, and L.F. Eastman, *IEEE Electron. Dev. Lett.,* Vol. EDL-2, 1981, p. 193.

47. B. Tell and K.F. Brown-Goebeler, unpublished.

48. R.A. Hopfel and N. Sawaki, *Appl. Phys. Lett.,* Vol. 55, 1989, p. 460.

49. R.J. Malik, T.R. Hayes, F. Capasso, K. Alavi, and A.Y. Cho, *IEEE Electron. Dev. Lett.,* Vol. EDL-4, 1983, p. 383.

50. W. Lee and C.G. Fonstad, *Appl. Phys. Lett.,* Vol. 50, 1987, p. 1278.

51. B. Tell, T.Y. Chang, K.F. Brown-Goebeler, J.M. Kuo, and N.J. Sauer, *J. Appl. Phys.,* Vol. 64, 1988, p. 3290.

52. B. Jalali, R.N. Nottenburg, Y.K. Chen, D. Sivco, and A.Y. Cho, *Appl. Phys. Lett.,* Vol. 54, 1989, p. 2333.

53. J.I. Davies, A.C. Marshall, M.D. Scott, and R.J.M. Griffiths, *J. Cryst. Growth,* Vol. 93, 1988, p. 782.

54. S.J. Pearton, W.S. Hobson, and U.K. Chakrabarti, *Appl. Phys. Lett.,* Vol. 55, 1989, p. 1786.

55. G.P. Agrawal and N.K. Dutta, *Long Wavelength Semiconductor Lasers,* New York: Van Nostrand, 1986.

56. S.M. Spitzer and J.C. North, *J. Appl. Phys.,* Vol. 44, 1973, p. 214.

57. J. Chevy, S.R. Forrest, B. Tell, B. Schwartz, and D.D. Wright, *J. Appl. Phys.,* Vol. 58, 1985, p. 1780.

58. S.A. Schwartz, B. Schwartz, T.T. Sheng, S. Singh, and B. Tell, *J. Appl. Phys.*, Vol. 58, 1985, p. 1698.

59. M. Gaunneau, H.L. Haridon, A. Rupert, and M. Salvi, *J. Appl. Phys.*, Vol. 53, 1982, p. 6823.

60. J.T. Hsieh, J.A. Rossi, and J.P. Donnelly, *Appl. Phys. Lett.*, Vol. 28, 1976, p. 709.

61. D.P. Wilt, B. Schwartz, B. Tell, E.D. Beebe, and R.J. Nelson, unpublished, 1983.

62. G. Nakajima, K. Nagata, Y. Yamauchi, H. Ho, and T. Ishibashi, *Electron. Lett.*, Vol. 22, 1986, p. 1317.

63. P.M. Asbeck, D.L. Miller, R.J. Andersen, and F.A. Eisen, *IEEE Electron. Dev. Lett.*, Vol. EDL-5, 1984, p. 310.

64. S. Adachi, *J. Appl. Phys.*, Vol. 60, 1986, p. 959.

65. W. Lee and L.G. Fonstad, *IEEE Electron Dev. Lett.*, Vol. EDL-8, 1987, p. 217.

66. S.J. Pearton, A. Katz, and M. Geva, *J. Appl. Phys.*, Vol. 68, 1990, p. 1081.

67. R. Singh, *J. Appl. Phys.*, Vol. 63, 1988, p. R59.

68. T.R. Hayes, this book.

69. C.I.H. Ashby, *Properties of GaAs, EMIS Dakareview*, RN 15422, London: IEEE, 1985.

70. S.W. Pang and W.J. Piancentini, *J. Vac. Sci. Technol.*, Vol. B1, 1983, p. 1334.

71. Y. Yaba, T. Ishida, K. Gamo, and S. Namba, *J. Vac. Sci. Technol.*, Vol. B6, 1988, p. 253.

72. P.C. Zalm, *Vacuum*, Vol. 36, 1986, p. 787.

73. G. Betz and G.K. Wehner, *Sputtering by Particle Bombardment*, ed. R. Behrisch, Vol. 2, No. 11, Berlin: Springer, 1983.

74. P. Kwan, K.N. Bhat, J.M. Borrego, and S.J. Ghandi, *Solid State Electron.*, Vol. 26, 1983, p. 125.

75. S.W. Pang, M.W. Geis, N. Efrememov, and G. Lincoln, *J. Vac. Sci. Tech.*, Vol. B3, 1985, p. 398.

76. Y. Yuba, K. Gamo, Y. Judai, and S. Namba, *The Physics of VLSI*, ed. J.C. Knights, AIP Conf. Ser. NU 1984, pp. 286–290.

77. S.J. Pearton, U.K. Chakrabarti, and F.A. Baiocchi, *Appl. Phys. Lett.*, Vol. 55, 1989, p. 1633.

78. J.W. Coburn and H.F. Winters, *J. Vac. Sci. Technol.*, Vol. 16, 1979, p. 391.

79. S. Somekh and H.C. Casey, Jr., *Appl. Opt.*, Vol. 16, 1977, p. 126.

80. S.J. Pearton, U.K. Chakrabarti, A.P. Perley, and K.S. Jones, *J. Appl. Phys.*, Vol. 68, 1990, p. 2760.

81. S.J. Pearton, U.K. Chakrabarti, A.P. Perley, C. Constantine, and D. Johnson, *Appl. Phys. Lett.*, Vol. 56, 1990, p. 1424.

82. S.K. Ghandi, P. Kwan, K.N. Bhat, and J.M. Borrego, *IEEE Electron. Dev. Lett.*, Vol. EDL-3, 1982, p. 50.

83. R.H. Burton, R.A. Gottscho, and G. Smolinsky, *Dry Etching for Microelectronics*, ed. R.A. Powell, New York: Elselvier, 1984.

84. D.L. Flamm, *Plasma Etching—An Introduction*, ed. D. Manes and D.L. Flamm, New York: Academic Press, 1989.

85. E.L. Hu and R.E. Howard, *Appl. Phys. Lett.*, Vol. 37, 1980, p. 1022.

86. U. Niggebrugge, M. Klug, and G. Garus, *Inst. Phys. Conf. Ser.*, Vol. 78, 1985, p. 367.

87. S.W. Pang, *J. Electrochem. Soc.*, Vol. 133, 1986, p. 784.

88. R. Cheung, S. Thomas, S.P. Beamont, G. Doughty, V. Law, and C.D.W. Wilkinson, *Electron. Lett.*, Vol. 23, 1987, p. 857.

89. H.E. Wong, D.L. Green, T.Y. Liv, D.G. Lishan, M. Bellis, E.L. Hu, P.M. Petroff, P.O. Holtz, and J.L. Merz, *J. Vac. Sci. Technol.*, Vol. B6, 1988, p. 1906.

90. W.C. Dautremont-Smith, J. Lopata, S.J. Pearton, L.A. Koszi, M. Stavola and V. Swaminathan, *J. Appl. Phys.*, Vol. 66, 1989, p. 1993.

91. J.P. Biersack and L.G. Haggmark, *Nucl. Insti. Meth.*, Vol. 174, 1980, p. 257.

92. J. Saito, K. Nanbu, T. Ishikawa, and K. Kondo, *J. Cryst. Growth*, Vol. 95, 1989, p. 322.

93. R.H.P. Chang, C.C. Chang, and S. Darack, *J. Vac. Sci. Technol.*, Vol. 20, 1982, p. 45.

94. S.J. Fonash, *Solid State Technol.*, Vol. 28, 1985, p. 150.

95. J.H. Thomas, III, G. Kaganowicz, and J.L. Robinson, *J. Electrochem. Soc.*, Vol. 135, 1988, p. 1201.

96. T.R. Hayes, M.A. Dreisbach, P.M. Thomas, W.C. Dautremont-Smith, and L.A. Heimbrook, *J. Vac. Sci. Technol.*, Vol. B7, 1989, p. 1130.

97. J. Asmussen, *J. Vac. Sci. Technol.*, Vol. A7, 1989, p. 883.

98. C. Constantine, D. Johnson, S.J. Pearton, U.K. Chakrabarti, A.B. Emerson, W.S. Hobson, and A.P. Kinsella, *J. Vac. Sci. Technol.*, Vol. B8, 1990, p. 596.

99. S.J. Pearton, W.S. Hobson, F.A. Baiocchi, A.B. Emerson, and K.S. Jones, *J. Vac. Sci. Technol.*, Vol. B8, 1990, p. 57.

100. D. Lecrosnier, L. Henry, A. LeCorre, and C. Vaudry, *Electron. Lett.*, Vol. 23, 1987, p. 1254.

101. R. Schutz, K. Matsushita, H.L. Hartunagel, T.Y. Longere, and S.K. Krawczyk, *Electron. Lett.*, Vol. 26, 1990, p. 564.

102. T. Sugino, A. Boonyasirikool, H. Hashimoto, and J. Shiratuji, *SPIE*, Vol. 1144, 1989, pp. 224–232.

103. V.M. Donnelly, D.L. Flamm, C.W. Tu, and D.E. Ibbotson, *J. Electrochem. Soc.*, Vol. 129, 1982, p. 253.

104. I. Adesida, C. Jones, N. Finnegan, and E. Andideh, *Proc. 2nd Int. Conf. InP and Related Mat.*, New Jersey: IEEE, 1990, p. 405.

Chapter 8
Dry Etching of In-Based Semiconductors

T.R. Hayes
AT&T Bell Laboratories

8.1 INTRODUCTION

The primary use of InP in the semiconductor industry is in the fabrication of opto-electronic devices such as lasers, photodetectors, solar cells, and, to a lesser extent, transistors to drive these devices. Applications for optoelectronic devices have become much more sophisticated in the last few years, placing greater demands on growth and processing to carefully control the dimensions of critical components within the devices. For example, distributed feedback (DFB) laser diodes used for high-speed digital communication contain active (lasing) regions whose width must be carefully controlled at about 1 μm. These lasers also contain a grating with a period of 0.20 to 0.24 μm near the active layer. Clearly, well-controlled microprocessing techniques are required to achieve these dimensions reproducibly.

While device dimensions in the direction perpendicular to the wafer plane are determined by controlled epitaxial growth, etch processes are typically used to define lateral dimensions. Wet etching has traditionally been used for this application. Wet etching is simple and inexpensive, but because most etch rates are isotropic (the same in all directions, such that the mask is undercut) or crystallographic (dependent on lattice orientation), adequate dimensional and profile control can be difficult to achieve. The rate of isotropic wet etching is often diffusion controlled, and non-uniform etching in both the vertical and horizontal directions can occur because of solution turbulence and reactant depletion effects. For some chemistries, III-V materials etch crystallographically because of differences in the atomic composition of different crystal planes. There are many varieties of crystallographic profiles, depending on the semiconductor, wafer/mask orientation, and etch chemistry. Figure

8.1 illustrates typical profiles for isotropic etching (a) and two varieties of crystallographic etching (b,c).

Dry etching is a low-pressure process that is used to remove material from a surface. Most dry-etch processes are based on sputtering by energetic ion or particle bombardment, evaporation of volatile compounds created through interactions with a reactive gas, or, most frequently, a combination of the two. Dry-etch processes that use particle bombardment usually yield vertical or angled sidewalls regardless of crystallographic orientation (Figure 1(d)). This is known as *anisotropic* etching because vertical etch rates are much higher than horizontal etch rates. As long as the etch selectivity between semiconductor and mask is high, the result is better lateral dimensional uniformity than is achieved with isotropic wet etching. The critical aspect of controlling lateral dimensions is then shifted in part to the lithographic (masking) process.

Photons can also be used as a directed energy component to give anisotropy, although less work has been done in this area. There are also dry-etch processes that do not use a directed energy component at all, and, as a result, etch isotropically or crystallographically. These are most useful for in situ (without breaking the process cycle) preparation of surfaces before epitaxial growth [1], or for in situ etch steps that require low damage but not stringent dimensional control. Because we are most interested in anisotropic etching, we will not concentrate on these processes; but we will discuss their etch mechanisms since they possess one mechanistic component of anisotropic processes.

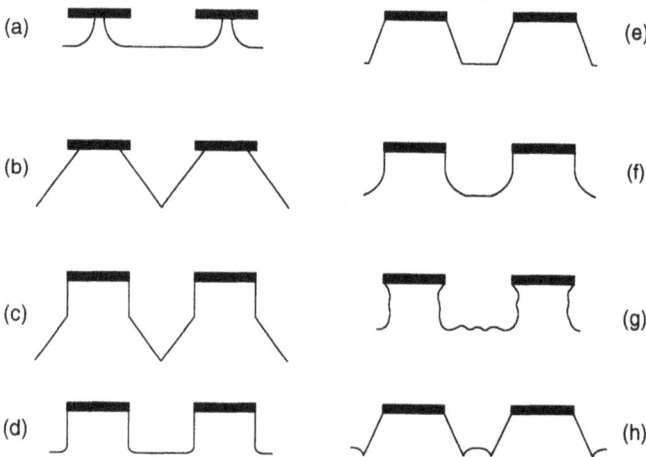

Figure 8.1 Examples of etch profiles for (100) InP surface: (a) isotropic: HBr-based etchant, mask oriented parallel to ⟨011⟩ direction; (b) crystallographic: HBr-based etchant, mask oriented parallel to ⟨01bar1⟩ direction; (c) crystallographic: HCl-based etchant, mask oriented parallel to ⟨011⟩ direction; (d) anisotropic.

In the mid to late 1980s there was a surge in the use of dry processing for In-based devices. This was partly because device performance requirements had pushed design rules to the point where the uniformity advantages of dry etching were found to increase yields, and because epitaxial growth techniques had matured to the point where both the processing and the material quality were limiting improvement in performance. DFB lasers and heterojunction bipolar transistors (HBT) are good examples. Perhaps another reason for this renewed interest was the discovery of a new dry-etch chemistry for III-V semiconductors, as discussed in Section 8.3.

This chapter reviews dry etching as applied to InP, InGaAsP, InGaAs, InAs, InAlAs, and InSb, compound semiconductors that are the heart of diode lasers, infrared detectors, and HBTs. We will discuss the dry-etch techniques and the chemistries used to etch these compounds. We will address issues of etch rate, anisotropy, semiconductor/mask selectivity, compositional selectivity, surface smoothness and contamination, etch mechanisms, and etch-induced damage. It is assumed that the reader has a general understanding of ion beam and plasma processing; if not, the reader is directed to reference sources to obtain this background [2–6]. We will also examine some in situ diagnostics for process monitoring and control, and discuss epitaxial growth on etched surfaces. Finally, we will review applications of dry etching in the fabrication of InP-based lasers. We will emphasize the performance improvements afforded by dry etching and issues of long-term reliability.

8.2 OVERVIEW OF ETCH TECHNIQUES

Dry-etch processes are by nature performed under vacuum and can be separated into two classes: those that etch isotropically and spontaneously by means of reactive chemistries, and those that use a directed energy component (e.g., energetic ions or photons) in order to etch anisotropically. They are further classified and described in the following paragraphs.

In *spontaneous chemical etching,* a wafer is exposed to a reactive gas such as chlorine which is in some cases dissociated in a plasma to create highly reactive atoms. Etching is isotropic or crystallographic, and temperature and reactant flux are used to adjust etch character. Usually the surface must be free of native oxide and other contaminants to etch uniformly.

Ion beam etching (IBE) uses a broad-area ion beam composed of a nonreactive gas such as Ar with energies on the order of 200 to 1000 eV. Focused ion beam etching (FIBE) is a serial writing process that requires no mask. Typically, Cs or Ga ions are used with energies of 10 to 20 keV (because of focusing requirements). In both cases, products are removed solely by sputtering. While GaAs can be successfully etched by IBE [7], preferential sputtering of phosphorus from InP surfaces can lead to excessive damage and, if extensive enough, the formation of In droplets [8,9]. Some progress has been made toward minimizing P loss, however, by using a liquid nitrogen-cooled sample holder [10]. Typical process variables for IBE are ion mass, energy, flux, angle of incidence, and wafer temperature.

Reactive ion beam etching (RIBE) uses an ion beam that consists of a reactive gas. Either a Kaufman or electron cyclotron resonance (ECR) ion source is used. If a halogen ion beam such as Cl^+ or Cl_2^+ is used, etching occurs by both chemical (formation of volatile products) and physical (sputtering) processes, and the process is typically very sensitive to wafer temperature. N_2 or O_2 [11,12] ion beams are also considered reactive because the formation of nitrides and oxides on the surface results in near-equal sputter rates for group III and V species, leading to smoother surfaces and less damage than IBE.

Ion-beam-assisted etching (IBAE), also known as *chemically-assisted ion beam etching* (CAIBE), uses a rare gas ion beam while a reactive gas such as Cl_2 is injected near the wafer. The etch mechanisms and results are similar to RIBE, but the ability to separate ion flux, energy, and reactant flux allows improved etch control. The local pressure and flow rate of the reactive gas are additional variables that can affect the process.

Reactive ion etching (RIE) uses radio frquency (RF) power to maintain a plasma between at least two electrodes, with the wafer placed on the powered electrode. The chamber walls and second electrode are usually grounded. In "plasma etching," the wafer is placed on the smaller (grounded) electrode, an inferior configuration for anisotropic etching because of the significantly lower plasma sheath voltage and ion energies. Although 13.56 MHz is most often used because of federal transmission regulations, frequencies ranging from 30 kHz to 20 MHz are also used. Reactive ions and neutrals are generated in the plasma and interact with the semiconductor surface to create volatile products. An electric sheath-field is formed at plasma-surface boundaries that accelerates positive ions to the wafer and enhances the etch rate by stimulating the formation and desorption of volatile products. A wider variety of chemistries have been studied for RIE than the other techniques. The interaction between experimental variables such as pressure, RF power, bias voltage (the sheath-field magnitude), flow rate, and wafer temperature and the resulting etch rate, anisotropy, and surface smoothness are much more complicated than for the other techniques.

There are also some emerging plasma-based processes that combine semiremote, magnetically confined plasma generation of ions and reactive species with RF biasing of the wafer for improved etch control. The most well known of these is ECR etching. Microwave power, typically at 2.45 GHz, is used to excite the plasma and, due to magnetic confinement, plasma densities are significantly higher than in RIE. This results in ion energies at the wafer that are much lower (typically tens of eletronvolts), creating the possibility of directed, low-damage etching. The application of an RF bias to the wafer allows more independent adjustment of the flux and energy of positive ions striking the wafer than in RIE. Helicon and helical resonator plasma sources also belong in this group.

8.3 ETCH CHEMISTRIES AND MECHANISMS

There are generally two classes of chemistries used for dry etching of III-V semi-conductors: halogen-based and hydrocarbon-based. We discuss each in detail and compare them in Section 8.3.3.

8.3.1 Halogen-Based Chemistries

The limited volatility of the group III fluorides obviates the use of F-based chemistries for etching III-V semiconductors. The higher halogens form sufficiently volatile compounds, however, and the first successful chemistries to be applied to III-V etching used Cl-containing gases. In fact, III-Vs etch spontaneously when exposed to Cl_2 or to Cl atoms created in an upstream plasma [14], with a strong temperature dependence. For example, in a study by Lishan *et al.* [14], the room temperature etch rate of GaAs in a Cl_2/Cl atom atmosphere was 300 Å/min, but InP did not etch at this rate until the wafer was heated to 130°C. As suggested by Table 8.1, the reason for the lower InP etch rate is that the indium chlorides ($InCl_3$ as well as InCl and $InCl_2$) [15] are significantly less volatile than the chlorides of Ga, P, and As [16] (see Table 8.1), such that spontaneous etching of In-based materials is usually limited by the $InCl_x$ desorption rate. Consistent with this point, Donnelly *et al.* [17] showed that in a low-power Cl plasma at 0.3 torr, the InP etch rate exhibits an

Table 8.1
Vapor Pressures for Group III and V Compounds

Compound	Vapor Pressure (torr) [16]	Compound	Vapor Pressure (torr) [139]
$InCl_3$*	$\sim 10^{-8}$ (100°C), $\sim 10^{-4}$ (180°C), $\sim 10^{-2}$ (250°C) [15]	$(CH_3)_3In$	7.2 (30°C)
$GaCl_3$	2 (50°C)	$(C_2H_6)_3In$	1.2 (44°C)
$AsCl_3$	290 (100°C)	$(CH_3)_3Ga$	65 (0°C)
PCl_3	400 (57°C)	$(CH_3)_3P$	380 (0°C)
PCl_5	1 (56°C)	PH_3	Gas at 0°C
$AlCl_3$	1 (100°C)	AsH_3	Gas at 0°C
$SbCl_3$	1 (49°C)		
AsF_3	750 (56°C)		
AlF_3	1 (1238°C)		

*InCl and $InCl_2$ vapor pressures are on the same order as $InCl_3$ [15].

Arrhenius dependence with temperature, with an activation energy nearly equivalent to the heat of vaporization of $InCl_{3(s)}$.

It is necessary to remove the low-reactivity oxides of InP and GaAs by hydrogen plasma treatment [18] or some other method before uniform, spontaneous etching can occur. This effect has been used to create an in situ lithographic process using FIBE (to remove the InP native oxide in localized areas) and thermal Cl_2 etching for pattern transfer [19].

Ion bombardment must be added to etch anisotropically. RIBE [20–23] has been most often performed with a Cl ion beam, and IBAE [24,25] has been most often performed with Cl as the dosing gas. Etch rates range between 500 and 5000 Å/min and are dependent on the wafer temperature, ion flux, ion energy, and, for IBAE, reactive gas flux. If sufficient reactant is available, the removal of $InCl_3$ from the surface is the rate-limiting step [26]. Near room temperature, sidewalls usually slope outward from the mask (known as *overcut profiles*) with an angle of 15° to 30° from normal, independent of crystallographic orientation (Figure 1(e)). Vertical sidewalls with rounded corners can be obtained by angling and spinning the sample during etching [20] (Figure 1(f)).

Etch selectivity with respect to SiO_2 and photoresist is poor at room temperature because of an inadequate chemical etch component for InP. To get around this, metal masks like Ni can be used [25], or strontium fluoride, which works well because it etches very slowly in Cl plasmas [27]. Operating at elevated temperature, however, yields more vertical sidewalls, increases InP etch rates, and can yield InP/SiO_2 selectivities on the order of 10 [28]. Optimum results are achieved at temperatures between 170°C and 200°C [21,25,28], which can be considered to be the effective sublimation temperature of $InCl_x$ products. At higher temperatures, profiles become undercut due to excessive spontaneous etching (Figure 1(g)), and surfaces become rougher. GaAs Cl-RIBE etch rates are higher than InP etch rates due to the higher volatility of gallium chlorides over indium chlorides. This leads to InGaAs and InGaAsP etch rates that are higher than InP by about a factor of two to three [28].

Iodine has also been used to etch InP. The use of ICl as a reactive gas for IBAE has enabled room-temperature InP etch rates of 1 μm/min, but leads to rough, undercut sidewalls [29]. Because of the high volatility of the indium iodides, room temperature IBAE using I_2 as the reactive gas [30] has been found to give smooth surfaces with no evidence of In enrichment, in contrast to Cl-based etching. Sidewalls were overcut, however, and etch rates were low (comparable to IBE). Iodine RIBE [31] has also been shown to allow preparation of In-based samples for transmission electron microscopy with smooth surfaces and no evidence of In droplets. It seems that iodine is a promising chemistry that has not yet been fully exploited.

Ar/O_2 [12], N_2, and N_2/O_2 chemistries [11] for RIBE have the advantage of being nontoxic and etch without excessive preferential sputtering of P. While volatile products are not formed, the formation of oxides and nitrides leads to similar sputter removal rates for group III and group V species. Compared to halogen chemistries, sputter/etch yields (atoms per incident ion) are much lower, and profiles tend to exhibit trenching (a groove at the bottom of the sidewall that occurs because of directional scattering of ions off the sidewall, Figure 1(h)). Reasonable InP/photoresist (PR) selectivities of 3 to 5:1 can be obtained, and so shallow structures such as gratings for DFB lasers have been successfully fabricated [32].

Ultraviolet or visible photons can be used instead of ions as a directed energy source for low-damage etching of InP [33–36]. Patterns can be created by projection lithography, writing with focused laser beams, or standard surface masking. We briefly describe here the results of masked surface studies that are relevant to anisotropic etching. While dosing an InP surface with Cl gas or atoms, we have used a pulsed excimer laser to heat the surface repeatedly above the sublimation temperature of $InCl_x$, removing monolayer(s) of products with each pulse [35]. While this process is capable of high etch rates (greater than 1 μm/min) and is expected to impart little damage, advances must still be made in broad-area etching, and, because of sidewalls that slope at 45°, improvements in vertical etching are necessary for applications requiring high anisotropy.

RIE has been applied to In-based compounds using Cl-based chemistries in such diverse forms as Cl_2 [37], CCl_4 [38,39], $SiCl_4$ [40,41], CCl_3F [42], CCl_2F_2 [43,44], and BCl_3/Cl_2 [45]. The latter two Freons have the advantage of being nontoxic, but their role in depletion of the ozone layer will limit their future availability. For most of these chemistries, O_2 and rare gases (Ar, He) have been used as additives. The addition of O_2 increases the concentration of Cl atoms in Cl-based discharges [38], tending to increase etch rates and improve surface smoothness. Rare gases tend to enhance sputtering and can increase etch rates.

Although the chemistries seem diverse, some general statements can be made about Cl-based RIE under optimized conditions. Room temperature etch rates for InP fall between 200 and 2000 Å/min, with typical values between 400 and 1000 Å/min. Surfaces tend to be somewhat rough, and sidewalls have overcut profiles. Just as with RIBE, the sidewall angle varies with etch conditions, but usually falls between 15° and 35° from the normal. This is illustrated in Figure 8.2, a secondary electron micrograph of a SiO_2-masked mesa etched by $SiCl_4$-RIE at 70°C [46]. The micrograph also shows roughness due to micromasking effects. Etched surfaces contain a Cl residue that can be completely removed by cleaning in water and organic solvents [41]. Because of the lower volatility of $InCl_x$ species compared to the chlorides of Ga, As, Al, and Sb, InP etches more slowly than the other compounds, with the exception of InAlAs, which tends to etch near the same rate as InP [41].

The overcut sidewalls observed with near-room temperature, Cl-based chemistries (not only for RIE but also for RIBE and IBAE [37,43,40,47] likely result from the low volatility of the indium chlorides. $InCl_x$ sputtered from horizontal surfaces is believed to deposit on mesa sidewalls and inhibit etching [40,48]. An angled sidewall probably begins to develop because sputter yields are lower on the angled surface (true for most materials at angles less than about 30° from normal [49]), such

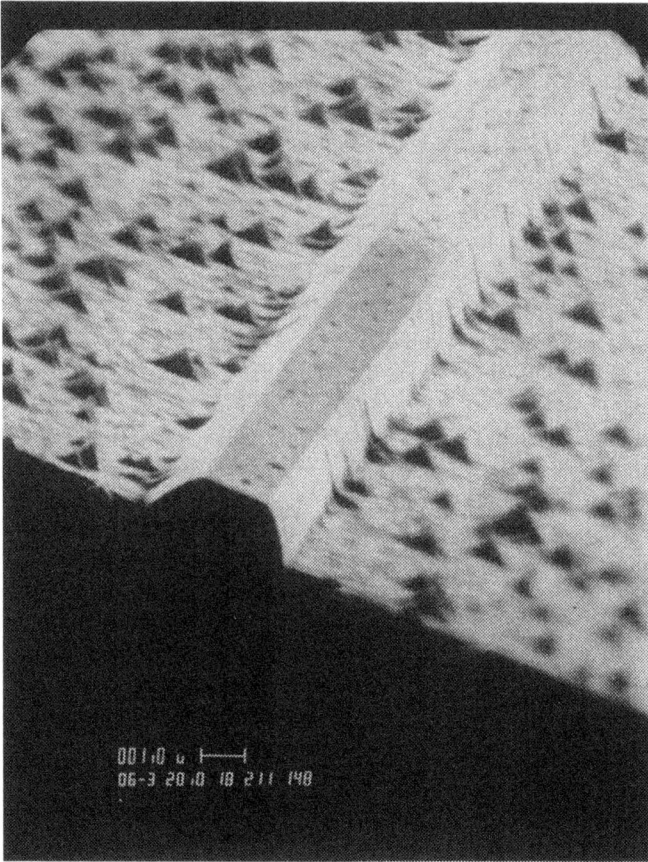

Figure 8.2 Mesa etched in a InGaAsP/InP double heterostructure using SiCl$_4$ RIE. Pressure = 30 mtorr, power density = 0.8 W/cm^2, bias voltage = 430 V, and electrode temperature = 70°C. The SiO$_2$ mask is still on the mesa.

that the etch rate and deposition rate are roughly balanced. The trend toward more vertical sidewalls at elevated temperatures, where the InCl$_x$ desorption rate is higher, supports this explanation. Overcut profiles are desirable if it is necessary to apply a conformal coating over a mesa (e.g., a metal line connecting two parts of an integrated circuit). Otherwise the result is a loss of dimensional control relative to perfectly vertical sidewalls. For example, when etching mesas (through the active layer) for buried heterostructure lasers, variations in the depth of the active layer, due to layer thickness nonuniformities, will result in variations in active width from the sloping sidewalls. As we have discussed, etching at a higher temperature yields more vertical sidewalls, but with increased surface roughness because of nonuniform volatilization.

A small amount of work has also been done on bromine- and iodine-based RIE of InP. Br_2/N_2 and Br_2/Ar mixtures have been shown to yield vertical InP sidewalls with slightly rough surfaces near room temperature [50]. The vertical sidewalls are in contrast with Cl-based etching, and they presumably result from the higher volatility of $InBr_x$ products. RIE with Ch_3I/O_2 has also been tried, although etch rates were low and sidewalls were overcut [30]. More work on Br- and I-based RIE is necessary to fully explore these chemistries.

8.3.2 Hydrocarbon-Based Chemistries

This second class of etchants for III-V semiconductors was introduced by Niggebrügge et al. [51] with an etch study of CH_4/H_2 mixtures, and this chemistry has since been heavily examined [52–54]. C_2H_6/H_2 mixtures perform with nearly identical results [55–57]. Hydrocarbon feedstocks have the advantage of being noncorrosive compared to Cl-based chemistries, and of low toxicity. A hydrocarbon polymer [53] deposits on less-reactive surfaces like PR, SiO_2, and chamber walls. This polymer can improve mask/semiconductor etch selectivity under the proper conditions, but also gives rise to particulates and an upper etch rate limitation. H_2, Ar, or O_2 are typically added as diluents to minimize deposition and enhance etch characteristics. Phosphine, detected by mass spectrometry [58,53], and organoindium species (e.g., $(CH_3)_xIn$) are believed to be the volatile products of InP. As shown in Table 8.1, organometallic and metal hydride compounds that might be considered to be etch products have high room temperature vapor pressures. While hydrocarbon chemistries have been applied to RIE and ECR processing, the low sticking coefficients of methane and ethane preclude their use in IBAE. Some authors have coined the term *metalorganic reactive ion etching* (MORIE) [59], a reference to the chemical parallels with metalorganic chemical vapor deposition (MOCVD).

We will use CH_4/H_2 reactive ion etching of InP to demonstrate general aspects of hydrocarbon-based etching. In Figure 8.3, InP etch rate and polymer deposition rate (on Si) have been plotted *versus* the percentage of CH_4 in H_2. The chamber has been previously "seasoned" by depositing polymer, and even for pure hydrogen discharges, a nonzero etch and deposition rate are observed. This illustrates the importance of wall reactions in generating organic etchants and of preparing the chamber for etching. While such a memory effect can be a nuisance, it poses no reproducibility problem for repetitive processing. At low methane fractions, etched surfaces are depleted of P (as measured by auger electron spectroscopy (AES)) because of the H-rich environment. In this regime, the flux of organic species to the surface limits the rate of In desorption and, thus the etch rate. As the fraction of methane in the discharge is increased, the etch rate increases and the surface becomes

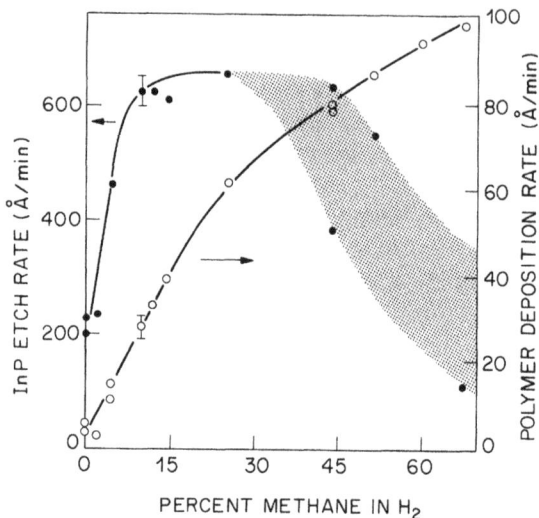

Figure 8.3 InP etch rate and polymer deposition rate (measured ellipsometrically) *versus* %CH$_4$ in H$_2$. Conditions: pressure = 90 mtorr, flow rate = 230 sccm, power density = 0.6 W/cm^2, self-bias = 270V.

less depleted of P. Past the plateau, the incident flux of organic reactants is so high that they tend to react with each other rather than with the semiconductor, resulting in the formation of an etch-inhibiting polymer film. Interestingly enough, until the etch rate begins to drop, carbon contamination of the surface is minimal, and x-ray photoelectron spectroscopy (XPS) shows that some of the C detected on the surface is bonded to In [60]. For CH$_4$/H$_2$ mixtures, maximum etch rates lie between 300 and 700 Å/min. Hydrocarbon RIE also exhibits a significant loading effect. For example, a 700-Å/min etch rate on a 0.5-cm^2 InP wafer can drop to 300 Å/min on a 5-cm^2 wafer [61].

Figure 8.4 is a secondary electron micrograph of a SiO$_2$-masked mesa etched by CH$_4$/H$_2$ RIE, showing the vertical sidewalls that can readily be achieved with this chemistry. Under some conditions (for example, higher pressure and methane fraction), undercut profiles are obtained. AES studies of the sidewall composition [53] have shown that there is minimal C contamination under typical optimum etch conditions (50–90 mtorr, 10% methane, bias voltage ~300V). This suggests that rather than an etch inhibitor mechanism of anisotropy (which for Cl chemistries gives overcut profiles), ion-driven enhancement of the etch rate in the direction of bombardment [62] is the primary cause of anisotropic etching. The fact that undercut sidewalls can be obtained with just an increase in methane fraction suggests that there is some small spontaneous etch effect and that the products are highly volatile.

Figure 8.4 Mesa etched in an InGaAsP/InP double heterostructure using CH_4/H_2 RIE. 10% CH_4 in H_2, pressure = 50 mtorr, power density = 0.4 W/cm^2, bias voltage = 340V. The SiO_2 mask is still on the mesa.

In fact, InP has been found to etch spontaneously in a downstream reactor producing methyl radicals and H atoms at a rate of 100 Å/min at 85°C [63]. Therefore, if the correct reactant mix is available, spontaneous etch rates can be substantial. Many authors have noted that for hydrocarbon RIE, however, ion bombardment is necessary to stimulate etching. This may indicate that the flux of reactive species to the surface in a hydrocarbon plasma is not the optimum for etching, and that ion bombardment is necessary to activate the surface and stimulate the formation or desorption of volatile products.

Etch rate is dependent on semiconductor composition for the hydrocarbon etchants as well, with the etch rate order being InP>InGaAsP>InGaAs>GaAs>AlGaAs [51,59], nearly the opposite of that observed for Cl-based etching. The difference between InP and GaAs vertical etch rates is a factor of three to six, depending on conditions. Etched InGaAsP surfaces are found to be depleted of both As and P, and GaAs surfaces are depleted of As because of the facile formation and desorption of arsine and phosphine. It is interesting to note that organogallium compounds and arsine have higher vapor pressures than their In and P analogs, yet these materials etch more slowly than InP. In order to successfully etch GaAs, it is necessary to etch under conditions (lower pressure and higher bias voltage) that favor more of a sputtering component than for etching InP. The mechanism responsible for these effects is not well understood. For both InP and GaAs, it has been found that the addition of Ar to the gas mixture increases etch rates and, in some cases, improves surface smoothness because of an increase in sputtering.

Microwave ECR etching has also been used to etch InP with $CH_4/H_2/(Ar)$ mixtures [64–67]. In order to avoid polymer deposition and obtain usable etch rates, an RF bias must be applied to the sample. It is extremely easy, however, to excessively deplete the surface of P, even to the point of In droplet formation, if operating under conditions that give excessive ion bombardment (both flux and energy) and excessive H atom partial pressure. General conditions to avoid include low pressure (<1 mT), high RF bias (>100V self-bias), and high microwave power (>150W–250W). Decreasing the H atom partial pressure and increasing sputtering by adding Ar to the discharge improves the etch characteristics significantly [64,65]. The carefully tuned process can yield etch rates that are in the range of those obtained with hydrocarbon RIE and similar anisotropic profiles and smooth surfaces [67,66]. ECR etching has the potential of causing very low damage, as discussed in Section 8.4.

8.3.3 Halogen and Hydrocarbon Mixtures

We now turn to a comparison of Cl- *versus* hydrocarbon-based etch chemistries. Pearton *et al.* have carried out a set of self-consistent etch experiments using CCl_2F_2/O_2 [44,56,68], $CClF_2H/O_2$ or CCl_2FH/O_2 [47], and C_2H_6/H_2 [68,44,56] to etch InP, $In_{0.53}Ga_{0.47}As$, $In_{0.52}Al_{0.48}As$, InAs, and InSb, which enables study of the transition from chlorofluorocarbon- to hydrocarbon-based chemistries. Typical process conditions were a pressure of 4 mtorr, a total flow rate of 20 sccm, power densities of 0.56 to 0.85 W/cm^2, and bias voltages of 350V to 450V, carried out at room temperature. The following trends were observed. Etch rates for all semiconductors decreased with increasing hydrogen/hydrocarbon composition: $CCl_2F_2>CCl_2FH$ or $CClF_2H>C_2H_6/H_2$, and surface smoothness improved with increasing hydrocarbon fraction: $CCl_2F_2<CCl_2FH$ or $CClF_2H\leq C_2H_6/H_2$. All the Cl-containing gases gave overcut sidewalls, while C_2H_6/H_2 gave vertical sidewalls. These trends are in agreement with all of the conclusions that have been reached up to this point. It has also been shown that the addition of SF_6 to Cl-based chemistries provides a natural etch stop for the removal of In-based layers over InAlAs [47,69], an application relevant to the fabrication of HBTs. The etch stop mechanism involves the formation of a nonvolatile AlF_3 layer on the InAlAs surface [70].

The main deficiencies of Cl-based etchants are that overcut sidewalls are obtained at room temperature, and even at higher temperatures that yield more vertical sidewalls, surfaces tend to be rough. The main deficiency of hydrocarbon etchants is that etch rates are low, possibly because reactive species responsible for etching are not efficiently generated in hydrocarbon plasmas, or because their lifetimes are short due to gas phase reactions. As evidence of this, mass spectrometric sampling of hydrocarbon plasmas detects an abundance of hydrocarbon species containing up to nine C atoms [53].

It seems, however, that a carefully composed mixture of hydrocarbon- and halogen-containing gases could give the optimum etch results. In addition to the RIE experiments with $CCl_xF_yH_z$ [47], studies of CH_3Cl/H_2 [57] and $Cl_2/CH_4/H_2/Ar$ [71]

mixtures have been reported. In particular, promising results have been achieved with RIE using a carefully balanced mixture of the latter chemistry at about 180°C. Etch rates on the order of 2000 to 5000 Å/min have been achieved, with a 10:1 InP:SiO$_2$ etch selectivity and vertical sidewalls. However, conditions had to be carefully adjusted to minimize "grass" formation due to micromasking by indium chlorides. It seems likely that the combination of Br or I and hydrocarbon chemistries will result in room temperature, vertical sidewall, high-rate etching without the micromasking problems that occur with chlorine chemistries.

8.4 ETCH-INDUCED DAMAGE

Unfortunately, anisotropic etching has its downside, and this is the damage that is almost inevitably imparted to the surface, bringing about degradation in the optical and electrical properties of the material. Dry etch-induced damage in III-V semiconductors takes the form of lattice damage (collision cascade mixing and preferential sputtering caused by ion bombardment) and "chemical" damage, such as dopant passivation for H-containing etch chemistries. The result is a reduction in photoluminescent (PL) lifetimes and carrier concentrations, which can potentially adversely affect the performance and reliability of optoelectronic devices. The depth of lattice (crystallographic) damage can be measured by channeling Rutherford backscattering spectrometry [68,53], and atomic composition changes can be measured by AES and XPS depth profiling [53,60]. While these techniques have high surface sensitivities, they are limited by their modest detection limits, 0.1% to 1%. Electrical and optical measurements are much more sensitive to low defect levels, 1 ppm or lower, and are useful for determining the overall depth of damage on a sensitive scale, one that is relevant to optoelectronic device performance. For example, electrochemical capacitance-voltage profiling can be used to measure changes in carrier concentrations arising from deep-level point defects. Schottky contact current-voltage measurements are sensitive to carrier levels and deep-level defects within the depletion region, and PL intensities are strongly affected by the presence of deep-level defects. When combined with wet etching, the latter two methods can yield the depth of near-surface damage by successively removing layers and monitoring the return to an undamaged condition. We are unable to meaningfully compare the relative sensitivities of these three diagnostics, however, since they are affected by doping levels, the kind of defects generated, band bending effects, and the details of the experimental methods. Here we assume similar sensitivities for the sake of comparison.

Table 8.2 summarizes electrical and optical measurements of the depth of damage induced by IBE, RIE, and ECR. Numbers in braces in the following discussion refer to study numbers in the table. Ar$^+$ IBE at 500 eV has been found to induce damage in InP to depths of 200Å {1} and 500Å {2} by PL/WE and Schottky contact/wet-etch (WE) methods, respectively. This factor of two variation in damage depth is probably typical of the scatter when comparing results from different authors with differing measurement methods and experimental configurations. Damage to depths of 1000Å to 2000Å occurs to both InP and GaAs when operating both Cl- and hydrocarbon-based RIE processes at low pressure (4 mtorr) and bias voltages between 280V and 550V {3–5}. Operating at higher pressure (e.g., 50 to 90 mtorr) with the attendant lower bias voltages {6,7} results in shallower damage, ~200Å deep. The predominant effect here is that charge-transfer neutralization limits the energy of ions and fast neutrals transported across the sheath more effectively at higher pressures, such that upon impact the energy is dissipated and defects are created much closer to the surface, bias voltage being roughly constant. In many applications wet-etch removal of such thin damaged layers does not compromise dimensional integrity. Lighter ions generate more damage than heavier ions [72,73] because of their higher velocity, and so while there is no dramatic difference in the depth of damage between Cl-based and hydrocarbon-based etchants, the extent of damage may be greater in the latter case due to the presence of low mass species, such as H and C.

$CH_4/H_2/Ar$ ECR {8} has been used to etch surfaces with extremely shallow damage (~20Å) by operating with no RF bias applied to the wafer (ion energies less than 50 eV) [64]. While the etch rate was low (50 Å/min), it should be possible to operate with an RF bias to enhance etch rates and anisotropy, and then remove the last few hundred angstroms of material under zero bias conditions to leave a nearly undamaged surface. Of course, this same mode can be used to optimize other methods, such as RIE for anisotropic etching, with minimum damage [70]. It has also been found that degradation-free ECR etching of InP can be obtained by adding a small amount of PCl_3 to the discharge when operating at low RF bias [67]. Photoluminescence measurements showed no loss of PL yield upon etching under optimized conditions, the improvement being attributed to the removal of P vacancies by the P overpressure. Etch rates are low under these conditions, so this may be a method for conditioning the surface after etching under a higher etch rate, which leads to more damage, and to obtain the structure required. The technique may be extendable to RIE insofar as conditions of ultralow damage etching can be found, with final surface conditioning with PCl_3 or another P-containing compound.

It has been observed by many authors that the measured maximum depth of etch-induced damage exceeds the predicted maximum range of ions implanted in the material, as calculated by Monte Carlo methods, by roughly an order of magnitude [74,73,68]. Several mechanisms may be responsible, including ion channeling effects and bombardment enhanced diffusion [75], and rapid propagation of defects such as dislocations. In addition to damage to horizontal surfaces, it is also important to consider damage to sidewalls (i.e., in the direction perpendicular to ion transport). Sidewall damage originates from the overlap of the spatial collision cascade distribution with the material under the mask that is not etched but is partially damaged, as well as from gas phase scattering of ions and energetic neutrals into exposed

Table 8.2

Optical and Electrical Studies of Etch-Induced Damage

Study, Method	Material	Analysis Method	Chemistry	Etch Conditions	Damage Depth (Å)	Damage Type	Ref.
{1}, IBE	n-InP, 1E16 cm^{-3}	PL/WE	Ar, <= 0°	E_{ion} = 500 eV / 1350 eV	200 / 500	Point defects	8
{2}, IBE	n-InP, 6E15 cm^{-3}	SC/WE	Ar, 10°C, <= 45°	E_{ion} = 500 eV	600	Point defects	9
{3}, RIE	n-InP, 1E17 cm^{-3}	El. C-V	1C$_2$H$_6$:H$_2$ / 19CCl$_2$F$_2$:1O$_2$	4 mT, 0.85 W cm^{-2}, 380V / 4 mT, 0.85 W cm^{-2}, 340V	1300 / 1000	Point defects	[68]
{4}, RIE	n-InP, 1E17 cm^{-3}	El. C-V	19CHCl$_2$F:1O$_2$	4 mT, 0.56 W cm^{-2}, 280V / 4 mT, 1.3 W cm^{-2}, 550 V	1500 / 2000	Point defects	[47]
{5}, RIE	n-GaAs, 1E17 cm^{-3}	El. C-V	19CCl$_2$F$_2$:1O$_2$	4 mT, 0.85 W cm^{-2}, 380V	1500	Point defects	[83]
{6}, RIE	n-InP, 3E17 cm^{-3}	SC/WE	1CH$_4$:9H$_2$	90 mT, 0.6 W cm^{-2}, 270V	200	Point defects	[60]
{7}, RIE	n-InP, 1E18 cm^{-3}	El. C-V	10SeCl$_4$:5Ar or 10Cl$_2$:5Ar	50 mT, 100V	<200	Point defects	[41]
{8}, ECR	n-InP, 6E15 cm^{-3}	PL/WE	5CH$_4$:15H$_2$:7Ar	1 mT, 250W, 0V bias	20	Point defects	[64]
{9}, RIE	p-InP, 1E18 cm^{-3}	El. C-V	1CH$_4$:9D$_2$	90 mT, 0.6 W cm^{-2}, 270V	2000	H passivation	[85]
{6}, RIE	p-InGaAsP, 1E18 cm^{-3}	El. C-V	1CH$_4$:9D$_2$	90 mT, 0.6 W cm^{-2}, 270V	1500	H passivation	[60]
{10}, RIE	InGaAs, 1E19 cm^{-3}	El. C-V	1CH$_4$:14H$_2$ / 1CH$_4$:14He / 19CHF$_3$:1H$_2$ CHF$_3$	75 mT, 0.4W cm^{-2}	800 / 700 / 500 / 400	H passivation	[86]

PL/WE Photoluminescent intensity measurements with wet-etch depth profiling
SC/WE Schottky contact current-voltage measurements with wet-etch depth profiling
El. C-V Capacitance-voltage measurements with electrochemical etching

sidewalls. Sidewall damage has been studied both for GaAs [76,77] and, to a lesser extent, InP structures. For GaAs quantum wires patterned by RIE and ECR, it was found that damaged layers were much thinner for ECR, and comparable to wet etching [78]. For InGaAs/InP quantum wire structures patterned by Ar/O$_2$ RIBE, analysis of PL intensities *versus* wire width yielded a "dead layer" thickness of 190Å [79], and separate magnetoresistance experiments led to an estimated dead layer thickness of 300Å [80]. For micropillars containing an InGaAs quantum well with InAlAs barriers etched in CH$_4$/H$_2$, peak current measurements made with a scanning tunneling microscope led to an estimated dead layer thickness of 300Å [81].

Hydrogen passivation of zinc acceptors occurs when etching InP {9}, InGaAsP {6}, and InGaAs {10} with H-containing feedstocks. For GaAs and AlGaAs, donors are most strongly affected [82,83]. This results in a significant decrease in carrier concentration within 500Å to 2000Å of the etched surface. For Figure 8.5, *p*-InP (1E18 cm^{-3}) was etched in a 9.7% CH$_4$/D$_2$ discharge under the conditions given in Table 2, {9}. Deuterium atom concentration measured by secondary ion mass spectrometry (SIMS) and carrier concentration measured by electrochemical C-V profiling are plotted *versus* depth. The inverse relationship between deuterium concentration and carrier concentration identifies deuterium (H) as a passivant for acceptors

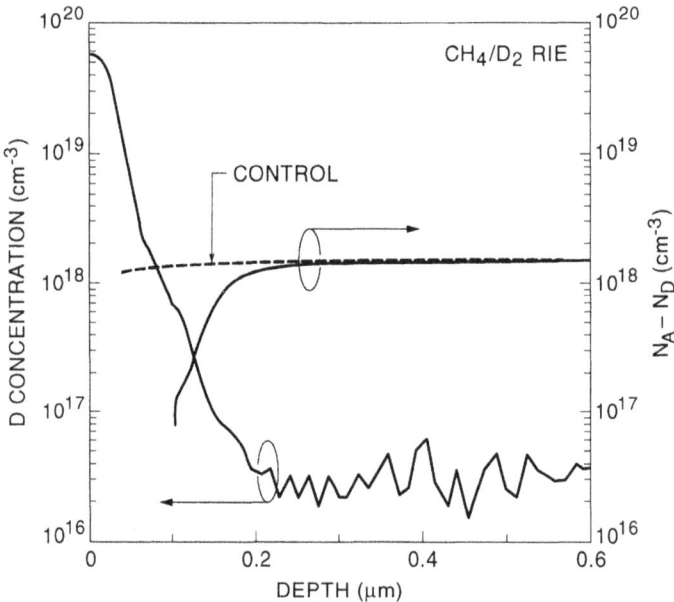

Figure 8.5 Deuterium atom concentration (by SIMS) and free carrier concentration (by eletrochemical C-V profiling) *versus* depth for *p*-InP etched in 9.7% CH$_4$/D$_2$.

in InP. The general mechanism appears to be diffusion of H into the material, followed by spontaneous reaction to form neutral complexes with acceptors [84]. *n*-InP shows little passivation effect. Fortunately, near-complete reactivation of acceptors can be accomplished by annealing at 350°C for 1 min [85], and the effect is completely reversed by heating for 20 min [86]. Processes that require regrowth on etched *p*-InP surfaces (at temperatures on the order of 600°C) will show no residual passivation.

8.5 ETCH DIAGNOSTICS

In situ etch diagnostics play an important role in process development and control. Typically a signal is monitored that identifies the layer being etched or allows determination of the thickness of the semiconductor removed, or allows measurement of the wafer temperature. We will discuss diagnostics that have been applied to InP/InGaAs(P) heterostructure etching that have the characteristics of simplicity, low cost, and high reliability, and, as a result, are particularily suitable for manufacturing as well as research and development.

Mass spectrometry can be used to sample species formed in the plasma whose concentrations vary depending on the layer being etched. The mass spectrometer must be differentially pumped to allow direct sampling from the comparatively high-pressure plasma environment. Schmid *et al.* [87] found that when etching heterostructures in a CH_4/H_2 plasma, the directly-sampled PH_4^+ signal varies depending on whether InP or InGaAs(P) is being etched, because of the difference in P concentration. A depth resolution of 50 nm or better (suitable for quantum well identification) and a compositional resolution of better than 15 at.% P was determined.

Optical emission spectroscopy (OES) is a simple, inexpensive diagnostic technique. Light from the plasma can be coupled into a moderate resolution spectrometer through a fiber, or if a narrow bandpass filter is used, directly in a photodetector. Indium atomic emission has been observed at 451.1 nm when etching InP in CCl_4 [39] and CH_4/H_2 [88] plasmas, and we have applied this to heterostructure identification [89]. In Figure 8.6(a), the In emission intensity is plotted *versus* time while etching through a double heterostructure wafer by CH_4/H_2 RIE. The signal intensity is clearly lower when etching through quaternary layers (Q_1–Q_3) which have a lower In concentration than InP. The technique has a depth resolution of less than 100Å and a compositional resolution of less than 12 at.% P, and so has comparable sensitivity to mass spectrometry, but at considerably lower cost and probably higher reliability.

While the above techniques are readily capable of identifying heterostructure layers during etching, no direct information is given about the etch rate. In situ etch rate measurements can expedite process development, and are useful for monitoring

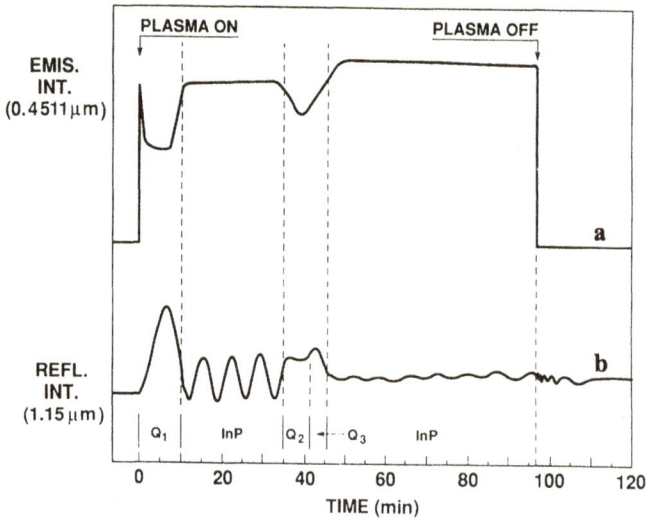

Figure 8.6 (a) Indium atom emission intensity (4511Å) *versus* etch time; (b) reflected intensity *versus* etch time, for $\lambda = 1.15$ μm laser, $\theta_i = 70°$. Heterostructure: Q_1: 1200Å InGaAsP ($\lambda_g = 1.3$ μm); 6000Å InP; Q_2: 900Å InGaAsP ($\lambda_g = 1.3$ μm); Q_3: 900Å InGaAsP ($\lambda_g = 1.1$ μm); ~300Å deep grating on InP substrate.

process reproducibility. Interferometry, the analysis of the intensity of coherent light reflected from a wafer during etching, has been used for more than a decade to measure etch rates and identify layers [90]. It has traditionally been applied using patterned samples and a HeNe laser operating at a wavelength of 6328Å (which is absorbed by semiconductors). Infrared interferometry, however, using light of a wavelength longer than the bandgap of the semiconductor, yields etch rate and layer identification even for unpatterned samples, and, in addition, allows direct measurement of the wafer temperature [91,92]. We have successfully used wavelengths of 1.15 μm (HeNe laser), 1.3 μm, and 1.55 μm (single-mode InP diode lasers).

Figure 8.6(b) shows the reflected intensity of a $\lambda = 1.15$ μm laser during etching, acquired at the same time as the OES signal. The signal modulation arises from interference between light reflected from the etching top surface, buried heterointerfaces, and the polished wafer bottom. The thickness of material Δd removed between sequential peaks is given by

$$\Delta d = \lambda/2(n^2 - \sin^2\theta_i)^{1/2} \tag{1.1}$$

where λ is the laser wavelength, n is the refractive index of the material, and θ_i is the angle of incidence with respect to the normal. Under the conditions of Figure 8.6, $\Delta d = 1850$Å and 1680Å for InP and $\lambda_g = 1.3$ μm InGaAsP, respectively.

Differences in refractive index lead to a change in slope of the reflected intensity at heterointerfaces, such that even the location of 100Å quantum wells and barriers can be identified [93]. The vertical lines identifying the interfaces in Figure 8.6 were verified by computer modeling [94]. The higher frequency oscillations observed after the plasma is extinguished arise from cooling of the wafer, and can be used to measure the process-induced change in wafer temperature [92]. This effect arises from the temperature dependence of the refractive index of the material, and to a lesser extent thermal expansion. A method that gives nearly identical information, based on infrared ellipsometry, has also been reported [95]. This method requires an elliptical polarizer, rotating analyzer, and computer analysis of the data, however, in addition to the infrared source and detector required for interferometry.

Measurement and control of the actual wafer temperature (not the electrode temperature) in a plasma is a vital issue, especially for Cl-based etchants whose etch characteristics show larger temperature sensitivities than hydrocarbon-based etchants. Plasma radiation has been found to increase the temperature of a poorly heat-sunk wafer 50°C to 75°C [96], with a strong pressure and power dependence. Wafers should be thermally bonded to the electrode to minimize this effect, but this is not always compatible with further processing, such as epitaxial growth. In addition to infrared interferometry, two other methods for actual wafer temperature measurement that have been put forth are the fluoroptic probe method [96] and in situ photoluminescence [97]. While the former two can be used regardless of the semiconductor, the latter requires a direct bandgap semiconductor. Photoluminescence also provides valuable information on the damage state of the etched surface, aiding in the development of low-damage processes.

8.6 EPITAXIAL GROWTH ON DRY-ETCHED SURFACES

Many device structures require epitaxial growth on etched surfaces and, in order to obtain epitaxy, it is usually necessary to remove from the surface any crystallographic damage resulting from ion bombardment. This can be accomplished by either brief wet chemical etching of the surface to remove the damaged layer, or by gas phase processing. The higher pressure RIE processes in Table 8.2, exhibiting damage no deeper than 200Å, are especially appropriate for this procedure.

Epitaxial growth on dry-etched InP surfaces can also be accomplished by annealing in a P-containing ambient. For example, hydrocarbon RIE-etched InP has been annealed and treated with P_2 (525°C) to remove the P-deficient surface layer before growth by gas-source molecular beam epitaxy (MBE) [98]. The surface carrier concentration and mobility values for 1-μm thick epitaxial layers were found to be highly dependent on etch conditions, and the best results, approaching those of InP grown on wet-etched surfaces, were obtained by etching with a mixture of $CH_4/H_2/Ar$. There are publications dealing with selective growth on dry-etched structures by metal organic molecular beam epitaxy (MOMBE) [99], MOCVD [100], and hydride vapor phase epitaxy (VPE) [101].

8.7 APPLICATIONS

In previous sections, general applications in device fabrication have been periodically mentioned; here we will cover specific applications that have taken advantage of the anisotropy and dimensional control of dry processing to fabricate gratings, active channels and mesas, mirrors, and facets for laser diodes. We concentrate on dry-etched/wet-etched device performance and reliability comparisons for buried heterostructure devices, and process improvements afforded by dry etching. This review of applications of dry etching in InP-based laser fabrication has been condensed from Hayes et al. [102].

8.7.1 Gratings for Distributed Feedback Lasers

Gratings are used in DFB and distributed Bragg reflector laser structures to obtain single-frequency operation. These devices are typically used in long-haul, high bit rate communication systems with stringent performance requirements, such as bit error rates from 10^{-11} to as low as 10^{-14}. In order to obtain a satisfactory spectral yield, the coupling of the light to the grating (which governs single mode operation and therefore bit error rate) must be well controlled.

Gratings are produced by fabricating a structure with a periodically varying refractive index within the optical field of the laser. One fabrication method involves patterning a PR grating mask (by holography or electron beam writing) on an InP substrate. The grating is then etched into the substrate by either a wet or dry process. Grating depths may range from 0.02 to 0.15 μm, depending on device and grating structure design, with periods from 0.20 to 0.24 μm ($\lambda = 1.3$ and 1.5 μm, respectively) in the first order. If standard Br-based wet etchants are used, a triangle-shaped grating results whose depth is a function of the PR duty cycle, serving to make grating character more difficult to control across a wafer.

By contrast, dry etching is capable of producing nearly vertical sidewalls and homogeneous depths. Chlorine-based RIE [103] and RIBE [23] processes have been used to fabricate InP gratings, with typical InP/PR selectivities of 2:1 to 4:1. These selectivities are about the same as those obtained with RIBE with N_2/O_2 and Ar/O_2 [104]. With these processes there is usually considerable evidence of linewidth narrowing during etching. The latter process was used to etch gratings, which were then successfully overgrown by an optimized liquid phase epitaxy (LPE) method with minimal mass transport and dislocation generation [105]. CH_4 or C_2H_6/H_2 [106,107] and CH_4/O_2 RIE [71] can yield higher selectivities than the above mentioned chemistries. This is in part due to the polymer film that deposits on masked surfaces. High selectivity is desirable due to the thin PR layers.

In order to evaluate the effects of ion damage on device behavior, we have compared the performance and reliability of wet- and CH_4/H_2 RIE-etched grating 1.55-μm DFB lasers [108], where the waveguide, active, and subsequent layers over the *n*-substrate grating were grown by LPE [109]. After LPE growth, wet etching was used to form ~1-μm-wide mesas, and semi-insulating InP blocking layers were grown by MOCVD. A planar *p*-InP cap was then grown on top of the structure, and standard metallizations were applied. These cleaved-facet, capped-mesa buried heterostructure (CMBH) [42] devices were then tested for lasing threshold and slope efficiency before bonding. The wet-etched (WE) and RIE grating devices performed equivalently in terms of both threshold and slope efficiency. After bonding and AR/HR facet coating, high-stress aging followed by burn-in and long-term aging measurements were made to examine device reliability. No statistical difference was observed between wet- and dry-etch degradation rates. Similarly, other measures of performance after purge and burn-in, such as peak power at 60°C and the change in threshold current were equivalent for wet- and dry-etched gratings. We conclude that the dry etch has no adverse effect on device performance and reliability, and we suggest that any residual ion damage is effectively removed or annealed out during LPE growth, possibly during the limited mass transport of the grating.

8.7.2 Active Channels and Mesas

Improved active width control is the primary motivation for this application of dry etching in laser fabrication. An example of this is the improvement of buried crescent laser performance by using RIBE with Cl_2 gas excited by an ECR plasma [110]. RIBE improved the width control of 1.3-μm-wide stripes opened in a top quaternary layer, which was followed by wet etching (conc. HCl) of an arrowhead groove with vertical sidewalls in the underlying InP base structure and LPE growth in the arrowhead to form λ = 1.3-μm buried crescent devices. The use of dry processing decreased thresholds by 25% and improved the single transverse mode device yield by over a factor of two [111].

The use of dry etching as demonstrated above to anisotropically etch through top-cap quaternary material followed by crystallographic WE that is selective for InP is a nondemanding but effective use of the dry etch. A similar process has been used by Liou *et al.* [112] to fabricate self-aligned, λ = 1.5-μm multi-quantum well (MQW) ridge waveguide lasers. CH_4/H_2 RIE hs also been used to etch vertically through the quaternary and well into the upper *p*-InP-clad layer of a MQW laser structure, but stopped before the active region to create λ = 1.3-μm gain-guided devices [113]. CAIBE has been used by Yap *et al.* to etch mesas [114] for λ = 1.3-μm buried heterostructure lasers. Cl_2 gas and Ar ion bombardment were used to fashion 3- to 4-μm-wide stripes, etching completely through the active layer. WE was then used to selectively narrow the quaternary active layer to 1.5 μm, and a 620°C mass transport (rather than regrowth step) was used to encase the active layer in InP. Because of the selective WE, ion-induced damage is of no concern, but by adding the isotropic etch step, some of the dimensional control benefits of RIE are probably lost. Nonetheless, the authors report improved uniformity of device performance over their previous WE process. In addition, the authors have used the process to fabricate mesas and facets concurrently in one etch step [115].

Perhaps the most demanding use of dry etching is to etch high mesas through the *p-n* junction to completely pattern the active region by dry etching. In Figure 8.7, we demonstrate the type of improvement in active width control possible with

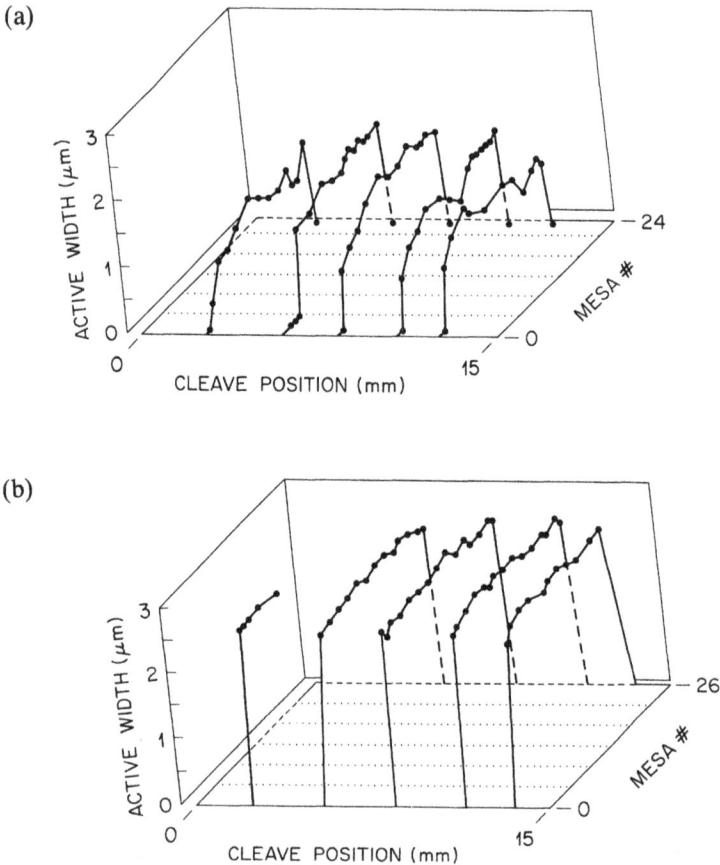

Figure 8.7 (a) Distribution of wet etched mesa widths across a wafer; (b) reactive ion-etched mesa width distribution.

RIE relative to WE [116]. Mesas for CMBH devices as described in Section 8.7.1 were etched either by an isotropic Br-based wet etch, or by CH_4/H_2 RIE. Active region widths were measured by growing blocking layers, cleaving the wafer perpendicular to the mesas, staining, and examining each active region by SEM. The WE process (Figure 8.7a) required undercutting a dielectric mask, resulting in active regions with an average width of 1.26 ± 0.27 μm. The mask used for RIE resulted in mesas wider than what would typically be used for high-performance lasers, 2.58 ± 0.09 μm (Figure 8.7b), but the factor of three reduction in standard deviation illustrates the greatly improved active width uniformity across the wafer. This should translate into more uniform device performance characteristics. There is a real challenge, however, to eliminate mask narrowing and maintain smooth sidewalls, because rough active sidewalls have been linked to poor far-field patterns [117].

In the only report of RIE mesa laser performance at the time of this writing, buried heterostructure lasers with C_2H_6/H_2 RIE mesas similar to those discussed above have been reported by Matsui et al. [118]. Two-micron high and wide mesas masked by SiO_2 were etched into an *n*-substrate, $\lambda = 1.5$-μm double heterostructure. Without additional wet chemical removal of any etch-induced damage, *p*- and *n*-InP blocking layers were grown by LPE, followed by planar *p*-InP-clad and *Q* cap layers. The authors report that RIE mesa devices had threshold currents of 15 to 20 mA, while WE mesa devices had thresholds above 25 mA, and it is stated that the RIE devices had generally superior performance characteristics. Although aging data were not reported, these results are encouraging and suggest that etch-induced damage does not limit device performance. High-temperature or current reliability data is necessary, however, to fully evaluate the damage issue.

8.7.3 Mirrors and Facets

There are more reports of the application of dry etching in InP laser facet fabrication than in either grating or mesa etching, probably due to the intense early interest in optoelectronic integrated circuits, and because much of the early work on InP-based lasers was on simple, gain-guided Fabry-Perot lasers not involving gratings or channel/mesa etching. The best metric of facet performance is comparison with cleaved facet devices, not only in terms of threshold and slope efficiency, but of reliability. To achieve comparable performance between etched and cleaved facet devices, the following are critical: a) vertical sidewalls, b) InP/InGaAsP selectivity near unity, c) high edge acuity masking, and d) high semiconductor/mask selectivity.

Work on etched InP-based laser facets originated at AT&T Bell Laboratories and used Cl-based RIE with mixtures of Cl_2/O_2 and Ti masks [119,120], in some cases in combination with WE to improve the vertical nature and smoothness of the sidewalls [121,122]. Sidewalls were made more vertical by inclining the sample in the plasma. In this early work, using immature dry processing, etched facet device performance was fair at best.

CCl$_4$/O$_2$ RIE [123] (SiO$_2$ masks) and Cl$_2$/Ar-inclined RIE [124] (TiO$_2$ masks) have been used at NTT to etch facets. The Cl$_2$/Ar work was furthered to include studies of aging [125,126] and facet reflectivity [67,69]. For $\lambda = 1.3$-μm double-channel planar buried heterostructure (DCPBH) devices with both facets etched, average pulsed thresholds were 12% higher (about 25 mA for 300-μm long devices) than for cleaved devices, with etched facet reflectivities 85% that of cleaved. The devices showed "no serious degradation" after aging at 50°C, 100 mA for 3200 hr, although there was no comparison to cleaved-aging results. Integrated laser/photo-diode (PD) pairs, such that the laser facet is vertical and the PD facet is inclined by 55° from the lasing plane, have also been fabricated by Cl$_2$/Ar-inclined RIE [127]. An angled mirror coated with a metallic reflector opposite a vertical etched facet has been used to fabricate surface-emitting lasers [128]. In addition to the mass-transported etched mesas and facets discussed in Section 8.7.2 [114,115], Liau *et al.* have also reported inclined mirror surface emitting lasers by CAIBE [129]. Elevated-temperature Cl RIBE (ECR ion source, perpendicular incidence) and hardbaked PR masking has been used to make devices with a single etched facet whose threshold and efficiency performance is equivalent to that of cleaved devices [130], but no statistics were given.

Br$_2$/Ar RIE with Ti/Si$_3$N$_4$ masks has been used to fabricate single- and double-etched facet devices [131]. The use of Br$_2$ was motivated by higher selectivities over Ti, TiO$_2$, Si$_3$N$_4$, and SiO$_2$ masks, and higher etch rates than Cl$_2$. Thresholds increased, however, by 30% and 80%, respectively, over dual-cleaved facet devices. The process was also used to create integrated laser/PD pairs [132]. Etched facet reflectivity was calculated to be only 3% due to nonvertical orientation and the presence of striations, compared to 32% for cleaved facets. Apparently, the Br$_2$ process holds little advantage over other chemistries yet. It must be reemphasized, however, that to obtain smooth facets, masks with excellent edge quality are required, in addition to anisotropy and selectivity of the etch chemistry.

C$_2$H$_6$/H$_2$ RIE and SiO$_2$ masking has been used to fabricate $\lambda = 1.5$-μm lasers with one or two etched facets [130]. The best etched/cleaved and etched/etched facet devices had thresholds 30% and 50% higher than cleaved/cleaved devices (300-μm cavities). The process was also used to integrate a laser and photodetector [134]. The facet reflectivity was only 6%, however, probably due to inferior masking.

Promising results in facet fabrication have been achieved with FIBE by Harriott *et al.* [135]. The advantage of FIBE over broad-area techniques is that no mask is required and sidewall striations can be minimized; the disadvantage is that one expects ion damage from perpendicularly-scattered ions to penetrate to a greater depth because of the high ($\sim$20 keV) ion energy. In addition, from a manufacturing perspective, FIBE is a serial writing method that suffers from throughput problems. Thresholds of channeled substrate buried heterostructure lasers increased by about 5% (to about 27 mA), and quantum efficiencies dropped by about 8%, for single-etched facet devices. An angled reflector was also successfully fabricated.

Two reports from CNET have described the application of IBE-to-facet fabrication. In the first, a broad-area Ar^+ beam, PR masks, and a tilted, liquid nitrogen-cooled sample holder were employed to fabricate single-etched facet BH devices [136]. A brief wet chemical etch was used to remove ion damage. Thresholds and efficiencies were "almost the same" as cleaved devices, and the etched mirror reflectivity was about 25%. A 600-hr, 50°C, 3-mW constant power aging test gave an average degradation rate of about 6%/khr, which seems somewhat high, perhaps due to ion-induced damage. In the second report, FIBE was used [137]. Single-etched facet devices had an increase in threshold of 13% with efficiencies equivalent to cleaved devices.

Finally, promising results have been reported by Philips [138,71] on DCPBH devices with both facets etched. $Cl_2/CH_4/Ar/H_2$ RIE (200°C) and PR masks fabricated by a trilevel process were used to obtain devices with threshold and slope efficiencies comparable to cleaved devices. Unfortunately, a statistical performance comparison was not given. Aging at 70°C, 5 mW constant power for 800 hr resulted in a degradation rate of about 2%/khr, which is quite low for this temperature. This combination of masking technology and etch chemistry appears capable of yielding the smooth sidewalls necessary for high-quality facets, the desired reliability, and the high etch rates (300 nm/min) necessary for high-throughput wafer processing.

From the above examples it is evident that, with careful integration and optimization of lithography, etching, and growth, device yields can be improved by using dry etching, while maintaining reliability equivalent to that of standard processing. This potential foretells substantial activity in this field in the future.

8.8 SUMMARY

Considering the large number of similar etching technologies and chemistries that have been discussed, which is the most appropriate for which application? IBE generally has inferior semiconductor/mask selectivity along with the potential for greater surface damage than techniques employing reactive chemistries, however, and so is not the optimum technology for etching of III–V semiconductors. However, FIB etching of facets may be a useful technology since it does not rely on masking, but reliable device operation must first be demonstrated. By virtue of the ease of angling the sample to adjust sidewall slopes, RIBE and CAIBE are uniquely suited for the fabrication of angled and parabolic mirrors. Room-temperature Cl chemistries, which inherently give angled sidewalls at normal incidence, can be used to advantage to fabricate integrated laser/photodetector pairs. By appropriately angling the wafer, a vertical laser facet can be formed opposite an angled photodetector facet (which improves coupling uniformity by reducing reflections back into the laser cavity). If high-rate (0.5–1 μm/min), normal incidence etching is required with vertical sidewalls, room-temperature ICl/Cl_2 IBAE can be used (with strontium fluoride masks), or RIBE at about 180°C. Surfaces are somewhat rough with these chemistries, however. Good temperature control or thermal contact of the wafer is important, especially at the low pressures at which these processes are carried out.

Due to the variety of chemistries available to RIE, there exists a greater opportunity to optimize mask/semiconductor selectivities and other etch parameters than for RIBE or IBAE, although RIE suffers from the difficulty of independently controlling plasma density or ion current and bias voltage. In addition, the sensitivity of etching to geometric design factors can make it tedious to reproduce an etch process in a reactor of different design. Still, RIE is currently the most common etch technique found in development and manufacturing. RIE with Cl-containing chemistries is very similar in character to RIBE and IBAE. Good temperature control of the wafer is important, and a bit easier at the higher pressures at which RIE is carried out. When smooth, shallow etching with vertical sidewalls and high mask selectivity is required, such as for laser gratings or mesas a few microns high or less, hydrocarbon RIE (or ECR) is the best choice. The process should be operated at the highest pressure and lowest bias voltage possible that will give satisfactory etch rate and anisotropy in order to keep ion damage as shallow as possible. Damage depths of less than 200Å can be obtained without difficulty. Halogen and hydrocarbon gas mixtures are a promising chemistry, capable of yielding high etch rates (0.2–0.5 μm/min), vertical sidewalls, and smooth surfaces when operated near 180°C. There is even greater hope for room temperature processes using Br or I and hydrocarbon mixtures, but most of this work remains to be done. Finally, $CH_4/H_2/Ar$ ECR shows substantial promise for anisotropic, low-damage etching, afforded by the ability to move to etch conditions of ultralow damage near the end of an etch.

REFERENCES

1.   P.D. Agnello and S.K. Ghandhi, *J. Crystal Growth,* Vol. 73, 1985, p. 453.
2.   B. Chapman, *Glow Discharge Processes,* New York: John Wiley & Sons, 1980.
3.   J.L. Vossen and W. Kern, eds., *Thin Film Processes,* New York: Academic Press, 1978.
4.   O. Auciello and R. Kelly, eds., *Beam Modification of Materials, 1. Ion Bombardment Modification of Surfaces, Fundamentals and Applications,* New York: Elsevier, 1984.
5.   D.E. Ibbotson and D.L. Flamm, *Solid State Technol.,* Vols. 31(10), p. 77 (1988) and 31(11), p. 105 (1988).
6.   W.C. Dautremont-Smith, R.A. Gottscho, and R.J. Schutz, in *Semiconductor Materials and Processing Technology Handbook,* ed. G.E. McGuire, Park Ridge, NJ: Noyes, 1988, p. 191.
7.   C.L. Chen and K.D. Wise, *IEEE J. Quantum Electron.,* Vol. QE-29, 1982, p. 1522.
8.   O. Wada, *J. Phys. D: Appl. Phys.,* Vol. 17, 1984, p. 2429.
9.   S.J. Pearton, U.K. Chakrabarti, and A.P. Perley, *J. Appl. Phys.,* Vol. 68, 1990, p. 2760.
10. N. Bouadma, P. Devoldere, B. Jusserand, and P. Ossart, *Appl. Phys. Lett.,* Vol. 48, 1986, p. 1285.
11. W. Katzschner, U. Niggebrügge, R. Löffler, and H. Schröter-Janssen, *Appl. Phys. Lett.,* Vol. 48, 1986, p. 230.

12. W. Katzschner, A. Steckenborn, R. Löffler, and N. Grote, *Appl. Phys. Lett.*, Vol. 44, 1984, p. 352.
13. N. Furuhata, H. Miyamoto, A. Okamoto, and K. Ohata, *J. Appl. Phys.*, Vol. 65, 1989, p. 168.
14. D.G. Lishan and E.L. Hu, *Appl. Phys. Lett.*, Vol. 56, 1990, p. 1667.
15. S.C. McNevin, *J. Vac. Sci. Technol. B*, Vol. 4, 1986, p. 1216.
16. *Chemical Rubber Company Handbook of Chemistry and Physics*, 51st and 57th editions, Cleveland: CRC Publishing Co., 1971 and 1976.
17. V.M. Donnelly, D.L. Flamm, C.W. Tu, and D.E. Ibbotson, *J. Electrochem. Soc.*, Vol. 129, 1982, p. 2533.
18. R.P.H. Chang, C.C. Chang, and S. Darack, *J. Vac. Sci. Technol.*, Vol. 20, 1982, p. 45.
19. Y.L. Wang, H. Temkin, L.R. Harriott, R.A. Hamm, and J.S. Weiner, *Appl. Phys. Lett.*, Vol. 57, 1990, p. 1672.
20. M.A. Bösch, L.A. Coldren, and E. Good, *Appl. Phys. Lett.*, Vol. 4, 1981, p. 264.
21. T. Tadokoro, F. Koyama, and K. Iga, *Jpn. J. Appl. Phys.*, Vol. 27, 1988, p. 389.
22. K. Mutoh, M. Nakajima, and M. Mihara, *Jpn. J. Appl. Phys.*, Vol. 29, 1990, p. 1022.
23. Y. Yuba, K. Gamo, X.G. He, Y.S. Zhang, and S. Namba, *Jpn. J. Appl. Phys.*, Vol. 22, 1983, p. 1211.
24. S.C. McNevin, *J. Vac. Sci. Technol. B*, Vol. 4, 1986, p. 1203.
25. N.L. DeMeo, J.P. Donnelly, F.J. O'Donnell, M.W. Geis and K.J. O'Connor, *Nuclear Instruments and Methods in Phys. Res.*, Vol. B7, 1985, p. 814.
26. R.A. Barker, T.M. Mayer, and R.H. Burton, *Appl. Phys. Lett.*, Vol. 40, 1982, p. 583.
27. A Scherer, H.G. Craighead, and E.D. Beebe, *J. Vac. Sci. Technol.*, Vol. B5, 1987, p. 1599.
28. H. Yamada, Y. Inomoto, M. Kitamura, I. Mito, and K. Asakawa, *Proc. 20th Intl. Conf. Solid State Devices and Materials*, Tokyo, 1988, p. 279.
29. A. Scherer, B.P. van der Gaag, K. Kash, H.G. Craighead, and P. Grabbe, *Mat. Res. Soc. Symp. Proc.*, Vol. 126, 1988, p. 245.
30. G.F. Doughty, S. Thoms, V. Law, and C.D.W. Wilkinson, *Vacuum*, Vol. 36, 1986, p. 803.
31. N.G. Chew and A.G. Cullis, *Appl. Phys. Lett.*, Vol. 44, 1984, p. 142.
32. M. Schilling and K. Wünstel, *Appl. Phys. Lett.*, Vol. 49, 1986, p. 710.
33. L. Ding, Q. Mingxin, and K. Zhong, *J. Electron. Materials*, Vol. 17, 1988, p. 29.
34. M. Takai, J. Tsuchimoto, J. Tokuda, H. Nakai, K. Gamo, and S. Namba, *Appl. Phys. A*, Vol. 45, 1988, p. 305.
35. V.M. Donnelly and T.R. Hayes, *Appl. Phys. Lett.*, Vol. 57, 1990, p. 701.
36. J.-L. Peyre, D. Rivière, C. Vannier, and G. Villela, *Hybrid Circuits*, Vol. 21, 1990, p. 6.
37. L.A. Coldren and J.A. Rentschler, *J. Vac. Sci. Technol.*, Vol. 19, 1981, p. 225.
38. R.H. Burton and G. Smolinsky, *J. Electrochem. Soc.*, Vol. 129, 1982, p. 1599.
39. R.A. Gottscho, G. Smolinsky, and R.H. Burton, *J. Appl. Phys.*, Vol. 53, 1982, p. 5908.
40. M.B. Stern and P.F. Liao, *J. Vac. Sci. Technol. B*, Vol. 1, 1983, p. 1053.
41. S.J. Pearton, U.K. Chakrabarti, W.S. Hobson, and A.P. Perley, *J. Electrochem. Soc.*, Vol. 137, 1990, p. 3188.
42. R.A. Burton, C.L. Hollien, L. Marchut, S.M. Abys, G. Smolinsky, and R.A. Gottscho, *J. Appl. Phys.*, Vol. 54, 1983, p. 6663.
43. E.L. Hu and R.E. Howard, *Appl. Phys. Lett.*, Vol. 37, 1980, p. 1022.
44. S.J. Pearton, W.S. Hobson, F.A. Baiocchi, A.B. Emerson, and K.S. Jones, *J. Vac. Sci. Technol.*, Vol. B8, 1990, p. 57.
45. R.J. Contolini, *J. Electrochem. Soc.*, Vol. 135, 1988, p. 929.
46. R.C. Wetzel and T.R. Hayes, previously unpublished.
47. S.J. Pearton, W.S. Hobson, U.K. Chakrabarti, G.E. Derkits, and A.P. Kinsella, *J. Vac. Sci. Technol. B*, Vol. 8, 1990, p. 1274.

48. G.A. Vawter and J.R. Wendt, *Appl. Phys. Lett.*, Vol. 58, 1991, p. 289.

49. R.E. Lee, *J. Vac. Sci. Technol.*, Vol. 16, 1979, p. 164.

50. K. Takimoto, K. Ohnaka, and J. Shibata, *Appl. Phys. Lett.*, Vol. 54, 1989, p. 1947.

51. U. Niggebrügge, M. Klug, and G. Garus, *Inst. Phys. Conf. Ser.*, Vol. 79, 1985, p. 367.

52. R. Cheung, S. Thoms, S.P. Beaumont, G. Doughty, V. Law, and C.D.W. Wilkinson, *Electron. Lett.*, Vol. 23, 1987, p. 857.

53. T.R. Hayes, M.A. Dreisbach, P.M. Thomas, W.C. Dautremont-Smith, and L.A. Heimbrook, *J. Vac. Sci. Technol. B*, Vol. 7, 1989, p. 1130.

54. J. Keskinen, J. Näppi, H. Asonen, and M. Pessa, *Electron. Lett.*, Vol. 26, 1990, p. 1371.

55. T. Matsui, H. Sugimoto, T. Ohishi, and H. Ogata, *Electron. Lett.*, Vol. 24, 1988, p. 798.

56. J. Pearton, W.S. Hobson, F.A. Baiocchi, A.B. Emerson, and K.S. Jones, *J. Vac. Sci. Technol. B*, Vol. 8, 1990, p. 57.

57. G. Franz, *J. Electrochem. Soc.*, Vol. 137, 1990, p. 2897.

58. H. Schmid, *Proc. 6th International Conference on Ion and Plasma-Assisted Techniques*, Brighton (UK), 1987, p. 98.

59. L. Henry, C. Vaudry, and P. Granjoux, *Electron. Lett.*, Vol. 23, 1987, p. 1254.

60. T.R. Hayes, U.K. Chakrabarti, F.A. Baiocchi, A.B. Emerson, H.S. Luftman, and W.C. Dautremont-Smith, *J. Appl. Phys.*, Vol. 68, 1990, p. 785.

61. T.R. Hayes, unpublished.

62. D.L. Flamm, V.M. Donnelly, and D.E. Ibbotson, *J. Vac. Sci. Technol. B*, Vol. 1, 1983, p. 23.

63. J.E. Spencer, T.R. Schimert, J.H. Dinan, D. Endres, and T.R. Hayes, *J. Vac. Sci. Technol. A*, Vol. 8, 1990, p. 1690.

64. S.J. Pearton, U.K. Chakrabarti, A.P. Kinsella, D. Johnson, and C. Constantine, *Appl. Phys. Lett.*, Vol. 56, 1990, p. 1424.

65. C. Constantine, D. Johnson, S.J. Pearton, U.K. Chakrabarti, A.B. Emerson, W.S. Hobson, and A.P. Kinsella, *J. Vac. Sci. Technol. B*, Vol. 8, 1990, p. 596.

66. S. Shah and T.R. Hayes, unpublished.

67. S.J. Pearton, U.K. Chakrabarti, and A.P. Perley, submitted to *Appl. Phys. Lett.*

68. S.J. Pearton, U.K. Chakrabarti, and F.A. Baiocchi, *Appl. Phys. Lett.*, Vol. 55, 1989, p. 1633.

69. S.J. Pearton and W.S. Hobson, *Appl. Phys. Lett.*, Vol. 56, 1990, p. 2186.

70. K.L. Seaward, N.J. Moll, D.J. Coulman, and W.F. Stickle, *J. Appl. Phys.*, Vol. 61, 1987, p. 2358.

71. G.J. van Gurp and J.M. Jacobs, *Philips J. Res.*, Vol. 44, 1989, p. 211.

72. S.W. Pang, *J. Electrochem. Soc.*, Vol. 133, 1986, p. 784.

73. H.F. Wong, D.L. Green, T.Y. Liu, D.G. Lishan, M. Bellis, E.L. Hu, P.M. Petroff, P.O. Holtz, and J.L. Merz, *J. Vac. Sci. Technol. B*, Vol. 6, 1988, p. 1906.

74. R. Germann, A. Forchel, and D. Grützmacher, *Appl. Phys. Lett.*, Vol. 55, 1989, p. 2196.

75. J.C. Bean, G.E. Becker, P.M. Petroff, and T.E. Seider, *J. Appl. Phys.*, Vol. 48, 1977, p. 902.

76. S.W. Pang, W.D. Goodhue, T.M. Lyszczarz, D.J. Ehrlich, R.B. Goodman, and G.D. Johnson, *J. Vac. Sci. Technol. B*, Vol. 6, 1988, p. 1916.

77. A. Scherer, H.G. Craighead, M.L. Roukes, and J.P. Harbison, *J. Vac. Sci. Technol. B*, Vol. 6, 1988, p. 277.

78. R. Cheung, Y.H. Lee, C.M. Knoedler, K.Y. Lee, T.P. Smith, III, and D.P. Kern, *Appl. Phys. Lett.*, Vol. 54, 1988, p. 2130.

79. B.E. Maile, A. Forchel, R. Germann, and D. Grützmacher, *Appl. Phys. Lett.*, Vol. 54, 1989, p. 1552.

80. A. Menschig, A. Forchel, B. Roos, R. Germann, W. Heuring, and D. Grützmacher, *Microelectronic Eng.*, Vol. 11, 1990, p. 11.

81. J.S. Weiner, H.F. Hess, R.B. Robinson, T.R. Hayes, D.L. Sivco, A.Y. Cho, and M. Ranada, submitted to *Appl. Phys. Lett.*

82. R. Cheung, S. Thoms, I. McIntyre, C.D.W. Wilkinson, and S.P. Beaumont, *J. Vac. Sci. Technol.*, Vol. B6, 1988, p. 1911.

83. S.J. Pearton, U.K. Chakrabarti, and W.S. Hobson, *J. Appl. Phys.*, vol. 66, 1989, p. 2061.

84. E.M. Omeljanovsky, A.V. Pakhomov, A.Y. Polyakov, O.M. Borodina, E.A. Kozhukhova, A.Y. Nashelskii, S.V. Yakobson, and V.V. Novikova, *Solid State Communiations*, Vol. 72, 1989, p. 409, and references within.

85. T.R. Hayes, W.C. Dautremont-Smith, H.S. Luftman, and J.W. Lee, *Appl. Phys. Lett.*, Vol. 55, 1989, p. 56.

86. M. Moehrle, *Appl. Phys. Lett.*, Vol. 56, 1990, p. 542.

87. H. Schmid, F. Fidorra, and D. Grützmacher, *Inst. Phys. Conf. Ser. No. 96*, 1988, p. 431.

88. D. Field, Y.P. Song, D.F. Klemperer, and A.P. Day, *Vacuum*, Vol. 40, 1990, p. 357.

89. T.R. Hayes, R. Pawelek, C.A. Green, K.E. Strege, V.R. McCrary, D.L. Coblenz, and P.M. Thomas, previously unpublished.

90. H.H. Busta, R.E. Lajos, and D.A. Kiewit, *Solid State Technol.*, Vol. 22(2), 1979, p. 61.

91. T.R. Hayes, P.A. Heimann, V.M. Donnelly, and K.E. Strege, *Appl. Phys. Lett.*, Vol. 57, 1990, p. 2817.

92. V.M. Donnelly and J.A. McCaulley, *J. Vac. Sci. Technol. A*, Vol. 8, 1990, p. 84.

93. K.E. Strege and T.R. Hayes, unpublished.

94. P.A. Heimann and T.R. Hayes, unpublished.

95. R. Müller, *Appl. Phys. Lett.*, Vol. 57, 1990, p. 1020.

96. R.J. Contolini and L.A. D'Asaro, *J. Vac. Sci. Technol. B*, Vol. 4, 1986, p. 706.

97. A. Mitchell, R.A. Gottscho, S.J. Pearton, and G.R. Scheller, *Appl. Phys. Lett.*, Vol. 56, 1990, p. 821.

98. L. Henry, C. Vaudry, A. Le Corre, D. Le Crosnier, P. Alnot, and J. Olivier, *Electron. Lett.*, Vol. 25, 1989, p. 1257.

99. M. Gailhanou, L. Goldstein, M. Lambert, M. Boulou, C. Starck, and L. Le Gouezigou, *Proc. 19th European Solid State Device Res. Conf.*, Berlin, 1989, p. 495.

100. B. Garrett and E.J. Thrush, *J. Crystal Growth*, Vol. 97, 1989, p. 273.

101. V.S. Ban, G.C. Erickson, S. Mason, and G.H. Olsen, *J. Electrochem. Soc.*, Vol. 137, 1990, p. 2904.

102. T.R. Hayes, S.J. Kim, and C.A. Green, *Proc SPIE Conf. No. 1418, Laser Diode Technology and Applications III*, Los Angeles, paper no. 1418-20, 1991.

103. K. Hirata, O. Mikami, and T. Saitoh, *J. Vac. Sci. Technol. B*, Vol. 2, 1984, p. 45.

104. M. Korn, M. Klingenstein, R. Germann, A. Forchel, H. Nickel, W. Schlapp, R. Lösch, K. Streubel, and F. Scholz, *J. Vac. Sci. Technol. B*, Vol. 7, 1989, p. 2057.

105. M. Schilling and K. Wünstel, *Appl. Phys. Lett.*, Vol. 49, 1986, p. 710.

106. E. Andideh, I. Adesida, T. Brock, C. Caneau, and V. Keramidas, *J. Vac. Sci. Technol. B*, Vol. 7, 1989, p. 1841.

107. J. Keskinen, J. Näppi, H. Asonen, and M. Pessa, *Electron. Lett.*, Vol. 26, 1990, p. 1369.

108. J.L. Zilko, L.J.P. Ketelsen, Y. Twu, D.P. Wilt, S.G. Naphlotz, J.P. Blaha, K.E. Strege, V.G. Riggs, D.L. Van Haren, S.Y. Leung, P.M. Nitzsche, J.A. Long, C.B. Roxlo, G. Pryzblek, J. Lopata, M.W. Focht, and L.A. Koszi, *IEEE J. Quant. Electron.*, Vol. 25, 1989, p. 2091.

109. T.R. Hayes, S.J. Kim, and C.A. Green.

110. A. Kasukawa, M. Iwase, Y. Hiratani, N. Matsumoto, Y. Ikegami, M. Irikawa, and S. Kashiwa, *Appl. Phys. Lett.*, Vol. 51, 1987, p. 1774.

111. A. Kasukawa, M. Iwase, N. Matsumoto, T. Makino, and S. Kashiwa, *J. Lightwave Technol,* Vol. 7, 1989, p. 2039.

112. K.Y. Liou, A.G. Dentai, E.C. Burrows, C.H. Joyner, C.A. Burrus, and G. Raybon, accepted by *Photonic Technology Letters.*

113. M.J. Ludowise, T.R. Ranganath, and A. Fischer-Colbrie, *Appl. Phys. Lett.*, Vol. 57, 1990, p. 1493.

114. D. Yap, Z.L. Liau, D.Z. Tsang, and J.N. Walpole, *Appl. Phys. Lett.*, Vol. 52, 1988, p. 1464.

115. D. Yap, J.N. Walpole, and Z.L. Liau, *Appl. Phys. Lett.*, Vol. 53, 1988, p. 1260.

116. T.R. Hayes, M.A. Dreisbach, W.C. Dautremont-Smith, E.K. Byrne, and V.R. McCrary, unpublished.

117. K. Junyaprasert, N. Ogasawara, R. Ito, and K. Aiki, *Jpn. J. Appl. Phys.*, Vol. 26, 1987, p. 1279.

118. T. Matsui, K. Ohtsuka, H. Sugimoto, Y. Abe, and T. Ohishi, *Appl. Phys. Lett.*, Vol. 56, 1990, p. 1641.

119. L.A. Coldren, K. Iga, B.I. Miller, and J.A. Rentschler, *Appl. Phys. Lett.*, Vol. 37, 1980, p. 681.

120. L.A. Coldren, B.I. Miller, K. Iga, and J.A. Rentschler, *Appl. Phys. Lett.*, Vol. 38, 1981, p. 315.

121. L.A. Coldren, K. Furuya, B.I. Miller, and J.A. Rentschler, *Electron. Lett.*, Vol. 18, 1982, p. 235.

122. L.A. Coldren, K. Furuya, B.I. Miller, and J.A. Rentschler, *IEEE Trans. Microwave Theory and Tech.*, Vol. MTT-30, 1982, p. 1667.

123. O. Mikami, H. Akiya, T. Saitoh, and H. Nakagome, *Electron. Lett.*, Vol. 19, 1983, p. 213.

124. H. Saito, Y. Noguchi, and H. Nagai, *Electron. Lett.*, Vol. 21, 1985, p. 748.

125. H. Saito, Y. Noguchi, and H. Nagai, *Electron. Lett.*, Vol. 22, 1986, p. 36.

126. H. Saito, Y. Noguchi, and H. Nagai, *Electron. Lett.*, vol. 22, 1986, p. 1157.

127. H. Saito and Y. Noguchi, *Electron. Lett.*, Vol. 25, 1989, p. 719.

128. H. Saito and Y. Noguchi, *Jpn. J. Appl. Phys.*, Vol. 28, 1989, p. L1239.

129. Z.L. Liau and J.N. Walpole, *Int. Electron Dev. Mtg. 1986, Tech. Dig.*, p. 622.

130. H. Yamada, Y. Inomoto, M. Kitamura, I. Mito, and K. Asakawa, *Ext. Abst. 20th Conf. Solid State Dev. and Mat.*, Tokyo, 1988, p. 279.

131. K. Takimoto, K. Ohnaka, and J. Shibata, *Appl. Phys. Lett.*, Vol. 54, 1989, p. 1947.

132. H. Tsujii, K. Ohnaka, and J. Shibata, *Electr. and Commun. in Japan, Part 2*, Vol. 72, 1989, p. 26.

133. T. Matsui, H. Sugimoto, T. Ohishi, Y. Abe, K. Ohtsuka, and H. Ogata, *Appl. Phys. Lett.*, Vol. 54, 1989, 1193.

134. T. Matsui, H. Sugimoto, K. Ohtsuka, Y. Abe, and H. Ogata, *Electron. Lett.*, Vol. 25, 1989, p. 954.

135. L.R. Harriott, R.E. Scotti, K.D. Cummings, and A.F. Ambrose, *Appl. Phys. Lett.*, vol. 48, 1986, p. 1704.

136. N. Bouadma, J.F. Hogrel, J. Charil, and M. Carré, *IEEE J. Quantum Electron.*, Vol. QE-23, 1987, p. 909.

137. G. Ben Assayag, P. Sudraud, J. Gierak, D. Remiens, L. Menigaux, and L. Dugrand, *Microelectron. Eng.*, Vol. 11, 1990, p. 413.

138. G.J. van Gurp, J.M. Jacobs, J.J.M. Binsma, and L.F. Tiemeijer, *Jpn. J. Appl. Phys.*, Vol. 28, 1989, p. L1236.

139. *Alfa Products* catalog, Morton Thiokol, Inc., 1986.

Chapter 9
Ohmic Contacts to InP and Related Materials

A. Katz
AT&T Bell Laboratories

9.1 INTRODUCTION

The issues involved in processing high-quality ohmic contacts to InP and related materials have been widely investigated and reported during the last ten years because of their essential role in the performance of electronic and photonic devices, which require operation over long periods. Ohmic contacts to InP-based materials are essential for electronic devices such as field effect transistors (FET) [1,2], junction field effect transistors (JFET) [3,4], high electron mobility transistors (HEMT), and heterojunction bipolar transistors (HBT) [5,6], as well as for photonic devices such as long-wavelength laser diodes, light-emitting diodes (LED) [7–9], and photoelectronic and solar cells [10]. The ohmic contact of choice performs as the electrical communication link between the active region of the semiconductor device and the external circuit, allowing a low-energy carrier transport mechanism through the defined contact zone geometry, and ensuring a negligible voltage drop across it. From the practical point of view, ohmic contacts are basically metal-semiconductor Schottky barriers, through which a majority of the carriers can tunnel because of the narrow barrier width, enhanced by high dopant concentration in the interface region. These interfaces are characterized by very low Schottky barrier heights (up to about 0.4 eV), which is reflected in almost a linear relationship between the measured current and the voltage applied across the contact [11,12]. The quality of the ohmic contact is frequently described in terms of contact resistivity ($\Omega \cdot$ mm) or specific contact resistance ($\Omega \cdot$ cm^2). The most conventional ways of forming a high semiconductor surface concentration of free carriers are by incorporating a high concentration of dopants into the initially grown epitaxial semiconductor layer, by introducing an external dopant diffusion source, from which the dopants will be driven

---

into the semiconductor by means of a heating process, and by ion implanting the semiconductor surface. The latter approach is rarely applied to InP and related materials because of the surface damage and loss of stoichiometry associated with bombarding the InP surface. The other two methods are widely used in the manufacturing process of ohmic contacts to InP-based devices. The former method does not require a heating cycle to enable the migration of the dopants into the semiconductors, and thus contacts that are processed onto such materials are frequently referred to as nonalloyed ohmic contacts. The second approach requires thermal activation in order to form the heavily doped intermixed interfacial layer, and thus contacts to these layers are termed alloyed contacts. These two approaches differ from each other by the kind and geometry of the metal-semiconductor interfacial related layers, and will be discussed in detail later.

A different approach to forming a metal-semiconductor ohmic contact is by applying a semiconductor with an extremely narrow bandgap, such as InAs ($\sim$0.35 eV), which can be epitaxially grown lattice matched to an InP substrate. This semiconductor has high electron and hole mobilities (33,000 and 460 cm^2/vs, respectively, at 300K), and has surface states pinned in the conduction bandgap; therefore, it forms an ohmic contact with almost every metal that is deposited onto it [13–15].

Two major concerns have to be considered when forming ohmic contacts to InP: (a) The unstable nature of the semiconductor, which tends to decompose through heat treatments at temperatures as low as 350°C while losing the group V volatile element. This is reflected in a degraded contact interface morphology, poor edge geometry definition, and uneven penetration of both the metal and the semiconductor elements. (b) The fundamental difference in the nature of the n-type and the p-type InP. The energy distribution of the surface states density (N_{ss}) in InP is parabolic, with its minimum positioned near the conduction band edge and increasing toward the valence band [16]. As a result, the surface of n-type InP is only slightly depleted. This is consistent with the measured low barrier height values of most metal contacts to n-InP (0.4–0.5 eV) [17] and the high barrier for p-type InP (0.7–0.8 eV). Therefore, it should be possible to realize low-resistance ohmic contacts to n-type InP, while it is a much more complicated task to identify a metal that would perform as a low barrier ohmic contact to p-type InP.

When designing an ohmic contact to a high-speed InP-based device, operated under a high current density and elevated-temperature conditions, a few constraints and requirements have to be met in order to provide a suitable contact technology to support both short- and long-term device operation. The main issues that have to be considered when selecting the metallization scheme and the process conditions are (a) optimizing the contact design to yield an ohmic contact with the lowest possible resistance, (b) providing stable morphology over a wide temperature range, which requires the presence of only limited intermetallic reactions, and thus requires an abrupt metal-semiconductor interface through the contact processing, (c) ensuring contact stability through the device electrical and thermal operating conditions, (d) causing no excessive stresses in the metal films, the underlying dielectric patterning layer, and the underlying semiconductor, (e) being compatible with the metal system used for the interconnection technology, and (f) fabricating with a robust process that fits as an integrated step into the overall device manufacturing scheme.

All the above-mentioned constraints are essential in the micron-size device technology currently applied for both electronic and photonic devices, but will be more pronounced in the future for submicron and deep submicron technologies. For this future miniaturization trend, the required morphological constraints on the low-resistance contacts will be even more crucial due to higher current densities present, and they will be harder to achieve because of the small contact geometry definition, which depends entirely on some other obstacles, such as lithography, etching, and testing processes. In addition, the more complicated and sophisticated the overall device geometry becomes, along with the more stringent market requirements, the more emphasis has to be put on producing integration within the device manufacturing sequence. Thus, the challenge is not only to design a contact metallization that will provide a sufficient stable ohmic contact, but also to choose a metallization scheme that will withstand the multistep processing sequence and that can be used to simplify the overall device fabrication process. One superior trend in designing such a process is to deposit a stable patterned metallization scheme in the initial stages of the device manufacturing sequence and use it to assist the realization of some other process later in the manufacturing sequence. As an example of this approach, we have recently developed a self-aligned etched-mesa buried heterostructure laser device structure, in which the metal contact tops the future mesa site and serves as a selective mask for both the mesa etching and regrowth of the blocking layers surrounding it [18,19].

In the current review we will discuss in detail recent work that has been carried out during the last two years, with emphasis on research that has elucidated the role of metallurgical phenomena in determining the electrical properties of the metal-InP and InP-based contacts. This work was aimed at providing the required technology for ohmic contacts to InP-based photonic devices, and, thus, some motivations in conjunction with this technology may be mentioned. It is recommended that the reader become familiar with some comprehensive reviews of the subject of ohmic contacts to InP, which have been published in the past [17,20–23]. This review contains mainly information on ohmic contacts to n-type and p-type InP, $In_{0.53}Ga_{0.47}As$, and InGaAsP. It is organized in three main sections reflecting the various needs for and approaches to producing ohmic contacts to InP and related materials. It provides a guideline of the advantages and deficiencies of contacts formed using the various techniques, discusses some ideas regarding present and future designs and choices of metallization schemes, and describes processes to fabricate ohmic contacts to InP and its related compound semiconductors.

9.2 FUNDAMENTALS OF METAL/InP OHMIC CONTACTS

9.2.1 Interfacial Reactions

The severe geometric design demands of advanced InP-based electronic and photonic devices calls for shallow and uniform interfacial layers in between the metallization scheme of choice and the semiconductor. The metal/InP ohmic contact design concepts are very similar to those suggested for metal/GaAs systems [21], with the exception that the InP binary system is more reactive than GaAs, as is reflected in the binary phase diagrams of those two systems shown in Figure 9.1. Capless InP begins to decompose at 350°C when heated at atmospheric pressure, as a result of out-diffusion of the group V atoms from the bulk to the surface [24]. Thus, InP tends to decompose at much lower temperatures than GaAs during the heating cycles applied for sintering the metal-semiconductor contact.

The thermodynamic, highly reactive nature of the InP in metal/InP interfaces has been widely observed, particularly in the commonly used Au-based contacts to InP. Analogous to the work on metal-GaAs, Au-based metallization schemes provide the advantage of introducing an external dopant (either p-type or n-type) source, which is alloyed into the adjacent semiconductor through solid or liquid phase reactions; however, this is done with the involvement of the Au as a stabilizer element, which eliminates dopant precipitate formation and interfacial defect nucleation. In order to drive the required alloying reactions that lead to the required semiconductor doping level, heat treatments in the temperature range of 350°C to 450°C are typically required, resulting in the formation of a thick, metal-semiconductor reactant, interfacial layer in between the metal and the InP substrate. This layer is typically about three times thicker than the original deposited metal layer [24].

Even with the lack of reported thermodynamic data for the simplest metal-InP ternary systems, a lot of information can be gained with regard to the nature of these systems by studying the metal-In and metal-P binary phase diagrams, referring to the metal system involved in the contact of interest. Figure 9.2 shows, as an example, the binary phase diagrams of Au-In and Be-In, which are the materials often chosen to form a p-type alloyed contact to InP. Due to a well-known thermodynamic theorem, a ternary system composed of three binary eutectic systems contains at least one ternary eutectic melting point at a temperature lower than the lowest binary eutectic melting point introduced in the system. Therefore, adding P, Be, or Au as the third element to each of those binary systems reduces the melting points to even lower temperatures and contributes to its reactive nature.

In summary, the issues involved in the metallurgical and thermal processing of high-quality ohmic contacts to InP and related materials are challenging and of great interest. The metallization systems of choice and the associated heat treatments have to enable and drive the required metal-semiconductor interfacial reactions, which

(a)

(b)

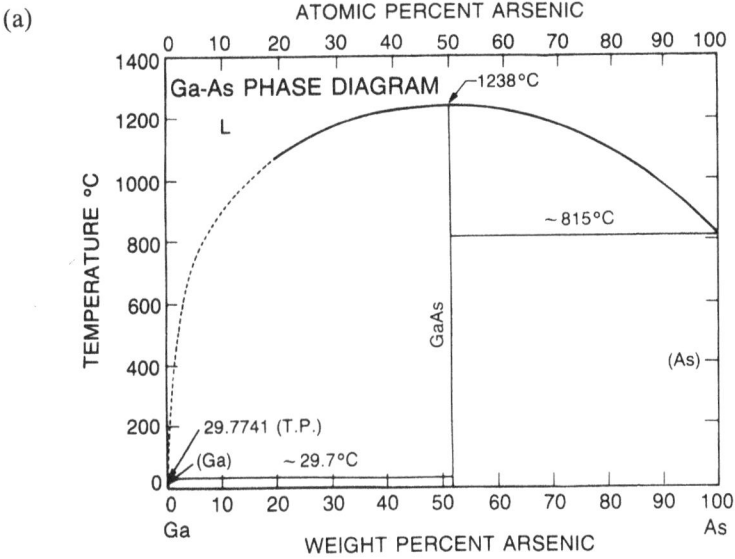

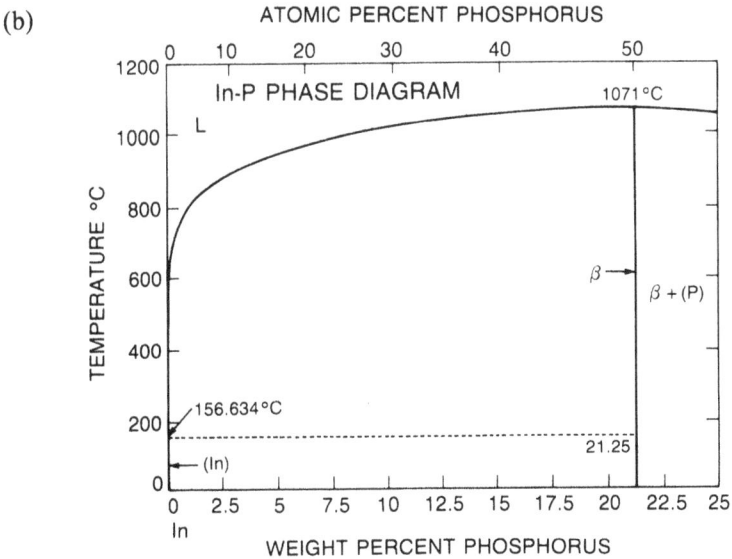

Figure 9.1 Binary phase diagram of (a) GaAs, and (b) InP system.

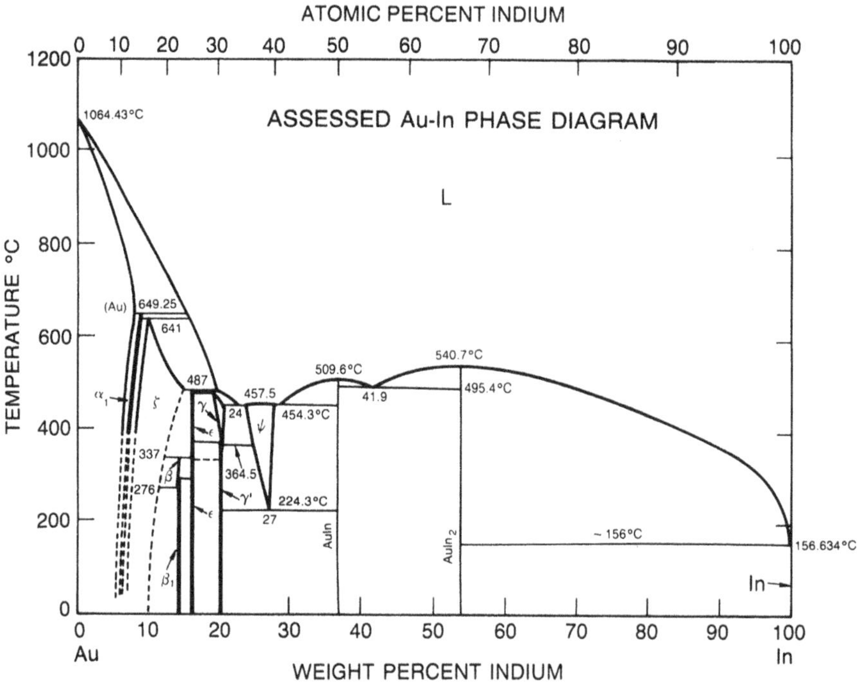

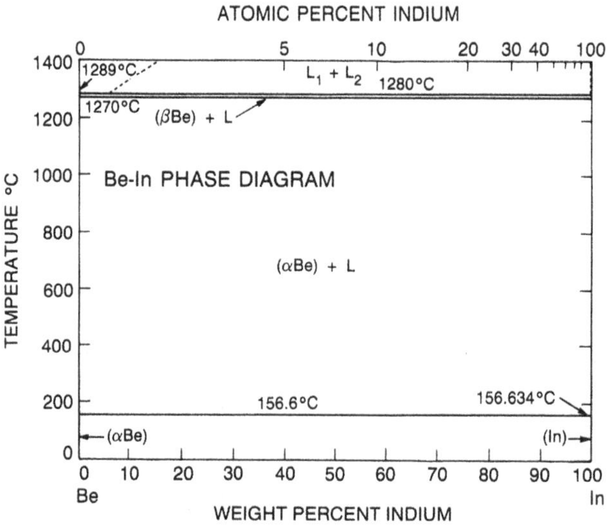

Figure 9.2 Binary phase diagrams of Au-In and Be-In systems.

accounts for the decomposition of the interfacial oxides and contaminations. In addition, this combination should provide sufficient adhesion between the metal layer and the semiconductor, intermix the dopants with the semiconductor interfacial layer, and then possibly form a variety of narrow bandgap interfacial compounds [25,26]. These reactions, however, have to be limited and controlled in order to eliminate the formation of spiky interfaces, which leads to nonuniform current density and, thus, to the evolution of local heated spots due to Joule effects. This is a twofold task, entirely dependent on the metallurgical system of choice and the contact process conditions (such as deposition technique and sintering conditions). Moreover, since the contact sintering process takes place at the final stage of the device manufacturing sequence, the contact sintering process may influence the entire device performance. Thus, a moderate heat treatment is essential for minimizing enhancement of various metallurgical reactions, such as spillover of dopants into the adjoining semiconductor layers (in the case of heterostructure devices), and reducing the occurrence of different interfacial reactions and diffusion processes in between those layers.

The main purpose of the following sections is to summarize the microstructure and electrical properties of a variety of contacts to InP, which are composed of Au-based alloys, near-noble transition metals such as Co, Ni, Pt, Ti, and refractory metals such as W. The influence of the processing conditions on these properties will also be discussed.

9.2.2 Electrical Properties

Modern high-speed electronic and photonic devices are operated under a high current density, with realistic values in the range of 4 to 10 kA/cm^2 in a narrow contact stripe geometry, typically in the range of 2 to 150 μm wide. They require a specific contact resistance lower than about 5×10^{-6} $\Omega \cdot cm^2$ in order to add only a negligible amount to the overall device resistance. Due to the relatively large energy bandgap of InP ($E_g = 1.3$ eV), some rectifying characteristic, as a result of the surface Schottky barrier, is always realized. High surface dopant concentration, higher than 5×10^{18} cm^{-3}, is occasionally applied, either by ion implantation or by incorporating the relevant dopant into an epitaxially grown structure, in order to narrow the associated Schottky barrier height and to allow for enhanced tunneling. When the device geometry allows for additional epitaxial layers, it is useful to cap the layered structure with lower energy bandgap materials, such as $In_{0.53}Ga_{0.47}As$ ($E_g = 0.75$ eV) [11] lattice matched to InP, or graded $In_xGa_{1-x}As$ completed with an InAs layer ($E_g = 0.35$ eV − 0.45 eV) [11,15,27]. A reasonable assumption is that with such a narrow bandgap semiconductor, the Schottky barrier values of the contacts are mainly determined by the composition of the layers, in particular when using a ternary layer. Thus, for the latter case, the specific contact resistance is a function of the InAs mole fraction of this layer [28]. The heavily doped ternary layer, formed by alloy regrowth [8], creating a shallow diffusing zone [29], ion implantation, or heavily doped epitaxially grown layer [11], provides significant advantages over highly doped *p*-type and *n*-type InP, because of the higher thermodynamic dopant solubility limits in the material. Furthermore, the InAs exhibits pinned surface states in the conduction band, and it is therefore used extensively as an *n*-type contacting layer for the formation of *n*-type ohmic contacts [15,28,30].

InP is an exception to Mead's rule, which suggests that the barrier height (Φ_{Bo}) of *n*-type semiconductors equals two-thirds of the energy bandgap (E_g), and that of *p*-type semiconductors equals one-third E_g [15,30]. The InP Φ_{Bo} as measured in *n*-type Schottky diodes was found to be 0.48 eV for a variety of metals, such as Au, Ti, and Al [17,31]. As mentioned before, Fritzche [16] has argued that the energy distribution of N_{ss} in InP is parabolic-like, but with its minimum positioned near the conduction band edge. Thus, the InP surface is only moderately depleted. It is therefore easy to achieve low-resistance ohmic contacts to *n*-type InP. On the other hand, formation of ohmic contacts to *p*-type InP is much more complicated. This, in fact, plays a major role in defining some of the device structures. While *n*-type contacts can be formed through alloying the contact metallization directly onto the InP substrate (typically doped to level of 2×10^{17} to 2×10^{18} cm^{-3}), the *p*-type contacts are frequently formed onto heavily doped epitaxial layers such as InGaAs or InGaAsP. Therefore, most two-sided contact devices, such as the majority of laser diodes, are grown on *n*-type InP substrates, with a simple *n*-type ohmic contact to the substrate and with a front *p*-type contact formed onto a *p*-type contacting semiconductor layer in the neighborhood of the active layer.

In contrast to the studies on the origin and physics of Schottky barrier heights in metal contacts to GaAs [32–36], no definitive model for the actual mechanism of the origin of Schottky barrier and the Fermi level pinning in the contacts to InP has yet been suggested. Some ideas have been suggested, such as the use of group IV or group II dopants to act as donor or acceptor, respectively, by substituting for the In to enhance the ohmic contact formation mechanism. However, some more fundamental issues, such as the Fermi-level pinning problem in InP, are rarely discussed. Since samples are invariably exposed to air prior to the metal deposition and contact processing, it may be presumed that the surface is initially pinned. However, it is not clear at this time whether subsequent metallurgical reactions taking place prior or during the contact formation eliminate the pinning of the Fermi level. It is agreed, however, that when the semiconductor surface is heavily doped, pinning of the Fermi-level does not dominate the ohmic nature and properties of the contact. Since heavily doping the semiconductor surface is currently the most common technology, the issues associated with Fermi-level pinning in the InP bandgap will be considered beyond the scope of this review, and, therefore, will not be discussed.

9.2.3 Correlation Between Ohmic Contact Microstructure and Electrical Properties

The accumulated experimental data show that the nature and properties of ohmic contacts to InP depend on several variables and imply that metallurgical reactions, which have a major role in defining these properties, can proceed even at room temperature. Fatemi and Weizer [37], for example, have clearly demonstrated that Au contacts to InP are chemically unstable at room temperature, due to the leaching action of indium nitrate islands formed in the interface. Tsai and Williams [38] discussed chemical reactions at the Au/InP interface and provided evidence for solid-state reactions which occurred at this interface even at room temperature. In addition, it has not been concluded yet whether the interfacial metallurgical reactions associated with the aging of a variety of InP-based devices through routine operating conditions are to the benefit or detriment of the ohmic contact performance. As will be detailed in the following sections, the metal/InP and metal/InP-related material (InGaAs, InGaAsP, *et cetera*) systems are extremely reactive, and complicated interfacial microstructures have been observed in all these systems, regardless of the applied metal, through the thermal processes essential in the contact processing sequence. It is our belief that the carrier-transport nature, particularly in the cases of ohmic contacts to InP, is solely determined by the interfacial microstructure. The specific contact resistance may be reduced by several orders of magnitude simply by optimizing the thermal processing conditions to achieve a low contact-resistance microstructure [24,39]. The metal/InP interfacial microstructure contains various different crystalline compounds of metal and groups III or V elements, each having different carrier-conduction properties. In addition, they exhibit various geometrical shapes and distributions at the interface, and are associated with massive crystalline defects generated because of the structural incompatibility. Thus, the electrical properties of this interface cannot be described by directly applying one of the classical approaches, such as the Schottky or tunneling barrier theories [11], which are based on the assumption of a chemically abrupt and structurally uniform interface. For these, the correlation between the contact resistance and the processing parameters, such as the semiconductor carrier concentration and the sintering temperature, was defined for the two basic carrier transport mechanisms at the metal-semiconductor contact interface.

For the thermionic-emission case

$$Rc \approx [K/(\alpha^* T)] \exp[(q\Phi_B)/(KT)] \tag{9.1}$$

and for the tunneling case

$$Rc \sim \exp\left[\left(2\sqrt{\varepsilon_s m^*}\right)\right]/\hbar\left(\Phi_B/\sqrt{N}\right) \tag{9.2}$$

where Φ_B is the Schottky barrier height at the metal-semiconductor interface, K is the Boltzmann constant, T is the absolute temperature, N is the carrier concentration, q is the magnitude of the electrical charge, ε is the semiconductor permittivity, m^* is the effective mass, α^* is the Richardson constant, and $\hbar$ is Planck's constant divided by 2π.

By evaluating the exponential dependence of the contact resistance as a function of either the reciprocal temperature or the doping concentration, the primary mode of the carrier transport mechanisms across the interfacial barrier may be defined. However, in order to properly describe the carrier transport across the interfaces of the metal/InP contacts, it is necessary to take into account the effect of the nonuniform, degraded, and complicated interfacial microstructure. Unless both the doped or alloyed interfacial semiconductor layers and the metallic contacted layers are grown in situ to complete the contact formation without exposing it to the atmosphere by means of, for example, molecular beam epitaxy (MBE) [40,43] or metalorganic chemical vapor deposition (MOCVD) [44], the above-mentioned approach is essential in order to understand the interfacial transport mechanisms. Thus, we have suggested [26,45,46] a phenomenological theory, based on the existence of simultaneous parallel carrier-conduction processes across the interface, and formulated it in order to explain, in a semiquantitative manner, the temperature dependence of the specific contact resistance and the interfacial microstructure of metal/InP-based materials, formed under different sintering conditions. An essential microstructure parameter was introduced that accounted for the effect of the interfacial microstructure on the carrier transport process. Based on multiple parallel carrier-transport mechanisms, the specific contact resistance of an ohmic contact is given by

$$Rc = \left[\left[\frac{\partial S}{\partial V} \right]_{v=0} \right]^{-1} = \left[\sum_{i=1}^{N} Xi\, ji(fi)ji(T) \left[\frac{\partial_{ji}(v)}{\partial V} \right]_{v=0} \right]^{-1} \quad (9.3)$$

where $Xi = Ai/A$ is the fraction for the interfacial area occupied by the "ith" compound, $ji(fi)$ is the structural parameter of the "ith" compound, and T is the absolute temperature. For a nearly ideal interface, where the carrier transport mechanism is dominant by a single thermionic emission process, Rc reduces to a single term, since $X_i = 0$, and thus

$$Rc = \frac{k}{j(f)\rho T} \exp \frac{\rho\phi_B}{kT} \quad (9.4)$$

or

$$\ln RcT = \ln \frac{k}{j(f)\rho} + \frac{\rho\phi B}{kT} \quad (9.5)$$

Thus, for a carrier transport process dominated by the thermionic emission mechanism [9], linear relationship between ln RcT and $1/T$ should be observed, as was measured, for example, for Pt/Ti/In$_{0.53}$Ga$_{0.47}$As as deposited contacts (see Figure 9.3). For this contact the barrier height was calculated to be 0.13 eV, and the corresponding microstructure factor to be 0.29A K^{-2} cm^{-2}.

The sharp decrease in the specific contact resistance of the Pt/Ti/InGaAs sample, with the increase in the rapid thermal processing (RTP) temperature up to 450°C [47], indicated the development of a new, low-contact resistance, interfacial micro-

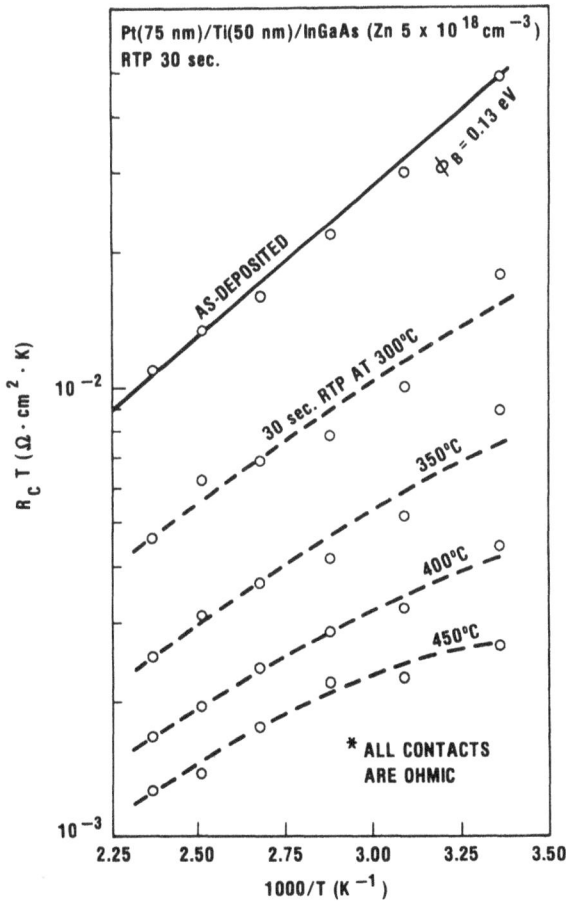

Figure 9.3 Natural logarithm (Ln) of specific contact resistance multiplied by temperature (R_cT) as a function of $1/T$ of as-deposited Pt/Ti/In$_{0.53}$Ga$_{0.47}$As (Zn doped 5×10^{18} cm^{-3}) sample and after RTP at different temperatures.

structure. The interface for elevated temperatures contains multiregime regions, and is therefore dominated by different carrier transport properties. Thus, the overall carrier transport mechanism could no longer be described by a single process, as for the as-deposited contact, and a deviation from the linearity in the $\ln R_c T$ versus $1/T$ curve was expected. Empirically we found that introducing only one temperature-independent component in Equation (9.3) was essential in order to fit the measured data to the theory. The implication was that the carrier transport mechanism across the newly developed interfacial microstructure was dominated by the field-emission process. For simplification, we have avoided the introduction of the possible existing thermionic-field emission process parameters. If (X) and $(1 - X)$ describe the fraction of interfacial areas occupied by the original and the newly developed microstructures, respectively, then Equation (9.3) may be written as

$$R_c = \left[\frac{X}{R_{th}} + \frac{1 - X}{R_{tu}} \right]^{-1} \qquad (9.6)$$

where

$$R_{th} = \frac{k}{j_1(f_1)\rho T} \exp \frac{\rho \phi_B}{kT} \qquad (9.7)$$

and

$$R_{tu} = \frac{\sqrt{N}}{j_2(f_2)C_1 C_2} \qquad (9.8)$$

As shown in Figure 9.4, all the measured R_c versus T values of all the samples sintered by RTP at different temperatures were accurately fitted by the values calculated using Equation (9.6). The parameters that were used for the best fit are listed in the inserted table. As expected, the fractional area, $1 - X$, of the low R_c field emission regions, increased from 7% after RTP at 300°C to about 30% after RTP at 450°C. A slight change in the original microstructure, which took place with the modification of the RTP temperature, was also reflected in the increase of the structure parameter $j(f)$. The same effect was likely to have caused the slight decrease in the R_{tu} values (from 6×10^{-6} $\Omega \cdot cm^2$ to 4×10^{-6} $\Omega \cdot cm^2$), which were used in fitting the measured data.

This phenomenological theory was applied to some other metal/InP-based material systems as well, and in all cases led to a good agreement with the measured values.

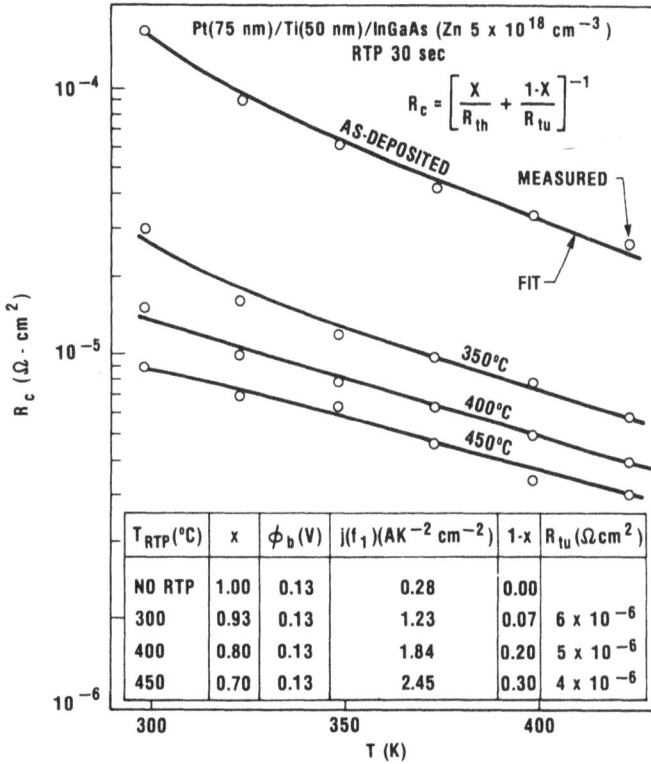

Figure 9.4 Specific contact resistance (R_c) as a function of the measuring temperature (T) of the as-deposited Pt/Ti/In$_{0.53}$Ga$_{0.47}$As (Zn doped 5×10^{18} cm^{-3}) sample and after RTP at different temperatures. The measured values (dots) and fitted values, calculated from the parallel conduction mechanism phenomenological theory (continuous line) are superimposed.

9.2.4 Intrinsic Thermal and Mechanical Stresses

The stress induced in the contact metallization, semiconductor and dieletric layers under it has an important influence on the quality of the contact properties, as well as an influence on the overall device performance. In general, this stress may be caused by a variety of intrinsic parameters, such as film deposition conditions, film mode of growth, nature and concentration of grown-in defects, and substrate temperature during depositions [48]. In addition, the contact quality is strongly dependent on the thermal stresses that are induced in the film during the contact sintering, which occurs as a result of one or more of the following reasons: the difference in the coefficient of thermal expansion between the film and the substrate, interfacial interdiffusion and interaction, compound formation, grain growth, recrystallization, grain reorientation, densification, and thermal gradients induced by the heat treatment source and nature [49]. High stress, particularly if tensile, may stimulate a number of contact and device failure modes through an immediate loss of adhesion between the film and the substrate, which leads to contact peeling, or (as a result of time-dependent phenomena such as enhancement of interdiffusion and interreaction phenomena at the metal-semiconductor interface) formation of dislocations in the substrate and the acceleration of defect migration [50].

Therefore, it is important to monitor and optimize the induced stress in the contact region, not only through the epitaxial layer growth geometry [51–53], but also through the large number of processing steps involved in the device fabrication, in particular through the contact sintering. In order to define the origin of stress through these various processes, in situ measurements are required [54]. Figure 9.5

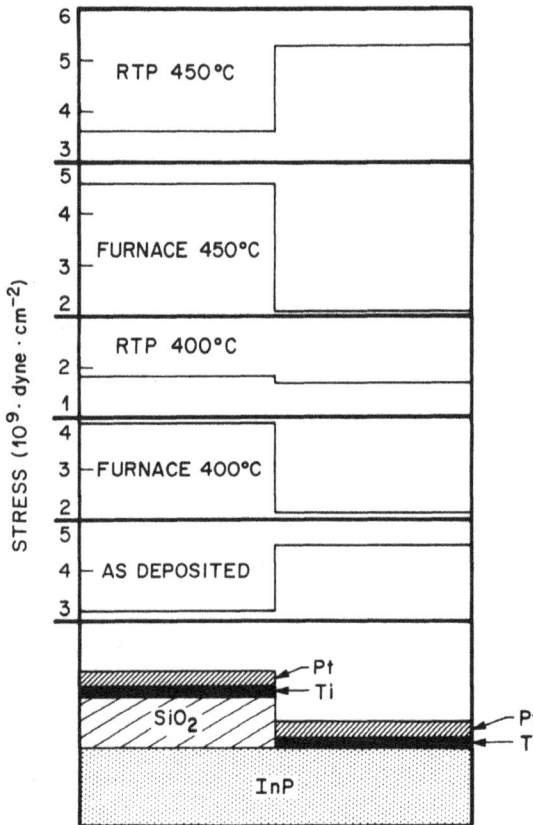

Figure 9.5 Schematic presentation of stress discontinuity in Pt/Ti/InP-Pt/Ti/SiO₂/InP contact edge, induced by RTP and conventional heat treatment at 400°C and 450°C.

provides an example of optimization of contact sintering conditions due to stress discontinuity induced in the various layers involved in the definition of an ohmic contact to some generic InP-based device. The figure shows in a schematic way the stresses induced in a typical Pt/Ti/InP contact edge. This contact was geometrically defined by a plasma-enhanced chemical vapor deposition (PECVD) SiO_2 layer, metallized with the same Pt/Ti scheme. Selecting a highly stressed metal or an incorrect sintering condition for the contact formation may lead to highly stressed structure, both in the immediate vicinity of the metal/InP contact and in the bonding pads in the peripheries, and to stress discontinuity at the contact edges. The latter may act as a core for three-dimensional dislocation propagation and as a preferred site for adhesion failures, and may, as a result, stimulate thermal failures by interdiffusion and interation behavior along these defects. Therefore, stress discontinuity in the contact borders must be minimized while optimizing the contact formation process. In all the contacts shown in Figure 9.5, processed through four different sets of conditions, stress was continuously measured in situ through the sintering cycle. The smallest stress discontinuity was measured in the sample that was sintered at 400°C by RTP, which may suggest some technological advantage for this set of processing conditions.

9.3 ELEMENTAL METAL-InP CONTACTS

9.3.1 Au-Based Contacts

A variety of Au alloys are traditionally suggested as the metallization scheme for the formation of ohmic contacts to InP and its related materials. The early success of these contacts is directly revealed by the thermodynamic driving forces of the Au-InP reaction (see, for example, Fig. 9.2). From the thermodynamic nature of the binary Au-In and Au-P systems, the ternary Au-InP system can be anticipated to be very reactive even at temperatures as low as 200°C. The dominant reaction is the one between the Au and the In, leading to the formation of Au-In compounds such as $AuIn_2$ and Au_7In_3 [46]. The propagation of this reaction is greatly assisted by the large entropy of phosphide sublimation.

These tunneling ohmic contact schemes contain the electronegative element, Au, which has a high solubility for In and the dopant element, and they therefore have to be alloyed at a temperature between 300°C and 450°C in order to form a thin, intermediate, heavily-doped layer between the metal and the semiconductor.

For contacts to *p*-type InP, Au-Zn [4,55–61] and Au-Be [62–66] evaporated by electron gun have been commonly used and annealed at 400°C to 430°C to yield specific contact resistance values lower than 1×10^{-5} Ω-cm^2. In addition to these

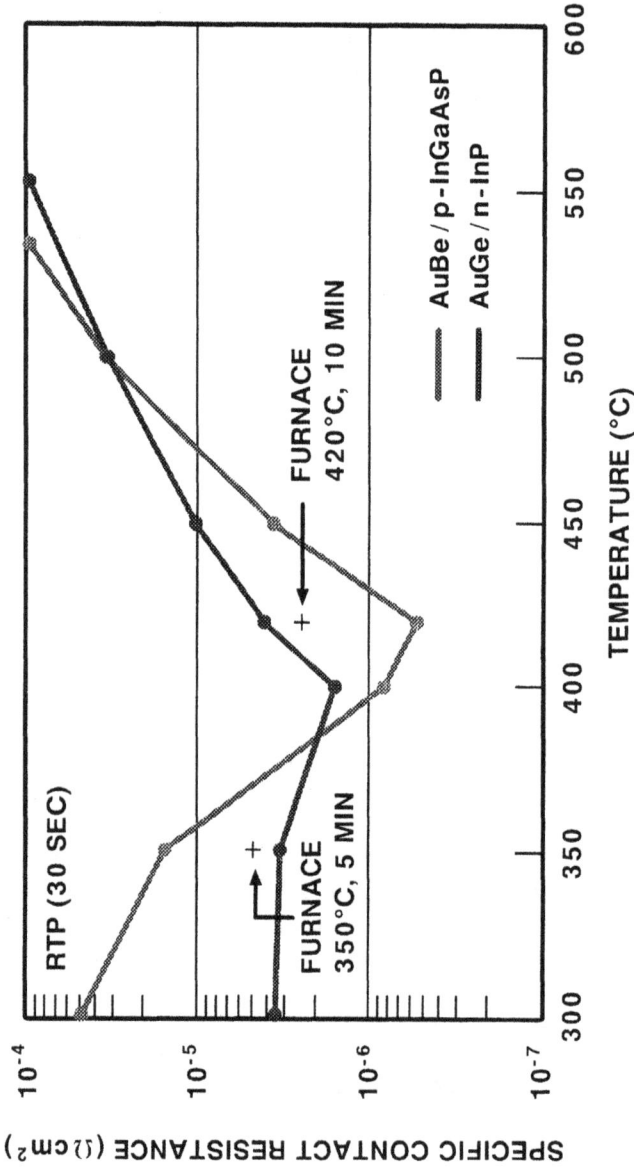

Figure 9.6 Specific contact resistance as a function of the processing temperature at the AuGe/n-InP and AuBe/p-InGaAsP systems.

metallization systems, Au-Mg [67], Au-Cd [68], Au-Ni [69], and some ternary al-loys, such as Au-Be-Cr [70] and Au-Zn-Ni [71], have been successfully applied to form an ohmic contact to *p*-type InP.

For contacts to *n*-InP, the most commonly employed scheme is the Au-Ge (12 wt % Ge) eutectic alloy [72–75]. Alloying this contact at the eutectic melting tem-perature of about 360°C drives the dissolution of InP and the resolidification of the phosphide-rich InP, as a result of the high solubility of indium in the Au-Ge melt adjacent to the substrate. Germanium is likely to incorporate preferentially on the In site in the precipitated InP. Alloying the contact at temperatures of 450°C and higher results in significant outdiffusion of P and enhanced formation of P vacancies. This leads to degradation of the contact electrical properties, as is reflected in the plots shown in Figure 9.6. Other metallization schemes often used to form ohmic contacts to *n*-type InP are Au-Ge-Ni ternary alloy [76–81] and Au-Sn [82–84].

All the Au-based contacts, which require an alloying thermal cycle in order to allow for the intermixing of the dopants in the metal/semiconductor interfacial layer, suffer some general deficiencies: (a) formation of wide interfacial reaction layer, typically 2 to 3 times the thickness of the initial deposited metal layer [85] (Figure 9.7 shows a typical example of the wide reaction zone of a AuBe contact on a laser device in which Au spikes, about 0.5 μm large, are observed), (b) nonuniform depth of interaction and spatial composition, leading to nonuniform current density across

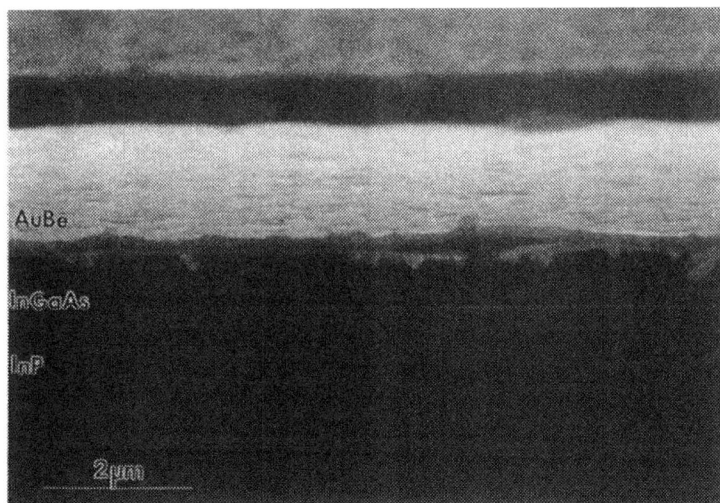

Figure 9.7 SEM cross-section micrograph of AuBe contact on CSBH laser device after furnace alloy at 420°C for 5 min.

the contact [63,85], and (c) lack of long-term metallurgical stability [37,86,87]. These deficiencies may be improved by alloying the Au-based contacts by means of RTP [24]. As an example, the improvement in the electrical properties, reflected in a reduction of one order of magnitude in the specific content resistance of the AuBe contact, is shown in Figure 9.6.

9.3.2 Near-Noble Transition Metal-Based Contacts

In order to overcome the deficiencies involved in the processing and performance of the Au-based alloyed contacts to both p-type and n-type InP-based materials, in particular their unstable nature and lateral nonuniformity, various metallization schemes have been suggested as alternative contacts. All of these solutions are based on the formation of an ohmic contact by driving Au-free solid phase reactions and by using higher eutectic melting point systems. These contacts use the near-noble transition metals as the metallization scheme of choice. Due to the more stable thermodynamic properties and less reactive nature of these metals, they provide a less reactive and less degraded contact microstructure through heat treatments at higher temperatures. The ohmic nature of these contacts is obtained by both heavily doping the semiconductor layer under the metals and by the formation of narrow-bandgap intermetallic phases in the metal/semiconductor interface through a limited reaction when sintering the contacts. Most commonly used are the Ni and Ni-P [88,89], and Ti and Ti/ Pt contacts [8,26,46–48,90–93]. Depositing the metal scheme, either by electron beam evaporation or by sputtering, onto highly doped p- and n-type InP-based material (typically Zn or S doped, respectively, to the levels of 1×10^{18} to 8×10^{18} cm^{-3}) and optimizing the process conditions may lead to the formation of contacts with specific resistance in the range of 1×10^{-6} to 3×10^{-8} $\Omega \cdot$cm^2. A typical example of the specific contact resistance dependence on the semiconductor doping level is shown in Figure 9.8. The electrolytic capacitance-voltage depth profiles as a function of depth in four different p-InGaAsP layers doped with Zn to different levels are given in Figure 8a, while the transmission line model (TLM)-derived resistance measurements at room temperature as a function of the RTP sintering temperature of the Pt/Ti contacts are given in Figure 8b. The RTP was carried out under N_2 ambient by means of the A.G. Associates' Heatpulse 410T model. The specific contact resistance values are clearly improved as a result of the increase in the dopant concentration. A substantial diminishment of about two orders of magnitude in the specific resistance is observed in the three higher doped specimens as a result of RTP at 450°C. Increasing the sintering temperature above 450°C increases the specific contact resistance values in all the samples, reflecting a degradation of the stable contact microstructure.

The limited reaction of these contacts to the various InP-based semiconductors was widely demonstrated, and a clear example of it is given in Figure 9.9 for the

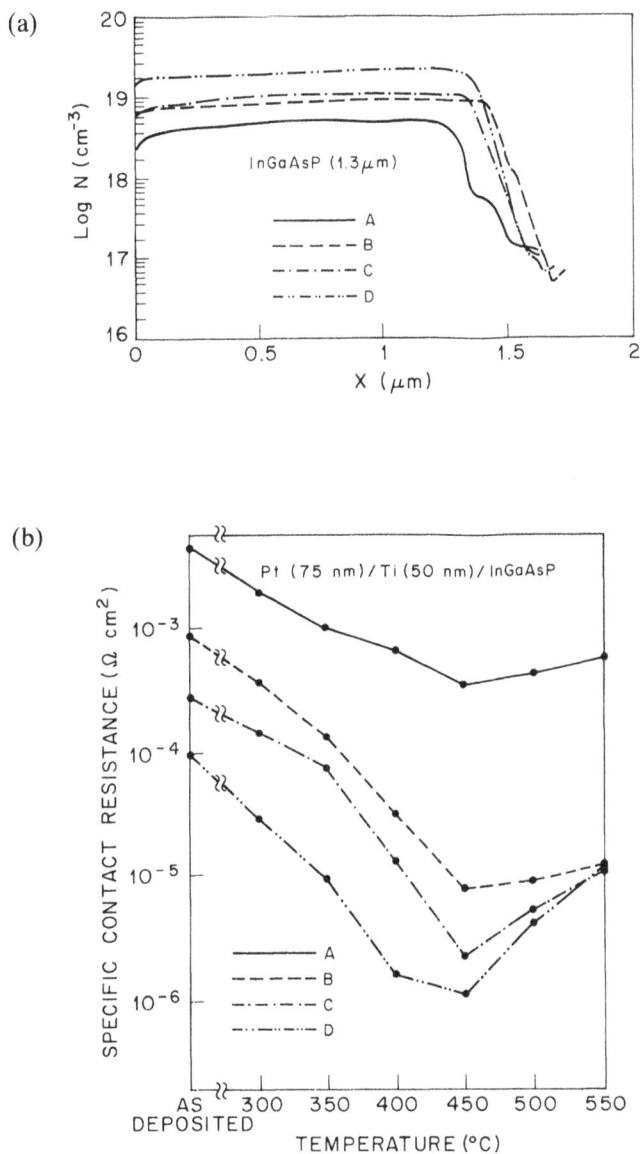

(a)

(b)

Figure 9.8 (a) Electrolytic capacitance-voltage depth profiles as a function of depth in four different Zn doped p-InGaAsP samples; (b) the specific contact resistance as a function of RTP temperature of Pt/Ti contacts to the four Zn doped p-InGaAsP layers.

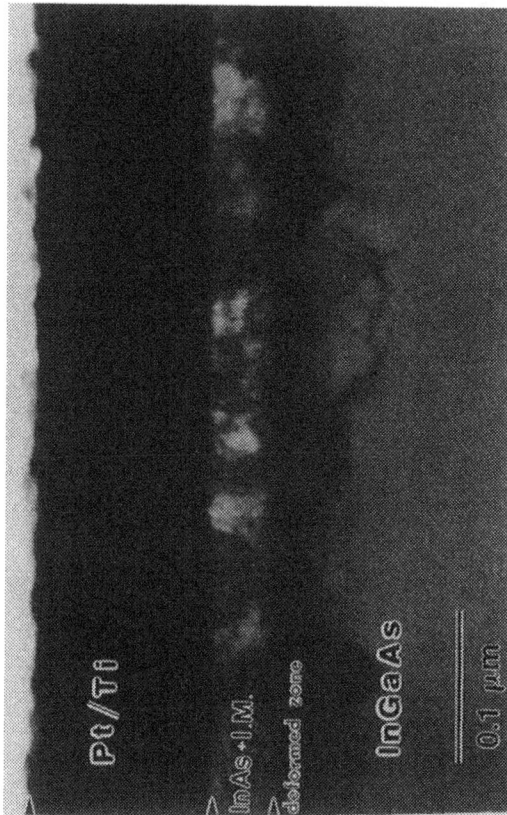

Figure 9.9 TEM cross-sectional micrograph of Pt/Ti/In$_{0.53}$Ga$_{0.47}$As contact sintered by RTP at 450°C for 30 sec.

case of a Pt/Ti contact to 1.5×10^{19} cm^{-3} Zn-doped In$_{0.53}$Ga$_{0.47}$As layers, sintered by means of RTP at 450°C for 30 sec. This figure shows a transmission electron microscope cross-sectional micrograph of the contact. A narrow interfacial reacted layer of about 40 nm thick, containing various intermetallics, is observed, as well as a deformed zone of about 30 nm thick, which was formed in the InGaAs layer adjacent to the metal/semiconductor interface. The latter was formed as a result of the observable outdiffusion of In and As, resulting in a significant depletion near the semiconductor surface, leading to decrease in the lattice constant in the InGaAs layer, and inducing misfit stresses. This reaction, however, is far more limited than the reactions formed in the Au-based alloys (see, for example, Figure 9.7) and demonstrates very clearly the limited reaction induced in the Pt/Ti contact, even for relatively high sintering temperatures.

9.3.3 Refractory Metal-Based (W, W-Alloys) Contacts

For high-speed devices operated at high temperature conditions (>600°C), the thermal stability offered by the transient metals may be not sufficient. These contacts cannot be exposed to temperatures higher than 450°C through the contact processing without suffering some severe degradation, the type of which may also be observed during operating under high-temperature and high-current density conditions. The highest level of thermodynamic stability may be achieved using refractory metal-based contacts, such as a variety of W-based alloys (W_xSi_y, $GeWSi_z$, WN, WSiN, WTiN, and WPtAg), suggested earlier as also a good metallization scheme for self-aligned metallization layers on either GaAs [94–102] or InP [18,19,103] substrates. These contacts exhibit an inert behavior through heating cycles at temperatures up to 700°C, preserving the almost abrupt metal/semiconductor interfaces, and have typical specific contact resistance values in the range of 1×10^{-6} to 1×10^{-7} $\Omega \cdot cm^2$, for highly doped semiconductors. As an example, the Rutherford backscattering spectrometry (RBS) spectra of the $W/In_{0.53}Ga_{0.47}As$ contact, as deposited and after RTP at various temperatures up to 800°C, are given in Figure 9.10. The RBS shows no difference in the shape of the W peak position for sintering in the temperature range of 300°C to 700°C, with an obvious deviation as a result of RTP at 800°C.

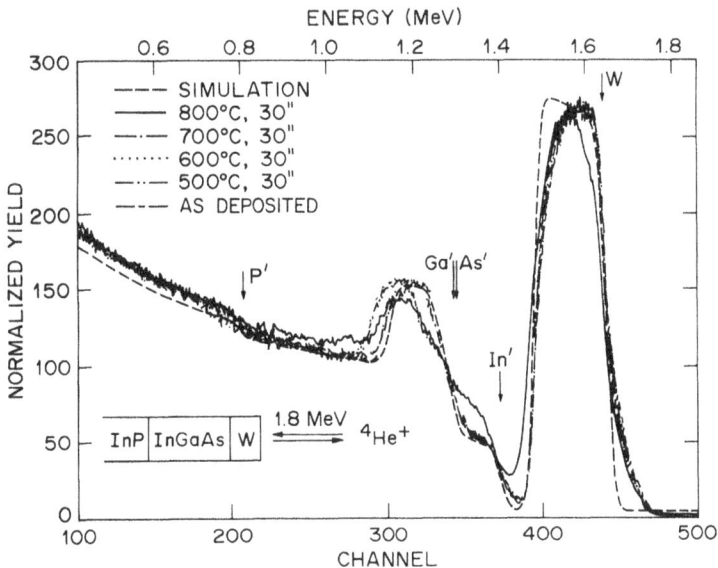

Figure 9.10 RBS spectra of W (100 nm)/p-InGaAs as-deposited contact and after RTP at different temperatures.

These results agree with other observations that clearly revealed the absolute stability of the W/InGaAs contacts at temperatures up to 800°C.

9.4 CONCLUSIONS, SUMMARY, AND FUTURE DIRECTIONS

Contacts to n-type and p-type InP and related materials are becoming a major field of concern and development, because these semiconductor materials are widely emloyed in high-speed electronic and photonic device technology. In contrast to contacts to GaAs and related material, this technology is still premature, and the understanding of the contact metallurgy and carrier transport mechanisms, and the correlation between the microstructure and electrical properties are far from satisfactory. Much more work has yet to be invested in these studies.

However, some principles and fundamentals have been demonstrated in this review, and the kind of metallization scheme and process conditions necessary for the required performance and device technology can be defined. The main issues that have to be considered when selecting the metallization scheme and process conditions are (a) optimization of the electrical properties to achieve the lowest possible contact resistance, (b) stable microstructure over a wide temperature range, (c) stability throughout the device electrical operating conditions, (d) minimization of stress in the underlying semiconductor and in the metal films, (e) compatibility with the metal system used for the interconnection technology, (f) fabrication through a robust process that fits as an integrated step into the overall device manufacturing scheme.

Figure 9.11 shows a rather complicated graph which is essential for optimizing a contact formation process. This figure gives both the specific contact resistance and the thin layer biaxial stress values of the common Pt/Ti contact to p-InGaAs and n-InP as a function of the RTP temperature. In addition, it presents the induced stresses in the Pt/Ti and Pt/Ti/SiO$_2$ structures on InP substrates. The specific resistance values of the Pt/Ti contacts to both p-InGaAs and n-InP are given as a function of the semiconductor doping level over the range of 5×10^{18} to 1.5×10^{19} cm^{-3} Zn doped for the former, and 5×10^{17} to 5×10^{18} cm^{-3} S doped for the latter. The vertical line at a temperature of 470°C shows the highest RTP temperature at which degradation of the contact microstructure has not yet occurred. The compilations of the above-mentioned properties enables the optimization of the contact formation processing conditions, namely, allowing for the lowest specific contact resistance, the lowest induced biaxial stress (both in the metallization films and in the bonding pad structure), and achieving a stable microstructure. For the Pt/Ti contacts, to both n-InP and p-InGaAs, it can be seen that to satisfy these three major requirements, the best sintering temperatures range from 400°C to 450°C. Sintering at the high end of this range leads to the formation of a contact with an extremely low specific contact resistant. Reducing the sintering temperature, however, to 400°C results in a considerable reduction in the biaxial induced stress of both the metal/

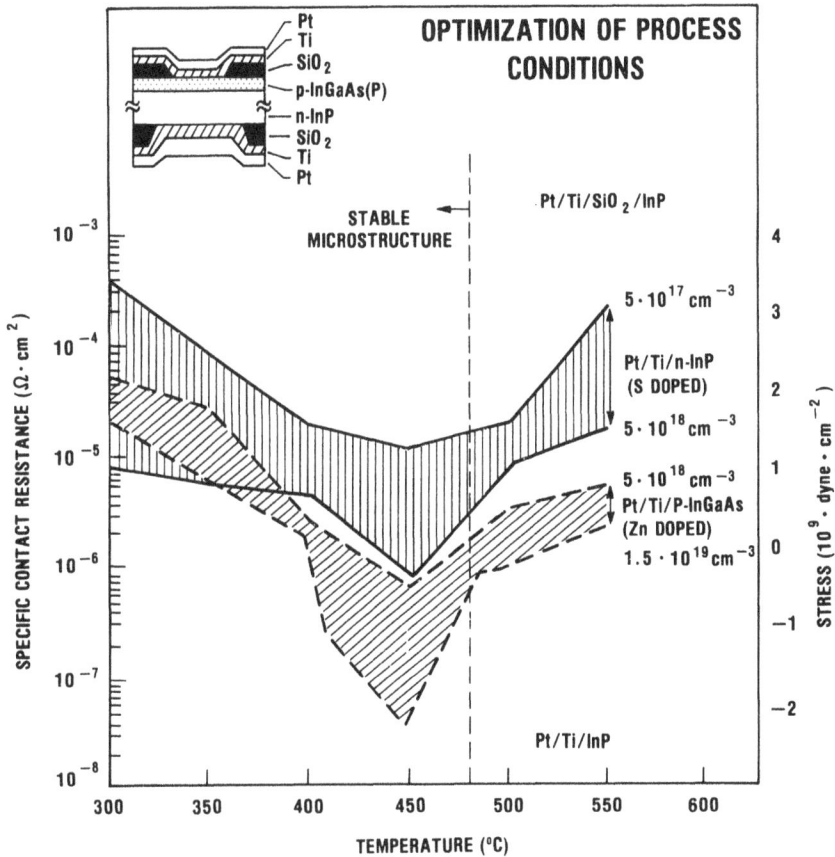

Figure 9.11 Specific contact assistance and induced biaxial stress as a function of the processing temperature of Pt/Ti contacts to Zn-doped $In_{0.53}G_{0.47}As$ and S-doped InP.

SiO_2 and metal/semiconductor systems. This widely available processing temperature window shows the attractive advantages of the Pt/Ti contact scheme, namely, its flexibility in processing, which enables the user to optimize the formation conditions according to individual specific needs and constraints.

The technology trend towards the fabrication of microstructures and submicron technology renders added significance to the understanding of the nature of the metal/InP-based interfaces. Improving the quality of these interfaces and controlling a very shallow interaction will provide the desired abrupt metal/semiconductor interfaces, leading to uniform current transport via the contact and enhanced reliability, both for short- and long-term operation. Further reduction in the interfacial defect densities is necessary in order to control the contact electrical properties and to achieve reproducible characteristics. Advances in this research will require an atomistic understanding of the interfacial reactions, in situ characterization capabilities of the process in real time, and the ability to absolutely control metal/semiconductor diffusion and interactions.

Elimination of interfacial defect and lattice mismatch in the metal/semiconductor system, for example, by fabricating stable epitaxial metal contacts or metal/semiconductor heterostructures, preferably by means of an in situ sequential process, is expected to improve the wetting and adhesion of the thin deposited layers, as well as the stability and electrical performance of the contacts and devices. Furthermore, kinetic effects, bulk and grain-boundaries diffusion processes, lateral and vertical spreading of the metallization during the depositions, and the subsequential sintering process would be effectively limited by using high eutectic melting point metal/III-V systems. Metals such as W and W-alloys are certainly the metals of choice, mainly due to their thermodynamic stability.

On top of these issues, it is necessary to consider the appropriate and suitable metallization processes for the future fabrication of multilevel devices with small (submicron) dimensions and high-aspect ratio contact windows and stripes, as well as the required shallow junctions and very thin cladding layers over the device active region.

The above mentioned discussion points out the need for debating the advantages and disadvantages of the variety of the metal deposition techniques, such as sputtering, evaporation, CVD, or MOCVD. Each of those techniques has its own merits and drawbacks, which have been widely discussed in conjunction with silicon technology [104]. However, the severe demands on the future metallization point to some advantages of deposition by means of CVD and MOCVD. Tungsten technology, both selectively deposited [105–108] or blanket deposited, followed by etchback to form the required geometry [109–110], was well developed and reported for applications in silicon device fabrication, and the fundamentals may be adopted for use in InP technology. The major attractive characteristics of the chemical deposition techniques are the properties of the films having a conformal coverage, high purity, easily doped and with controlled stress, and the processing benefits, such as low-temperature deposition, radiation-damage free, deposition selectivity, and excellent step-coverage.

CVD metallization in a rapid thermal processor (RTP-CVD) with load locks and cold walls, provides an attractive technique due to the improvement in the deposition selectivity, controlled deposition operation modes, availability of both gas flow- or lamp intensity-switched, and a short and reproducible contact sintering sequence [111–113].

Beyond the ohmic contact applications and issues partially covered in this review, there is a wide range of exciting research and development work necessary to provide currently needed and future projected metallization processes for InP-based electronic and photonic device manufacturing. As the understanding of the electrical, optical, mechanical, and morphological properties (and the correlations between them) of these thin metal films on InP and related materials, are improved, more exciting and advanced applications will be explored and realized to benefit device manufacturing technology.

ACKNOWLEDGMENTS

I gratefully acknowledge the many contributions and very close and friendly help of a lot of my AT&T Bell Laboratories colleagues. In particular, I would like to express my appreciation of the work provided by C.R. Abernathy, S.N.G. Chu, W.C. Dautremont-Smith, M. Geva, W. Hobson, J.W. Lee, S.G. Napholtz, S.J. Pearton, P.M. Thomas, and B.G. Weir, as well as my colleague W. Savin from NJIT.

REFERENCES

1. T.C. Eschbich, R.D. Carroll, R.N. Sacks, and W.J. Tanski, *IEEE Trans. Electron. Devices*, Vol. 36, 1989, pp. 1213–1215.
2. J.A. Del Alamo and J. Mizutani, *Solid State Electron.*, Vol. 31, 1988, pp. 1635–9.
3. P.E. Hallali, P. Blanconnier, L. Bricard and J.C. Renaud, *J. Phys. Colloq.*, Vol. 49, 1988, pp. 453–456.
4. J.B. Boos and W. Kruppa, *J. Vac. Sci. Technol. B*, Vol. 7, 1989, pp. 502–504.
5. H. Schumacher, J.R. Hayes, R. Bhat, and M. Koza, *International Electron. Devices Meeting, IEDM, Technical Digest*, 1987, pp. 852–853.
6. A. Yoshida, H. Tamura, T. Fujii, and S. Hasuo, *Extended Abstracts of 1987 International Superconductivity Electronic Conference*—ISEC, 1987, pp. 368–371.
7. H. Temkin, R.A. Logan, R.K. Karlicek, Jr., K.E. Strege, J.P. Blaha, and P.M. Gabla, *Appl. Phys. Lett.*, Vol. 53, 1988, pp. 1156–1158.
8. R. Kaumanns, N. Grote, H.G. Bach, and F. Fidorra, *Inst. Phys. Conf. Ser.*, Vol. 91, 1987, pp. 501–506.
9. M. Fukada, O. Fujita, and S. Uehara, *J. Lightwave Technol. G.*, 1987, pp. 1808–1811.
10. D.L. Meier and D.R. Schroeder, *IEEE Trans. Electron. Dev.*, Vol. Ed-31, 1984, pp. 647–651.
11. S.M. Sze, *Physics of Semiconductor Devices*, John Wiley & Son, 1981.
12. S.H. Rhoderich and R.H. Williams, *Metal-Semiconductor Contacts*, Clarendon, 1988.
13. K. Kojiyama, Y. Mizushima, and S. Sakata, *Appl. Phys. Lett.*, Vol. 23, 1973, pp. 458–60.
14. T.E. Fischer, F.G. Allen, and G.W. Giobeli, *Phys. Rev.*, Vol. 163, 1967, p. 703.
15. C.A. Mead and W.G. Spitzer, *Phys. Rev. Lett.*, Vol. 10, 1963, pp. 471–472.
16. D. Fritzche, *Inst. Phys. Conf. Ser.*, Vol. 50, 1980, pp. 258–268.
17. E. Kuphal, *Solid State Electronics*, Vol. 24, 1981, pp. 69–78.
18. A. Katz, B.E. Weir, D.M. Maher, P.M. Thomas, M. Soler, W.C. Dautremont-Smith, R.R. Karlicek, Jr., J.D. Wynn, and L.C. Kimerling, *Appl. Phys. Lett.*, Vol. 55, 1989, pp. 2220–2223.
19. A. Katz, S.J. Pearton, and M. Geva, *J. Appl. Phys.*, Vol. 68, 1990, pp. 3110–3113.
20. A. Piotrowska, A. Guivafch, and G. Pelous, *Solid State Electron.*, Vol. 26, 1983, pp. 179–183.

21. T. Sands, *Materials Science and Eng.*, Vol. B1, 1989, pp. 289–312.
22. R.P. Gupta, R.C. Dubey, and W.S. Khokle, *J. Pure Appl. Phys.*, Vol. 28, 1990, pp. 126–135.
23. A. Katz, S.N.G. Chu, B.E. Weir, C.R. Abernathy, W.S. Hobson, S.J. Pearton, and W. Savin, *IEEE Trans. Elect. Dev.*, 1991.
24. A. Katz, P.M. Thomas, S.N.G. Chu, J.W. Lee, and W.C. Dautremont-Smith, *J. Appl. Phys.*, Vol. 66, 1989, pp. 2056–2062.
25. R. Singh, R.P.S. Thakur, A. Katz, A.J. Nelson, S.C. Gebhard, and A.B. Swartzlander, *Appl. Phys. Lett.*, Vol. 57, 1990, pp. 1239–1241.
26. S.N.G. Chu, A. Katz, T. Boone, P.M. Thomas, V.G. Riggs, W.C. Dautremont-Smith, and W.D. Johnston, Jr., *J. Appl. Phys.*, Vol. 67, 199, pp. 3754–3760.
27. J.N. Walpole and K.W. Nill, *J. Appl. Phys.*, Vol. 42, 1971, pp. 5609–5614.
28. J.M. Woodall, J.L. Freeouf, G.D. Pettit, T.J. Jackson, and P. Kirchner, *J. Vac. Sci. Technol.*, Vol. 19, 1981, pp. 626–634.
29. M.C. Amaan and G. Franz, *J. Appl. Phys.*, Vol. 62, 1987, pp. 1541–1544.
30. C.A. Mead and W.G. Spitzer, *Phys. Rev.*, Vol. 134, 1964, pp. A713–716.
31. A.M. White, A.J. Grant, and G. Day, *Electron. Lett.*, Vol. 14, 1978, 409–411.
32. V. Heine, *Phys. Rev. A*, Vol. 138, 1965, pp. 1689–1694.
33. J. Tersoff, *Phys. Rev. Lett.*, Vol. 52, 1984, pp. 465–467.
34. W.E. Spicer, I. Lindau, P. Skeath, C.Y. Su, and P. Chye, *Phys. Rev. Lett.*, Vol. 44, 1980, pp. 420–422.
35. L.J. Brillson, *International Electron. Device Meeting, IEDM, Technical Digest*, 1983, pp. 111–114.
36. L.J. Brillson, R.E. Viturro, S. Chang, J.L. Shaw, C. Maichiot, R. Zanoni, Y. Hwu, G. Maryaritondo, P. Kirchner, and J.M. Woodal, *Mat. Res. Soc. Symp. Proc.*, Vol. 148, 1989, pp. 103–115.
37. N.S. Fatemi and V.G. Weizer, *Appl. Phys. Lett.*, Vol. 57, 1990, pp. 500–502.
38. C.T. Tsai and R.S. Williams, *J. Mater. Res.*, Vol. 1, 1986, pp. 1–7.
39. A. Katz, W.C. Dautremont-Smith, P.M. Thomas, L.A. Koszi, J.W. Lee, V.G. Riggs, R.L. Brown, J.L. Zilko, and A. Lahav, *U. Appl. Phys.*, Vol. 65, 1989, pp. 4319–4323.
40. P.D. Kirchner, T.N. Jackson, G.D. Pettit, and J.M. Woodal, *Appl. Phys. Lett.*, Vol. 47, 1985, pp. 26–28.
41. R. Stall, C.E.C. Wood, K. Board, and L.F. Eastman, *Electron. Lett.*, Vol. 15, 1979, pp. 800–802.
42. M. Eizenberg, M. Heiblum, M.I. Nathan, N. Braslau, and P.M. Mooney, *J. Appl. Phys.*, Vol. 61, 1987, pp. 1516–1522.
43. C.J. Palmstrom and G.J. Galvin, *Appl. Phys. Lett.*, Vol. 47, 1985, pp. 815–817.
44. A. Katz, Feingold, S.J. Pearton, and U.K. Chakrabarti, *Appl. Phys. Lett.*, Vol. 59, 1991, pp. 579–581.
45. A. Katz, S.N.G. Chu, B.E. Weir, W. Savin, D.W. Harris, W.C. Dautremont-Smith, R.A. Logan, and T.T. Tabun-Ek, *J. Appl. Phys.*, Vol. 68, 1990, pp. 4141–4149.
46. A. Katz, S.N.G. Chu, W.C. Dautremont-Smith, M. Soler, B.E. Weir, and P.M. Thomas, *Proceedings of the SPIE Meeting, J. SPIE*, Vol. 1189, 1989, pp. 142–158.
47. A. Katz, W.C. Dautremont-Smith, S.N.G. Chu, P.M. Thomas, L.A. Koszi, J.W. Lee, V.G. Riggs, R.L. Brown, S.G. Napholtz, and J.L. Zilko, *Appl. Phys. Lett.*, Vol. 54, 1989, pp. 2306–2308.
48. A. Katz and W.C. Dautremont-Smith, *J. Appl. Phys.*, Vol. 67, 1990, pp. 6237–6246.
49. G.E. Henien and W.R. Wagner, *J. Appl. Phys.*, Vol. 54, 1983, pp. 6395–6404.
50. A.K. Chin, M.A. DiGiuseppe, and W.A. Bonner, *Mat. Lett.*, Vol. 1, 1982, pp. 19–25.
51. K. Ishida, T. Kamejima, Y. Matsumoto, and K. Endo, *Appl. Phys. Lett.*, Vol. 40, 1982, pp. 16–18.

52. R.L. Brown and R.G. Sobers, *J. Appl. Phys.*, Vol. 45, 1974, pp. 4735–4739.

53. M.A. Aframowitz and D.L. Rode, *J. Appl. Phys.*, Vol. 45, 1974, pp. 4738–4746.

54. J.T. Pan and I.A. Blech, *J. Appl. Phys.*, Vol. 55, 1984, pp. 2874–2880.

55. L. Wang and Q. Zhang, *Chin. Phys.*, Vol. 7, 1987, pp. 1139–1143.

56. F. Schulte and H. Beneking, *Solid State Electron.*, Vol. 30, 1987, pp. 1039–1042.

57. O. Oparaku, L.L. Dargan, N.M. Pearsall, and R. Hill, *Semicond. Sci. Technol.*, 1990, pp. 65–68.

58. P. Auvray, A. Guivarc'h, H. L'Haridon, J.P. Bercier, and P. Henol, *Thin Solid Films*, Vol. 127, 1985, pp. 39–43.

59. E. Kaminska, A. Piotrowska, A. Barcz, J. Adamczewska, and A. Turos, *Solid State Electron.*, Vol. 29, 1986, pp. 274–283.

60. A. Piotrowska, E. Kaminska, A. Bercz, J. Adamczewska, and A. Turos, *Thin Solid Films*, Vol. 130, 1985, pp. 231–236.

61. W.X. Chen, S.C. Hasueh, P.K.L. Yu, and S.S. Lau, *Electron. Dev. Lett.*, Vol. 7, 1986, pp. 471–476.

62. H. Temkin, R.J. McCoy, V.G. Karamidas, and W.A. Bonner, *Appl. Phys. Lett.*, Vol. 36, 1980, pp. 444–446.

63. H. Temkin and S. Mahajan, *Inst. Phys. Conf. Ser.*, Vol. 56, 1981, pp. 293–296.

64. B.P. Segner, L.A. Koszi, H. Temkin, E.J. Flynn, L.J.P. Ketelsen, S.G. Napholtz and G.J. Przybylek, *J. Appl. Phys.*, Vol. 64, 1988, pp. 3718–3721.

65. D.R. Myers, E.D. Jones, I.J. Fritz, L.R. Dawson, T.E. Zipperian, R.M. Biefeld, M.C. Smith, and J.E. Schirber, *J. Electron. Mat.*, Vol. 18, 1989, pp. 465–472.

66. N.K. Dutta, T. Cella, A.B. Piccirilli, and R.L. Brown, *Appl. Phys. Lett.*, Vol. 49, 1986, pp. 1227–1229.

67. L.P. Erickson, A. Waseen, and G.Y. Robinson, *Thin Solid Films*, Vol. 64, 1979, pp. 421–425.

68. J.J. Kelly, J.M.G. Rikken, J.W.M. Jocobs, and A. Vacster, *J. Vac. Sci. Technol. B*, Vol. 6, 1988, pp. 48–52.

69. D.G. Ivey, R. Bruce, and C.R. Piercy, *J. Electronic Materials*, Vol. 17, 1988, pp. 373–377.

70. T.C. Hasenberg and E. Garmir, *J. Appl. Phys.*, Vol. 61, 1987, pp. 808–809.

71. V. Malinka, K. Vogel, and J. Zelinka, *Semicond. Sci. Technol.*, Vol. 3, 1988, pp. 1015–1021.

72. H. Morkoc, T.J. Drummond, and C.M. Stancha, *IEEE Trans. Electron. Devices*, Vol. 28, 1981, pp. 1–4.

73. D.V. Morgan, J. Frey, and W.J. Devlin, *J. Electrochem. Soc.*, Vol. 127, 1980, pp. 1202–1205.

74. J. Dunn and G.B. Stringfellow, *J. Electron. Mat.*, Vol. 19, 1990, pp. 1–3.

75. S.C. Binari and J.B. Boos, *Electron. Lett.*, Vol. 25, 1989, pp. 1207–1209.

76. D.A. Anderson, R.J. Graham, and J.W. Steeds, *Semicond. Sci. Technol.*, Vol. 3, 1988, pp. 63–76.

77. J.L. Merz, J.R. Abelson, and T.W. Sigmon, *S. Electron. Mat.*, Vol. 16, 1987, pp. 257–262.

78. K.P. Pande, E. Martin, D. Gutierrez, and O. Aina, *Solid State Electron.*, Vol. 30, 1987, pp. 253–259.

79. M.F.J. O'Keefe, R.E. Miles, and M.J. Howes, *Proc. SPIE—Int. Soc. Opt. Eng.*, Vol. 1144 1989, pp. 361–367.

80. P.J. Mellor and J. Herminan, *Proc. SPIE—Int. Soc. Opt. Eng.*, Vol. 1144, 1989, pp. 347–353.

81. I. Mehdi, U.K. Reddy, J. Oh, J.R. East, and G.I. Haddad, *J. Appl. Phys.*, Vol. 65, 1989, pp. 867–869.

82. H. Schumacher, J.R. Hayes, R. Bhat, and M. Koza, *1987 Int. Electron Device Meeting, IEDM, Technical Digest*, 1987, pp. 852–853.

83. P.A. Barnes and R.S. Williams, *Solid State Electron.*, Vol. 24, 1981, pp. 907–910.
84. I. Camlibel, A.K. Chin, F. Ermanis, M.A. Digiuseppe, J.A. Lourenco, W.A. Bonner, *J. Electrochem. Soc.*, Vol. 129, 1982, pp. 2585–2590.
85. J.M. Vandenberg, H. Temkin, R.A. Hamm, and M.A. Digiuseppe, *J. Appl. Phys.*, Vol. 53, 1982, pp. 7385–7389.
86. A.K. Chin, C.L. Zipfel, F. Ermanis, L. Marchut, L. Camlibel, M.A. Digiuseppe, and B.H. Chin, *IEEE Trans. Electron. Devices,* Vol. 30, 1983, pp. 304–309.
87. A.K. Chin, C.L. Zipfel, M. Geva, I. Camlibel, P. Skeath, and B.H. Chin, *Appl. Phys. Lett.,* Vol. 54, 1984, pp. 37–39.
88. A. Appelbaum, M. Robins, and R. Schrey, *IEEE Trans. Electron. Devices,* Vol. 34, 1987, pp. 1026–1030.
89. A. Appelbaum, L.C. Feldman, L.A. Koszi, P.M. Thomas, and P.A. Barnes, *AT&T Bell Labs. Int. Technical Report,* Vol. 20, 1985.
90. A. Katz, P.M. Thomas, S.N.G. Chu, W.C. Dautremont-Smith, R.G. Sobers, and S.G. Napholtz, *J. Appl. Phys.*, Vol. 67, 1990, pp. 884–889.
91. A. Katz, S.N.G. Chu, P.M. Thomas, and W.C. Dautremont-Smith, *Proc. SPIE—Int. Soc. Opt. Eng.,* Vol. 1144, 1989, pp. 321–331.
92. A. Katz, B.E. Weir, and W.C. Dautremont-Smith, *J. Appl. Phys.,* Vol. 68, 1990, pp. 1123–1128.
93. W.C. Dautremont-Smith, P.A. Barnes, J.W. Stayt, Jr., *J. Vac. Sci. Technol. B,* Vol. 2, 1984, pp. 620–625.
94. T. Oshnishi, N. Yokoyama, H. Onodera, S. Suzuki, A. Shibatomi, *Appl. Phys. Lett.,* Vol. 43, 1983, pp. 600–602.
95. K. Ishii, T. Oshima, T. Futatsugi, T. Fujii, N. Yokoyama, and A. Shibatomi, *International Electron Devices Meeting—IEDM Proc.,* Vol. 11, 1986, pp. 274–278.
96. A. Lahav, C.S. Wu, and F.A. Baiocchi, *J. Vac. Sci. Technol. B,* Vol. 6, 1988, pp. 1785–1792.
97. R.P. Gupta, W.S. Khokle, J. Wuerfl, and H.L. Hartnagel, *J. Electrochem. Soc.,* Vol. 137, 1990, pp. 631–635.
98. N. Uchitomi, M. Nagaoka, K. Shimada, T. Mizoguchi, and N. Toyoda, *J. Vac. Sci. Technol. B,* Vol. 4, 1986, pp. 1392–1396.
99. K.M. Yu, J.M. Jaklevic, E.E. Haller, S.K. Cheung, and S.K. Kwok, *J. Appl. Phys.,* Vol. 64, 1988, pp. 1284–1289.
100. K. Asai, H. Sugahara, Y. Matuoka, and M. Tokumitsu, *J. Vac. Sci. Technol. B,* Vol. 6, 1988, pp. 1526–1529.
101. A.E. Geissberger, R.A. Sadler, and M.L. Balzman, J.W. Crites, *J. Vac. Sci. Technol. B,* Vol. 5, 1987, pp. 1701–1706.
102. A. Lahav, F. Ren, and R.F. Kopf, *Appl. Phys. Lett.,* Vol. 54, 1989, pp. 1963–1965.
103. A. Katz, S.J. Pearton, and M. Geva, *Appl. Phys. Lett.* Vol. 59, 1991, pp. 286–288.
104. J.L. Vossen and W. Kern, *Thin Film Processes,* New York: Academic Press, 1978.
105. W.A. Metz and E.A. Beam in *Tungsten and Other Refractory Metals for VLSI Applications,* Pittsburgh: MRS Publications, 1986, p. 249.
106. C. Fuhs, E.J. McInerny, L. Watson, and N. Zettequist in *Tungsten and Other Refractory Metals for VLSI Applications,* Pittsburgh: MRS Publications, 1986, p. 257.
107. D.W. Woodruff, R.H. Wilson, and R.A. Sanchez-Martines in *Tungsten and Other Refractory Metals for VLSI Applications,* Pittsburgh: MRS Publications, 1986, p. 173.
108. K.C. Ray and N.E. Zetterquist in *Tungsten and Other Refractory Metals for VLSI Applications II,* Pittsburgh: MRS Publications, 1987, p. 177.
109. N. Tsuzuki, M. Ichikawa, K. Kurita, K. Watanable, and K. Inayoshi, in *Tungsten and Other Refractory Metals for VLSI Applications,* Pittsburgh: MRS Publications, 1987, p. 257.

110. R.A. Wilson, R.W. Stoll, and M.A. Calacone in *Tungsten and Other Refractory Metals for VLSI Applications,* Pittsburgh: MRS Publications, 1986, p. 35.

111. A. Katz, A. Feingold, S. Nakahara, M. Geva, E. Lane, S. Z. Dearton, and K. S. Jones, *J. Appl. Phys.,* Vol. 69, 1991, pp. 7664–7669.

112. A. Katz, A. Feingold, M. Geva, E. Lane, S. J. Pearson, M. Ellington, and U. K. Chakabarti, *J. Appl. Phys.,* Vol. 70, 1991, pp. 1–11.

113. A. Katz, A. Feingold, S. Nakahara, E. Cane, M. Geva, S. J. Dearton, F. A. Stevie, and K. Jones, *J. Appl. Phys.,* 1991, to be published.

Chapter 10
Issues for Dielectric Technology for InP MISFETs

U.K. Chakrabarti

AT&T Bell Laboratories

10.1 INTRODUCTION

For high-speed electronics, a semiconductor with high electron mobility and high saturation velocity is desirable. High thermal conductivity and high breakdown fields are also desirable properties. All of these requirements are well met in InP [1,2]. However, in the recent past, InP gained importance as a semiconductor for long-wavelength optoelectronic devices (lasers, LEDs, photo detectors, *et cetera*) [3], but not for transistor technology. The field of high-speed electronics is dominated by the GaAs and AlGaAs/GaAs material system [4]. In GaAs, the metal-semiconductor field effect transistor (MESFET) technology has been pursued and very successfully implemented; for example, the Japanese had announced, by 1985, a 16K static RAM containing 102,000 transistors [5]. Because of the inherent problems of the material system, metal-insulator-semiconductor FET (MISFET) technology could not be implemented very successfully in GaAs [6]. However, there is a constant effort going on to implement MISFET approach in the GaAs-based material system [7,8]. The motivating factors are that MISFET technology offers the advantages of low power dissipation, large dynamic range, and better noise margins than can be achieved by MESFET technology.

Although efforts started in the mid-1970s [10], the progress in transistor technology in InP has been lagging behind GaAs. There are several factors responsible for this. The early efforts to follow GaAs MESFET technology in InP were not successful. The reason is fundamental in nature; the Fermi-level pinning in the vicinity of the conduction band minimum (CBM) produces a low Schottky barrier height in InP and related alloy systems [11]. The barrier height of most metals on InP falls in the range of 0.4 to 0.5 eV [12]. This low barrier height gives rise to a large leakage current when the gate is reverse biased [13]. The other factor is that the quality of semi-insulating and low-doped InP substrates has been inferior to those of GaAs. However, the MESFET technology in GaAs has attained a reasonable degree of maturity and its simplicity in implementation has always been an attractive feature. Hence, there has been considerable effort put in to realize a reliable MESFET technology for InP. The major effort in this direction has been to enhance the Schottky barrier height on InP. Two approaches have been attempted: (a) to use a thin layer of insulator on the semiconductor [14–17], and (b) to produce a shallow p^+ layer on InP [13,18,19]. However, a reliable MESFET technology in InP still remains elusive. But the very fact that surface Fermi-level pinning is near the CBM, which gives rise to low Schottky barrier height, is fortuitous for MISFET applications.

10.2 InP MISFET

The operation of a field effect transistor (FET) depends on the electrostatic modulation of a current which flows between its source and drain electrodes. Control of the source-drain conductance is effected by the modulation of the space charge region under a control gate located between the source and drain electrodes. In the case of MISFET, the space charge region is due to a dielectrically isolated metal gate. Most of the relevant properties of a MISFET are governed by the nature of the dielectric-semiconductor interface. Extensive research has been conducted over the past decade to understand the electronic nature of the dielectric-InP interface and exellent reviews are available in Wieder [11] and Wilmsen [20]. Despite the fact that various types of InP-MISFETs and circuits have been reported [21], this technology still suffers from a number of deficiencies. Lack of uniformity and reproducibility of device data is a continuing problem, exemplified by seemingly random variation in transconductance (g_m) and drain saturation current (I_{DSS}) between supposedly identical devices across a wafer as well as between wafers. This has generally been believed to be due to uncontrolled variations in both dielectric and interface properties, as well as to nonuniformity in the starting bulk InP substrate. It is expected that the quality of InP substrates will improve with time as crystal growth technology advances.

The more vexing problem has been the slow progress in understanding and overcoming the drift in the current-voltage characteristics of an InP-MISFET. This phenomenon results from injection of electrons from the channel of the MISFET into the localized interface states either in the near surface region of the semiconductor or the dielectric. This charge trapping effect will show up very clearly in the form of hysteresis in capacitance-voltage (C-V) measurement and in the looping and drift of FET characteristics. The latter effect is the most dangerous one, particularly for analog circuit applications and has been the major reason for the slow progress in InP-MISFET technology.

10.3 DIELECTRIC TECHNOLOGY FOR InP MISFET

As mentioned earlier, the FET characteristics are controlled by the electronic charge interaction at the dielectric-InP interface, and the drift in the drain current is presumably due to the presence of charge-trapping states at the dielectric-InP interface. Despite the fact that significant research has been done to understand the origin of trapping sites, their energy distribution, and the kinetics of charge trapping, many questions remain unanswered [21,22]. The situation is further complicated by the fact that there is always a transition region between InP and the deposited dielectric. The exact chemical composition and electrical properties of this transition region are relatively unknown.

As a contrast to the Si-SiO$_2$ system, it has been found [23] that thermally grown native oxide on InP produces a very bad electrical interface. This had encouraged researchers to use deposited dielectric as the gate insulator for MISFET. Various insulators, namely SiO$_2$, Si$_3$N$_4$, Al$_2$O$_3$, Ge$_3$N$_4$, P$_3$N$_5$, PAS$_{1-x}$Nx, AlPO$_4$, and InPxOy, have been tried. However, there is a definite preference for SiO$_2$ and Al$_2$O$_3$ in the published literature.

The major problem arises during the deposition of the dielectric. Depending upon the surface cleaning procedure adopted, one would start with an InP surface that is relatively devoid of native oxide, or the surface may be covered with chemically grown oxide of variable composition. In the growth chamber where the dielectric is deposited, a severe "interaction" [24] takes place on the InP surface, the extent of which will depend on various factors: (a) the nature of reactants used, (b) the temperature of the substrate during deposition, and (c) the physical nature of the environment (e.g., presence of energetic charged particles, *et cetera*).

The above picture points out the complicated nature of the situation. Furthermore, no clear-cut technology has been established. However, by scanning the literature, a guideline for future work to achieve a reliable dielectric technology for InP can be established [24].

10.3.1 Requirements of the Dielectric

(a) The dielectric should adhere very well to the semiconductor. It should be chemically and physically stable and be a good barrier against environmental contaminants.

(b) The insulator should have high electrical resistivity, typically $>10^{12}$ $\Omega \cdot$cm. To obtain high resistivity, the insulator should have wide bandgap, the impurity content should be low, and the composition should be stoichiometric. Structurally the dielectric should be amorphous, devoid of pinholes, and the surface morphology should be smooth.

(c) The dielectric constant of the insulator should be frequency independent. The dielectric should not be heterogeneous, and ionic impurities should be low.

(d) The dielectric should have high breakdown voltage. All the criteria mentioned above will ensure high breakdown voltage.

(e) The bulk traps in the insulator should be low. Traps in the insulators have significant control over the electrical conduction and, as a consequence, have a profound effect on the electrical resistivity, breakdown field, and dielectric dispersion. Also, the bulk traps are the major source of electrical instabilities at the dielectric-semiconductor interface. To obtain a stable FET operation, it is essential to reduce bulk-trap density in the dielectric. However, traps are inherent in dielectrics, and their densities and energy distributions vary among different dielectrics. Also, the physical and electrical characteristics of the traps depend upon impurities, nonstoichiometry, physical defects, and strain in chemical bonding. Therefore, it is imperative to understand that the same dielectric prepared in different ways would have traps of different characteristics.

(f) The chemical composition of the dielectric should be such that it does not promote interfacial reaction with InP (i.e., the particular dielectric is capable of forming a sharp interface under the appropriate deposition conditions).

In summary, it is important to choose an insulator that does not have intrinsic traps energetically located within the bandgap of the semiconductor. This will minimize charge interaction at the interface and lead to a stable FET operation. From this viewpoint, SiO_2 is the best choice, followed by Al_2O_3 [24]. Silicon nitride, which is a very good dielectric, is particularly unsuitable as a gate dielectric in an InP MISFET because of the unfavorable energy distribution of traps [24] and because the reducing ambient under which it must be deposited promotes loss of P from an InP surface.

10.3.2 Requirements of Dielectric Deposition Technology

10.3.2.1 Interfacial Considerations

It has been mentioned earlier and is reemphasized here that the operation of an MISFET will be controlled by the electronic interaction at the dielectric-semiconductor interface. Native oxide grown on InP, particularly by thermal oxidation, does not produce a stable interface [23]. The gate insulator for the InP MISFET is invariably a non-native type. It is important to notice that the interface between the deposited dielectric and the semiconductor is not abrupt. The dielectric and the semiconductor are separated by a thin transition region. The physico-chemical and electrical properties of this region will depend upon the following:

(a) The dielectric being deposited.

(b) The chemical composition of the semiconductor surface before deposition.

(c) The chemical reaction and the different chemical moieties at the surface of the semiconductor during the dielectric deposition.

(d) The mode of providing energy to synthesize the dielectric (e.g., thermal energy or excitation of reactants by plasma).

A combination of these factors will decide the nature of the dielectric-semiconductor interface including, of course, the transition region. Significant effort has been made to select a dielectric deposition process that produces either (a) a very abrupt dielectric-semiconductor interface or (b) a graded transition region with reproducible properties.

10.3.2.2 Process Considerations

Non-native dielectrics on InP can be deposited by various techniques. The literature on deposition of thin-film dielectrics is large [25,26]. However, for the particular applications, such as gate dielectric for InP MISFETs, the choice is limited by the process considerations in the following paragraphs.

Substrate Temperature

It had been realized early on that the substrate temperature of InP during dielectric deposition has to be kept low. At about 365°C, phosphorus evaporates from the surface, preferentially leaving behind an indium-rich surface [27]. If this process is not restricted in time and temperature, In droplets appear and thermally etch the surface resulting in a morphologically degraded surface [28–30]. Even before morphological defects are observable, electrical properties of InP surface are significantly altered [31,32].

The preferential loss of P from the surface creates anion vacancies and has been directly correlated with drift problems in InP MISFETs [21,33]. Hence, the substrate temperature during deposition has to be kept low. A typical example is 300°C or lower.

Energy Transfer During Deposition

Theoretical studies have predicted that both vacancies [34] and antisite defects [35] can give rise to traps within the bandgap of the semiconductor. The estimates of the enthalpy of formation of various types of defects show that only 0.42 eV is needed for creating anion-on-cation antisite defect, while 3.04 eV is needed for producing neutral cation vacancy. The enthalpy values for producing other types of defects lie in between [24,36,37]. It should be borne in mind that these values are for bulk defects, and it is expected that energies required for creating surface/interface defects are even smaller [38]. Different kinds of defects may give rise to interface states and, therefore, it is important to use a deposition technology that produces the least amount of defects. Because of this consideration, a large number of physical deposition techniques (ion-beam, DC, RF, and magnetron sputtering, *et cetera*), although low-temperature processes (no intentional heating of substrate required), are ruled out.

Thermal evaporation is a low-energy deposition process. However, it is difficult to deposit stoichiometric dielectric thin films by thermal evaporation. The dielectric materials usually have very high melting points, and significant dissociation of the starting material takes place in the deposition chamber producing a nonstoichiometric material [26,38]. One way to get around this problem is to use electron-beam (e-beam) heating and bleed a reactive gas into the vacuum chamber to compensate for the loss of the anion from a deposited thin film [26,38]. However, the e-beam deposition process is not a low-energy technique, and it is well documented that e-beam deposition produces large amounts of surface states at the interface because of x-ray exposure of the interface [39].

Chemical vapor deposition (CVD) is another low-energy process that can be used to deposit dielectrics at low substrate temperature. However, dielectrics produced by low-temperature ($<300°C$) CVD do not have desirable properties and do not produce stable InP MISFETs [40]. In a normal plasma CVD (PECVD) process, the substrate is exposed to a glow discharge plasma where it can be bombarded by energetic ions, atoms, and electrons with an energy of ≈ 10 eV. As mentioned before, these energies are sufficient to create electronic defects at the interface, so normal PECVD is not a very desirable process for gate dielectric deposition for InP MISFETs.

To circumvent the electronic damage due to plasma, the normal CVD plasma process has been modified to keep the substrate away from the main volume of the plasma; that is, the plasma is employed to enhance gas phase dissociation of reactants, but is not involved in heterogeneous chemistry. This process has been variously called indirect [41] or isolated [42] plasma CVD, and has been very successfully employed in fabricating reasonably stable InP MISFETs [43]. In one approach [41], silicon dioxide was deposited by reacting silane and oxygen, the latter being dissociated by a RF discharge chamber physically removed from the deposition zone. It has been argued that the major excited species in the deposition zone is atomic oxygen. In the other approach [42], plasma was generated using a DC potential and the film deposition was achieved at a low plasma power (0.3 W/cm^2). The plasma is struck between a pedestal holding the substrate and an annular electrode made of an aluminum tube placed below the pedestal. The oxidizer, in this case N_2O, was fed into the plasma region through the annular electrode, which had gas dispersing holes. Silane was introduced from the top, near the substrate surface where the reaction took place. In this configuration, radiation damage was minimized and, indeed, reasonably well behaved InP MISFETs were fabricated [43].

Protection of Surface from Phosphorus Loss

In a recent result [44] it has been shown that if the surface of InP is covered with a thin layer of P before the deposition of the dielectric, the drain current drift of InP MISFET can be made insignificant. A comparison of the results of C-V measurements on InP MIS capacitors fabricated in an isolated plasma system with [44] and without [43] the P layer reveals that the drift in the drain current of InP MISFET is possibly due to the defects generated by loss of surface P atoms. As mentioned earlier, the energies required to create surface defects are very small [37], and even indirect plasma systems are not safe enough. With this technique of prepassivation with a thin P layer, gigahertz logic gates based on InP-MISFETs with minimal drain current have been recently demonstrated [44]. It had also been shown earlier that P can be used as an insulator in an InP MIS capacitor where the C-V curves have extremely small hystereses and the interface has a very low surface state density [45]. However, use of P as the gate dielectric in a MISFET is not technologically appealing.

Another important point meriting consideration and exemplified by the well-known SiO_2-Si system is the chemical structural compatibility between the dielectric and the semiconductor. It has been shown [46] that thin layers of $AlPO_4$ can be built on InP atom by atom to completely saturate the bonds with In and P at the surface, similar to forming $InPO_4$. Significantly low values of surface state density have been obtained by using AlPxOy as the insulator in an InP MIS capacitor. In a similar approach, InPxOy has also been used as a dielectric on InP, and C-V curves with 0.2V hysteresis were obtained. Fast surface state densities as measured from the C-V curves exhibited minima as low as 10^{10} cm^{-2} eV [47]. Although encouraging C-V results have been obtained by using ternary dielectrics, FET results are not yet available. Furthermore, technologically speaking, ternary dielectrics may not be very desirable. Indeed, it has been mentioned [47] that the electrical properties of InPxOy layers are a sensitive function of chemical composition, and it may be difficult to control the stoichiometry reproducibly from one deposition run to the next. Also, results are not available at this time on the nature of bulk traps in these materials.

The Guidelines for Dielectric Technology

The guidelines for the dielectric technology for InP MISFETs are as follows:

(a) The gate dielectric should have very high electrical resistivity, high breakdown strength, low dielectric dispersion, and low intrinsic traps. SiO_2 appears to be a good choice. However, as described in the next section, PN_x and BaF_2 hold promise as suitable dielectrics.

(b) The deposition technique should use low substrate temperature, minimize energy transfer to the substrate surface, and should also produce abrupt interface. An indirect plasma CVD is a good choice among the reported techniques. Thermal evaporation can also be used to deposit fluoride-based dielectrics. Low-temperature photochemical CVD also holds promise.

(c) There is a universal agreement that the surface should be protected against P loss, at least during the initial stages of dielectric deposition. Furthermore, surface treatment of InP before dielectric deposition can alleviate this problem.

Recent Advances

Starting from the early 1970s and extending to the middle 1980s, the bulk of research work centered around developing dielectrics and deposition techniques along the guidelines presented in Section 10.4. A major achievement was the introduction of the remote plasma CVD technique to deposit high-quality dielectrics [41]. This process was later supplemented by downstream microwave plasma to deposit silicon nitride at a low temperature ($\approx 250°C$) [48]. In the latter half of the 1980s, some researchers increased their attention to surface preparation techniques to make the InP surface more "rugged" toward subsequent processing (e.g., dielectric deposition) [43,49]. Based on earlier work [50,51], a significant breakthrough was reported in which the surface of InP was treated in an aqueous solution of ammonium sulfide prior to SiO_2 deposition [52]. This treatment leaves a crusty residue on the surface which, by chemical analysis, was shown to be elemental sulfur. This layer can easily be removed by sublimination at $\sim 110°C$ in a remote plasma CVD machine, and following immediately with SiO_2 deposition at $\sim 250°C$ [53]. The electrical quality was studied by C-V analysis of an MIS structure and is reproduced in Figure 10.1(a). An excellent C-V characteristic is immediately evident. However, to fit the high-frequency curve, the authors had to assume a doping density of a factor of ~ 20 below the Hall data values for the starting wafers. This fact is yet to be explained. However, the analysis of the C-V curve as shown in Figure 10.1(b) yields the lowest surface state density at a surface potential of -0.4 eV, and the value is in the range of 10^{10} to a few 10^{11} cm^{-2} eV^{-1}. Enhancement mode FETs on *p*-type and semi-insulating InP were fabricated using this process, and drain current drifts of <5% over a 12-hr test period were reported [53].

It was also reported that an annealing step subsequent to SiO_2 deposition was critical in obtaining good results, and a 1-hr anneal at 350°C in N_2/H_2 ambient was adequate. In addition, it was observed that anneal at a temperature of $\sim 400°C$ led to no degradation of the C-V characteristics. This is very important information when considering postgate fabrication steps in making devices, circuits, and the packaging of these devices.

The aforementioned successes led to other researchers looking at the S-treated InP surfaces for chemical information. Auger/x-ray photoelectron spectroscopy was

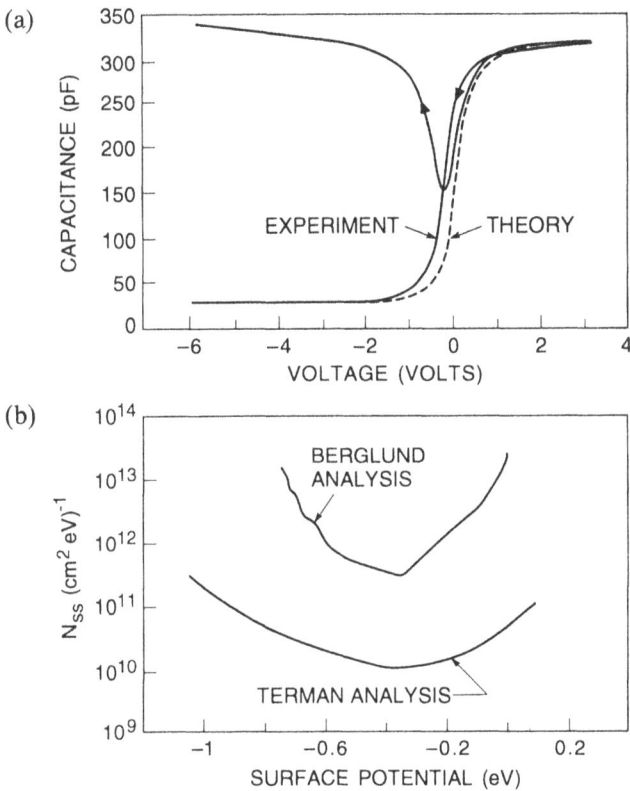

Figure 10.1 (a) High-frequency (1 MHz) and quasistatic C-V behavior of sulfurized n-type InP with IPCVD SiO_2 as dielectric. The high-frequency C-V measurement exhibits a clockwise hysteresis (not shown) of 200 mV. Sweep rate: 100 mV/s for high-frequency and 64 mV/s for quasistatic C-V. Theoretical curve calculated for N_D of 5×10^{14} cm^{-3} and C_{ox} of 72 nF/cm^2 [53]. (b) Surface state density of sulfurized n-type InP [53].

used to determine the nature of bonding [54]. It was observed that S is bonded to In. There was no indication of elemental S or S bonded to P. The authors concluded that the S has replaced P on the surface and has filled the P vacancies. Similar conclusions were also arrived at from low-temperature photoluminescence study of S-treated InP [55]. Furthermore, it was concluded that S treatment of InP reduced the P escape rate during thermal annealing [55]. This observation, together with point (c) in Section 10.4, can help explain the improved performance of MISFETs.

The issue of P loss from the surface has been approached from another angle: to deposit phosphorus nitride as the dielectric. In one study [56], surface preparation was done by in situ etching using PCl_3 and followed by phosphorus nitride deposition by adding ammonia. The smallest drain current drift ($<2\%$ in 10^3 sec) was obtained at 450°C for etching and deposition. However, a significant drop in channel electron mobility was observed at 450°C.

The problems associated with thermal CVD of phosphorus nitride using PCl_3 and NH_3 have been eliminated by using a low-temperature photochemical vapor deposition technique [57]. A typical 1-MHz C-V curve for the $Al/P_3N_5/n$-InP structure is shown in Figure 10.2(a), and the Terman analysis is presented in Figure 10.2(b) [57]. After in situ postannealing, the minimum interface trap state density was about $3 \cdot 6 \times 10^{10}$ cm^{-2} eV^{-1}, and the traps were found to be near the midgap of InP. In this work no MISFET results were published. However, C-V characteristics are very similar to those obtained by S treatment of InP surfaces.

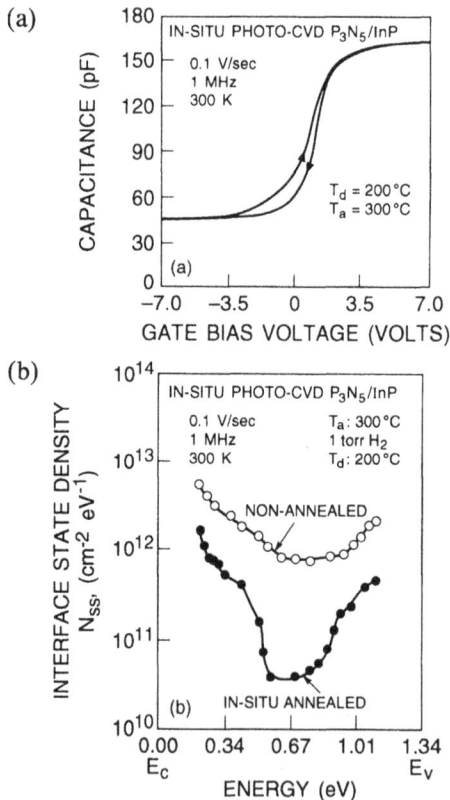

Figure 10.2 (a) High-frequency (1 MHz) C-V plots for $Al/P_3N_5/n$-InP MIS structure measured at room temperature. (T_d) deposition temperature, (T_a) annealing temperature [57]. (b) The in situ postdeposition annealing on density distribution of interface trap states in the energy gap of InP from high-frequency C-V data [57].

Another attractive approach to producing clean InP-dielectric interface would be to grow InP surface and cover the surface in situ by a suitable dielectric, adhering, however, to the guidelines developed in Section 10.4. Dielectrics based on fluorides seems ideal for this application. BaF_2, SrF_2, and CaF_2 and their solid solutions have attracted attention for a variety of reasons. These dielectrics can be deposited by thermal evaporation, which is a very low energy deposition process [26]. The depositon of dielectric-based fluorides on InP has been known for some time [58]; however, only recently detailed MIS properties of the fluoride-InP system have been reported [53]. Films of $Ba_{1-x}Sr_xF_2$ (x = 0.0, 0.5, 0.83, and 1.0) were deposited thermally at 10^{-5} torr pressure onto an n-InP substrate at 300K. Tin was used as the gate metal. A typical C-V characteristic of an MIS diode with unannealed BaF_2 as an insulator is shown in Figure 10.3(a), and Figure 10.3(b) shows the interface state density distribution (N_{ss}) obtained by Terman's method. The C-V curve showed small hysteresis as observed by others [53,57], and N_{ss} had the common U-shaped distribution with a minimum density near the upper half of the bandgap. In case of annealed BaF_2 (250°C for 15 min in H_2 ambient), N_{ss} was found to be 5×10^{10} cm^{-2} eV^{-1} at an energy 0.47 eV below E_c. Low N_{ss} values with fluoride gate dielectrics are considered to be due to partial compensation of P vacancies at the surface by F atoms, which are strongly attached to the InP surface. In this work, no FET results

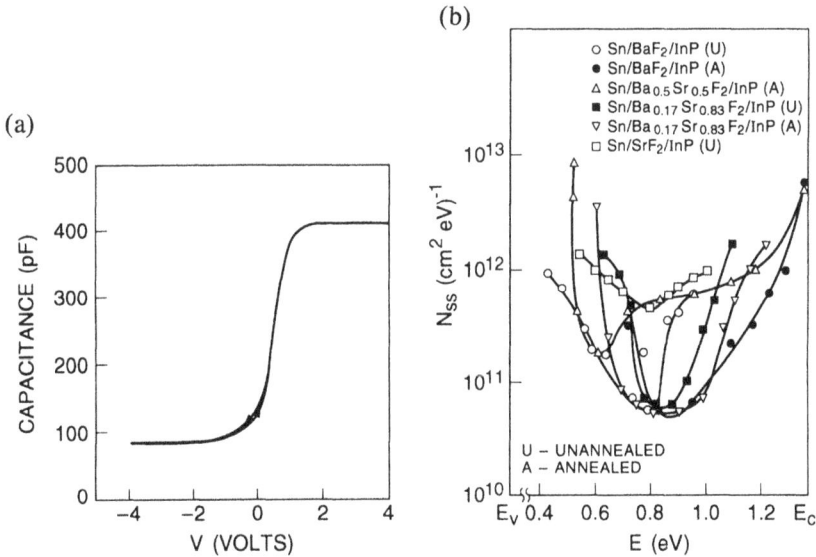

Figure 10.3 (a) C-V (1 MHz) characteristic of an MIS diode with an unannealed BaF_2 film on InP showing negligible hysteresis [59]. (b) Interface state density distribution for different MIS systems. The minimum value of N_{ss} obtained is 5.0×10^{10} cm^2 eV^{-1} in the upper part of the bandgap for annealed BaF_2 films.

were reported. However, considering the fact that no elegant surface preparation was used before depositing the dielectric, the C-V characteristics of the BaF_2-InP structure are good enough to merit further investigation.

REFERENCES

1. J.M. Golio and R.J. Trew, "Optimum Semiconductor for High Frequency and Low Noise MESFET Applications," *IEEE Trans. Electron. Devices,* Vol. ED-30, 1983, p. 1411.

2. H. Morkoc, J.T. Andrews, and S.B. Hyder, "Effects of an n^- Layer Under the Gate on the Performance of InP MESFETs," *IEEE Trans. Electron. Devices,* Vol. ED-26, 1979, p. 238.

3. T.P. Pearsall, ed., *GaInAsP Alloy Semiconductors,* New York: John Wiley & Sons, 1982.

4. For example, see the special issue of *IEEE Trans. Electron Devices,* Vol. ED-33, 1986.

5. D.K. Ferry, ed., *Gallium Arsenide Technology,* Indiana: Howard W. Sams & Co., Inc., 1985.

6. D.L. Lile, "Interfacial Constraints on III-V Compound MIS Devices," Chapter 6, in *Physics and Chemistry of III-V Compound Semiconductor Interfaces,* ed., C.W. Wilmsen, New York: Plenum Press, 1985, p. 360.

7. K. Matsumoto, M. Ogura, T. Wada, and N. Hashizume, "Accumulation-Mode GaAs MIS-Like FET Self-Aligned by Ion-Implantian," paper no. IIA-5, presented at 42nd Annual Device Research Conference, June 18–20, 1984.

8. P.M. Solomon, C.M. Knoedler, and S.L. Wright, "A GaAs Gate Heterojunction FET," paper no. IIA-7, presented at 42nd Annual Device Research Conference, June 18–20, 1984.

9. J.S. Barrera and R.J. Archer, *IEEE Trans. Electron. Devices,* Vol. ED-22, 1975, p. 1023.

10. L. Messick, D.L. Lile, and A.R. Clawson, "A Microwave InP/SiO_2 MISFET," *Appl. Phys. Lett.,* Vol. 32, 1978, p. 494.

11. H.H. Wieder, "Interface Constraints on MESFET and MISFET," Chapter 5, in *VLSI Electronics,* Vol. 10, treatise ed., N.G. Einsbruch, New York, 1985, p. 174 and 226.

12. C.A. Mead and W.G. Spitzer, "Fermi Level Position at Metal-Semiconductor Interface," *Phys. Rev.,* Vol. 134, 1964, p. A173.

13. T.P. Pearsall and M.A. Digiuseppe, "InP Metal Barrier Junctions with Improved I-V Characteristics," *IEEE Trans. Electron. Devices,* Vol. EDL-7, 1986, p. 317.

14. H. Morkoc, T.J. Drummond, and C.L. Stanchak, "Schottky Barriers and Ohmic Contacts on *n*-Type InP Base Compound Semiconductors for Microwave FETs," *IEEE Trans. Electron. Devices,* Vol. ED-28, 1981, p. 1.

15. Y. Imai, T. Ishibashi, and M. Ida, "Characteristics of InP MIS Schottky Diodes Prepared by Plasma Oxidation," *J. Electrochem. Soc.,* Vol. 129, 1982, p. 221.

16. O. Wada and A. Majerfeld, "Low Leakage Nearly Ideal Schottky Barriers to *n*-InP," *Electron. Lett.,* Vol. 14, 1978, p. 125.

17. K. Kamimura, T. Suzuki, and A. Kunioka, "InP Metal-Insulated-Semiconductor Schottky Contacts Using Surface Oxide Layers Prepared with Bromine Water," *J. Appl. Phys.,* Vol. 51, 1980, p. 4905.

18. G.P. Schwartz, G.L. Gualtieri, and W.A. Bonner, "Metal-p^+-*n* Enhanced Schottky Barrier Structures on (100) InP," *J. Electrochem. Soc.,* Vol. 133, 1986, p. 1021.

19. G.P. Schwartz and G.L. Gualtieri, "Schottky Barrier Height Enhancement on M-p^+-N Structures Including Free-Carriers," *J. Electrochem. Soc.,* Vol. 133, 1986, p. 1266.

20. C.W. Wilmsen, ed., *Physics and Chemistry of III-V Compound Semiconductor Interface,* New York: Plenum Press, 1985.

21. K.P. Pande, D. Gutierrez, and L. Messick, "Advances in InP-MISFET Technology," paper presented at the 168th Meeting of the Electrochemical Society, Las Vegas, October 13-18, 1985.

22. D.L. Lile and M.J. Taylor, "The Effect of Interfacial Traps on the Stability of Insulated Gate Devices on InP," *J. Appl. Phys.*, Vol. 54, 1983, p. 260.

23. M. Okamura and T. Kobayashi, "Current Drifting Behaviors in InP MISFET with Thermally Oxidized InP/InP Interface," *Electron. Lett.*, Vol. 17, 1981, p. 941.

24. J.F. Wager and C.W. Wilmsen, "The Deposited Insulator/III-V Semiconductor Interface," Chapter 3 in *Physics and Chemistry of III-V Compound Semiconductor Interfaces*, ed., C.W. Wilmsen, New York: Plenum Press, 1985, p. 165.

25. J.L. Vossen and W. Kern, ed., *Thin Film Processes*, New York: Academic Press, 1978.

26. R.F. Bunshah, ed., *Deposition Technologies for Films and Coatings*, New Jersey: Noyes Publication, 1982.

27. C.T. Foxon, J.A. Harvey, and B.A. Joyce, "The Evaporation of InP Under Kundsen (Equilibrium) and Langmuir (Free) Evaporation Conditions," *J. Phys. Chem. Solids*, Vol. 34, 1973, p. 2436.

28. L.G. Van Uitert, P.K. Gallagher, S. Singh, and G.J. Zydzik, "Time and Temperature Dependence of Phosphorus Evolution from InP," *J. Vac. Sci. Technol.*, Vol. B1, 1983, p. 825.

29. S.N.G. Chu, C.M. Jodlauk, and W.D. Johnston Jr., "Morphological Study of Thermal Decomposition of InP Surfaces," *J. Electrochem. Soc.*, Vol. 130, 1983, p. 2398.

30. G.A. Antypas, "Prevention of InP Surface Decomposition in Liquid Phase Epitaxial Growth," *Appl. Phys. Lett.*, Vol. 37, 1980, p. 64.

31. S. Guha and F. Hasegawa, "Effect of Heat Treatment on n-Type Bulk Grown and Vapour Phase Epitaxial Indium Phosphide," *Solid State Electron.*, Vol. 20, 1977, p. 27.

32. J.D. Oberstar and B.G. Streetman, "Annealing Encapsulants for InP II: Photoluminescence Studies," *Thin Solid Films*, Vol. 94, 1982, p. 161.

33. J.A. Van Vechten and J.F. Wager, "Consequence of Anion Vacancy Nearest-Neighbor Hopping in III-V Compound Semiconductors: Drift in InP Metal-Insulator-Semiconductor Field Effect Transistor," *J. Appl. Phys.*, Vol. 57, 1985, p. 1956.

34. M.S. Dow and D.L. Smith, "Energy Levels of Semiconductor Surface Vacancies," *J. Vac. Sci. Technol.*, Vol. 17, 1980, p. 1028.

35. J.D. Dow and R.E. Allen, "Surface Defects and Fermi-level Pinning in InP," *J. Vac. Sci. Technol.*, Vol. 20, 1982, p. 659.

36. J.A. Van Vecten, "Simple Theoretical Estimates of the Schottky Constants and Virtual-Enthalpies of Single Vacancy Formation in Zinc-Blend and Wurzite Type Semiconductors," *J. Electrochem. Soc.*, Vol. 122, 1975, p. 419.

37. P.W. Chye, I. Lindau, P. Pianetta, C.M. Carner, C.Y. Su, and W.E. Spicer, "Photoemission Study of Au Schottky-Barrier Formation on GaSb, GaAs, and InP Using Synchrotron Radiation," *Phys. Rev.*, Vol. B18, 1978, p. 5545.

38. L.I. Maissel and R. Glang, ed., *Handbook of Thin Film Technology*, New York: McGraw-Hill, 1970.

39. E.H. Nicollian and J.R. Brews, Chapter 11 of *MOS (Metal · Oxide Semiconductor) Physics and Technology*, New York: John Wiley & Sons, 1982, p. 549.

40. W.F. Tseng, M.L. Bark, H.B. Dietrich, A. Christou, R.L. Henry, W.A. Schmidt, and N.S. Saks, *IEDM Technical Digest*, 1981, p. 111.

41. L.G. Meiners, "Indirect Plasma Deposition of Silicon Dioxide," *J. Vac. Sci. Technol.*, Vol. 21, 1982, p. 655.

42. K.P. Pande and V.K.R. Nair, "High Mobility n-Channel Metal-Oxide-Semiconductor Field-Effect Transistors Based on SiO_2-InP Interface," *J. Appl. Phys.*, Vol. 55, 1984, p. 3109.

43. K.P. Pande and D. Gutierrez, "Channel Mobility Enhancement in InP Metal-Insulator-Semiconductor Field-Effect Transistor," *Appl. Phys. Lett.*, Vol. 46, 1985, p. 416.

44. K.P. Pande, M.A. Fathimula, D. Gutierrez, and L. Messick, "Gigahertz Logic Gates Based on InP-MISFETs with Minimal Drain Current Drift," *IEEE Electron Device Lett.*, Vol. EDL-7, 1986, p. 407.

45. R. Schachter, D.J. Olego, J.A. Baumann, L.A. Bunz, P.M. Raccah, and W.E. Spicer, "Interfacial Properties of InP and Phosphorus Deposited at Low Temperature," *Appl. Phys. Lett.*, Vol. 47, 1985, p. 272.

46. L.G. Meiners, "Electrical Properties of Al_2O_3 and AlPxOy Dielectric Layers on InP," *Thin Solid Films,* Vol. 113, 1984, p. 85.

47. H.L. Chang, L.G. Meiners, and C.J. Sa, "Preparation and Electrial Properties of InPxOy Gate Insulator on InP," *Appl. Phys. Lett.*, Vol. 48, 1986, p. 375.

48. S. Dzioba, S. Meikle, and R.W. Streater, "Downstream Plasma Induced Deposition of, SiN_x on Si, InP and InGaAs," *J. Electrochem. Soc.*, Vol. 134, 1987, p. 2599.

49. J. Chave, A. Choujaa, C. Santinelli, R. Blanchet, and P. Viktorovitch, "Arsenic Passivation of InP Surface for Metal-Insulator-Semiconductor Devices Based on Both Ultra-High Vacuum Technique and Chemical Procedure," *J. Appl. Phys.*, Vol. 61, 1987, p. 257.

50. G. Post, P. Dimitriou, A. Scavennec, N. Duhamel, and A. Mircea, "InP MIS Transistors with Grown-in Sulphur Dielectric," *Electron. Lett.*, Vol. 19, 1983, p. 459.

51. M. Gendry, J. Durand, J.M. Villeneuve, and L. Cot, "Étude du diagramme des phases condensées du système In-P-S," *Thin Solid Films,* Vol. 139, 1986, p. 53.

52. R. Iyer, R.R. Chang, and D.L. Lile, "The Growth and In Situ Characterization of Chemical Vapor Deposited SiO_2," *J. Cryst. Growth,* Vol. 83, 1987, p. 290.

53. R. Iyer, R.R. Chang, and D.L. Lile, "Sulfur as a Surface Passivation for InP," *Appl. Phys. Lett.*, Vol. 53, 1988, p. 134.

54. C.W. Wilmsen, K.M. Geib, J. Shin, R. Iyer, D.L. Lile, and J.J. Pouch, "The Sulfurized InP Surface," *J. Vac. Sci. Technol. B*, Vol. 7, 1989, p. 851.

55. R. Leonelli, C.S. Sundararaman, and J.F. Currie, "Photoluminescence Study of Sulfide Layers on *p*-Type InP," *Appl. Phys. Lett.*, Vol. 57, 1990, p. 2678.

56. Y. Iwase, F. Arai, and T. Sugano, "Effects of Deposition Temperature of Insulator Films on the Electrical Characteristics of InP Metal-Insulator-Semiconductor Field-Effect Transistors," *Appl. Phys. Lett.*, Vol. 52, 1988, p. 1437.

57. Y-H Jeong, J-H. Lee, Y-H. Bae, and Y-T. Hong, "Composition of Phosphorus-Nitride Film Deposited on InP Surfaces by a Photochemical Vapor Deposition Technique and Electrical Properties of the Interface," *Appl. Phys. Lett.*, Vol. 57, 1990, p. 2680.

58. R.F.C. Farrow, P.W. Sullivan, G.M. Williams, G.R. Jones, and D.C. Cameron, "MBE-Grown Fluoride Films: A New Class of Epitaxial Dielectrics," *J. Vac. Sci. Technol.*, Vol. 19, 1981, p. 415.

59. T.K. Paul and D.N. Bose, "Flouride Dielectric Films on InP for Metal-Insulator-Semiconductor Applications," *J. Appl. Phys.*, Vol. 67, 1990, p. 3744.

Chapter 11
InGaAsP Quantum Well Lasers

N.K. Dutta
AT&T Bell Laboratories

11.1 INTRODUCTION

A double heterostructure laser consists of an active layer, which emits the light, sandwiched between two higher gap cladding layers. When the thickness of the active layer becomes comparable to the de Broglie wavelength ($\lambda \sim h/p$), quantum mechanical effects are expected to occur. In double heterostructure lasers with a very thin active region ($\sim 100 \text{Å}$), the carrier (electron or hole) motion normal to the active layer is restricted. As a result, the kinetic energy of the carriers moving in that direction are quantized into discrete energy levels similar to the well-known quantum mechanical problem of the one-dimensional potential well. This effect is observed in the absorption and emission (including laser action) characteristics and transport characteristics, including phenomena such as tunneling. Double heterostructure lasers with a very thin active region or multiple thin active regions are known as quantum well (QW) lasers.

The optical properties of semiconductor QW double heterostructures were initially studied by Dingle *et al.* [1]. Since then, extensive work on GaAlAs QW lasers has been done by Holonyak *et al.* [2,3], Tsang [4,5], and Hersee *et al.* [6]. Much of the optical and transport properties of QWs using GaAlAs material system is reviewed in *Heterojunction Band Discontinuation: Physics and Applications* [7].

QW lasers fabricated using the GaAlAs material system have improved performance characteristics, such as lower threshold current and higher quantum efficiency, compared to regular double heterostructure lasers. These improved performance characteristics of QW lasers relative to regular double heterostructure lasers have been demonstrated for the InGaAsP material system using the metal organic chemical vapor deposition (MOCVD) epitaxial growth technique. Early work on

InGaAsP multi-quantum well (MQW) lasers was done using liquid phase epitaxy (LPE) growth [8–11].

The physical principle, fabrication, and performance characteristics of QW lasers fabricated using the InGaAsP material system will be described in this chapter. One advantage of QW lasers over regular double heterostructure lasers is that the emission wavelength of the former can be varied simply by varying the width of the QWs that form the active region. This phenomenon will be discussed in Section 11.2. The restriction of the carrier motion normal to the well leads to a modification of the density of states in a QW, which changes the radiative and nonradiative recombination rates of electrons and holes in a QW heterostructure compared to that of a regular double heterostructure. These changes may result in several desirable characteristics, such as lower threshold current density, higher efficiency, and lower temperature dependence of threshold current of suitably designed QW lasers compared to regular double heterostructure lasers. The calculation of optical gain, radiative recombination rate, and nonradiative Auger recombination rates will be discussed in Section 11.3. The various types of QW structures, such as the single quantum well (SQW), MQW, modified MQW, and the graded index SQW, and their principal performance characteristics, such as threshold current and efficiency, will be discussed in Section 11.4.

The performance characteristics of QW lasers of different structures will be described in Section 11.5. QW lasers are found to have lower linewidth under modulation. Measurements and calculation of this effect, along with the fabrication of distributed feedback and distributed Bragg reflector-type single-frequency lasers, will be described in Section 11.6. The recent results on strained QW lasers will be described in Section 11.7.

11.2 ENERGY LEVELS

The schematic of a QW structure, along with the band diagram, is shown in Figure 11.1. L_Z and L_B are well and barrier thicknesses, respectively.

A carrier (electron or hole) in a double heterostructure is confined in a three-dimensional potential well. The energy levels of such carriers are obtained by separating the Hamiltonian into three parts, corresponding to the kinetic energies in x, y, and z directions, each of which form a continuum of states. When the thickness of the heterostructure (L_z) is comparable to the de Broglie wavelength, the kinetic energy corresponding to the particle motion along the z direction is quantized. The energy levels can be obtained by separating the Hamiltonian into energies corresponding to x, y, and z directions. For the x and y directions, the energy levels form a continuum of states given by

$$E = \frac{\hbar^2}{2m}(k_x^2 + k_y^2) \tag{11.1}$$

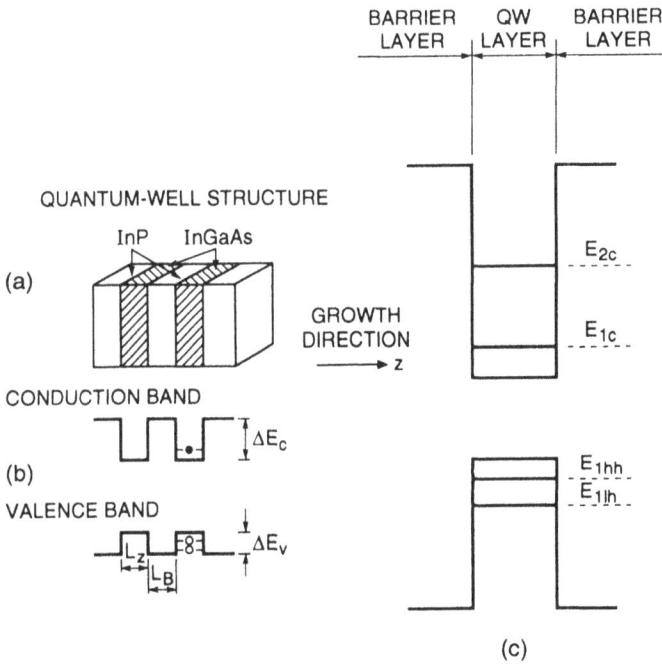

Figure 11.1 InGaAs/InP QW structure (a) layer structure, z direction is the growth direction; (b) band structure for conduction and valence band; (c) energy levels in the conduction and valence band.

where m is the effective mass of the carrier, and k_x and k_y are the wave vectors along the x and y directions, respectively. Thus, the electrons or holes in a QW may be viewed as forming a two-dimensional Fermi gas.

The energy levels in the z direction are obtained by solving the Schrödinger equation for a one-dimensional potential well. It is given by

$$-\frac{\hbar^2}{2m}\frac{d\Psi^2}{dz^2} = E\Psi \qquad \text{in the well} \qquad (0 \leq z \leq L_z),$$

$$\text{(11.2)}$$

$$-\frac{\hbar^2}{2m}\frac{d^2\Psi}{dz^2} + V\Psi = E\Psi \quad \text{outside the well} \quad (z \geq L_z; z \leq 0),$$

where Ψ is the Schrödinger wavefunction, and V is the depth of the potential well. For the limiting case of an infinite well, the energy levels and the wavefunctions are

$$E_n = \frac{\hbar^2}{2m} \left(\frac{n\pi}{L_z}\right)^2 \quad \text{and} \quad \Psi_n = A \sin \frac{n\pi z}{L_z} \quad (n = 1, 2, 3), \qquad (11.3)$$

where A is a normalization constant. For very large L_z, Equation (11.3) yields a continuum of states, and the system no longer exhibits quantum effects.

For a finite well, the energy levels and wavefunction can be obtained from Equation (11.2) using the boundary conditions that Ψ and $d\Psi/dz$ are continuous at the interfaces $z = 0$ and $z = L$.

The potential well for electrons (ΔE_c) and holes (ΔE_v) in a double heterostructure depends on the materials involved. A knowledge of ΔE_c and ΔE_v is necessary in order to accurately calculate the energy levels. For an InGaAsP-InP double heterostructure, the values obtained by Forrest et al. [12] are

$$\Delta E_c/\Delta E = 0.39 \pm 0.01, \quad \Delta E_v/\Delta E = 0.16 \pm 0.01. \qquad (11.4)$$

where ΔE is the bandgap difference between the confining layers and active region.

The energy eigenvalues for a particle confined in the QW are

$$E(n, k_z, k_y) = E_n + \frac{\hbar^2}{2m_n^*}(k_x^2 + k_y^2), \qquad (11.5)$$

where E_n is the nth confined-particle energy level for carrier motion normal to the well, and m_n^* is the effective mass. Figure 11.2 shows schematically the energy levels E_n of the electrons and holes confined in a QW. The confined-particle energy levels (E_n) are denoted by E_{1c}, E_{2c} for electrons, E_{1hh} for heavy holes, and E_{1lh} for light holes. These quantities can be calculated by solving the eigenvalue equations, equation (11.2), for a given potential barrier (ΔE_c or ΔE_v), as described earlier.

Electron-hole recombination in a QW follows the selection rule $\Delta n = 0$; that is, the electrons in states $E_{1c}(E_{2c}, E_{3c},$ et cetera) can combine with the heavy holes E_{1hh} (E_{2hh}, E_{3hh}, et cetera) and with light holes $E_{1lh}(E_{2lh}, E_{3lh},$ et cetera).

The separation between the lowest conduction band level and the highest valence band level for an infinite well is given by

$$E \simeq E_g + \frac{h^2}{8L_z^2}\left(\frac{1}{m_c} + \frac{1}{m_{hh}}\right) \qquad (11.6)$$

where m_c and m_{hh} are the conduction band masses and heavy hole mass, respectively. Thus, the energy of the emitted photons can be varied by simply varying the well width L_z. Figure 11.2 shows the experimental results for InGaAs QW lasers with different well thicknesses bounded by InP cladding layers [13]. As well thickness is reduced, the laser emission shifts to higher energies.

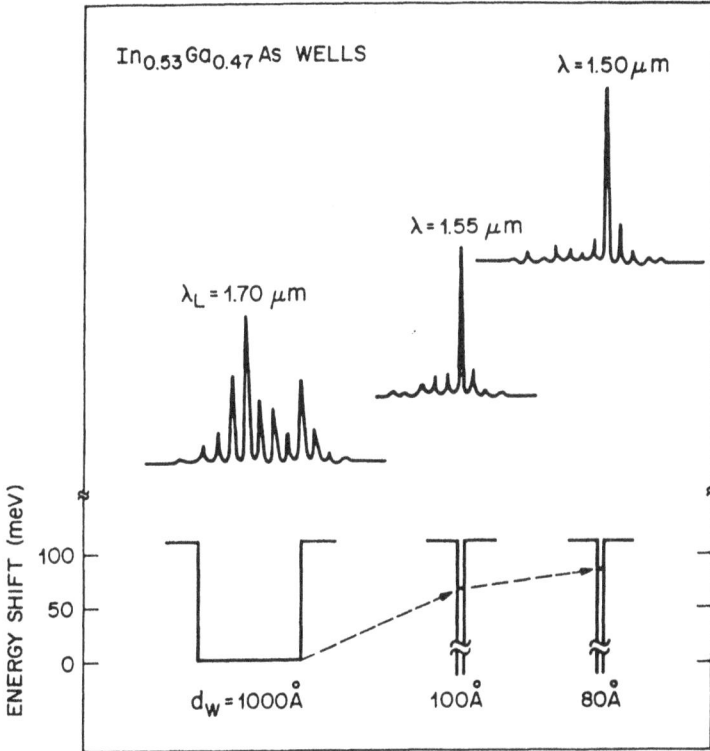

Figure 11.2 Lasing wavelength for different well thicknesses of an InGaAs/InP MQW well structure.

11.3 GAIN AND RECOMBINATION

The basis of light emission in semiconductors is the recombination of an electron in the conduction band with a hole from the valence band, and the excess energy is emitted as a photon (light quantum). The process is called radiative recombination. Sufficient numbers of electrons and holes must be excited in the semiconductor for stimulated emission or net optical gain. The condition for net gain at a photon energy E is given by [14]

$$E_{fc} + E_{fv} = E - E_g \qquad (11.7)$$

where E_{fc} and E_{fv} are the quasi-Fermi levels of electrons and holes, respectively, measured from the respective band edges (positive into the band), and E_g is the bandgap of the semiconductor. The quasi-Fermi levels can be calculated from a knowledge of the density of states of electrons and holes in the QW heterostructure.

11.3.1 Density of States

In a QW structure, a series of energy levels and associated subbands are formed because of the quantization of electron energy in the direction normal to the well. The density of states of such electrons in a single quantum well (SQW) is given by [15,16]

$$g_c(E) = \sum_{n=1}^{\infty} \frac{m_c}{L_z \pi \hbar^2} H[E - E_n] \qquad (11.8a)$$

where $H(x)$, m_c, $\hbar$, and E_n are the Heaviside function, the effective mass of electrons, Planck's constant (h) divided by 2π and the energy level for n_{th} subband. A similar equation holds for the holes in the valence band. For the regular three-dimensional case, the density of states is given by [15]

$$g(E) = 2(2\pi m_c k_B T/h^2)^{3/2} E^{1/2} \qquad (11.8b)$$

where k_B is the Boltzmann constant and T is the temperature. A comparison of Equations (11.8a) and (11.8b) shows that the density of states in a QW is independent of carrier energy and temperature. The modification of the density of states in a QW is sketched in Figure 11.3. This modification can significantly alter the recombination rates in a QW relative to a regular double heterostructure.

If we use a MQW structure instead of the SQW, the density of states is modified. When the barrier layers between the wells are thick, each well is independent, and the density of states for the entire MQW structure is simply N times the density of states for a SQW (equation (11.8a)), where N is the number of wells. However, if the barrier is thin or the barrier height is small, the energy levels in the adjacent

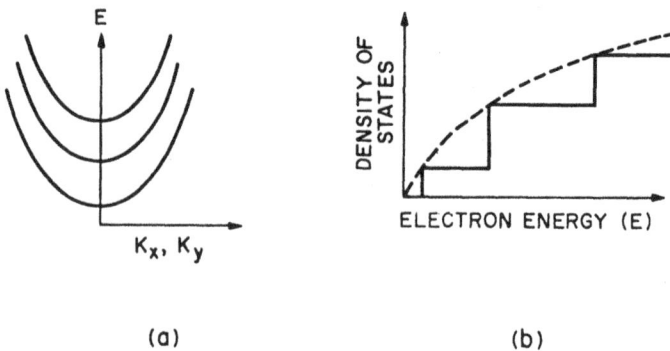

(a)   (b)

Figure 11.3 (a) Energy *versus* wave vector for each subband; (b) schematic representation of the density of states in a QW.

wells are coupled, which splits each single well level into N different energy levels. In this case, the density of states is given by

$$g_c(E) = \sum_{n=1}^{\infty} \sum_{k=1}^{\infty} \frac{m_c}{L_z \pi \hbar^2} H[E - E_{nk}] \qquad (11.9)$$

where E_{nk} $(k = 1, \ldots, N)$ are the energy levels that split from a single well energy level. The difference between the maximum and minimum value of E_{nk} indicates the broadening of each QW level due to coupling. For the step-like density of states (Figure 11.3) to be preserved in a MQW structure, the broadening due to coupling between wells must be smaller than the broadening due to intraband relaxation. For the discussion of recombination rates, we assume that the interwell coupling in a MQW structure is weak enough so that the density of states for each well is described by Equation (11.8a).

11.3.2 Gain and Radiative Recombination

The optical gain and radiative recombination rate in a QW structure has been calculated using a single-level approximation and assuming no broadening of each level [16]. In order to explain the saturation of gain at high currents, we need to take into account the broadening of each level due to intraband relaxation [17–19]. The calculated model gain as a function of current density is shown in Figure 11.4. N denotes the number of wells in the MQW structure. The calculation takes into account

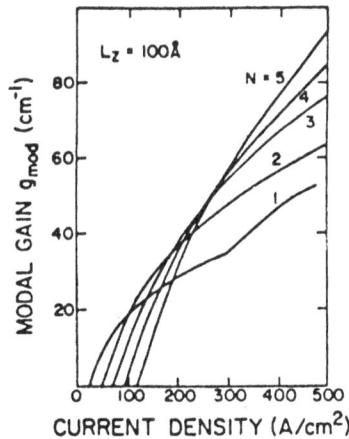

Figure 11.4 The modal gain (Γ_g) as a function of injected current density for different numbers of QWs [after 17].

level broadening. The modal gain (g_{mod}) is defined as $g_{mod} = \Gamma_g$ where Γ is the confinement factor and g is the optical gain. The confinement factor for the calculation in Figure 11.4 is $\Gamma = 0.03N$, which is a typical value for a 100Å thick well.

11.3.3 Nonradiative Recombination

An electron-hole pair can recombine nonradiatively, meaning that the recombination can occur through any process that does not emit a photon. In many semiconductors (e.g., pure germanium or silicon) the nonradiative recombination dominates radiative recombination.

The effect of nonradiative recombination on the performance of injection lasers is to increase the threshold current. If τ_{nr} is the carrier lifetime associated with the nonradiative process, the increase in threshold current density is given approximately by

$$J_{nr} = \frac{en_{th}d}{\tau_{nr}} \tag{11.10}$$

where n_{th} is the carrier density at threshold, d is the active layer thickness, and e is the electron charge.

It is generally believed that the Auger effect plays a significant role in determining the observed high-temperature sensitivity of the threshold current of InGaAsP lasers emitting near 1.3 μm and 1.55 μm [20–27]. The active region of the InGaAsP MQW laser is nominally undoped. Under high injection the Auger rate varies approximately as

$$R_A = Cn^3 \tag{11.11}$$

and the current lost to Auger recombination at threshold is given by

$$J_A = edCn^3 \tag{11.12}$$

The Auger recombination rate in QW structures has been calculated [24,25]. The value for a $\lambda \sim 1.3$-μm InGaAsP QW is $\sim 3 \times 10^{29}$ cm^6 sec^{-1}, which is close to the measured value. The Auger lifetime τ_A is given by

$$\tau_A = \frac{n}{R_A} = \frac{1}{Cn^2} \tag{11.13}$$

The calculated Auger lifetimes for an InGaAsP-InP QW structure is shown in Figure

11.5. The calculation assumes the same matrix elements as that for bulk semiconductors. Because of the uncertainties of both the matrix element and effective masses of the bands away from the band edge, only an order of magnitude estimate of the Auger coefficient in a QW structure can be obtained.

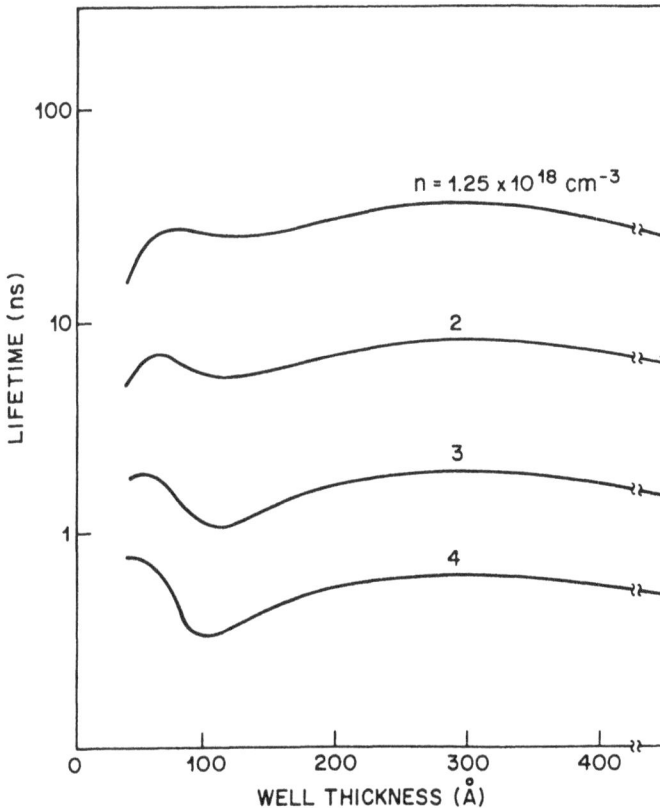

Figure 11.5 The calculated auger lifetime in an undoped InGaAsP/InP QW as a function of well thickness. The parameter n is the injected carrier density [after 19].

11.4 SINGLE QW AND MQW LASERS

QW injection lasers with both single and multiple active layers have been fabricated. QW lasers with one active region are called SQW lasers, and those with multiple active regions are called MQW lasers. The layers separating the active layers in a MQW structure are called barrier layers. The energy band diagram of these laser structures are shown schematically in Figure 11.6. MQW lasers with the bandgap

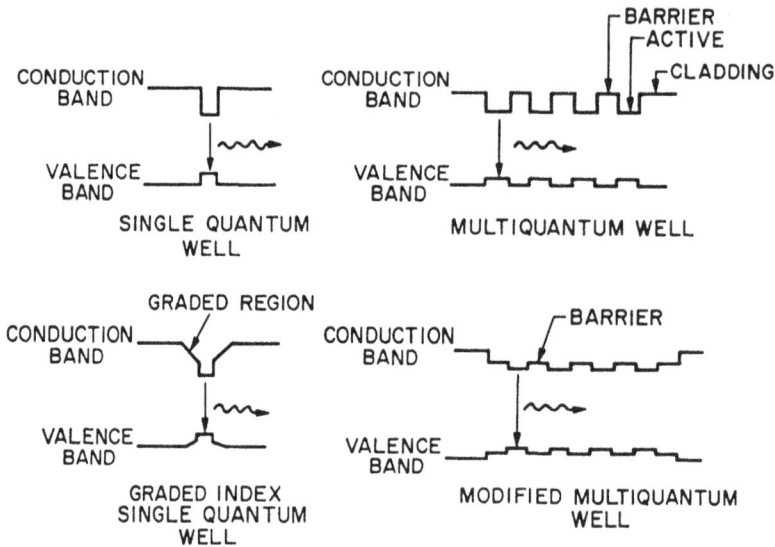

Figure 11.6 Single QW and MQW laser structures are shown schematically.

of the barrier layers different from that of the cladding layers are sometimes referred to as modified MQW lasers [27].

One of the main differences between the SQW and MQW lasers is that the confinement factor (Γ) of the optical mode is significantly smaller for the former than that for the latter. This can result in higher threshold carrier density and in higher threshold current density of SQW lasers when compared with MQW lasers. The confinement factor of a SQW heterostructure can be significantly increased using a graded-index cladding layer (see Figure 11.6). This layer allows the use of the intrinsic advantage of the QW structure (high gain at low carrier density) without the penalty of a small mode-confinement factor. Threshold current densities as low as 400 A/cm^2 have been reported for a InGaAs graded-index (GRIN) QW laser [28].

We now discuss the calculation of the mode-confinement factor in SQW and MQW structures. For small active layer thickness, such as that of a SQW, the expression for Γ is [29]

$$\Gamma \simeq 2\pi^2(\mu_a^2 - \mu_c^2)d^2/\lambda_0^2 \qquad (11.14)$$

where μ_a and μ_c are the refractive indices of the active and cladding layer, respectively, d is the active layer thickness, and λ_0 is the free space wavelength. For typical values, $\mu_a = 3.5$, $\mu_c = 3.2$, $d = 200$Å, and $\lambda = 1.3$ μm, the calculated $\Gamma \simeq 0.016$.

The mode confinement in MQW structures can be analyzed by solving the electromagnetic wave equations for each of the layers with appropriate boundary conditions. This procedure is quite tedious because of a large number of layers involved. Streifer *et al.* [30] have shown that the following simple formula gives reasonably accurate results

$$\Gamma(\text{MQW}) = \gamma \frac{N_a d_a}{N_a d_a + N_b d_b} \tag{11.15}$$

where

$$\gamma = 2\pi^2 (N_a d_a + N_b d_b)^2 (\bar{\mu}^2 = \mu_c^2) \quad \text{and} \quad \bar{\mu} = \frac{N_a d_a \mu_a + N_b d_b \mu_b}{N_a d_a + N_b d_b}$$

N_a and N_b are the number of active and barrier layers in the MQW structure and d_a and d_b (and μ_a and μ_b) are the thicknesses (and refractive indices) of the active and barrier layers, respectively. The formulation of equation (11.15) may be seen as follows; $\bar{\mu}$ is the average refractive index of a uniform optical mode in the MQW active region (i.e., the active and barrier layers) of total thickness $N_a d_a + N_b d_b$. Hence γ is the confinement factor of the optical mode (same as equation (11.14)) in both active and barrier layers. $\Gamma(\text{MQW})$ is thus obtained by multiplying γ by the ratio of the total active thickness, $N_a d_a$, to the total thickness of the active and barrier layers, $N_a d_a + N_b d_b$. For a 1.3-μm InGaAsP-InP MQW structure with four active layer wells (150Å thick) and three barrier layers (InP, 150Å thick), the calculated $\Gamma \approx 0.2$. This is considerably larger than the typical Γ values for SQW lasers with active layer thicknesses in the range of 100Å to 150Å.

Let us now discuss the effect of an Auger process on the threshold current of a single QW laser vis-a-vis a MQW laser. For a SQW laser, the confinement factor is small, and hence the threshold gain is very large ($g_{th} \gtrsim 10^3$ cm^{-1}). This makes the carrier density at threshold large ($n_{th} \sim 2.5 \times 10^{18}$ cm^{-3}). For a MQW laser, the effective confinement factor is large, and hence g_{th} is smaller ($g_{th} \approx 300$ cm^{-1}). The carrier density at threshold is $n_{th} \approx 8 \times 10^{17}$ cm^{-3} at this gain. The Auger rate varies as n^3; thus, it follows that, compared to the SQW laser, the Auger rate will be significantly smaller in MQW lasers.

11.5 LASER STRUCTURES

In QW lasers, the optical mode is confined perpendicular to the junction plane because the cladding layers have a lower index of refraction compared to that of the active region [30,31]. For stable fundamental mode operation with a low threshold current, additional confinement of the optical mode along the junction plane is required. In the absence of this lateral mode confinement (which may be induced by a region of well-defined optical gain or refractive index step), the laser behaves as a broad-area laser.

11.5.1 Broad-Area Lasers

A broad-area laser does not employ any scheme for current confinement. The processing of these devices is relatively simple. The wafer with the MQW active region is grown by MOCVD growth technique on a n-InP substrate. The layers grown are (a) n-InP cladding layer, (b) MQW active region, (c) p-InP cladding layer and (d) p-InGaAs contact layer. The as-grown wafer, typically 250 to 400 μm thick, is thinned down to a thickness of 75 to 100 μm, and metallic contacts are deposited on the substrate side (typically AuSn or AuGe alloy) and the epitaxially grown side (typically CrAu or AuBe alloy). The thinning of the wafer is necessary to facilitate cleaving along a crystallographic plane.

The threshold current density J_{th} depends on the confinement of the optical mode and therefore varies with the design parameters of the MQW active region, such as well thickness, number of wells, etc. The optical gain g in a QW is approximately related to the current J by [24]

$$g = a(J - J_0) \tag{11.16}$$

where a is the gain constant and J_0 is the transparency current density. The threshold gain g_{th} is given by [7,37]

$$\Gamma g_{th} = \alpha + \frac{1}{L} \ln\left(\frac{1}{R}\right) \tag{11.17}$$

where L is the cavity length, α is the total loss, and R is the facet reflectivity. From equations (11.16) and (11.17), the threshold current J_{th} is given by

$$J_{th} = J_0 + \left[\alpha + \frac{1}{L} \ln\left(\frac{1}{R}\right) \right] \bigg/ a\Gamma \tag{11.18}$$

Equation (11.18) shows that J_{th} decreases with increasing cavity length, and for small losses it equals the transparency current density J_0 as $L \to \infty$. The measured threshold current density for InGaAs MQW lasers grown by MOCVD and CBE (chemical beam epitaxy) growth technique is shown in Figure 11.7. The threshold current density is found to decrease with increasing cavity length, with the lowest value being 380 A/cm^2 for 2-mm cavity length. The measured temperature dependence of threshold current is represented by the usual expression $J_{th} \sim J \exp(T/T_0)$ with $T_0 \sim 60$K in the temperature range 20°C to 70°C.

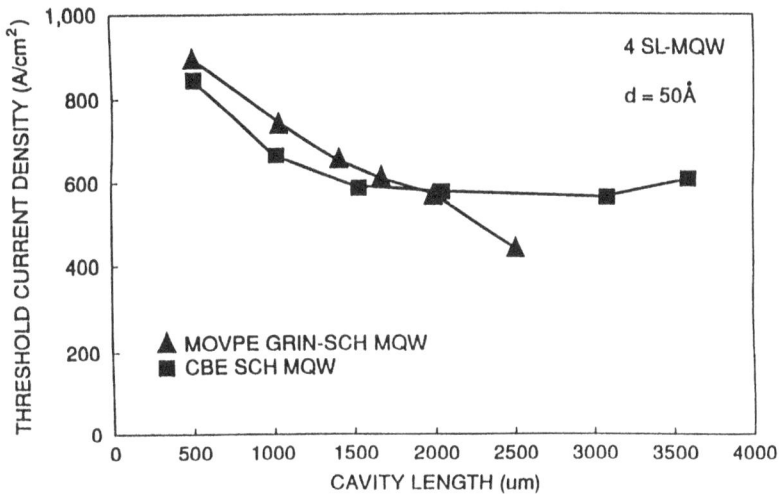

Figure 11.7 Broad-area threshold current density as a function of cavity length.

11.5.2 Index-Guided Lasers

Index guiding along the junction plane is necessary in order to control the light emission pattern of a semiconductor laser [32]. The schematic of an index-guided buried heterostructure (BH) laser is shown in Figure 11.8. The active region width is ~2 μm. The laser uses Fe-doped InP layers to confine the current to the active region and to provide lateral index guiding. The light *versus* current characteristics of a MQW BH laser at different temperatures is shown in Figure 11.9.

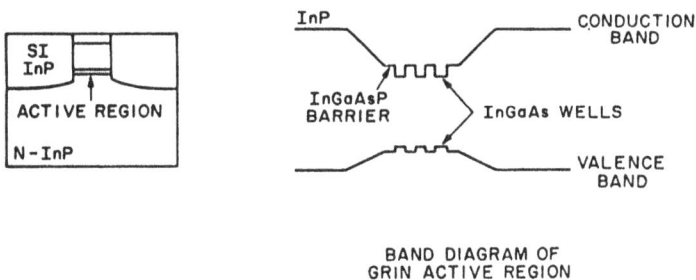

Figure 11.8 Schematic of a buried heterostructure laser using semi-insulating Fe doped InP current confining layer.

For low threshold current and high efficiency, it is necessary to optimize the MQW structure. Temkin *et al.* [33] performed such an optimization study. BH lasers of the type shown in Figure 11.8 were fabricated using three different MQW structure designs. The MQW structures have four 100Å-thick InGaAs wells and 100Å-thick barrier layers and is confined between two step GRIN regions. The schematic of three MQW designs studied is shown in Figure 11.10. These designs are identical except for the composition of the barrier layer.

Figure 11.11 shows plots of (a) quantum efficiency and (b) the threshold current of the three GRIN MQW laser designs. The solid lines in Figure 11.11a represent fits to the expression describing the differential quantum efficiency (η_D) as a function of cavity length, L:

$$\eta_D^{-1} = \eta_i^{-1}\left[1 + \alpha L \Big/ \ln\left(\frac{1}{2}\right)\right] \tag{11.19}$$

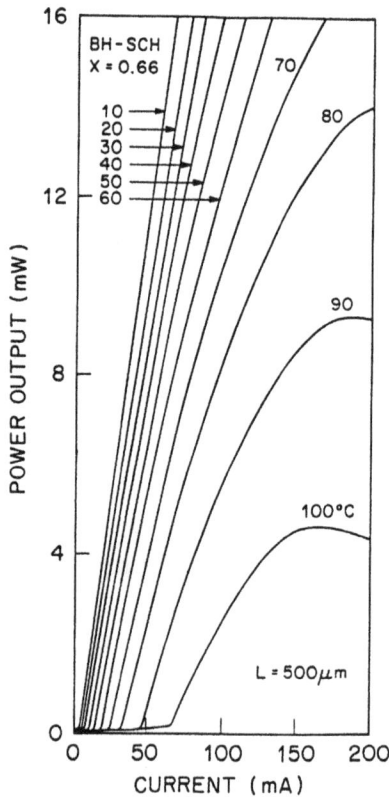

Figure 11.9 Light *versus* current characteristics of a MQW BH laser at different temperatures.

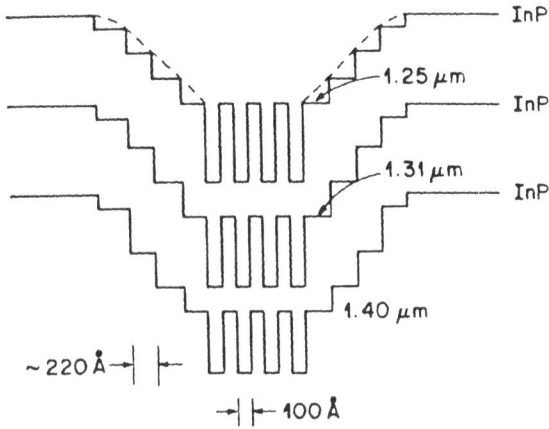

Figure 11.10 Schematic of three GRIN MQW designs. The bandgap of the barrier layer is varied from 1.25 μm to 1.40 μm [after 33].

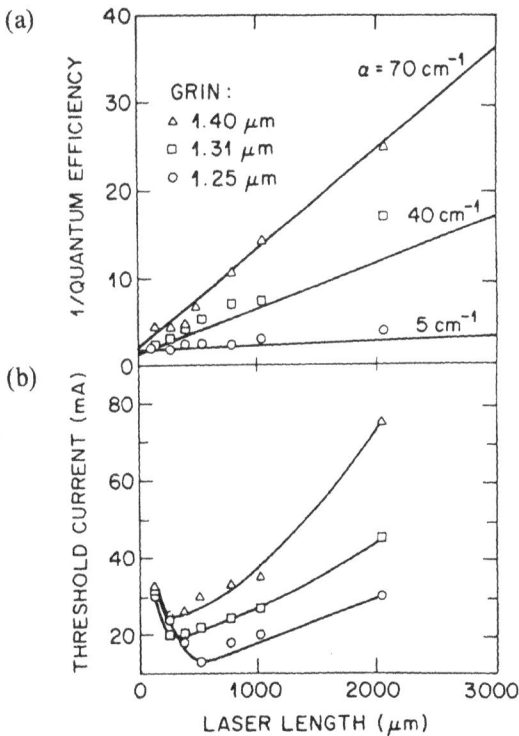

Figure 11.11 (a) Inverse of the quantum efficiency is plotted as a function of cavity length. (b) Threshold current is plotted as a function of cavity length [after 33].

where η_i is the internal quantum efficiency. The optical loss α is found to decrease dramatically for larger gap barriers. Larger gap barriers also produce lasers with the lowest threshold current (Figure 11.11b). The results of Figure 11.11 show that there is an optimum cavity length and a GRIN MQW design for lasers with low threshold and high efficiency. The strong effect of compositional changes of the barrier layers suggest that the waveguide loss is dominated by the absorption in the barrier layers forming the MQW structure.

11.5.3 Low-Threshold Lasers

Equation (5.3) shows that the threshold current density of a laser is primarily determined by two components: (a) transparency current density, J_0, and (b) current needed to overcome losses. The latter is small if α is small and can be significantly reduced using high reflectivity coatings. For a QW structure, the transparency current density is small ~ 50 A/cm^2 (Figure 11.3) [34,35] compared to a value of ~ 1000 A/cm^2 for a regular double heterostructure laser. This allows the possibility of very low threshold operation [34]. The threshold current of a BH laser is given as $I_{th} = J_{th}WL$ where W is the width of the active region. The calculated threshold current as a function of cavity length for different mirror reflectivities is shown in Figure 11.12 for a 1-μm-wide active region. The calculation shows that InGaAs/InP MQW lasers can have submilliampere threshold current for cavity lengths of ~ 150 μm with high reflectivity coatings [36].

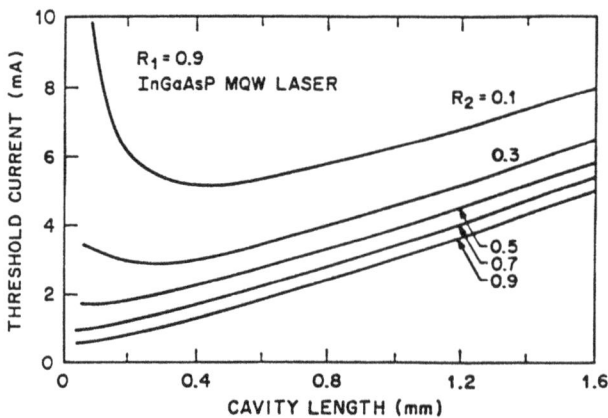

Figure 11.12 Calculated threshold current as a function of cavity length for different mirror reflectivities [after 36].

The light *versus* current characteristics of a MQW BH laser with 90% and 70% facet coatings, respectively, is shown in Figure 11.13 [36]. The laser has a threshold current of 1.1 mA at 25°C and 0.9 mA at 10°C.

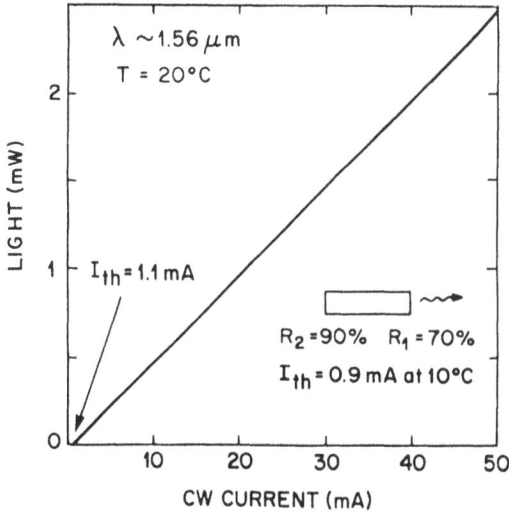

Figure 11.13 Light *versus* current characteristics of a MQW BH laser with 90% and 70% facet coatings. The cavity length is 200 μm [after 36].

11.6 SINGLE-FREQUENCY LASERS

A conventional semiconductor laser with cleaved facets does not emit in a single frequency (single longitudinal mode). It emits in a few longitudinal modes centered around the gain peak. Single-frequency lasers are important as sources for optical fiber communication systems for reducing pulse broadening due to fiber dispersion. Two principle types of single frequency lasers are the distributed feedback laser (DFB) and the distributed Bragg reflector (DBR) laser. A DFB laser has a grating built into the laser structure close to the active region. The grating provides frequency-selective feedback to the optical mode, which makes the laser emit in the single frequency. For a DBR laser the grating is external to the active region and acts as a single-frequency reflector to the optical mode.

11.6.1 DFB Lasers

DFB lasers fabricated using the MQW active region have been reported by several investigators. The transmission electron micrograph (TEM) of the active region of such a device is shown in Figure 11.14. The grating has a periodicity of 2350Å and is fabricated using optical holography followed by wet chemical etching on a (100) oriented *n*-InP substrate. The MQW active, cladding, and contact regions are then grown on the wafer using the MOCVD growth technique. The wafer was processed following a second growth to fabricate BH lasers of the type shown in Figure 11.8.

The light *versus* current characteristics of a BH MQW DFB laser is shown in Figure 11.15. More than 80 mW of power from such lasers have been coupled into a single mode fiber.

A key characteristic, which is very important for lightwave transmission systems, of MQW DFB lasers is that these devices have lower spectral width under modulation than regular double heterostructure lasers. The spectral width under modulation is also known as frequency chirp. The measured chirp width as a function of modulation current for MQW and regular double heterostructure lasers is shown in Figure 11.16. The MQW laser has about a factor of two lower chirp than double

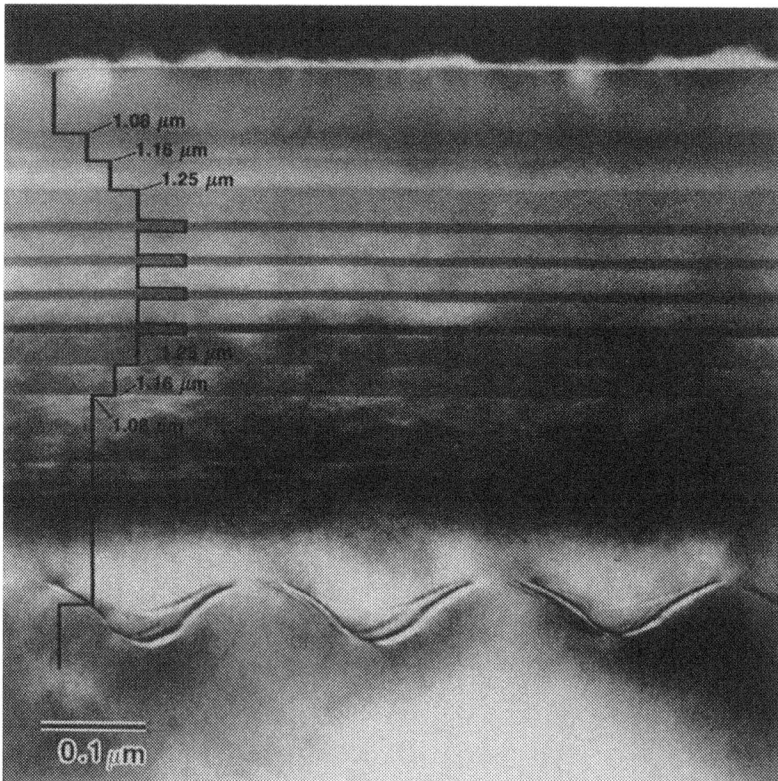

Figure 11.14 Transmission electron microscope (TEM) cross-section of a MQW DFB laser wafer.

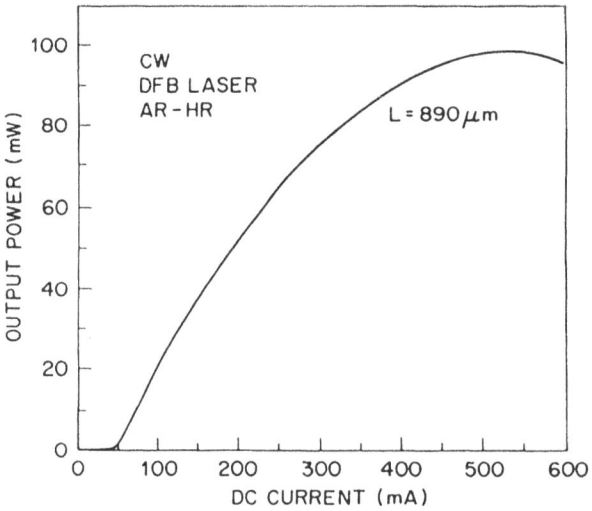

Figure 11.15 Light *versus* current characteristics of a buried heterostructure MQW DFB laser.

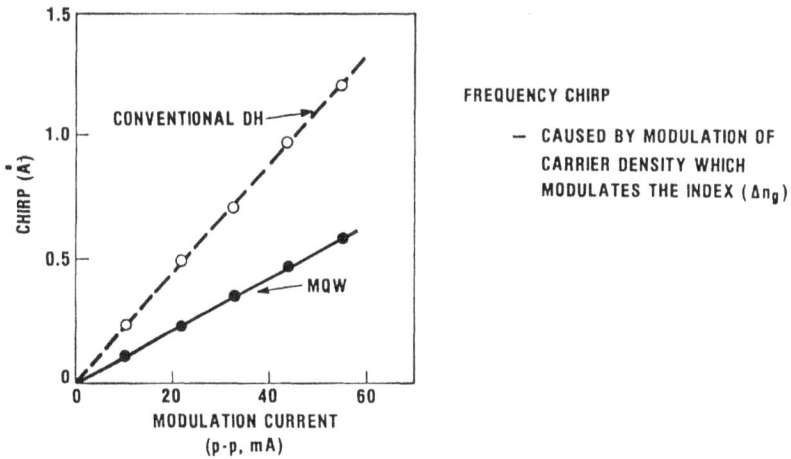

FREQUENCY CHIRP

— CAUSED BY MODULATION OF CARRIER DENSITY WHICH MODULATES THE INDEX (Δn_g)

Figure 11.16 Measured chirp plotted as a function of modulation current for regular DH and MQW lasers.

heterostructure lasers. Lower chirp implies that the MQW laser source will cause much less dispersion penalty than regular double heterostructure for high data rate fiber transmission systems.

The lower chirp of MQW lasers results from the lower linewidth enhancement factor (α) of MQW structures relative to double heterostructure [37–39]. Chirp is proportional to α, where α is defined as the ratio of the real to the imaginary part of the change in index with respect to change in carrier density. The smaller α in QW lasers is primarily due to the modification of the density of states. The measured α values as a function of photon energy for a regular double heterostructure and MQW structure are shown in Figure 11.17 [39]. Note that α equals 3.0 for a MQW laser and 5.5 for a regular DH laser at the lasing wavelength.

MQW DFB lasers also exhibit narrow CW spectral linewidth compared to a regular DH DFB laser. The spectral linewidth $\Delta\nu$ varies as $1 + \alpha^2$. Thus, small α will result in narrow linewidth. The measured CW linewidth is plotted on a function of the inverse of the output power for a MQW DFB laser in Figure 11.18. The smallest linewidth observed is 400 KHz at 30 mW of output power.

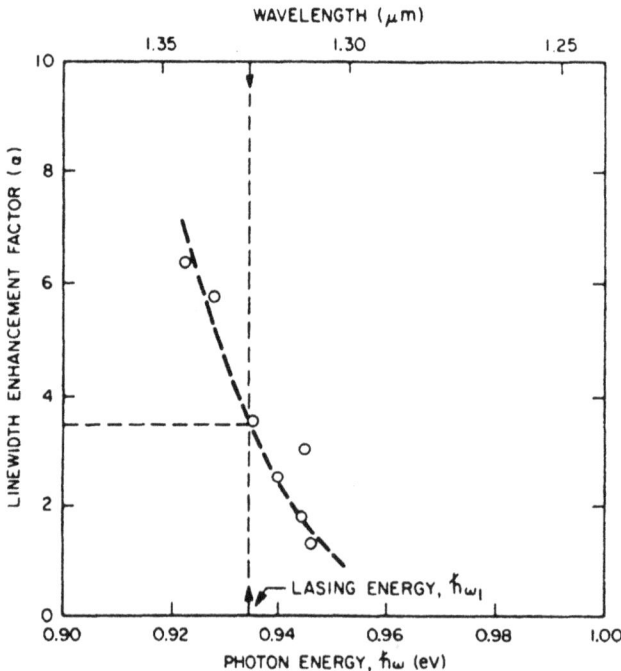

Figure 11.17 Measured linewidth enhancement factor (α) as a function of photon energy [after 39].

CW LINEWIDTH (MHz)

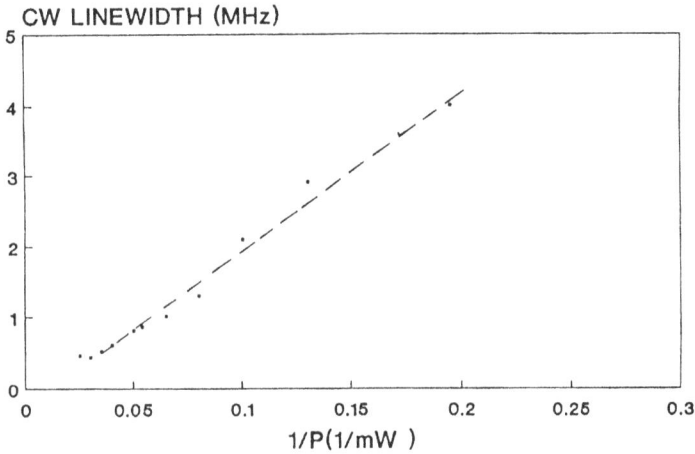

Figure 11.18 CW linewidth as a function of inverse power for a MQW DFB laser. Cavity length = 800 μm.

11.6.2 DBR Lasers

As mentioned previously, the DBR laser provides an alternative scheme in which the frequency dependence of the DFB mechanism is used to select a single frequency. In contrast to DFB lasers, the grating in DBR lasers is etched outside the active region.

Historically, DFB and DBR semiconductor lasers were investigated simultaneously for the InGaAsP material system. The performance of a DBR laser is comparable to that of a DFB laser as far as the spectral and dynamic properties are concerned.

The frequency selective wavelength (λ) of a DBR laser is given by

$$\lambda = 2n \wedge \qquad (11.20)$$

where $\wedge$ is the grating period and n is the effective refractive index of the lasing optical mode in the DBR waveguide. Since the refractive index depends on the carrier density, it is possible to vary index and hence the emission wavelength by injecting carriers (current) in the external DBR waveguide. The schematic of such a tunable DBR laser [40] is shown in Figure 11.19. It has three sections, the light emitting section with the active region, the index tuning (wavelength tuning) section, and the phase tuning section. The latter is necessary in order to ensure continuous tuning. The emission wavelength characteristics of a MQW DBR laser is illustrated in Figure 11.20 as a function of the phase and the frequency tuning section currents.

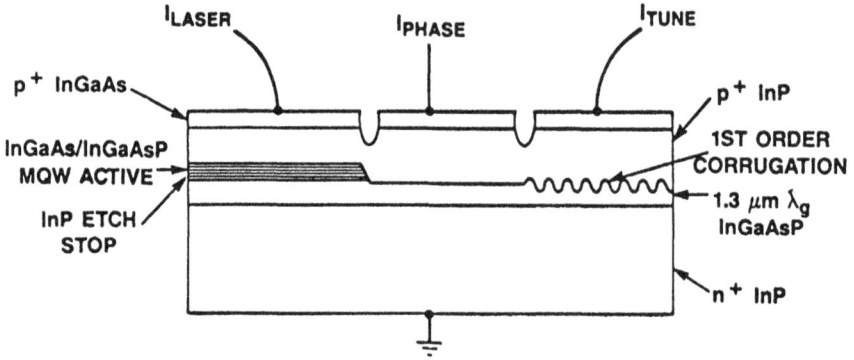

Figure 11.19 Schematic of a tunable three-section DBR laser [after 40].

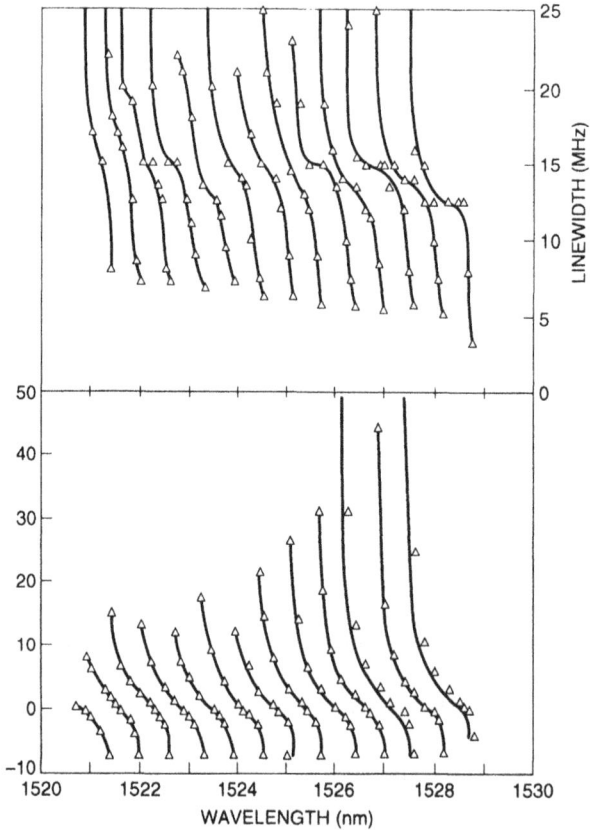

Figure 11.20 Tuning characteristics of a three-section DBR laser [after 40].

11.7 STRAINED QW LASERS

Recently, there has been considerable amount of interest in the fabrication of novel electro-optic devices using strained layer QWs [41–48]. It has been shown theoretically that strain can alter the band structure parameters significantly to produce many desirable features in laser characteristics, such as better high-temperature performance due to reduced Auger recombination [41,46] and small chirp [45], high bandwidth due to larger gain coefficient [41,44].

Strained QW lasers have been fabricated using $In_xGa_{1-x}As$ QW and GaAs barrier layers for $x \leq 0.2$ over a GaAs substrate using a MOCVD or MBE (molecular beam epitaxy) growth technique [42,47,48]. These lasers exhibit threshold current densities as low as 50 A/cm^2 and high CW and pulsed output powers have been reported. These lasers emit in the wavelength range of 0.85 to 1 μm.

11.7.1 InGaAs Strained QW Lasers

The alloy $In_{0.53}Ga_{0.47}As$ has the same lattice constant as InP. Lasers with an $In_{0.53}Ga_{0.47}As$ active region have been grown on InP by the MOCVD growth technique. Recently, it has been shown that excellent material quality is also obtained for $In_xGa_{1-x}As$ alloys grown over InP by MOCVD for non-lattice-matched compositions. In this case, the laser structure generally consists of one or many $In_xGa_{1-x}As$ QW layers with a InGaAsP barrier layer whose composition lattice matches that of InP. For $x < 0.53$, the active layer in these lasers is under tensile stress, and for $x > 0.53$, the active layer is under compressive stress.

Superlattice structures of InGaAs/InGaAsP with tensile and compressive stress have been grown by both MOCVD and CBE growth techniques over a n-InP substrate. Figure 11.21 shows the broad-area threshold current density as a function of cavity length for a MQW laser with four $In_{0.65}Ga_{0.35}As$ quantum wells. The active region in this laser is under 0.8% compressive strain. Note that the threshold current density is lower than the value obtained for a similar lattice-matched MQW structure (Figure 11.7).

BH lasers have been fabricated using compressive and tensile strained MQW lasers. The threshold current of these lasers as a function of In concentration (x) is shown in Figure 11.22. Lasers with compressive strain have a lower threshold current than lasers with tensile strain. This can be explained by the shifting of the light hole and heavy hole bands under stress.

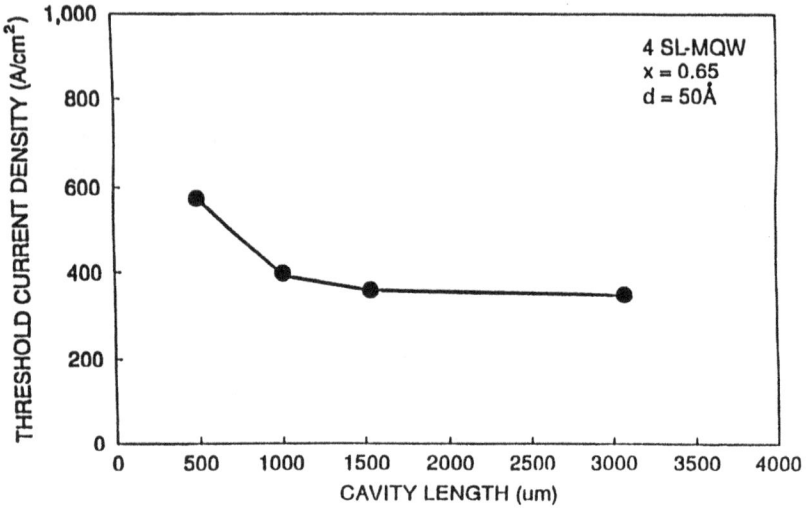

Figure 11.21 Broad area threshold current density as a function of cavity length for 1.5 μm strained MQW lasers. The active region composition is In$_{0.65}$Ga$_{0.35}$As [after 28].

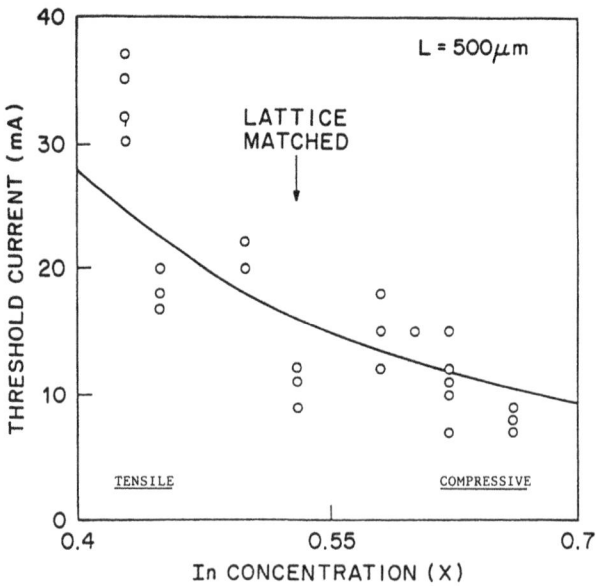

Figure 11.22 Threshold current of strained In$_x$Ga$_{1-x}$As/InP MQW lasers as a function of In concentration x.
(Courtesy H. Temkin)

11.7.2 Dynamic Characteristics

It has been predicted theoretically that strained MQW lasers can have higher bandwidth due to larger gain coefficient. Experimental results on the small signal bandwidth of $\lambda \sim 1.5$-μm strained MQW lasers have been reported. The measured maximum value of 15 GHz at 20 mW of output power is comparable to that for regular lasers. This suggests further optimization in device structure or MQW design may be necessary to achieve higher bandwidth.

The frequency chirp under modulation of strained MQW lasers has been measured. The lasers exhibit somewhat smaller chirp than unstrained MQW lasers. The smaller chirp is due to the smaller linewidth enhancement factor (α) of these strained MQW lasers. The measured α as a function of photon energy of a compressively strained MQW laser is shown in Figure 11.23. The α value of 2.0 at the lasing wavelength is smaller than the value observed for MQW lasers (Figure 11.17). The α value depends on the injected carrier density in both unstrained and strained MQW lasers [45], and, therefore, a small variation of α with cavity length and MQW active region design is expected.

The InGaAsP MQW laser technology is well advanced. Lasers with good reliability have been fabricated. These layers are being used in lightwave transmission systems operating at data rates of 2.4 Gb/s and higher. Even the lower chirp possible with strained MQW lasers makes them potentially useful for systems with larger regenerator spacing.

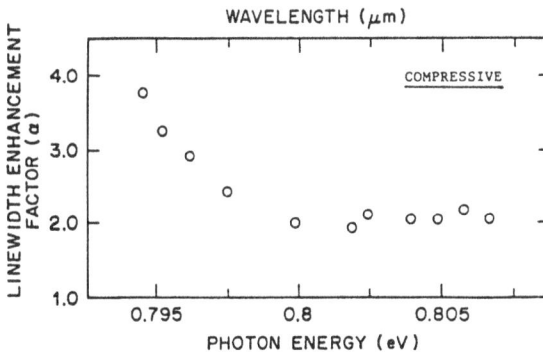

Figure 11.23 Measured α as a function of photon energy for a compressively strained $In_{0.66}Ga_{0.34}As/$ InP MQW laser.

REFERENCES

1.   R. Dingle, W. Wiegmann, and C.H. Henry, *Phys. Rev. Lett.*, Vol. 33, 1974, R. Dingle and C.H. Henry, U.S. Patent 3982207, September 21, 1976, p. 827.

2. N. Holonyak, Jr., R.M. Kolbas, R.D. Dupuis, and P.D. Dapkus, *IEEE J. Quantum Electron.,* Vol. QE-16, 1980, p. 170.

3. N. Holonyak, Jr., R.M. Kolbas, W.D. Laidig, B.A. Vojak, K. Hess, R.D. Dupuis, and P.D. Dapkus, *J. Appl. Phys.,* Vol. 51, 1980, p. 1328.

4. W.T. Tsang, *Appl. Phys. Lett.,* Vol. 39, 1981, p. 786.

5. W.T. Tsang, *IEEE J. Quantum Electron.,* Vol. QE-20, 1986, p. 1119.

6. S.D. Hersee, B. DeCremoux, and J.P. Duchemin, *Appl. Phys. Lett.,* Vol. 44, 1984, p. 476.

7. F. Capasso and G. Margaritondo, eds., *Heterojunction Band Discontinuation: Physics and Applications,* Amsterdam: North Holland, 1987.

8. N.K. Dutta, S.G. Napholtz, R. Yen, R.L. Brown, T.M. Shen, N.A. Olsson, and D.C. Craft, *Appl. Phys. Lett.,* Vol. 46, 1985, p. 19.

9. N.K. Dutta, S.G. Napholtz, R. Yen, T. Wessel, and N.A. Olsson, *Appl. Phys. Lett.,* Vol. 46, 1985, p. 1036.

10. N.K. Dutta, T. Wessel, N.A. Olsson, R.A. Logan, R. Yen, and P.J. Anthony, *Electron. Lett.,* Vol. 21, 1985, p. 571.

11. N.K. Dutta, T. Wessel, N.A. Olsson, R.A. Logan, and R. Yen, *Appl. Phys. Lett.,* Vol. 46, 1985, p. 525.

12. S.R. Forrest, P.H. Schmidt, R.B. Wilson, and M.L. Kaplan, *Appl. Phys. Lett.,* Vol. 45, 1984, p. 1199.

13. H. Temkin, M.B. Panish, P.M. Petroff, R.A. Hamm, J.M. Vandenberg, and S. Sumski, *Appl. Phys. Lett.,* Vol. 47, 1985, p. 394.

14. M.G.A. Bernard and G. Duraffourg, *Phys. Status Solidus,* Vol. 1, 1961, p. 699.

15. G.P. Agrawal and N.K. Dutta, *Long Wavelength Semiconductor Lasers,* New York: Van Nostrand, 1986.

16. Y. Arakawa and A. Yariv, *IEEE J. Quantum Electron.,* Vol. QE-21, 1985, p. 1666.

17. Y. Arakawa and A. Yariv, *IEEE J. Quantum Electron.,* Vol. QE-22, 1986, p. 1887.

18. A. Yariv, C. Lindsey, and V. Sivan, *J. Appl. Phys.,* Vol. 58, 1985, p. 3669.

19. A. Sugimura, *IEEE J. Quantum Electron.,* Vol. QE-20, 1984, p. 336.

20. A. Sugimura, *Appl. Phys. Lett.,* Vol. 43, 1983, p. 728.

21. N.K. Dutta and R.J. Nelson, *J. Appl. Phys.,* Vol. 53, 1982, p. 74.

22. R.J. Nelson and N.K. Dutta, Chapter 1 in *Semiconductors and Semimetals,* Vol. 22, Part C, ed., W.T. Tsang, New York: Academic Press, 1985.

23. L.C. Chiu and A. Yariv, *IEEE J. Quantum Electron.,* Vol. QE-18, 1982, p. 1406.

24. N.K. Dutta, *J. Appl. Phys.,* Vol. 54, 1983, p. 1236.

25. A. Sugimura, *IEEE J. Quantum Electron.,* Vol. QE-19, 1983, p. 923.

26. C. Smith, R.A. Abram, and M.G. Burt, *J. Phys. C,* Vol. 16, 1983, p. L171.

27. W.T. Tsang, *Appl. Phys. Lett.,* Vol. 39, 1981, p. 134.

28. W.T. Tsang, L. Yang, M.C. Wu, Y.K. Chen, and A.M. Sergent, *Electron. Lett. 2033,* 1990.

29. W.P. Dumke, *IEEE J. Quantum Electron.,* Vol. QE011, 1975, p. 400.

30. W. Streifer, D.R. Scifres, and R.D. Burnham, *Appl. Opt.,* Vol. 18, 1979, p. 3547.

31. H.C. Casey, Jr., and M.B. Panish, Chapter 3 of *Heterostructure Lasers, Part A,* New York: Academic Press, 1978.

32. Several index guided laser structures are described in Chap. 5 of Agrawa and Dutla [15].

33. H. Temkin, T. Tanbun-Ek, R.A. Logan, J.A. Lewis, and N.K. Dutta, *Appl. Phys. Lett.,* Vol. 56, 1990, p. 1222.

34. P.L. Derry, A. Yariv, K.Y. Lau, N. Bar-Chaim, and J. Rosenberg, *Appl. Phys. Lett.,* Vol. 50, 1987, p. 1773.

35. P.S. Zory, A.R. Reisinger, L.J. Marost, G. Costrini, C.A. Zmudzinski, M.A. Emanud, M.E. Givens, and J.J. Coleman, *Electron. Lett.,* Vol. 22, 1986, p. 475.

36. H. Temkin, N.K. Dutta, T. Tanbun-Ek, R.A. Logan, and A.M. Sergent, *Appl. Phys. Lett.,* Vol. 57, 1990, p. 1610.

37. M.G. Burt, *Electron. Lett.,* Vol. 20, 1984, p. 27.

38. Y. Arakawa, K. Vahala, and A. Yariv, *Appl. Phys. Lett.,* Vol. 45, 1984, p. 950.

39. C.A. Green, N.K. Dutta, and W. Watson, *Appl. Phys. Lett.,* Vol. 50, 1987, p. 1409.

40. T.L. Koch and U. Koren, *IEEE J. Lightwave Tech.,* Vol. 8, 1990, p. 274, and references therein.

41. A.R. Adams, *Electron. Lett.,* Vol. 22, 1986, p. 249.

42. S.E. Fischer, D. Fekete, G.B. Feak, and J.A. Ballantyne, *Appl. Phys. Lett.,* Vol. 50, 1987, p. 714.

43. K.Y. Lau, S. Xin, W.I. Wang, N. Bar-Chaim, and M. Mittelstein, *Appl. Phys. Lett.,* Vol. 55, 1989, p. 1173.

44. I. Suemune, L.A. Coldren, M. Yamanishi, and Y. Kan, *Appl. Phys. Lett.,* Vol. 53, 1988, p. 1378.

45. T. Ohtoshi and N. Chinone, *Photon. Technol. Lett.,* Vol. 1, 1989, p. 117.

46. E. Yablonovitch and E.O. Kane, *IEEE J. Lightwave Technol.,* Vol. LT-4, 1986, p. 504.

47. M.C. Wu, N.A. Olsson, D.L. Sivco, and A.Y. Cho, *Appl. Phys. Lett.,* Vol. 56, 1990, p. 221.

48. N.K. Dutta, J.D. Wynn, D.L. Sivco, and A.Y. Cho, *Appl. Phys. Lett.,* Vol. 56, 1990, p. 2293.

Chapter 12

Heterostructure Bipolar Transistors

B. Jalali, R.N. Nottenburg

AT&T Bell Laboratories

12.1 INTRODUCTION

Electronic devices and circuits fabricated from GaAs have carved out a small niche in applications that demand high speed, low noise, and cannot be satisfied by silicon technology. There has been significant success with GaAs metal semiconductor field effect transistors (MESFET) and high electron mobility transistor (HEMT); in particular, gate array circuits with a large integration level have been realized using GaAs MESFETs [1]. Heterostructure bipolar transistors (HBT) fabricated from III-V materials have also attracted a significant amount of attention [2,3]. These devices are uniquely suited for ultra-high-speed applications because of their large transconductance, inherent in bipolar transistors, and because of the excellent transport properties of III-V materials. With continuing advances in silicon technology, the performance gap between Si and GaAs is closing [4]. Therefore, interest has grown in InP-based electronics, where the superior transport properties of InGaAs can be exploited. This article presents an overview of InGaAs HBT technology for applications in high-speed electronics.

12.2 BASIC DEVICE CONCEPTS

The concept of HBTs, developed by Shockley and Kroemer [5], is based on the use of a wide-bandgap emitter to remove the tradeoff between current gain and base doping level in a bipolar transistor. Figure 12.1a shows the energy band diagram for

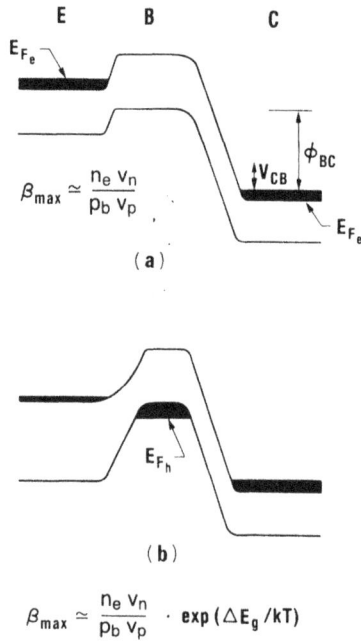

Figure 12.1 Energy band diagram for (a) homojunction (top), and (b) heterojunction (bottom) bipolar
transistor.

a homojunction transistor. In a homojunction NPN (negative-positive-negative) transistor with a sufficiently thin base, where recombination in the neutral base is negligible, the current gain is given by the ratio of electron current in the base to the
hole current in the emitter

$$\beta \approx n_e v_n / p_b v_p. \tag{12.1}$$

The quantities n_e and p_b are majority carrier densities in the emitter and base, v_n is
the electron velocity in the base, and v_p is the hole velocity in the emitter. To increase
the speed of bipolar transistor circuits, it is desirable to reduce the base resistance
by increasing the base doping concentration. However, this results in reduction in
current gain given by equation (12.1). This limitation may be removed by use of a
wide-bandgap emitter material as shown in Figure 12.1b. If we assume that the base
current is governed by hole diffusion in the emitter, similar to its behavior in a
homojunction transistor, then its magnitude will be reduced because of a smaller
equilibrium hole concentration in the wide-bandgap emitter. The resulting increase
in current gain is given by

$$\beta = \beta_o \exp(\Delta Eg/kT), \tag{12.2}$$

where β_o is the corresponding gain for the homojunction, ΔE_g is the bandgap difference, and kT is the thermal energy.

The analysis presented above is for a simplified case. The operation of an actual HBT may differ in several respects. For example, if the band discontinuity in Figure 12.1b is abrupt and sufficiently large, then the base current will be limited by thermionic emission of holes over the barrier, and the condition for thermal equilibrium of minority carrier density at the emitter side of the *p-n* junction does not hold. In general, thermionic emission will govern charge transport in any region where the potential barrier changes by kT over a distance that is less than the scattering mean-free path [6]. Another deviation from the simple picture in Figure 12.1b stems from the discontinuity in the conduction band. In this case (Figure 12.2), the electron current may also be limited by thermionic emission over the conduction band barrier rather than by diffusion in the base.

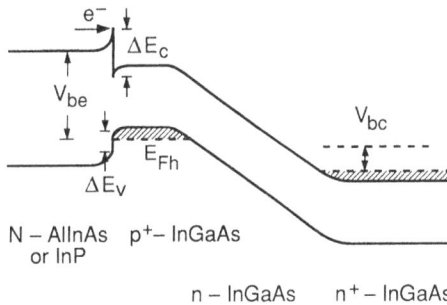

Figure 12.2 Band diagram of a HBT with abrupt emitter-base heterostructure.

12.3 AVAILABLE HETEROSTRUCTURE SYSTEMS

Compound semiconductor pairs such as AlGaAs/GaAs, InP/In$_{0.53}$Ga$_{0.47}$As, and Al$_{0.48}$In$_{0.52}$As/In$_{0.53}$Ga$_{0.47}$As are particularly attractive for HBT devices because of favorable band lineups and the ability to achieve high-quality heterostructures lattice matched to a semi-insulating substrate. In the above material systems, ΔE_v is large enough to permit the realization of HBTs with base doping levels well beyond $p_b = 1 \times 10^{19}$ cm^{-3} while maintaining a useful current gain for most applications.

The AlGaAs/GaAs material system is attractive for HBTs because high-quality heterostructures may be realized over a wide range of ternary compositions. This permits precise tailoring of band profiles, and transistors with impressive microwave performance were obtained in the 1980s [7,8]. In addition, small- to medium-scale digital, analog, and millimeter-wave integrated circuits with record performance have been achieved [9–12]. However, an area of major concern is that of power dissipation. The difficulty of laterally scaling these transistors, and the large bandgap of GaAs (E_g = 1.43 eV), has restricted integration level in practical high-speed digital circuits to a few hundred gates. Because of the large emitter-base turn-on voltage (approximately equal to the bandgap of GaAs), digital circuits require a supply voltage of 7V to 10V. In addition, submicrometer emitter dimensions are required to capture the total device bandwidth in a large-scale integrated circuit. For example, to achieve a power dissipation of 2 mW/gate for an ECL gate operating at 5V and using a transistor whose cutoff frequency, f_T, peaks at 50 kA/cm^{-2}, requires an emitter area of less than 1 μm^2.

Heterostructures of Al$_{0.48}$In$_{0.52}$As/In$_{0.53}$Ga$_{0.47}$As and InP/In$_{0.53}$Ga$_{0.47}$As are lattice matched to the InP substrates on which they are grown. Although both heterostructure systems benefit from the excellent physical properties of InGaAs, differences in energy band discontinuities (Table 12.1) have an impact on device properties. In the AlInAs/InGaAs heterostructure, most of the discontinuity occurs in the conduction band ($\Delta E_c/\Delta E_v$ = 67/33), whereas in the InP/InGaAs system it is mostly in the valence band ($\Delta E_c/\Delta E_v$ = 42/58) [14]. As will be seen later, the large ΔE_v and small ΔE_c in the InP/InGaAs system results in lower base resistance and lower power dissipation, respectively. Therefore, the InP/InGaAs is a more desirable material system for HBTs. However, it should be noted that growth of P-containing compounds requires gas-source molecular beam epitaxy (GSMBE) or metalorganic MBE (MOMBE) [15], whereas AlInAs/InGaAs devices can be grown with MBE using conventional elemental sources (ESMBE). Although both material systems can also be grown by metalorganic chemical vapor deposition (MOCVD) [16,17], better control over doping and compositional profiles is usually obtained by the MBE methods.

Heterostructure bipolar transistors using InGaAs have a number of advantages compared with GaAs devices. Because of the small bandgap of the InGaAs base

Table 12.1
Bandgap and Band Offset Values for Al$_{0.48}$In$_{0.52}$As/In$_{0.53}$Ga$_{0.47}$As and InP/In$_{0.53}$Ga$_{0.47}$As [14], Along with Al$_{0.30}$Ga$_{0.70}$As/GaAs [13] Heterostructure Systems

	Al$_{0.48}$In$_{0.52}$As/In$_{0.53}$Ga$_{0.47}$As	InP/In$_{0.53}$Ga$_{0.47}$As	Al$_{0.30}$Ga$_{0.70}$As/GaAs
E_{ge} (eV)	1.48	1.35	1.86
E_{gb} (eV)	0.76	0.76	1.43
ΔE_c (eV)	0.48	0.25	0.28
ΔE_v (eV)	0.24	0.34	0.15
$\Delta E_c/\Delta E_v$	67/33	42/58	65/35

region, these HBTs have a lower turn-on voltage than AlGaAs/GaAs HBTs. This means that ultra-high-speed integrated circuits that operate from a supply voltage of 5V or less may be readily realized. In addition, the large conduction band Γ-X and Γ-L intervalley separation in InGaAs permits nonequilibrium electron transport to be realized in practical HBTs. A low surface recombination velocity (10^3 cm/s), together with nonequilibrium electron base transport, allows devices to be scaled to submicrometer lateral dimensions without the use of passivating layers or a complicated process technology. The small InGaAs bandgap allows nonalloyed ohmic contacts to the emitter, base, and collector. Completely selective wet chemical etching between InGaAs and InP, as well as InGaAs and AlInAs, layers is possible. Dry etching using $C_2H_4/H_2/SF_6$ has been developed and looks promising for the fabrication of InGaAs HBTs [18]. While reactive ion etching (RIE) requires a subsequent removal of a damaged layer ($\sim$500Å) by wet etching, electron cyclotron resonance (ECR) plasma etching results in much reduced damage and looks even more promising.

12.4 CURRENT VOLTAGE CHARACTERISTICS

Figure 12.3 shows the layer structure of a typical InGaAs HBT. The $Al_{0.48}In_{0.52}As/In_{0.53}Ga_{0.47}As$ and $InP/In_{0.53}Ga_{0.47}As$ HBTs were grown lattice matched on semi-insulating InP substrates using ESMBE and GSMBE, respectively. The collector consisted of a 5000Å-thick heavily doped n^+-InGaAs contact layer and a 3000Å-thick InGaAs active region doped to $n = 1 \times 10^{16}$ cm^{-3}. The base region consisted of an InGaAs active region 500Å to 700Å thick and doped to $p = 1 \times 10^{19}$ to 5×10^{20} cm^{-3}. An undoped InGaAs spacer layer was incorporated between base and emitter. The thickness chosen, in the range 25Å to 100Å, depended on the growth temperature, Be concentration, and base thickness. The wide-bandgap emitter consisted of a 2000Å-thick InP or AlInAs layer doped to $N = 1 \times 10^{18}$ cm^{-3}, and a 1000Å-thick InGaAs contact layer doped to $n > 1 \times 10^{19}$ cm^{-3}.

n^+	$In_{.53}Ga_{.47}As$	1000Å
N	$Al_{.48}In_{.52}As$ or InP	2000Å
U	$In_{.53}Ga_{.47}As$	
p^+	$In_{.53}Ga_{.47}As$	500Å
n	$In_{.53}Ga_{.47}As$	3000Å
n^+	$In_{.53}Ga_{.47}As$	5000Å
	S.I. InP Substrate	

Figure 12.3 Layer structure of a typical InGaAs HBT.

Figure 12.4 shows the common-emitter output curves, $I_c - V_{CE}$, (a) for an AlInAs/InGaAs HBT, and (b) for an InP/InGaAs HBT. The emitter-base junction for both devices is abrupt and a spacer layer thickness of 100Å was used. In both transistors the observed offset voltage, which is due to the difference in turn-on voltages of the emitter-base junction ($E_{gb} + \Delta E_c$) and the collector-base junction (E_{gb}), is equal to the conduction band discontinuity, ΔE_c, at the emitter-base heterointerface, and agrees with values for ΔE_c given in Table 12.1. The output conductance, observed in Figure 12.4, is a result of avalanche multiplication and device heating, both of which are important in the small bandgap $In_{0.53}Ga_{0.47}As$ collector.

Figure 12.5 shows the collector current, I_c, and base current, I_b, as a function of base-emitter voltage, V_{be} (Gummel plot), for an InP/InGaAs HBT. For simulation purposes, the collector current can be modeled simply as

$$I_c = I_s \exp(qV_{be}/n_c kT - 1). \tag{12.3}$$

Although this expression is similar to the expected thermionic emission equation, the measured saturation current, I_s, is much larger than that calculated from a simple thermionic emission model. Furthermore, unlike an ideal thermionic emission process, the measured collector ideality factor, n_c is greater than unity, as shown in Figure 12.5 ($n_c = 1.3$). Our detailed study of collector current in these HBTs indicates that the high saturation current is a result of electron tunneling through the conduction band barrier from states above the Fermi energy of the emitter into the conduction band of the base. In these HBTs, a collector current ideality factor, $n_c > 1.0$, is due to modulation of the electron barrier height with applied base-emitter voltage. This occurs because a decrease of the notch on the base side of the junction (see Figure 12.2) results in an increase in barrier height with applied forward bias. The ideality factor thus depends on the base doping level near the junction and can be analytically expressed as

$$n_c = 1 + (\varepsilon_c n_e / \varepsilon_b p_b) \tag{12.4}$$

where ε_e and ε_b are the dielectric constants in the emitter and the base, respectively. This was first predicted by Ankri and Eastman [19].

The base current in a bipolar transistor consists of two components: recombination in the base, and back injection of holes into the emitter. In HBTs, hole injection into the emitter is negligible when the valence band is discontinuous, $\Delta E_v >> kT$. In this case, the base current is controlled by minority carrier recombination in the base and may be modeled as

$$I_b = I_s/\beta_f[\exp(V_{be}/n_c kT) - 1] + I_l[\exp(V_{be}/2kT) - 1] \tag{12.5}$$

where β_f is the forward current gain and I_l is a constant. The first term in equation

(a)

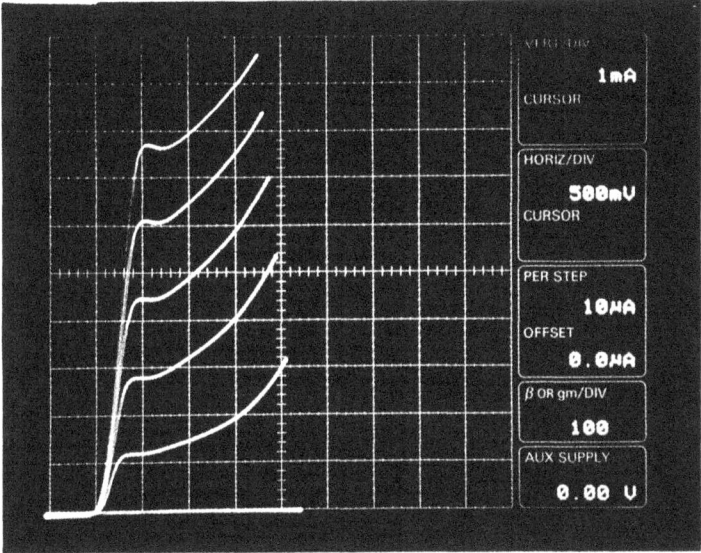

(b)

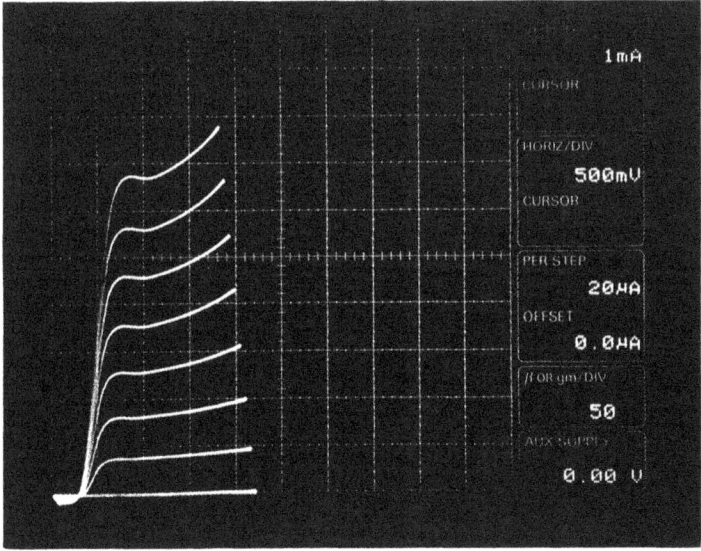

Figure 12.4 Output characteristics of a typical (a) AlInAs/InGaAs HBT and (b) InP/InGaAs HBT.

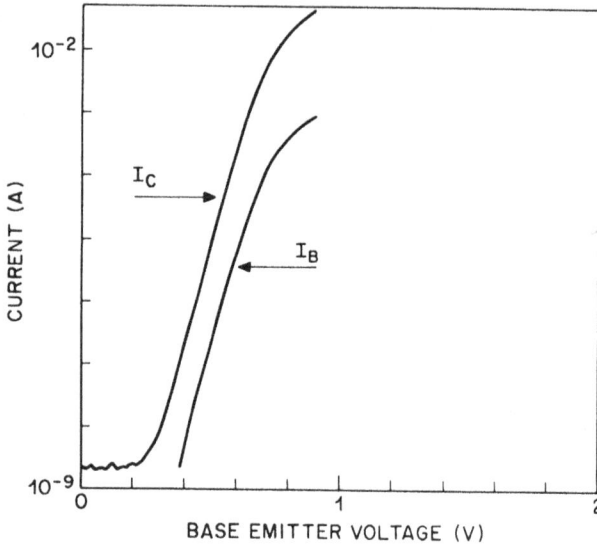

Figure 12.5 Collector and base currents as a function of base-emitter voltage.

(12.5) represents recombination in the neutral base region, while the second term describes the nonideal 2-kT recombination process in the intrinsic depletion layer and emitter periphery. Because of a high surface recombination velocity in GaAs (10^6 cm/s), the base current is dominated by the second term in equation (12.5) and has an ideality factor very close to two. The base current in InGaAs HBTs is mostly determined by the first term, and thus has an ideality factor close to that of the collector. The smaller nonideal component of base current in InGaAs HBTs is due to a smaller surface recombination velocity (10^3 cm/s) and confinement of injected electrons to the intrinsic base.

12.5 BASE TRANSPORT

The abrupt emitter-base junction, shown in Figure 12.2, injects electrons into the base with energies that are considerably higher than the ambient thermal energy, kT. Because of the high electron injection energy (given approximately by ΔE_c) at the AlInAs/InGaAs emitter-base junction, and conservation of the electron wave vector parallel to the hetero-interface, electrons are launched in a small angular cone perpendicular to the junction [20,21]. Since extrinsic base recombination depends on the lateral transport of carriers, a high perpendicular-to-parallel electron velocity ratio throughout the base results in better scaling. In bipolar transistors, where carriers are injected with the ambient thermal energy, the velocity ratio is small so that a small recombination velocity in the extrinsic base becomes a prerequisite for good lateral scaling. Therefore, an abrupt emitter with large injection energy can be used to improve lateral scaling in a HBT. Figure 12.6 shows the current gain β as a function of collector current for transistors with different emitter dimensions. The maximum current gain for the 0.5×12 μm^2 device was 122 at an emitter current density of 2.4×10^4 A/cm^2. The interesting feature is the high current gain at low current density and the small change in β with different emitter dimensions [21]. The current gain characteristics were the same for emitter stripe widths ranging from 3.5 to 50 μm.

Figure 12.7 shows the group velocity of an electron as a function of injection energy in the conduction band of the $In_{0.53}Ga_{0.47}As$ base in the $\langle 100 \rangle$ direction. The initial velocity of electrons in the base, with both AlInAs and InP emitters, is $\sim 10^8$ cm/s, and the electrons maintain a high average velocity over a distance of about 700Å. The base transit time in these structures is about 100 fs and is negligible compared to that of a transistor that employs equilibrium-diffusive transport. This reduces charge storage in the base. Charge storage is often the dominant delay in a conventional diffusive bipolar transistor.

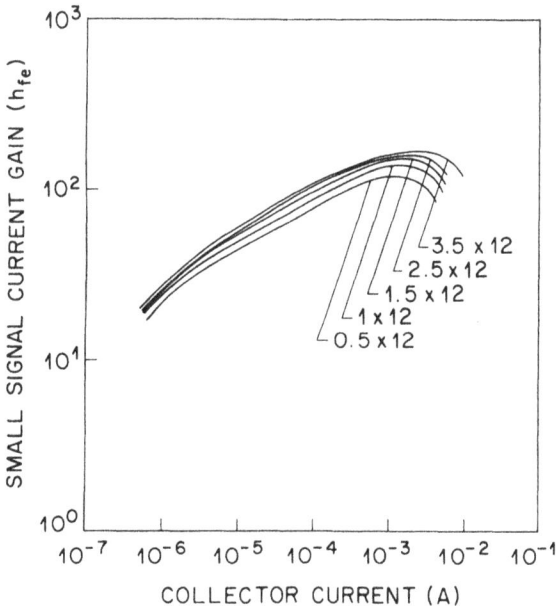

Figure 12.6 Common emitter current gain, β, versus the collector current for an AlInAs/InGaAs HBT with different emitter dimensions.

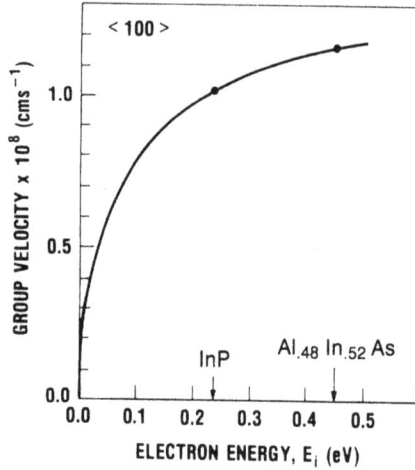

Figure 12.7 Group velocity of electrons as a function of their energy in the conduction band of $In_{0.53}Ga_{0.47}As$.

12.6 COLLECTOR TRANSPORT AND HIGH-FREQUENCY CHARACTERISTICS

The total forward delay time in a bipolar transistor is the sum of the parasitic charging times and the intrinsic transit times

$$\tau_{ec} = \tau_f + R_e C_e + (R_{ee} + R_e + R_c)C_c. \tag{12.6}$$

In the above relation, τ_{ec} is the total emitter-to-collector delay time and is related to the common emitter short circuit current gain cutoff frequency, $f_T = 1/2\pi\tau_{ec}$. The quantity $R_e = n_c kT/qI_e$ is the emitter dynamic resistance, R_{ee} is the emitter series resistance, R_c is the collector series resistance, and C_e and C_c are the emitter-base and collector-base depletion capacitances, respectively. The forward transit time, τ_f, is the sum of the base, τ_b, and the collector, τ_c, transit times. The power gain of the transistor can be expressed in terms of the real part of the input and output admittance and the current gain, and is characterized by the cutoff frequency f_{max}, where

$$f_{max} = (f_T/8\pi R_b C_c)^{1/2} \tag{12.7}$$

and R_b is the base resistance. The maximum frequency of oscillation, f_{max}, is defined as the frequency at which the unilateral power gain is reduced to unity in a device where the feedback is neutralized using a lossless network, and both input and output are conjugately matched [22].

One of the most useful equivalent circuit models of the transistor is the small signal hybrid-pi model. Figure 12.8 shows this circuit with physical device parameters. The delay associated with injection of the electrons (i.e., charging time of the base) is modeled as a "diffusion-like" capacitance, $C_d = g_m \cdot \tau_b$, which is added to the depletion layer capacitance of the emitter-base junction. After injection into the base, the carriers encounter delays due to base and collector transit processes, resulting in both an attenuation in amplitude and a shift in the phase of the output signal at the collector terminal. These effects are included in the current generator term in the equivalent circuit model of Figure 12.8. The frequency dependence of the current generator, shown in Figure 12.8, is the product of transfer functions for the base and the collector. In a conventional silicon bipolar transistor, the attenuation term associated with collector transport, $\text{Sinc}(\omega \tau_c)$, is often neglected, since the collector transit time is a small fraction of the total forward transit time.

As was mentioned above, in conventional diffusive transistors, τ_f is dominated by the base transit time. However, in abrupt InGaAs HBTs with nonequilibrium electron base transport, τ_b is negligible, and the collector transit time, τ_c, dominates. It is useful to understand the physics controlling the collector transit time of these vertically scaled transistors. Figure 12.9 shows the energy band structure of InGaAs. In III-V HBTs, τ_c is limited by electron-phonon scattering and phonon-assisted intervalley scattering. Because of the finite Γ-L and Γ-X intervalley separations, (Table

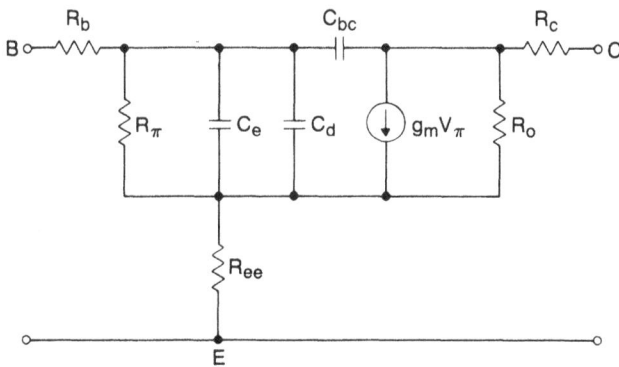

$$R_\pi = \frac{\beta}{g_{mo}}$$

$$g_m = g_{mo} \cdot \frac{\text{Sinc}(\omega \tau_c)}{1 + j\omega \tau_B} \cdot e^{-j\omega(\tau_B + \tau_c)}$$

$$C_d = g_{mo}\tau_B$$

$$g_{mo} = \frac{qI_c}{n_c kT}$$

$$\text{Sinc}(\omega \tau_c) = \frac{\text{Sin}(\omega \tau_c)}{\omega \tau_c}$$

Figure 12.8 An approximate equivalent circuit for the HBT.

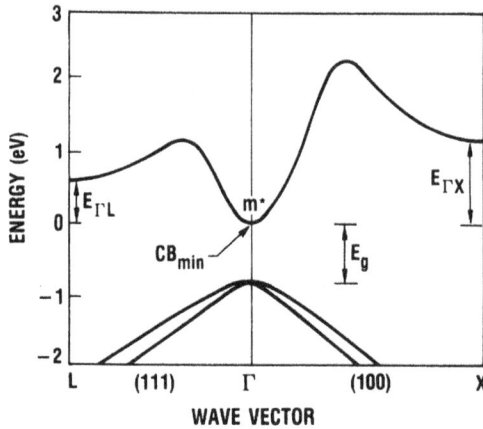

Figure 12.9 Energy band structure for InGaAs.

Table 12.2
Band Structure Parameters That Are Relevant for Charge Transport in the Ternary $In_{0.53}Ga_{0.47}As$ and Its Binary Parents GaAs and InAs

	GaAs	$In_{0.53}Ga_{0.47}As$	InAs
E_g (eV)	1.42	0.76	0.36
$E_{\Gamma L}$ (eV)	0.33	0.55	0.9
$E_{\Gamma X}$ (eV)	0.5	1.15	1.8
m^*/m_0	0.067	0.051	0.021

12.2), τ_c, and, consequently, f_T may vary with the applied collector-base voltage, V_{cb}. In AlGaAs/GaAs HBTs the cutoff frequency, f_T, decreases dramatically with increasing collector electric field and f_T actually peaks with $V_{cb} < 0$V. However, as shown in Figure 12.10, the cutoff frequency f_T for InGaAs HBTs is well over 100 Ghz for $-0.2 < V_{cb} < 1.5$V [23]. The different behavior for the InGaAs and GaAs HBTs reflects the relatively large Γ-L and Γ-X intervalley separations in $In_{0.53}Ga_{0.47}As$ and the small built-in electric field in the InGaAs p-n junction. The more ideal dependence of f_T on base-collector voltage in InGaAs HBTs is particularly useful in circuit applications, where a large output voltage swing is needed. The average electron velocity through the base and collector of these InGaAs HBTs, obtained from high-frequency measurements, is 4×10^7 cm/s. Because of coupling of base and collector transport, it is difficult to separate the individual components of the total

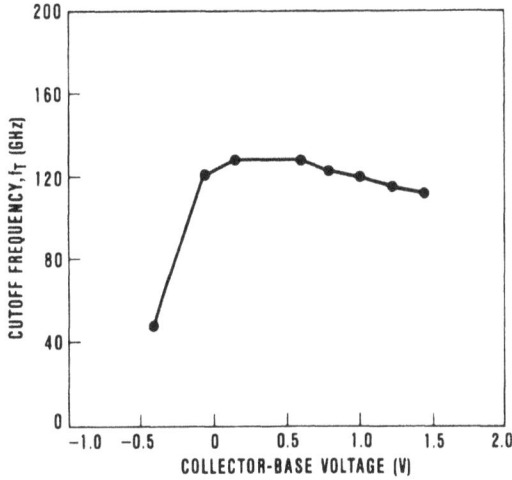

Figure 12.10 Cutoff frequency, f_T, versus the base collector voltage for an InP/InGaAs HBT.

transit time in HBTs, which employ extreme nonequilibrium electron base transport. Nevertheless, the small intrinsic delay, $\tau_f = 0.5$ ps, in such devices is an unambiguous demonstration of nonequilibrium electron transport [23]. Even in AlInAs/InGaAs HBTs with a base thickness of 1500Å, and collector thickness of 5000Å, an $f_T = 51$ Ghz corresponding to $\tau_{ec} = 3.1$ ps is measured, as shown in Figure 12.11. Assuming a maximum equilibrium velocity of 1×10^7 cm/s in both base and collector (the diffusion velocity in the base is actually much smaller than this value), the transit times in the base and collector are $\tau_b = X_b/v_e = 1.5$ ps and $\tau_c = X_c/2v_e = 2.5$ ps, respectively. This total intrinsic delay time of 4.0 ps is larger than the measured extrinsic (intrinsic plus parasitic) delay time of 3.1 ps. Parameter extraction from the measured S-parameters shows that the intrinsic part (τ_f) of the total delay is 1.7 ps, corresponding to an average electron velocity of 2.4×10^7 cm/s through 1500Å of base and 5000Å of collector. Therefore, even in these HBTs with relatively thick base and collector, nonequilibrium electron transport plays an important role.

The common emitter characteristics of vertically scaled InGaAs HBTs are characterized by high output conductance, as shown in Figure 12.4. This may affect the collector breakdown voltage and cause instability at high current density. One apparent reason is the small bandgap of $In_{0.53}Ga_{0.47}As$ ($E_g = 0.76$ eV), resulting in large base-collector junction reverse leakage current reminiscent of Ge ($E_g = 0.67$ eV) diodes. In addition, nonequilibrium electron base transport will influence collector breakdown voltage.

Figure 12.12 shows the measured multiplication constant in the collector versus base thickness for an AlInAs/InGaAs HBT [24]. The multiplication constant is defined as $\gamma = \Delta I_{av}/I_c$, and is independent of current gain, which also influences the

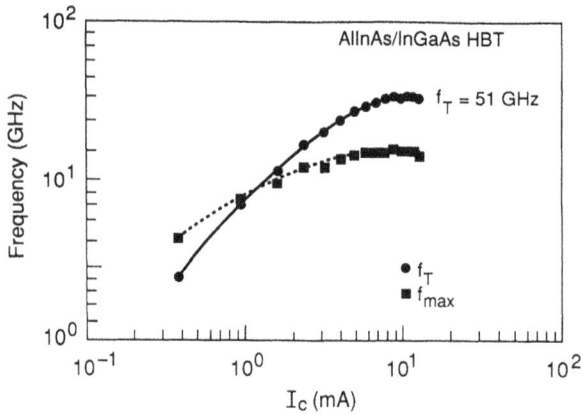

Figure 12.11 Cutoff frequency, f_T, as a function of collector current for an AlInAs/InGaAs HBT with 1500Å base and 5000Å collector.

common-emitter breakdown voltage. The measurements were performed in the common-emitter configuration. The common-base breakdown voltage was the same for all devices. When the base thickness was less than the mean-free path for energy relaxation in the base ($\sim$800Å), the avalanche processes in the collector were enhanced by high-energy injection from the emitter. On the other hand, no such dependence was observed for long base transistors with equilibrium base transport. These effects are expected as the emitter injection energy of 0.48 eV is appreciable compared to the impact ionization threshold of 0.83 eV in the $\langle 100 \rangle$ direction in the InGaAs collector. The multiplication process is shown in Figure 12.13 [24]. Because of high-energy injection from the emitter and nonequilibrium base transport, electrons arrive at the collector with excess energy and initiate impact ionization after acquiring additional kinetic energy from the electric field. In a conventional bipolar transistor, electron motion in the base is characterized by diffusive or drift-diffusive transport, and significantly more energy must be gained from the collector electric field to initiate impact ionization. The data in Figure 12.12 indicates that the mean-free path for energy relaxation in these devices is as long as 800Å. This value is in excellent agreement with independent experiments in which magnetic field was used to probe the electron transport in these devices [25].

The coupling of base and collector transports poses a tradeoff between the need for short base transit time and adequate output breakdown voltage, and must be considered in the design of practical HBTs with abrupt emitter-base junctions. Figure 12.14 shows a comparison of output characteristics for transistors with 200Å and 4000Å base widths. The high output conductance in ultra-thin base transistors sets a limit on minimum base thickness in practical devices.

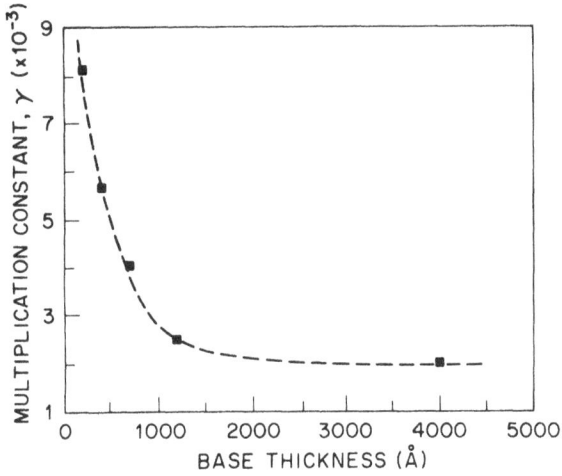

Figure 12.12 Multiplication constant, γ, versus base thickness for an AlInAs/InGaAs HBTs.

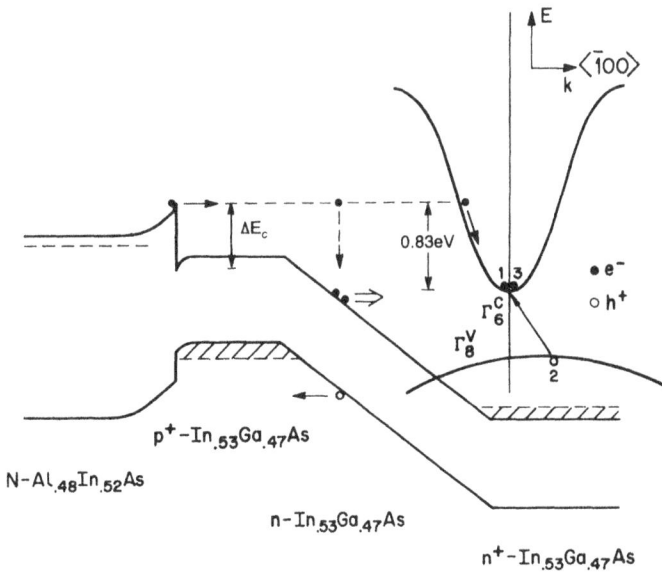

Figure 12.13 Energy band diagram of the AlInAs/InGaAs HBT showing the enhancement of avalanche in the collector by high energy injection from the emitter.

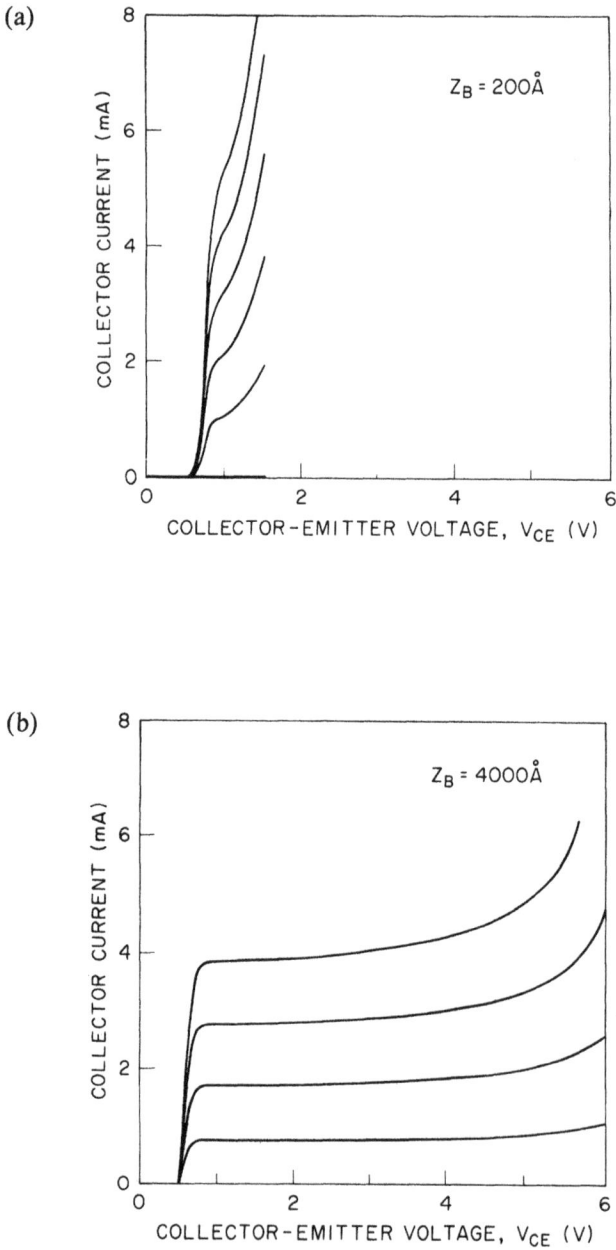

(a)

COLLECTOR CURRENT (mA)

$Z_B = 200 \mathring{A}$

COLLECTOR-EMITTER VOLTAGE, V_{CE} (V)

(b)

COLLECTOR CURRENT (mA)

$Z_B = 4000 \mathring{A}$

COLLECTOR-EMITTER VOLTAGE, V_{CE} (V)

Figure 12.14 Comparison of the output characteristics of abrupt AlInAs/InGaAs HBTs with base thicknesses of (a) 200Å and (b) 4000Å.

12.7 HIGH DOPING LIMIT IN THE BASE

The switching time of a bipolar transistor circuit is determined by the forward transit time, τ_f, and parasitic charging times. For example, the total delay time in an ECL gate, driven by a voltage input signal, may be approximated by [26]

$$\tau_{ECL} = \tau_f + C_c[R_L + R_b(1 + R_L/R_c) + R_c] + (C_d + C_e)R_b \quad (12.8)$$

where R_L is the load resistance. Because of the nonlinear nature of diffusion capacitance (C_d is proportional to the collector current), $(C_d + C_e)R_b$ may dominate the delay of an ECL gate, particularly at high collector current. When the last term in equation (12.8) is important, reduction of the total base resistance, R_b, is the only effective way to minimize τ_{ECL}. The base resistance may be reduced by lateral scaling of the device to small dimensions. However, the ultimate value of R_b may only be obtained by reducing the base sheet resistance. Because the majority carrier mobility decreases sublinearly with increasing acceptor concentration, it is always beneficial to reduce the base sheet resistance by increasing the p-type impurity concentration. The optimum device structure has a relatively thin, heavily doped base and maintains a useful current gain. For example, InGaAs/InP HBTs with current gain $\beta = 50$ and a base sheet resistance of 400 $\Omega/\square$ have been realized in a device structure with a base doping and a base thickness of $p_b = 1 \times 10^{20}$ cm^{-3} and $W_b = 500$Å, respectively [27].

The current gain in these and other HBT structures with high base doping is limited by band filling [14]. At high doping concentrations, the Fermi energy in the p^+ base moves into the valence band as a result of filling of available hole states. This effect is similar to the Burstein-Moss shift observed in semiconductor lasers [28]. In a HBT, band filling reduces the effective confinement barrier against hole back injection into the emitter. To estimate the magnitude of this effect, one may calculate the Fermi energy, E_f, shown in Figure 12.2, and compare that value with the valence band offset, ΔE_v, listed in Table 12.1. Assuming that the valence bands of InGaAs are isotropic around the zone center, $k = 0$, and treating the nonparabolicity in the light hole band with a $K \cdot P$ approach, the relation between hole concentration and Fermi energy can be written [14] as

$$p = 2(2\pi kT/h^2)^{3/2}[m_{hh}^{3/2}F_{1/2}(W) + m_{lh}^{3/2}[F_{1/2}(W) - (15\gamma kT/4E_g)F_{3/2}(W)] \quad (12.9)$$

where m_{hh} and m_{lh} are the $k = 0$ heavy hole and light hole masses, respectively, and $F_j(W)$ is the Fermi-Dirac integral for the reduced energy $W = (E - E_f)/kT$. The factor $\gamma = -(1 + E_g/2\Delta_{so})/(1 - E_g/2\chi)^2$ is the nonparabolicity parameter, where Δ_{so} is the spin-orbit splitting, and χ is the momentum matrix element corresponding to the interaction between different bands. It can be estimated from the band-edge effective mass of the light hole band ($m_{lh} = 0.051 \, m_o$).

Figure 12.15 shows the calculated room-temperature hole Fermi energy as a function of carrier concentration in the range $10^{18} < p < 10^{21}$ cm^{-3}. The effective density of states in the valence bands of InGaAs is $\sim 7 \times 10^{18}$ cm^{-3}. For hole concentrations above this value, the Fermi energy moves inside the valence band. At $p = 2 \times 10^{19}$ cm^{-3} and $p = 1 \times 10^{20}$ cm^{-3}, we obtain $E_{fh} = 40$ and 150 meV, respectively. At $p = 1 \times 10^{20}$ cm^{-3}, the effective valence band barrier of an AlInAs/InGaAs device is reduced to 90 meV. Because of a larger valence band offset, an InP/InGaAs HBT more effectively confines carriers to a heavily doped p-type base region, and a lower base sheet resistance may be realized in these structures. It should be noted that, in calculating the Fermi energy at very high carrier concentrations, it is necessary to include terms higher than fourth order in the energy dispersion relation, and thus the $K \cdot P$ approximation becomes less accurate. However, since the underlying physics, namely the restriction imposed on electron population by the exclusion principle, remains the same, the phase space filling will still be severe.

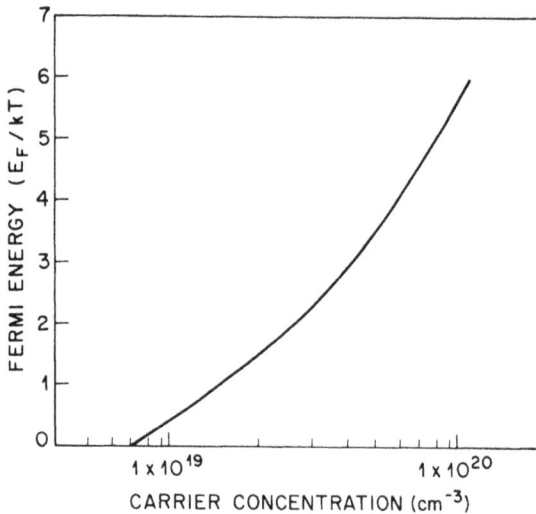

Figure 12.15 Fermi energy as a function of hole concentration in the base of an InGaAs HBT.

12.8 INTEGRATED CIRCUITS

The important issues in small-scale digital integrated circuits are speed and power. To obtain a small power-delay product in a high-speed circuit, it is desirable to employ a supply voltage of 5V or less. The minimum supply voltage required by a circuit depends on the turn-on voltage of individual transistors, internal headroom, and output voltage swing. Figure 12.16 shows a comparison of threshold character-istics of three well-known abrupt HBTs. The turn-on voltage of an abrupt HBT is $V_{be} \approx E_{gb} + \Delta E_c$ and is significantly lower in InGaAs HBTs than in GaAs HBTs because of the difference in bandgap (see Table 12.2). This difference is also apparent in HBTs with graded emitter-base junctions where $V_{be} \approx E_{gb}$. The larger turn-on voltage of AlInAs/InGaAs devices relative to InP/InGaAs transistors in Figure 12.16 is due to the 230-meV difference in the conduction band offsets (see Table 12.1).

Consider, for example, the impact of turn-on voltage on power dissipation for the commonly used current mode logic (CML) latch shown in Figure 12.17. The minimum voltage required for operation is

$$V_{ee} = 3V_{be} + 2V_s \tag{12.10}$$

where V_s is the voltage swing. Table 12.3 shows a comparison of power dissipation in a CML latch operating at a gate current of 5 mA and $V_s = 250$ mV. The smaller turn-on voltage in InGaAs HBTs results in circuits with less power dissipation compared to GaAs HBTs. This improvement is as much as 40% for the abrupt InP/InGaAs technology.

The choice for circuit architecture is dictated by device characteristics. For example, power dissipation, saturation voltage, breakdown voltage, and electric field dependence of the electron velocity in the collector make ECL logic an attractive choice for InGaAs digital circuits.

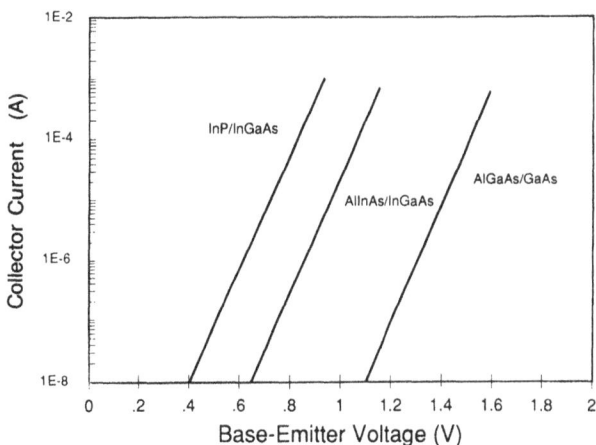

Figure 12.16 Comparison of the threshold voltage in different III-V HBT technologies.

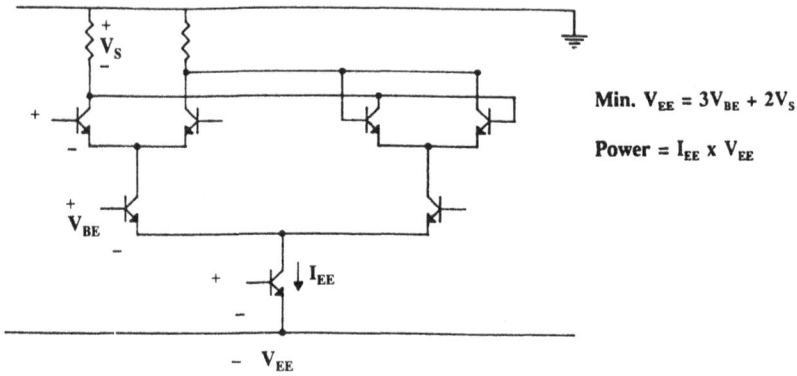

Figure 12.17 Circuit diagram of a commonly used CML latch.

Table 12.3
Comparison of Power Dissipation in a CML Latch for the Three Different HBT Technologies

	AlGaAs/GaAs	AlInAs/InGaAs	InP/InGaAs
V_{BE}	1.6V	1.1V	0.9V
V_{EE} ($V_s = 0.25$V)	5.3V	3.8V	3.2V
Power ($I_{EE} = 5$ mA)	26.5 mW	19 mW	16 mW

An important application for integrated circuits with high-bandwidth and low-power dissipation is in lightwave communication systems. In this section, some examples of such circuits realized with InGaAs HBTs are presented.

Figure 12.18a shows the circuit diagram of a master slave flip-flop used to demonstrate a decision circuit. The emitter followers have two principal functions: first, to prevent the transistors in the second differential pair from going into saturation, and second, to buffer the input capacitance of those transistors. Current mirrors are used to set precisely the operating current of the gate. The output circuit consists of an open collector driver with an independent current source. The chip photograph of the complete circuit using $Al_{0.48}In_{0.52}As/In_{0.47}Ga_{0.53}As$ HBTs is shown in Figure 12.18b. The circuit occupies an area of 0.85×1.2 mm^2. Transistors with emitter sizes 3×5 μm^2 and 3×11 μm^2 were used. The choice of size depended on their function inside the circuit. The high-speed input lines are terminated with on-chip 50Ω resistors. The circuit uses Ni/Cr resistors and a SiO$_2$ dielectric with two levels of interconnect metal. Figure 12.18c shows the eye diagram of the decision circuit with 10 Gb/s pseudorandom input signal. The circuit operates from a

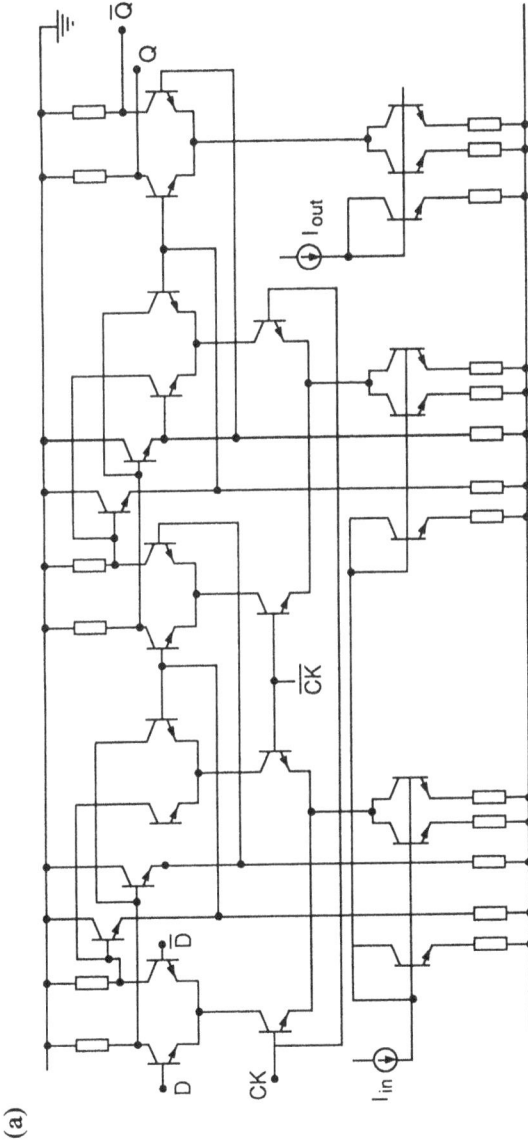

(a)

Figure 12.18 (a) circuit diagram, (b) chip photograph, and (c) single-ended output eye diagram at 10 Gb/s for the AlInAs/InGaAs D-flip flop.

(b)

(c)

30ps   ↕ 240mV

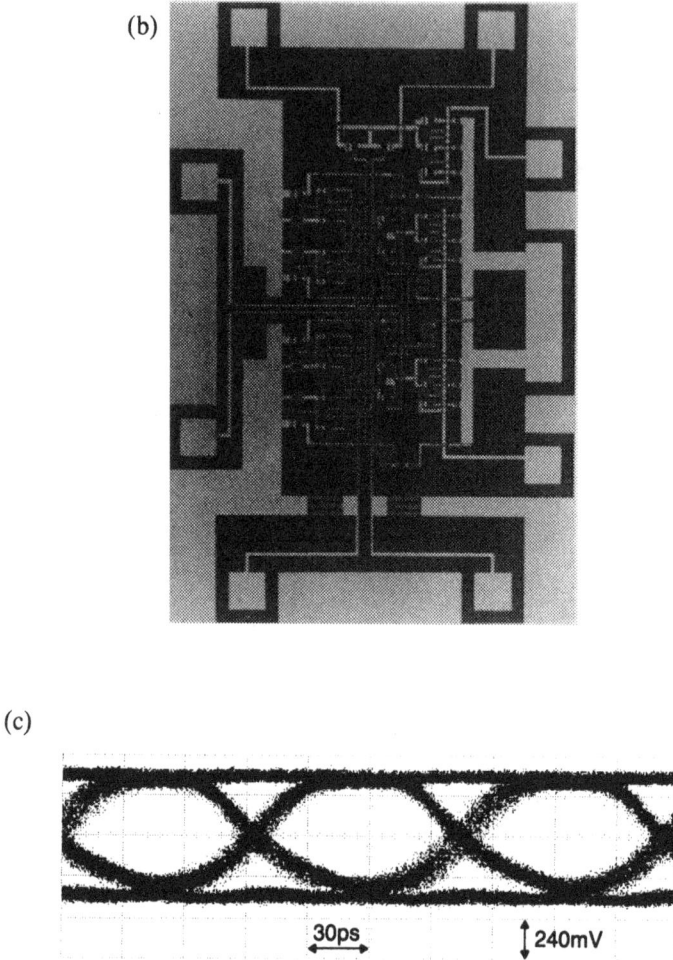

Figure 12.18 continued.

5V power supply voltage and has a differential output with a single-ended swing of 750 mV at 10 Gb/s.

Figure 12.19a shows the block diagram of a 2:1 bit interlever circuit using InP/InGaAs HBTs. This circuit employs emitter follower input buffers with resistor current sources and 50Ω termination. The interlever circuit consists of differential pairs and a current mirror. An open collector output driver allows compatibility between input and output signal levels. The circuit was tested with a 1100-input data pattern at 10 Gb/s and a 10-Ghz differential clock. Figure 12.19b shows the resulting

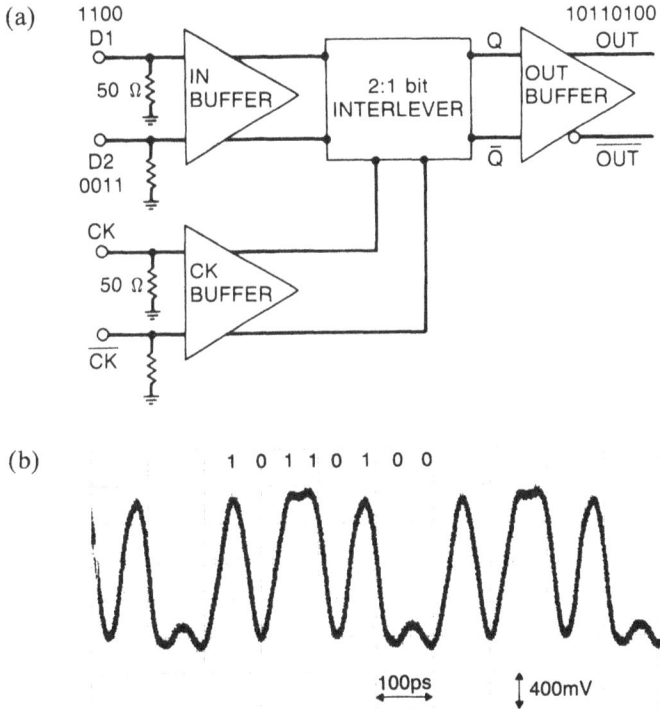

Figure 12.19 (a) Block diagram and (b) output data pattern at 20 Gb/s for the InP/InGaAs 2:1 bit interlever circuit.

10100101 differential output pattern at 20 Gb/s with 1.6V amplitude. This circuit also operates from 4V to 5V with a single supply voltage.

Another important high-speed circuit in a lightwave system is a laser driver. This circuit is particularly demanding on device technology since large currents must be switched at high speed. InGaAs HBTs are well suited for this application because of their wide bandwidth and high current capability, which result from a high average electron velocity. Figure 12.20a shows the block diagram of a laser driver designed in InP/InGaAs HBT technology [29]. The circuit consists of a differential current switch, an input buffer, and current mirror circuits. The current mirror circuits used matched devices and base current compensation techniques. To realize a modulation current of 100 mA, the current switch consisted of four HBTs with an emitter size of $2.5 \times 11 \ \mu m^2$. The large transconductance-per-unit capacitance, g_m/C, is a characteristic of these HBTs and is extremely important for such applications. The input buffer, designed to drive the 0.5 pf input capacitance, consisted of emitter followers

(a)

(b)

50 mV

50 ps

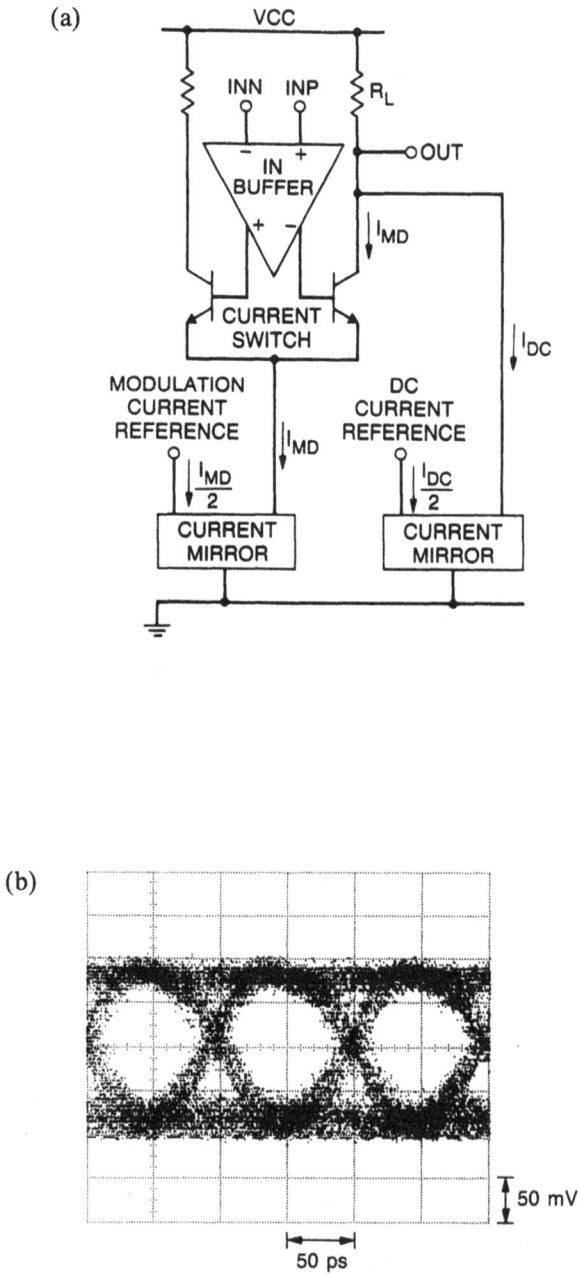

Figure 12.20 (a) Block diagram and (b) output eye diagram for the InP/InGaAs laser driver circuit.

biased with cascoded current sources. A chip load resistor, R_L, was used so that the testing could be done at wafer level.

Figure 12.20b shows the output eye diagram of the laser driver circuit at 10 Gb/s. The circuit modulates 100 mA of current and operates from a single 4V supply voltage. The rise and fall times of the output signal are 35 and 40 ps, respectively, and are limited by the measurement setup. Cascode amplifiers fabricated on the same wafer showed a DC gain of 17.3 dB and a -3-dB frequency of 12.2 GHz, corresponding to a gain-bandwidth product of over 90 GHz [30].

The circuits described above demonstrate ultra-high speed capability with supply voltages that are compatible with those used in standard silicon technology. This is a significant improvement over the conventional AlGaAs/GaAs HBT technology where similar circuits require a 7V to 10V supply voltage, resulting in significantly higher power dissipation and the obvious incompatibility with existing circuit standards. In addition to the circuits discussed above, which have been realized at AT&T Bell Laboratories, there are other examples from Rockwell International and Hughes Laboratories. Divide-by-four frequency dividers, which operate up to 17.1 GHz with a minimum power delay product of 66 fJ, have been fabricated using AlInAs/InGaAs technology [31]. These circuits, when operated at 15 GHz consume only one-fifth the power of the best AlGaAs/GaAs divide-by-four circuits at the same frequency. In addition, dual-modulus prescalers with a clock frequency of 9 GHz and 700 mV of power dissipation have been realized using AlInAs/InGaAs HBTs [32].

12.9 SUMMARY

In summary, we have seen that ultra-high speed bipolar transistors can be realized by using nonequilibrium electron injection from a wide-bandgap InP or AlInAs emitter into a thin, heavily doped InGaAs base. High velocity Γ-valley nonequilibrium electron transport throughout the InGaAs base and collector reduces the transistor intrinsic transit delay to less than 0.5 ps at room temperature. Because of the "one-dimensional" nature of electron transport in these HBTs, they may also be laterally scaled to submicrometer dimensions without effects due to the classical recombination in the extrinsic base region. Base doping levels up to $p = 1 \times 10^{20}$ cm^{-3} may be readily achieved in practical HBTs with current gains of 50 and a base sheet resistance of a few hundred $\Omega/\square$. Beyond a base doping level of 1×10^{20} cm^{-3}, the current gain in thin base structures is limited by band filling. The long mean-free path for nonequilibrium transport in the base results in the coupling of base and collector transport, and manifests itself in nonclassical events, such as the dependence of avalanche multiplication in the collector on base thickness.

Integrated circuits fabricated with AlInAs/InGaAs and InP/InGaAs show that both high-speed and low-power dissipation is possible. Typical digital and analog circuits can be designed to operate from silicon standard supply voltages of 5V or less. A simple cascode amplifier has a gain bandwidth product of more than 90 GHz and digital circuits operating beyond 20 Gbit/s have been realized. However, small integration levels restrict this technology to special niche applications. With modest advances in processing and crystal growth, other applications, such as high-speed analog/digital conversion, should be possible.

ACKNOWLEDGMENTS

The authors would like to acknowledge their collaborators, who have made significant contributions to this work: M. Banu, A.Y. Cho, R.A. Hamm, D.A. Humphrey, A.F.J. Levi, R.K. Montgomery, M.B. Panish, and D. Sivco. Special thanks to M.B. Panish for reviewing the manuscript.

REFERENCES

1. See for example: G. Lee, S. Canaga, B. Terrell, and I. Deyhimy, "A High Performance GaAs Gate Array Family," *GaAs IC Symposium Technical Digest*, San Diego, Calif., Oct. 22–25, 1989, pp. 33–36.

2. K.C. Wang, P.M. Asbeck, M.F. Chang, R.B. Nubling, R.L. Pierson, N.H. Sheng, G.J. Sullivan, J.Y. Chen, D. Clement, T.C. Tsen, H.F. Basit, J.D. George, and R. Young, "A 15GHz Gate Array Implemented with AlGaAs/GaAs Heterojunction Bipolar Transistors," *IEEE International Solid State Circuits Conference (ISSCC) Technical Digest*, San Francisco, Calif., Feb. 13–15, 1991, pp. 154–155.

3. H.T. Yuan, H.D. Shih, J. Delaney, and C. Fuller, "The Development of Heterojunction Integrated Injection Logic," *IEEE Trans. Electron. Devices*, Vol. ED-36, No. 10, Oct. 1989, pp. 2083–2092.

4. See for example: J. Warnock, J.D. Cressler, K.A. Jenkins, T.C. Chen, J.Y.C. Sun, and D.D. Tang, "50GHz Self-Aligned Silicon Bipolar Transistors with Ion-Implanted Base Profiles," *IEEE Elec. Dev. Lett.*, Vol. 11, No. 10, Oct. 1990, pp. 475–477.

5. W. Shockley, U.S. patent no. 2569347, and H. Kroemer, *Proc. IRE*, Vol. 45, 1957, p. 1535.

6. H.A. Bethe, "Theory of the Boundary Layer of Crystal Rectifiers," *MIT Rad. Lab. Rep.*, 1942, p. 43-12.

7. P.M. Asbeck, M.F. Chang, K.C. Wang, D.L. Miller, G.J. Sullivan, N.H. Sheng, E. Sovero, and J.A. Higgins, "Heterojunction Bipolar Transistors for Microwave and Millimeter-Wave Integrated Circuits," *IEEE Trans. Electron. Dev.*, Vol. ED-34, pp. 1987, 2571–2579.

8. O. Nakajima, K. Nagata, Y. Yamachi, H. Ito, and T. Ishibashi, "High Performance AlGaAs/GaAs HBT's Utilizing Proton-Implanted Buried Layers and Highly Doped Base Layers," *IEEE Trans. Electron Dev.*, Vol. ED-34, 1987, pp. 2393–2397.

9. Y. Yamauchi, O. Nakajima, K. Nagata, H. Ito, and T. Ishibashi, "A 34.8GHz 1/4 Static Frequency Divider Using AlGaAs/GaAs HBTs," *GaAs IC Symposium Technical Digest*, San Diego, Calif., Oct. 22–25, 1989, pp. 121–124.

10. R.B. Nubling, N.H. Sheng, K.C. Wang, M.F. Chang, W.J. Ho, G.J. Sullivan, C.W. Farley, and P.M. Asbeck, "25GHz HBT Frequency Dividers," *GaAs IC Symposium Technical Digest*, San Diego, Calif., Oct. 22–25, 1989, pp. 125–128.

11. K. Kobayashi, R. Esfandiari, A. Oki, D. Umemoto, J. Camou, and M. Kim, "GaAs Heterojunction Bipolar Transistor MMIC DC to 10GHz Direct-Coupled Feedback Amplifier," *GaAs IC Symposium Technical Digest*, San Diego, Calif., Oct. 22–25, 1989, pp. 87–91.

12. Bayraktaroglu, M. Khatibzadeh, and R. Hudgens, "Monolithic X-Band Heterojunction Bipolar Transistor Power Amplifiers," *GaAs IC Symposium Technical Digest*, San Diego, Calif., Oct. 22–25, 1989, pp. 271–275.

13. T.F. Kuech, D.J. Wolford, R. Potemski, J.A. Bradley, K.H. Kelleher, D. Yan, J. Paul Farrel, P.M.S. Lesser, and F.H. Pollak, "Dependence of the $Al_xGa_{1-x}As$ Band Edge on Alloy Composition Based on the Absolute Measurement of x," *Appl. Phys. Lett.*, Vol. 51, No. 7, Aug. 1987, pp. 505–507.

14. B. Jalali, R.N. Nottenburg, A.F.J. Levi, R.A. Hamm, M.B. Panish, D. Sivco, and A.Y. Cho, "Base Doping Limits in Heterostructure Bipolar Transistors," *Appl. Phys. Lett.*, Vol. 56, No. 15, April 1990, pp. 1460–1462.

15. M.B. Panish and H. Temkin, "Gas-Source Molecular Beam Epitaxy," *Ann. Rev. Mat. Sci.*, Vol. 19, 1989, pp. 209–229.

16. R.N. Nottenburg, Y.K. Chen, T. Tanbun-Ek, R.A. Logan, and D.A. Humphrey, "High Performance InP/InGaAs Heterostructure Bipolar Transistors Grown by Metalorganic Vapor Phase Epitaxy," *Appl. Phys. Lett.*, Vol. 55, No. 2, July 1989, pp. 171–172.

17. B. Jalali, R.N. Nottenburg, W.S. Hobson, Y.K. Chen, R. Fullowan, S.J. Pearton, and A.S. Jordan, "AlInAs.GaInAs Heterostructure Bipolar Transistors Grown by Metalorganic Chemical Vapour Deposition," *Electron. Lett.*, Vol. 25, No. 22, Oct. 1989, pp. 1496–1497.

18. See chapter by Pearton, and Chakrabarti on "Ion Beam Processing of InP and Related Materials" in this book.

19. D. Ankri and L.F. Eastman, "GaAlAs-GaAs Ballistic Heterojunction Bipolar Transistor," *Elec. Lett.*, Vol. 18, No. 17, August 1982, pp. 750–751.

20. R.N. Nottenburg, Y.K. Chen, M.B. Panish, R.A. Hamm, and D.A. Humphrey, "High Current Gain Submicrometer InP/InGaAs Heterostructure Bipolar Transistors," *IEEE Electron Dev. Lett.*, Vol. EDL-9, No. 10, Oct. 1988, pp. 524–526.

21. B. Jalali, R.N. Nottenburg, Y.K. Chen, M.B. Panish, R.A. Hamm, and D.A. Humphrey, "Near-Ideal Lateral Scaling in Abrupt AlInAs/InGaAs Heterostructure Bipolar Transistors Prepared by Molecular Beam Epitaxy," *Appl. Phys. Lett.*, Vol. 54, No. 23, June 1989, pp. 2333–2335.

22. H.F. Cook, "Microwave Transistors: Theory and Design," *Proc. of IEEE*, Vol. 59, No. 8, Aug. 1971, pp. 1163–1181.

23. R.N. Nottenburg, Y.K. Chen, M.B. Panish, D.A. Humphrey, and R.A. Hamm, "Hot-Electron InGaAs/InP Heterostructure Bipolar Transistors with f_T of 110GHz," *IEEE Electron Dev. Lett.*, Vol. EDL-10, No. 1, Jan. 1989, pp. 30–32; Y.K. Chen, R.N. Nottenburg, M.B. Panish, R.A. Hamm, and D.A. Humphrey, "Subpicosecond InP/InGaAs Heterostructure Bipolar Transistors," *IEEE Electron Dev. Lett.*, Vol. EDL-10, No. 6, June 1989, pp. 267–269.

24. B. Jalali, Y.K. Chen, R.N. Nottenburg, D. Sivco, D.A. Humphrey, and A.Y. Cho, "Influence of Base Thickness on Collector Breakdown in Abrupt AlInAs/InGaAs Heterostructure Bipolar Transistors," *IEEE Electron Dev. Lett.*, Vol. EDL-11, No. 9, Sept. 1990, pp. 400–402.

25. R.N. Nottenburg, A.F.J. Levi, B. Jalali, D. Sivco, D.A. Humphrey, and A.Y. Cho, "Nonequilibrium Electron Transport in Heterostructure Bipolar Transistors Probed by Magnetic Field," *Appl. Phys. Lett.*, Vol. 56, No. 26, June 1990, pp. 2660–2662.

26. K.G. Ashar, "The Method of Estimating Delay in Switching Circuits and the Figure of Merit of a Switching Transistor," *IEEE Trans. Electron Devices*, 1964, p. 497.

27. R.A. Hamm, M.B. Panish, R.N. Nottenburg, Y.K. Chen, and D.A. Humphrey, "Ultrahigh Be Doping of GaInAs by Low-Temperature Molecular Beam Epitaxy," *Appl. Phys. Lett.*, Vol. 54, No. 25, June 1989, pp. 2586–2588.

28. E. Burstein, "Anomalous Optical Absorption Limit in InSb," *Phys. Rev.*, Vol. 93, No. 3, Feb. 1954, pp. 632–633; and T.S. Moss, *Proc. Phys. Soc.*, Vol. B76, London, 1954, p. 775.

29.   M. Banu, B. Jalali, R.N. Nottenburg, D.A. Humphrey, R.K. Montgomery, R.A. Hamm, and M.B. Panish, "10 Gbit/s Bipolar Laser Driver," *Elec. Lett.*, Vol. 27, No. 3, Jan. 1991, pp. 278–280.

30.   R.N. Nottenburg, M. Banu, B. Jalali, D.A. Humphrey, R.K. Montgomery, R.A. Hamm, and M.B. Panish, *Elec. Lett.*, Vol. 26, No. 24, Nov. 1990, pp. 2016–2018.

31.   C.W. Farley, M.F. Chang, P.M. Asbeck, K.C. Wang, N.H. Sheng, R. Pierson, and G.J. Sullivan, "A High-Speed, Low-Power Divide-by-4 Frequency Divider Implemented with AlInAs/InGaAs HBTs," *IEEE Electron Dev. Lett.*, Vol. 10, No. 8, Aug. 1989, pp. 377–379.

32.   J.F. Jensen, W.E. Stanchina, R.A. Metzger, D.B. Rensch, Y.K. Allen, M.W. Pierce, and T.V. Kargodorian, "High Speed Duel Modulus Dividers Using AlInAs-GaInAs HBT Technology," *GaAs IC Symposium Technical Digest*, New Orleans, La., Oct. 7–10, 1990, pp. 41–44.

Chapter 13
Optimization and Fabrication of Metal Contacts for Photovoltaic Solar Cells

T.A. Gessert and T.J. Coutts
Solar Energy Research Institute

13.1 INTRODUCTION

A photovoltaic (PV) solar cell exploits the PV effect and delivers electrical power to an external load when illuminated by a suitable light source. Since sunlight is a relatively diffuse source of energy at the surface of the earth, the electrical power output per unit area of a PV device is quite small; thus, large areas of solar cells are normally required to generate significant amounts of power. At present, the maximum reported conversion efficiency for one-sun,* Si-based solar cells is slightly over 24% [1]. These devices rely on very sophisticated metallization involving laser grooving and other techniques derived from the microelectronics industry. Although Si-based solar cells are becoming increasingly cost-effective for many power generation applications, the most efficient solar cells (although not the most commonly used), with efficiencies presently approaching 35% under optical concentration, are those based on the III-V binary, ternary, and quaternary alloys involving GaAs and InP [2,3]. Because these compound semiconductor materials are expensive, when compared with Si, the relative cost of generating the energy economically becomes an even more subtle issue. Indeed, in order to achieve cost-effectiveness, these compound semiconductor solar cells must operate at very high optical concentration, thereby further increasing the demands on the metallization.

---

*A "one-sun" solar cell is operated without optical concentration and thus has an incident energy flux (at the surface of the earth) standardized to 1000 W/m^2. A "concentrator" solar cell operates under optical concentration and can therefore experience an incident energy flux several orders of magnitude higher than a one-sun cell.

One way of distinguishing between different types of PV devices is to consider how much energy-converting area is used in each individual application. Typical small-area applications include wrist watches and pocket calculators, while large applications include traditional grid-connected power production, with some installations producing as much as 6.5 MW with an array size of 160 acres.** The cost of energy produced by PV is presently about four times that of conventionally produced grid-environmental concerns associated with the burning of fossil fuels, have given PV solar cells a future market of considerable potential growth. Indeed, few competing energy-producing technologies can match many of the long-term and/or life-cycle benefits of PV. For these reasons, it is believed that the slow, yet steady increase of PV use will continue for many years.

Regardless of how large or small the application, all PV installations can still be reduced to the basic PV element, most commonly called the "cell." One of the main differences between a cell and larger PV units (panels, modules, arrays, *et cetera*) is that the electrical contacts are an inseparable part of the cell. Thus, unlike other aspects of the completed PV installation, such as those providing electrical interconnection between the cells, the metallization used to form the electrical contacts to the cell must be studied along with the junction during the research and design stages. Indeed, unless these metallization issues are correctly addressed, the cell performance may not be a clear indication of junction quality. In this chapter, the issues necessary to produce PV solar cells with minimum power losses associated with metallization are identified and discussed. Both front (grid) and back metal contacts are considered, emphasizing, where possible, the way in which process-related aspects can affect the metallization and thus the power loss. Although the discussion of PV solar cell contacts will be general, solar cells based on the particular material InP will be used as an example. This material is of importance for application in space because of its radiation hardness; thus, it has been of major interest to the space industry for several years [4]. Additionally, if the cost of these devices could be decreased, perhaps through techniques such as heteroepitaxial growth on cheaper substrates such as Si, these InP-based solar cells could also have a considerable number of terrestrial applications.

For both one-sun and concentrator solar cell designs, inherent junction and optical losses limit achievable efficiency. However, a significant part of the available power is consumed by series resistance losses associated with the electrical contacts to the cell. These losses not only limit the amount of collectable power available from a typical one-sun cell, but they also determine the maximum intensity of concentrated sunlight that a "concentrator cell" can effectively use. For the III-V materials mentioned, achieving the relatively low values of the series resistance required

---

**Carrissa Plain Power Plant, San Luis Obispo County, CA. Completed in 1984. Owned and operated by ARCO Solar Power Production with a 30-year agreement to sell the power produced to Pacific Gas & Electric.

for concentrator operation is by no means a simple matter, particularly when considering that the grid line contacts must have many of the qualities required of very-large-scale integration (VLSI) contacts (e.g., reliability, ease of manufacture, durability, high-temperature stability, *et cetera*). It must also be noted that, unlike VLSI structures, these contacts may be exposed to degrading environmental conditions, such as moisture, temperature extremes, particulates, and ultraviolet irradiation. Although the required values of the series resistance parameters have been achieved for some materials under laboratory conditions (e.g., the contact resistance for heavily doped GaAs), this is not the case for InP since there are additional restrictions on the thermal treatments that may be used.

To investigate solar cell contacts adequately, many separate disciplines must be assembled. Although perhaps foremost of these will be an understanding of the general physics of metal-semiconductor interfaces and the associated resistance loss mechanism, it is also necessary to become familiar with the related processing and measurement aspects of metal deposition and patterning. These include an understanding of the parameters and suitability of the type of deposition used, techniques for measurement of contact and bulk resistance, and ways to assess and improve the adhesion of metal to semiconductor surfaces. Often, the contact will be subjected to temperature variations due to concentration or environmental effects. For example, solar cells operating on space satellites are subject to particularly severe thermal stresses, with temperatures varying from that of liquid nitrogen to ~90° C. Additionally, processes often considered to be only indirectly related to metallization must be understood (such as those related to photolithography and computer grid modeling). Finally, although often considered an entirely different discipline, an understanding of relevant chemical and physical etching processes is often essential to determine whether a particular metallization process will be compatible with the etching processes necessary to isolate the active cell structures and/or pattern metallizations.

The purpose of this chapter is to discuss the requirements necessary for making good-quality contacts to the previously mentioned III-V-based solar cell devices, and to discuss how these metal-semiconductor contacts are formed for laboratory cell use. Also discussed are some practical guidelines for the effective modeling, design, and fabrication of solar cell collector grids, focusing primarily on the determination of realistic parameters necessary for effective grid modeling. Additionally, the causes and effects of variance in these parameters are presented, illustrating how unexpected variations often lead to sources of resistive losses in an otherwise well-designed solar cell grid. Finally, it should be noted that many of the suggestions and considerations that will be proposed to improve PV solar cell metallizations are not necessarily original. Indeed, many of the individual guidelines and processes were developed long ago to support what is now a mature VLSI industry. However, because funding for material studies in the PV research environment has typically preceded that for device studies, the benefits of many now-standard practices of VLSI metallization have not been incorporated (or often, even attempted) in PV solar cell research. This is true even though, as will be shown, the operating criteria for several types of PV devices can be very similar to those encountered in VLSI structures. For these reasons, the following chapter is structured to assist the solar cell designer in recognizing critical metallization issues and identifying some situations where useful VLSI processing practices may already exist. With this in mind, it should not be inferred that the processes described here are necessarily consistent with state-of-the-art VLSI practices. Rather, they are meant to illustrate processes (and related problems) that may improve PV solar cell metallizations (and therefore device performance) in a typical research environment.

13.2 THE PHYSICS OF SOLAR CELL GRID MODELING

13.2.1 Photovoltaic Solar Cell Equations

The current-voltage characteristics of an ideal solar cell junction (i.e., one in which all resistance terms have been neglected) are described by the basic diode equation with a light generation component (See, for example, Green [5]):

$$J = J_o[\exp(eV/nkT) - 1] - J_g \tag{13.1}$$

where J_o is the reverse saturation current density, J_g is the minority-carrier, light-generated current density, n is the diode ideality factor, k is Boltzmann's constant, and T is the temperature in kelvins. For a good cell, it is reasonable to put J_g equal to the measured short-circuit current density (J_{sc}). In this case, the open-circuit voltage (V_{oc}) of the device is

$$V_{oc} = nkT/e \ln[(J_{sc}/J_o) + 1]. \tag{13.2}$$

On the current-voltage characteristic, there is a point at which the power ($V \cdot J \cdot$ Area) that may be supplied to an external load reaches a maximum. The coordinates of this point are known as the maximum power point current density and voltage (J_{mpp} and V_{mpp}). The output power at this point, divided by the product of V_{oc} and J_{sc}, is the fill factor (FF) of the solar cell and, for typical good-quality cells, is equal to ~80% to 85%. Since the power generated by a PV solar cell will be dependent not only upon the total amount of input power but also on the specific spectral content, for cell calibration purposes, the incident illumination (for terrestrial applications) is standardized to both 1000 W/m² and a specific spectral content [6]. The final conversion efficiency of the cell is equal to the product of the three junction parameters (V_{oc}, J_{sc}, and FF) divided by the standardized incident energy.

In the presence of a significant series resistance (R_s), the V_{mpp} measured across the external terminals of the cell is reduced as illustrated in Figure 13.1. This voltage

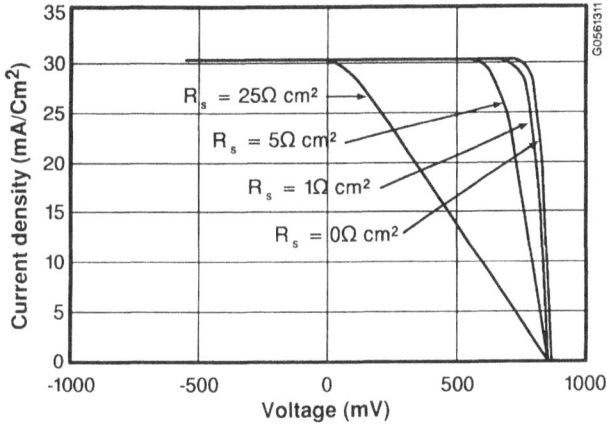

Figure 13.1 Diagram illustrating the effect of increasing series resistance on the illuminated current-voltage characteristics of a PV solar cell. Note that in this diagram, the typical convention of inverting the J-V characteristics from the fourth quadrant into the first quadrant has been used.

reduction can be qualitatively represented by modifying the ideal diode equation (Eq. 13.1), resulting in

$$J = J_o\{\exp[e(V + JR_s)/nkT] - 1\} - J_g \tag{13.3}$$

13.2.2 Top Contact Grid Modeling Equations

The methodology of optimizing the design of solar cell grids is well established (see, for example, Green [5]), but it is necessary to indicate the approach used here to establish the relevant points. As in all optimizations, effective grid design involves balancing conflicting criteria. Specifically, the grid contact of a cell simultaneously increases the cell efficiency by reducing R_s; however, the presence of the grid shadows the junction, thereby decreasing the input energy. To optimize the grid design, standard formalisms for the component power losses have been established and a computer program has been developed. This computer program is similar to those developed by others but includes some additional capabilities [7,8]. The first of these accounts for the effect of current crowding, which, although usually associated with the fine line conductors and contacts of VLSI structures, is also relevant to the design of certain types of solar cells [9]. The second involves the capability to calculate the total and component power losses of nonoptimum designs—a highly useful tool when assessing the effects on nonideal metallizations and variations in the design parameters. The program assumes that no current flows directly from the emitter to the bus-bar (i.e., one-dimensional current flow; see Fig. 13.2) and is currently configured for use with rectangular cells. Following is the mathematical formalism used to compute the total fractional power losses ($\Delta P/P$) for a grid with a single center bus-bar, with each of the individual component power loss terms indicated.

$$\Delta P/P = J_{mpp}SR_FB^2/(TW_FV_{mpp}) \qquad \text{(Power loss in grid fingers)} \qquad (13.4)$$

$$+J_{mpp}BR_BA^2/(TW_BV_{mpp}) \qquad \text{(Power loss in bus-bar)} \qquad (13.5)$$

$$+ J_{mpp}S[r_cR_E]^{1/2} \coth[W_F/L_T]/(2\ V_{mpp}) \qquad \text{(Power loss due to} \qquad \text{contact resistance)} \qquad (13.6)$$

$$+ J_{mpp}S^2R_E/(12\ V_{mpp}) \qquad \text{(Power loss due to} \qquad \text{emitter sheet resistance)} \qquad (13.7)$$

$$+ W_F/S + W_B/B \qquad \text{(Power loss due to finger} \qquad \text{and bus-bar shadowing)} \qquad (13.8)$$

The value of the parameters in Table 13.1 must first be simultaneously determined or assigned before a solar cell grid can be modeled. However, often accurate values of some of these basic parameters are unknown and measurement techniques for determining them must first be developed and tested. Since this is a long process, initially only a few of the parameters can be known with certainty and, although the subsequent design will only be as good as the parameters used to model it, the unknown parameters must initially be estimated. Nevertheless, the process of grid modeling should begin, since the results of a well-considered modeling study will not

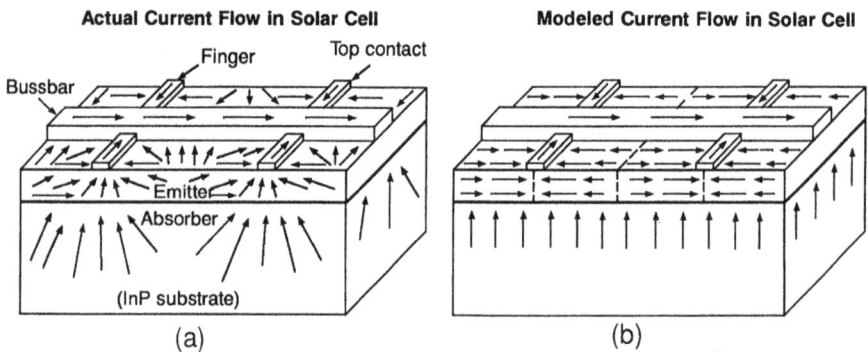

Figure 13.2 Schematic of the simple rectangular grid used for the modeling study illustrating the (a) actual and (b) modeled current flow of a PV solar cell. Source: After G.C. DeSalvo, and A.M. Barnett, "An Optimized Top Contact for Solar Cell Concentrators," *Proc. 18th IEEE Photovoltaic Specialists Conference*, New York: IEEE, 1985, pp. 435–440.

Table 13.1

List of Assigned and Calculated Parameters Used in Solar Cell Grid Modeling

V_{mpp}	Voltage at maximum power point (mV)
J_{mpp}	Active-area current density at maximum power point (mA/cm^2)
r_c	Specific contact resistance between metal and emitter ($\Omega \cdot$ cm^2) [10]
R_E	Sheet resistance of the emitter ($\Omega/\square$)
R_B	Sheet resistance of the bus-bar ($\Omega/\square$)
R_F	Sheet resistance of the grid fingers ($\Omega/\square$)
W_F	Width of the grid fingers (cm)
A	Length of the cell parallel to bus-bar (cm)
B	Half-width of the cell parallel to fingers (cm)
T	Taper factor $=3$ ($=4$ if fingers [bus-bar] are tapered)
L_T	Calculated transfer length, $L_T = (r_c/R_s)^{1/2}$ (cm) [9].
W_B	Calculated half-width of the bus-bar (cm)
S	Calculated spacings between the finger centers (cm)

only give insight into the probable areas of dominant cell power loss but also indicate which of the parameter measurement techniques should be developed first. Of the ten parameters listed in Table 13.1, there can normally only be control of the first seven, and two of these are often controlled by the same process-related aspects (R_B and R_F). This leaves six parameters to be determined or assigned: (a) V_{mpp}, (b) J_{mpp} of the photo-active area (i.e., the total solar cell area minus that obscured by metal-lization), (c) R_E, (d) r_c, (e) R_F and R_B, and (f) W_F (which can be processed with no electrical open circuits and with sufficient adhesion to withstand subsequent measurements and field testing). The above six parameters are presented in the order of the most difficult to the easiest to modify; (a) and (b) are essentially impossible to modify since they are governed by the materials of which the solar cell is fabricated, (c) is possible but often difficult, and (d), (e), and (f) are relatively easy to modify and/or demonstrate the greatest degree of process-related variability. For this reason, the grid modeling addressed here centers on variation in parameters (d), (e), and (f), with some mention of the benefits of varying parameter (c).

One final aspect of these procedures relates to the fact that, in a laboratory environment, the material, electrical and process parameters are often continuously changing. Thus, a grid will seldom be perfectly optimum for the solar cell it is being used with. For this reason, the design of the grid should take into account reasonable variation in the grid-related parameters. This is normally done by considering not a single, but a range of parameters, thereby taking into account not only the uncertainty in an "estimated" parameter(s) but revealing what effect future changes in the material/junction quality may have on the power losses of the metallization. Although this guideline may seem obvious, it again implies the requirement of interaction between metallization-related and material/junction-related activities.

13.2.3 Concentrator Solar Cells

Operation under optical concentration will likely be essential for the more expensive materials, such as those related to GaAs and InP. Because the underlying reason for concentration is to acquire an economic advantage by increasing the device efficiency (through increases in V_{oc} and FF), the cost of the additional equipment (concentrating optics, frames, alignment mechanics and electronics, *et cetera*) must be balanced against the benefit of increased efficiency and additional power output due to the greater illumination. Under these constraints, a concentrator cell is normally redesigned as follows: Since there will be a considerable amount of nonilluminated area beneath the optics (99.9% at 1000 suns concentration), the active area of the concentrator cell includes only that area which is actually illuminated by the equipment. Thus, the bus-bar is normally widened and moved outside the illumination area; the associated shadow and resistance losses are removed from the actual solar cell and, thus, from the calculation of efficiency. The final differences between these cells are that the concentrator cell normally has a smaller area (than a typical one-sun device) and the emitter is much thicker (to decrease emitter resistance losses, as will be discussed). To reflect these differences, the concentrator solar cell used here for modeling purposes is assigned a total area of 0.5 cm^2 (compared with the area assigned to the model one-sun cell of 4 cm^2) and an R_E of 100 $\Omega/\square$ (see Table 13.2).

In presenting concentrator cell data, it must be noted that, since the J_{sc} increases linearly with concentration, both the V_{oc} and the FF will also (to a point) increase, thereby causing the measured efficiency to increase. It is important to note that this efficiency increase is due to V_{oc} and FF increases, and not increases in J_{sc} (which is

Table 13.2
List of Typical InP Shallow Homojunction Parameters Used in Modeling

R_E	600 $\Omega/\square$ (100 Ω/sq for concentrator cells)
r_c	10^{-2}, 10^{-3} or 10^{-4} $\Omega \cdot$ cm^2
J_o	9×10^{-13} mA/cm^2
J_{mpp} (1 sun)	30 mA/cm^2
V_{mpp} (1 sun)	800 mV
Cell area	4 cm^2 (0.5 cm^2 for concentrator cells)
Minimum grid linewidth	10 μm
Grid line resistivity	2×10^{-6} (pure Au)–2×10^{-5} $\Omega \cdot$ cm
Concentration ratios	1–1000 suns

normalized for each concentration ratio). Because of this increasing junction efficiency with concentration, the effects of resistance on a cell is more clearly viewed when both the calculated efficiency of the model cell and the efficiency of an ideal junction (i.e., those characteristics which neglect resistance losses) are viewed together as a function of concentration.

13.3 SOLAR CELL METALLIZATION AND MODELING STUDIES

Although developing and understanding a methodology of resistance loss is a critical first step in optimizing solar cell metallization, one quickly discovers that the seemingly simple metallizations involved in even rectangular grid modeling are really not as straightforward to produce as the mathematics may indicate. In the next sections, several reasons for this will be discussed. These reasons not only include the need for the contacts to be reliable and durable, but also involve issues stemming from the environmental conditions in which the cells must operate. Additionally, although not discussed in detail here, solar cells are very large compared to individual VLSI structures, and thus the cost of metallization could become an important issue. Therefore, the choice of metallization technique and material may not only be a matter of identifying the process which leads to the best metallization parameters (resistivity, adhesion, diffusion, oxidation properties, *et cetera*), but also one of identifying and comparing the process availability and/or cost.

Since one is often either not sure of the exact parameters to use or is unsure of what degree of parameter variation is expected, the following modeling studies will consider a range of input parameters. This procedure not only indicates to the designer which parameters will be dominant in the grid loss, but also how suspected variation may affect the cell performance. Additionally, although multiple-junction solar cells are quickly becoming more commonly researched, for purposes of this discussion of solar cell metallization, only single-junction cells will be modeled. This single-junction cell will have two electrical contacts. Generally, one of these contacts covers the entire surface of the cell facing away from the light source and is therefore called the "back" contact; the other is on the illuminated side of the cell in the form of grids and is known as the "top" or "grid" contact.

For both one-sun and concentrator cells, increasing series resistance decreases the measured FF of the cell. This effect is easily visualized (see Fig. 13.1) and therefore promotes an intuitive understanding of the effect of R_s on the I-V characteristics of the cell. However, in actual grid optimization, increases in J_{sc} (from decreased active area metallization) tend to offset some FF loss. Therefore, it is better to represent grid performance by the parameter of total (or component) power loss rather than FF variation. It is in this fashion that the modeling results for one-sun grids will be presented. As a general guideline, a well-designed one-sun grid employing typical metallizations will have a total power loss of 5% to 10%.

13.3.1 Contact Resistance of the Back and Front Contact

The r_c of the metal-to-semiconductor interface is one of the most overlooked yet critical parameters of the grid design and modeling process. This is probably because the accurate measurement of r_c is difficult, often requiring specialized test structures and techniques. Additionally, if measurements yield a tolerably low value of r_c, often the highly process-dependent nature of r_c is not appreciated and, thus, power loss due to r_c variation is not suspected nor allowed for in the modeling. For these reasons, significant power loss due to r_c (or r_c variation as a cell is exposed to environmental conditions) is often considered improbable until it can be proven to be a severe deterrent to the performance of a cell. An example of this is that of presently fabricated contacts to CdTe solar cells [11].

Although it is generally believed that r_c is primarily only a function of the choice of metal and semiconductor (material and doping), experience will often show that the ideal metal-semiconductor values of r_c are seldom achieved with normal laboratory procedures. This is because, in addition to the specific metal and semiconductor chosen, many other process-related aspects are involved. Relative to the metal, these aspects include its purity, porosity, adhesion, and the interdiffusion of the metal layers. Relative to the semiconductor, they include the extent of surface cleaning prior to metal deposition, the type of deposition used (e.g., vacuum evaporation, sputtering, and electrolytic plating) and any intentional or unintentional thermal treatments experienced by the interface. Indeed, it has been observed in our own work that relatively slight changes in any one of these process parameters can cause a change in r_c of several orders of magnitude.

As mentioned previously, an accurate determination of the r_c is generally one of the more difficult of the measurement techniques to develop and qualify. Within the grid modeling studies presented here, several methods have been used, including the methods of Keramidas [12] and Cox and Strack [13]. Often, however, it is more interesting to describe the variation of the specific contact resistance as a function of some other parameter, such as predeposition cleaning procedures or postdeposition heat treatments, rather than the absolute value of r_c. In these cases of trend analysis, a simpler r_c measurement technique is often desirable, such as the four-bar method [14]. Although this method lacks the precision of the previously mentioned methods, studies have shown that the four-bar method will consistently give about an order of magnitude higher measure of r_c compared with that obtained using the Cox and Strack method.

Although fabricating a solar cell back contact is often considered to be relatively simple compared to that of fabricating the top contact, since solar cells are minority carrier devices, this is not always the case. For example, in Si solar cells, high resistivity (low doping, long diffusion length) substrate material is often employed. Thus, the back contact must be formed very carefully to avoid excessive heat treatments which may decrease the minority carrier lifetime, yet still provide a metal-semiconductor contact with low r_c. For compound semiconductor solar cells, the situation is similar; however, even lower-temperature contact formation is often necessary. For example, in our own work, both indium tin oxide (ITO)/InP and InP shallow homojunction cells require the formation of a back contact to p-type InP substrates. Generally, unless heat treatments are used, all metals form rectifying barriers when deposited onto p-type InP. This has been found to be true for the range of carrier concentration from 10^{16} cm^{-3} to $\sim 3 \times 10^{18}$ cm^{-3} and is probably a result of strong pinning of the surface Fermi level at about 0.4 eV below the conduction band minimum. This causes considerable band bending, resulting in the formation of a rectifying barrier that is too wide to permit tunneling. This can be contrasted with n-type InP where, for most metals, an ohmic contact forms if the carrier concentration is greater than about 5×10^{17} cm^{-3}. Shown in Figure 13.3 is the temperature-dependent r_c of several different metallizations used for back contacts on p-type InP. Here, the four-bar method had been used primarily to determine the temperature at which the contact will become ohmic. The demands of the back contact resistance become even more severe if the cell is to be used under concentration. For example, for a one-sun cell operating with a J_{mpp} of 30 mA/cm^2, a back r_c of 3×10^{-2} $\Omega \cdot$ cm^2 will result in a voltage loss (at V_{mpp}) of about 1 mV, which would be completely

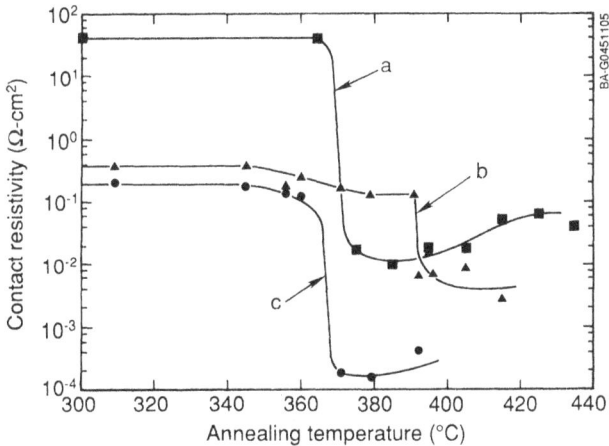

Figure 13.3 Contact resistance as a function of annealing temperature (in 10% H$_2$:90% N$_2$ forming gas) for metallizations used for the back contact of p-type InP of different doping levels. (a) Vacuum evaporated AuBe (1 wt. % Be) contact on p^-InP(1.6 $\times$ 10^{16}/cm^3[Zn]). (b) Plated Au on p^+ InP(1 to 4 $\times$ 10^{18}/cm^3[Zn]). (c) Plated Au/Zn/Au on same InP as in (b). Note that even for the p^+ material, significant temperature is required before the contact becomes ohmic.

acceptable for most applications. However, if the cell were to be used at a concentration of 500 suns, the corresponding voltage drop would be 450 mV, totally dominating the expected voltage gain (through the logarithmic dependence on J_{sc}; see (Eq. 13.2)) of 190 mV.

Since the contact area of the top grid contact is much smaller, even at one sun, a contact resistance of $3 \times 10^{-2}\ \Omega \cdot cm^2$ will dominate the grid losses for an otherwise optimum grid. Additionally, unlike the back contact, which experiences relatively uniform interfacial current density and thus the power loss is dependent only on r_c, the front grid contact collects current predominantly from lateral current flow, resulting in highly nonuniform interfacial current density. To account for this, the power loss associated with contact resistance of the grid contact depends on two (related) process aspects: (1) the value of r_c and (2) the grid linewidth (see Eq. (13.6) and Berger [9]). Figure 13.4 illustrates how two different values of r_c will affect the total power loss of the model one-sun solar cell. As can be seen, for linewidths greater than ~30 μm, it is often believed that an r_c of $1 \times 10^{-2}\ \Omega \cdot cm^2$ is sufficient for one-sun grids. However, for the narrower linewidths modeled here, this value of r_c will cause the contact resistance to be the dominant loss mechanism, and a substantial loss reduction could be achieved if r_c could be reduced to $1 \times 10^{-4}\ \Omega \cdot cm^2$. Also shown in Figure 13.4 is the effect that current crowding has on the calculation of power loss. Current crowding appears when the width of the grid finger (W_F) becomes less than about twice the calculated transfer length (L_T). Physically, if W_F is shorter than the natural length required by the current to flow into the contact, current crowding results with associated power loss proportional to $\coth(W_F / L_T)$. In

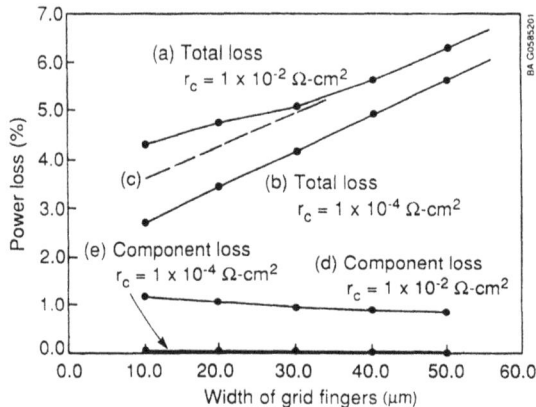

Figure 13.4 Effect of contact resistance for a model one-sun cell. Note that for thin grid lines and $r_c \geq 1 \times 10^{-2}\ \Omega \cdot cm^2$, contact resistance will dominate the total power loss. Shown in (c) is the effect of neglecting current crowding.

Figure 13.5 are shown similar calculations, but for a model concentrator cell operated at 1000 suns. Note here that, unlike the case of the one-sun design, an r_c of even 1 × 10^{-4} $\Omega \cdot cm^2$ will still have a significant (although not dominant) effect on the power loss. For this case, it is seen that r_c values approaching those usually required in VLSI would be advantageous. Finally, in Figure 13.6 is shown what effect increasing r_c from 1 × 10^{-4} $\Omega \cdot cm^2$ to 1 × 10^{-3} $\Omega \cdot cm^2$ would have on the cell as a function of increasing concentration. Here it is seen that even a slight nonuniformity or instability in r_c (as might be caused by progressive adhesion loss or interfacial oxide growth) would have a considerable effect on not only the efficiency but also the maximum concentration ratio.

In developing methods to decrease r_c one must note that r_c will likely be interrelated with other process parameters, which must be considered during a successful modeling study. For example, since the emitter is often very thin, excessive heat treatment of the contacts may cause junction shorting and must therefore be avoided. Indeed, our own work has demonstrated that annealing with a simple strip heater, at relatively low temperature (~400°C) is a sensitive process that requires considerable temperature uniformity and control [15]. Plasma treatments of n-type materials have been shown to achieve notably low r_c through the creation of a heavily doped n^+ surface [16] and improved adhesion through (predeposition) removal of surface contamination. Although, if carefully controlled, plasma processes can be beneficial to a solar cell [17], they may also adversely affect the junction by incorporating plasma species into the junction region. In light of these constraints, non-annealed yet temperature-stable contacts are often desirable. Examples of this include

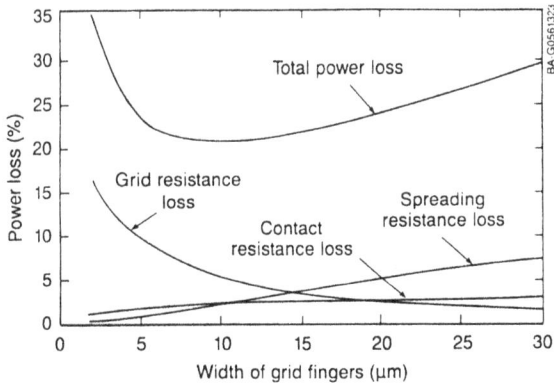

Figure 13.5 Comparison of r_c loss with other resistance losses for the model concentrator cell operated at 1000 suns as a function of finger width, where finger width/height = 2. Note that the spreading resistance loss is due to R_E, and that shadowing losses are not shown.

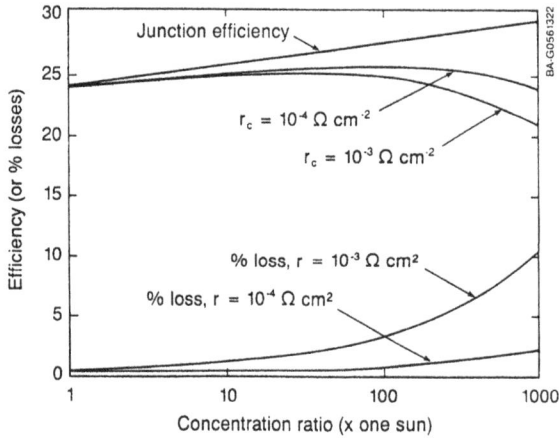

Figure 13.6 Comparison of $r_c = 10^{-3}$ *versus* $r_c = 10^{-4}$ $\Omega \cdot cm^2$ as a function of concentration for the model concentrator cell. Note that efficiency gains would be expected if r_c were reduced below 10^{-4} $\Omega \cdot cm^2$.

the use of plated, nonannealed Au onto bulk n^+-InP; measurements have demonstrated r_c values of $\sim 10^{-3}$ to 10^{-4} $\Omega \cdot cm^2$. Additionally, recent work has incorporated similar InP material and involved a combination of low-power Ar plasma surface cleaning and Cr/Pd/Ag deposition. With this metallization, results indicate that specific contact resistance values of less than 10^{-5} $\Omega \cdot cm^2$ and a level of adhesion necessary for even narrow (~ 6 μm) linewidths can be routinely achieved. Additionally, devices with thicker emitters are currently being investigated and these have the advantage of allowing higher temperature processing without junction shorting. Finally, although it has not yet been attempted in the present context, rapid thermal annealing processing may be of considerable benefit [18].

13.3.2 Grid Metal Resistivity

The metal grid lines not only transfer useful energy from the junction region but are also perhaps one of the most characteristic aspects of a PV solar cell. Nevertheless, subtle, yet important aspects of the metal are often overlooked. The parameter of metal sheet resistance of the grid lines is an area of solar cell grid modeling and design technology where the causes and effects of normal process variations are often not fully appreciated and, therefore, not accounted for in the grid design. Although the value of the metal resistance stems primarily from the choice of metal, like specific contact resistance it tends to vary greatly with the grid line fabrication process.

The selection of a metallization process involves not only acquiring the best resultant electrical and mechanical parameters for the solar cell device, but often must take into account issues of availability and process cost. For these reasons, relatively cost-effective metallizations such as metal plating, screen printing, or solder dipping must often be considered in lieu of more elegant methods such as vacuum evaporation and sputtering. Although a grid with acceptable performance can be designed from almost any metallization process, this can only be done if process-related variations in the metallization properties, such as resistivity, adhesion, and minimum linewidth, are accounted for in the modeling. Often, it is helpful to employ a measurement that specifically tests these process-dependent parameters. A typical test structure for this is shown in Figure 13.7. When a metallization is formed using this pattern, it is possible to measure the resistivity as a function of linewidth, measure the linewidth as a function of process and, often indirectly, measure the adhesion as a function of linewidth. Finally, if the linelength on the test pattern is representative of the linelength that will actually be used on a cell, any length-dependent and/or stress-related issues of the metal can also be identified.

Another area of concern when defining metal resistivity is that the grid modeling program assumes that the grid lines are perfectly rectangular and homogeneous, as shown in Figure 13.8(a). However, experience reveals that this is seldom the case. Figure 13.8(b-i) illustrate that the metal cross-section can vary greatly in geometry and degree of resistance homogeneity, depending on process-related aspects. As will be discussed in the following sections, these cross-sectional variations are often due to the particular photolithography and/or metal deposition technique used. Since it is difficult to predetermine how these material and/or process aspects will combine to affect the sheet resistance (and the degree of obscuration, as will be discussed), the use of a test pattern becomes invaluable. One method to define the measured resistivity is to use the average metal thickness (as measured, for example, with a stylus profilometer) and the average width of line obscuration (as measured, for example, by transmission optical or IR microscopy). This definition allows the modeling program to incorporate all of the previously mentioned aspects of cross-section and homogeneity accurately, while essentially ignoring the less practical, yet normally quoted, resistivity of a microscopic region within the metal line. Once this measured value of the effective metal resistivity is determined, it is often helpful to indicate the difference between the ideal resistance value (i.e., that determined from the pure, bulk value assuming perfectly rectangular grid lines) and the measured value (as defined above) as the resistance ratio (RR), where RR = measured sheet resistance/ideal sheet resistance.

Even for a well-established metallization processes with reliable edge definition, there are many mechanisms which can greatly affect the actual sheet resistance. Consider, for example, the process of high-purity plated Au contacts. In this case, it takes less than 1% of Fe contamination to increase the RR to 10 [18]. Increases in resistivity also occur when Au is contaminated by other metals (such as Co, Sn, Ni, Sn, and In), although to a lesser extent. Additionally, our own research has shown that the plating current has a substantial effect on both the porosity of the Au

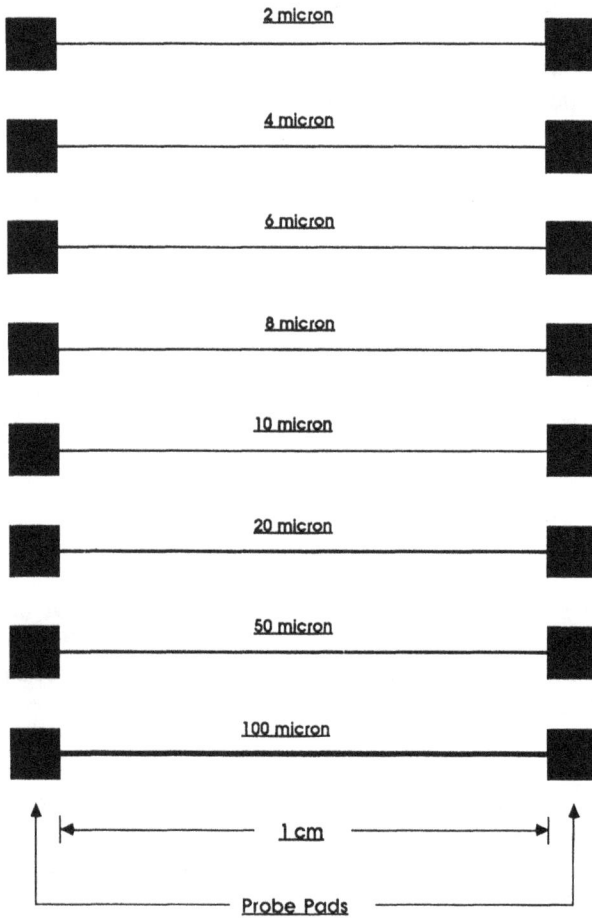

Figure 13.7 Metallization pattern used to measure metal resistivity necessary to calculate the resistance ratio between the actual and ideal resistance of a grid line of known width and thickness.

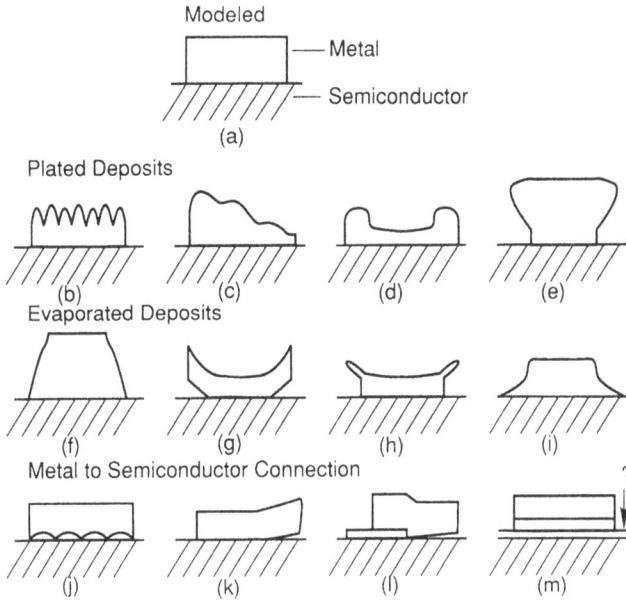

Figure 13.8 (a) Rectangular grid-line cross-section used in modeling program. (b-i) Grid-line cross-sections possible from different deposition techniques. Note that, due to edge effects, many of these geometries may obscure a significant amount of incident light but still not provide sufficient conduction cross-section. For this reason, observed line broadening often does not significantly reduce grid-line resistance losses. (j-m) Illustration of typical adhesion problems. (j) Spot adhesion due to poor surface cleaning. (k) Poor adhesion due to metallic stress. (l) Poorly-aligned, low-adhesion conduction layer on adhesion-promoting layer. (m) Well-aligned conduction and adhesion layer on insulating surface oxide or other unknown contamination.

deposit and the measured resistivity (see Figs. 13.9 and 13.10). In particular, standard bus-bar patterns have been electroplated onto ITO films (100 nm thick) to a thickness of 1 to 4 μm, and current-voltage measurements performed, suggesting that the RR = ~2 to 8. The higher values of the RR are associated with higher current densities and a darker deposit appearance due to surface roughness, resulting in a deposit similar to that illustrated in Figure 13.8(b). Furthermore, the greater porosity of the deposit in Figure 13.9(a) could enable oxidizing substances to enter the metal, causing the grid resistivity (and perhaps r_c) to be unstable. Finally, for electrolytically plated contacts, the plating rate (and thus the resistivity) may be highly dependent on the direction of current flow, yielding higher plating rates in regions where the electric field is the strongest (see Figs. 13.8(c, d)). On actual cells where the current density has been carefully controlled to produce a uniform, bright Au-colored deposit, measurements indicate that RR = ~2 to 3. As will be shown, this variation is sufficient to necessitate consideration in the grid design.

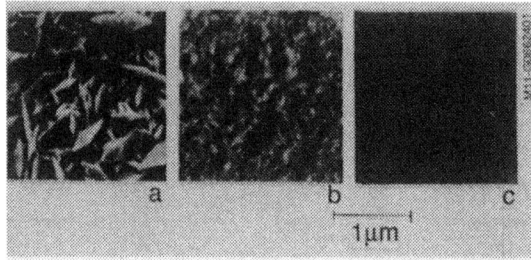

Figure 13.9 (a, b) SEM micrographs of Au plated onto ITO. (a) 5-μm Au plated at high current density (~0.5 μm/min), (b) 1-μm Au plated at low current (~0.1 μm/min), (c) 120-nm evaporated Au. Deposit (a) had a resistivity ratio ~6 to 8 times that of pure gold and (b) had a resistivity ratio of ~2 to 3.

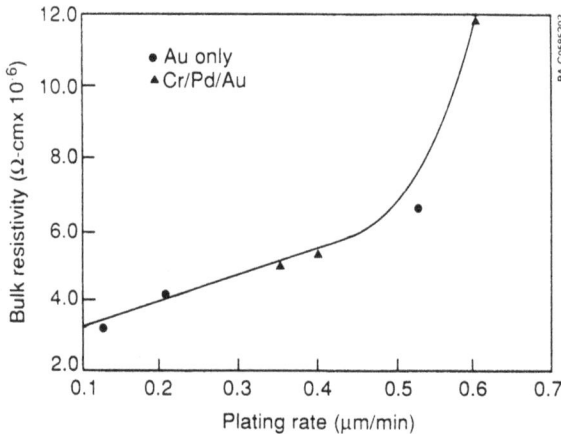

Figure 13.10 Plot of plated-Au resistivity as a function of plating current. As indicated, these results were obtained for grid lines plated both directly on an ITO film and onto Cr/Pd adhesion layers. Since the sheet resistance of the Au is low compared with that of the ITO and the Cr/Pd layers, a reasonable estimate of the resistivity may be made.

In contrast to plated metals, vacuum-evaporated contacts (made by photolithographic lift-off process, as will be discussed) appear to suffer less from sheet resistance variations. Indeed, measurements of electron-beam-evaporated Ag contacts deposited onto cleaned glass substrates indicate that, without great effort or control, RR values of 1.7 to 1.8 are routinely possible over a fairly wide range of deposition rates (1–5 nm/s). Similarly, electron-beam-evaporated Al demonstrates an even lower RR of 1.6 (2 nm/s). It should be noted, however, that concerns of electromigration, as will be discussed, may exclude this material for concentrator cell applications in favor of CuAl$_2$ alloys [20]. Thus, when choosing a metal to use for a grid line, it is not only important to choose one with a relatively low ideal resistivity, but to know how the resistivity is expected to change as a function of deposition, environmental, and/or operating conditions.

Figure 13.11 shows the percentage power loss as a function of RR for a one-sun cell illustrating the effect of resistivity variation. For each of the three deposit thicknesses shown, two different situations are considered: The upper curve of each pair assumes that the designer was unaware of the higher grid-line resistivity, and the lower curve assumes that the designer had been aware that the grid lines were of higher resistivity and had accounted for this. Obviously, the first situation is more likely. With the information that resistivities may be up to eight times the ideal value, modeling shows that if higher resistivities are not assumed, then even at one sun, these losses will certainly dominate. However, as also shown, a prudent design incorporating thicker and/or wider grid lines can be used to insure against these variations. The same effect of increasing the grid-line resistivity is modeled for the concentrator cell in Figure 13.12. Here, it is shown that if R_E can be reduced from 600 to 100 $\Omega/\square$, and if near-ideal metal resistivity can be achieved, then the efficiency will increase up to a concentration ratio of about 100 suns. Although concentrator systems may be designed for operation at up to 1000 suns, still beyond the efficiency maximum for the above design, this exercise illustrates the advantages of knowing which cell parameters to address first to achieve the greatest benefit. Additionally, although not studied here, minimizing the resistance-related power loss of the concentrator grid will require conforming the geometry of the grid design (rectangular or circular) to that produced by whatever focusing apparatus is used.

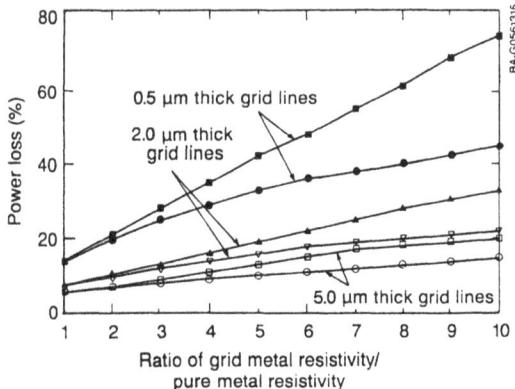

Figure 13.11 Percentage power loss as a function of grid-line resistance for three metal thicknesses for a one-sun cell. The upper curve in each pair assumes that the designer was unaware of the higher resistivity; the lower curve assumes that higher resistivity was expected and accounted for in the model.

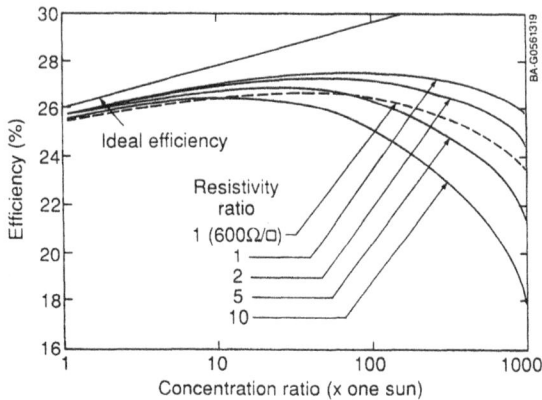

Figure 13.12 Efficiency *versus* concentration ratio for a concentrator cell with R_E equal to 100 $\Omega/\square$, in which the grid metal resistivity ratio is treated parametrically. Also shown is the modeled result if R_E had not been decreased from 600 $\Omega/\square$, illustrating this requirement for a concentrator design.

13.3.3 Grid Metal Linewidth Considerations

The parameter of minimum linewidth often represents the least optimized parameter of an existing grid design and thus an excellent avenue for improved cell performance. In this section, two separate aspects of this parameter are addressed. The first describes the practical guideline that, for typical metallizations, it is generally better to design a grid using the minimum linewidth that can be fabricated. This guideline suggests that many existing solar cell grid designs are probably not optimum, especially if state-of-the art techniques currently used in the microelectronics industry have not been incorporated. The second aspect, on the other hand, cautions that fabricating optimized grids with thinner lines also implies a grid with a higher density of grid fingers; therefore, several process-related problems are likely to occur. Additionally, if the cells are to be used under very high optical concentration (~1000 suns), the effects of electromigration should be considered [21]. Indeed, if these problems are not properly addressed, designing for the minimum linewidth will, in itself, probably not yield a better-performing solar cell.

Often, the geometry of the metal grid line is considered to be an unchangeable parameter in the grid design, dependent on the type of metallization and pattering processes used. However, a well considered grid design study will likely identify that process changes which lead to metallizations with narrower linewidths will likely improve the performance of a grid. For example, it can be seen intuitively that as the linewidth decreases, more grid fingers can be put on the cell, therefore reducing the emitter losses; the result is a new (lower) total power loss design. Although this is mathematically correct, as many more lines appear per unit area of the cell surface, small variations in the linewidth, unfortunately, will have a much greater effect on the shadow loss. Additionally, although it may at first be thought that the decreases in line resistance due to line "broadening" will offset the associated increasing shadow loss, this is seldom the case: the total power loss normally increases, especially for one-sun cells. Thus, in effect, the grid fabrication process has two rather difficult tasks: (1) to produce the narrowest yet thickest lines possible, and (2) to prevent process-related deviations in the linewidth (be it narrow or wide) from the designed width and thickness. As will be seen, although a well-considered design can minimize some of the effects of linewidth variation, it is one of the parameters where even a moderate variation can result in considerable power loss.

As implied in Figure 13.4, for one-sun cells, the total power loss will, to a point, decrease substantially as the linewidth decreases. However, for the cell modeled in this figure, the point of minimum power loss will be dependent upon the process parameter leading to the limiting aspect ratio of the narrow grid line. For example, in this modeled grid, the metal thickness is fixed at 5 μm. Thus, at some linewidth less than 10 μm (at 10 μm, the aspect ratio of linewidth/height = 2), the metal thickness will likely have to be decreased (or else problems such as adhesion loss likely will result due to the increased metallic stress which cannot be accommodated), and this will lead to a minimum in the power loss at a linewidth less than 10 μm. Thus, there is a linewidth at which the power loss will reach a minimum. However, for one-sun cells and typical metallizations, this width is often submicron and therefore not practical for solar cells at this time because the more demanding fabrication procedures would likely lead to a prohibitive process cost. The general guideline of using the minimum linewidth is also illustrated for a one-sun cell in Figure 13.13; the series of curves representing different metal thicknesses shows not only how the total power loss decreases as the linewidth decreases, but also the power loss reduction as the metal thickness increases.

Although the optimum minimum linewidth condition is not usually considered for one-sun grids, for concentrator cells operating a high current density, it must be addressed. This fact is illustrated in Figure 13.14 (and Fig. 13.15). Here, it is seen that when the aspect ratio is treated parametrically and the efficiency is calculated as a function of the grid linewidth, there is an optimum linewidth in the 5-to 10-μm range (dependent on aspect ratio). Since this range of linewidth can readily be fabricated with optical photolithography, it should be used in a concentrator design. Also shown in Figure 13.14 is the modeled effect of a technique developed by Entech, Inc., in which the power loss due to grid line shadowing is eliminated. This technique is based on the use of a silicone plastic cover, the top surface of which is patterned to form an array of corrugations (prisms) which act to focus the light away from the grid lines [22]. As can be seen, when this cover is incorporated into cell construction, the influence on the modeled performance is quite spectacular. However, it should be noted that in this modeling, additional optical effects, such

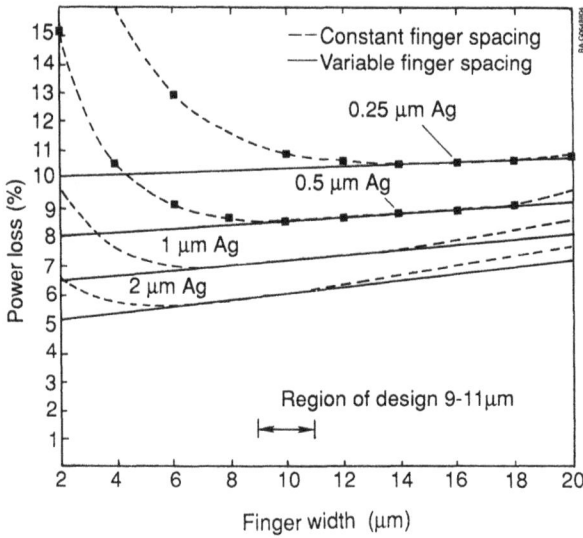

Figure 13.13 Diagram illustrating the percentage power loss for a 4-cm² grid as a function of linewidth where the grid metal thickness (Au, RR = 1) has been treated parametrically. Also shown (dashed lines) is the effect grid-line thickness will have on a cell where the grid spacing is constrained to a constant value (in this case to 508 μm; a typical spacing for Entech prismatic covers).

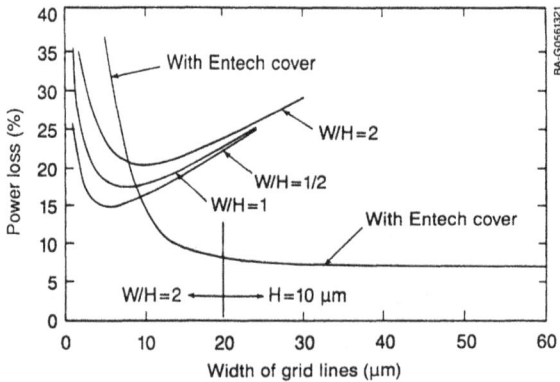

Figure 13.14 Percentage power loss as a function of the grid linewidth and aspect ratio at 1000 suns. Note that there is a well-defined minimum position for each aspect ratio. The grid-line resistivity ratio is taken as unity. Also shown is the effect of an Entech prismatic cover.

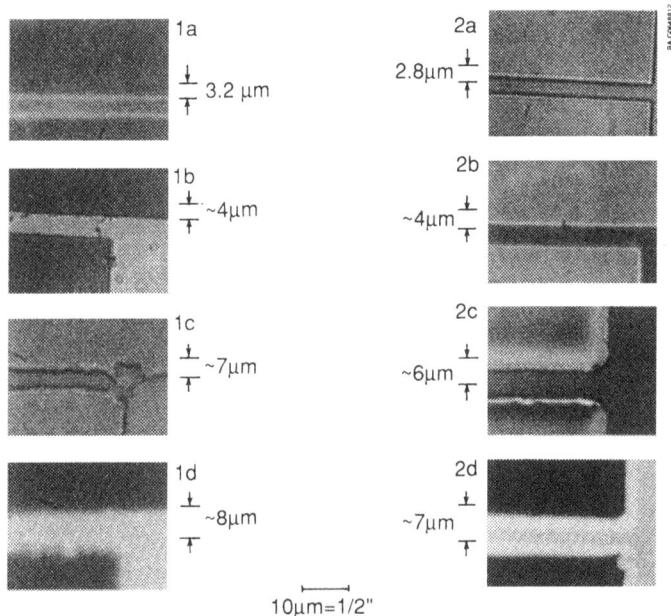

Figure 13.15 Two columns of photos illustrating how a grid line can increase in width from 2 to 8 μm as a metal grid line goes through the processing stages of a photolithographic lift-off procedure. (1a) After pattern exposure and development using a MANN 3600 Pattern Generator; (1b) after etching of FeO_2 (photomask) layer and subsequent removal of patterning photoresist; (1c) after photomask is used to pattern photoresist on PV device; (1d) resultant metallization (Cr/Pd/Ag, 1.5 μm total thickness). Column 2 shows same steps, except that improved etching and photoresist removal has been used in (2b), resulting in improved final metallization definition (2d).

as changes in the antireflection coating due to optical effects of the covers, have not been considered.

The preceding arguments suggest that narrower grid lines are very often a simple path to better solar cell grid performance. Although from the modeling standpoint this is true, in practice many process-related problems can arise with these narrow-line grid designs. These problems can more than offset the gains projected by the mathematical modeling. The primary reason for this is that the relative size (and thus the effect) of a variation in linewidth increases dramatically as the linewidth narrows. In other words, a narrow-line design will have many more, and very closely spaced, grid fingers; therefore, obscuration losses caused by process-related line broadening will tend to dominate any related gains due to the associated decreases in line resistance. As an example of this point, consider the following situation. Suppose a grid line is designed with fingers nominally 50 μm wide. After processing, an additional 2-μm-wide region appears on each edge of the line where irregularities of some sort are observed (see, for example, Fig. 13.8(g-i)). It is believed that this 2-μm-wide region has suspect resistance and/or adhesion quality. However, since it represents only 8% of the width of the grid line, these uncertainties can usually be incorporated into the model through appropriate assignment of the related parameters (as would be done by using a test pattern such as Fig. 13.7). However, if in a later design, the linewidth is decreased to 6 μm and no improvements are incorporated into the metallization process, the region of suspect quality will increase dramatically to 67% of the metallization area and the line resistance and/or adhesion will almost certainly suffer.

Another problem arises in fabricating narrow-line grid designs: when a new (narrower) linewidth is incorporated into a design, there is often uncertainty concerning how the initial linewidth, assigned at the photomask development stage, will compare to the postprocessing width and/or what degree of variation can be expected during normal processing. This point is illustrated in Figure 13.15, where two sets of photographs are compared to show how linewidth broadening propagates through a photolithographic lift-off process. Note that although the initial linewidth specified during photomask manufacture is 2 μm, the final metallized width is closer to 8 μm. Additionally, not only does the linewidth expand (causing increased obscuration), but, unless the photolithographic processing is performed very carefully, the edge definition is poor, causing the effective metal sheet resistance to increase more than would be assumed. This point is also illustrated in Figure 13.15, where, the first column of photographs shows how irregularities in the metallization (Fig. 13.15, 1(d)) can be traced back to residual photoresist remaining after incomplete removal during photomask manufacture (Fig. 13.15, 1(b)). When this photoresist removal processing is optimized (Fig. 13.15, 2(b)), the final metallized edge definition (Fig. 13.15, 2(d)) is also improved.

As mentioned earlier, if the cell is to be used at high optical concentrations, effects due to electromigration should be considered [23]. Electromigration occurs in metals stressed at high current densities; it results in electrical "opens" due to material voids caused by metal ion movement. Electromigration is a relatively common engineering concern in VLSI technology. However, it should be noted that an optimized solar cell grid, operating at 1000-sun concentration, will have a grid finger current density of about 1×10^5 A/cm^2. Hypothetically, this level of current density should not be problematic for solar cell research (e.g., for pure Al contacts operating continuously at 150°C and the above current density, the projected mean time to failure would be ~2 years). However, if the metallization is not of uniform cross-section, or if the cell were to undergo long-term field testing, the metallization alloy and/or the cross-section may need to be changed [20]. Additionally, if the bus-bar of the cell is operating at a much lower current density than are the fingers, as would be the case in a typical concentrator design, then account must be taken of temperature-influenced failure at locations where the narrow grid fingers attach to the bus-bar.

13.3.4 Adhesion, Intermetallic Stress, and Diffusion of the Grid Line

In addition to the metallization issues already discussed, the design of solar cell metallizations must also address some more subtle issues which, due to the relative size of a PV cell and the subsequent environmental conditions in which it must operate, are not often of concern in VLSI technology. Perhaps the most important of these is the adhesion of the metal to the semiconductor emitter. The study of metal-semiconductor adhesion can be one of the most frustrating areas of semiconductor research. Nevertheless, if not properly identified and corrected, adhesion problems will affect not only the obvious mechanical integrity of the solar cell contact but also, through decreasing contact area, the power loss due to contact resistance (i.e., $R_c = r_c/\text{area}$ $[\Omega \cdot cm^2/cm^2]$). Some typical adhesion problems are illustrated in Figure 13.8(j-m).

Many of the techniques used to measure the state of adhesion are very material-specific and highly dependent on the apparatus used [24]. Adhesion testing techniques used for solar cell contacts often include pull testing of contact tabs affixed to the metallization. Although this technique can give a quantitative value for adhesion strength, the processing used to affix the tabs may affect the metallization (and thus the adhesion) in the area beneath and surrounding the tab, resulting in a measurement of adhesion which may not be representative of the nontabbed region. Also, although subjective and not very elegant, the traditional method of pull testing with adhesive tape has been found to be very useful as a qualitative guide for adhesion testing.

Some basic guidelines can be used to improve adhesion. The first of these is that poor adhesion is often caused by surface contamination prior to metal deposition [25]. These contaminants may include chemically bound oxides and sulphides, organic deposits from related processes, and/or particulates from the surrounding environment. Many procedures can be used to clean the surface, and these involve both chemical and physical processes. In our own work with vacuum deposited metals, the use of adhesion layers has been very beneficial. Specifically, thin layers (80 nm) of Cr are used to support thick layers of Ag when a metal/ITO contact must be produced. As mentioned earlier, the necessary degree of adhesion becomes more critical as the linewidth narrows. For example, even for the Cr metallization mentioned here, RF plasma cleaning with Ar must be used if the linewidth decreases to less than about 20 μm. With plasma cleaning, adherent Cr/Ag metallizations have been produced which are 1.5 μm thick and ~6 μm wide.

Adhesion-promoting metals, such as Cr and Ti, are chosen because they are highly reactive and form adherent ionic species with residual surface oxides (e.g., CrO_x). However, these reactions may not stop with the surface oxides but may continue into the underlying semiconductor region. For example, our own work has indicated that Cr may diffuse into InP to a depth of several tens of nanometers during typical electron-beam depositions. Since these junctions are typically very shallow (40–80 nm), this degree of diffusion may lead to junction degradation at elevated temperatures. Thus, a different adhesion-promoting metal which may not diffuse as readily (such as Ti) may eventually have to be used. In addition to diffusion into the semiconductor, the adhesion layers are very susceptible to reaction with any oxygen allowed to permeate the conduction layer(s). Indeed, it has been shown that if the top surface of a Ti/Ag contact is not capped with a thick layer of soft solder, oxygen will diffuse through the Ag to form TiO_x at the Ti-Ag interface, resulting in greatly increased resistance and adhesion loss [26]. However, since solder dipping is not conducive to thin-line grid finger designs, an alternative "solderless" approach incorporates a "diffusion barrier" layer composed of a passivating metal such as Pd [26]. Indeed, even for the Cr/Ag metallization mentioned previously, a 40-nm-thick Pd diffusion layer is normally used to form a Cr/Pd/Ag contact.

13.3.5 Photolithographic Considerations

The photolithographic process used on a particular solar cell is the anchor to which all subsequent grid metallization aspects are tied. The type of photolithography used will not only indicate the minimum linewidth and thickness attainable, but will greatly influence the cross-sectional geometry of the metal line and therefore the RR and often the R_c. For example, if the photoresist application method inadvertently produces thickness variations in the preexposed photoresist, the resulting linewidth irregularity will be duplicated through the process, leading to variations in the final linewidth with associated increase in RR and/or shadow loss. Although a detailed discussion of photolithographic processing [27,28] is far too broad to be addressed here, the following section will present a brief description of some of the more common processes used in solar grid cell metallization research. Also, suggestions will be made concerning how specific photolithography procedures can be advantageously incorporated into a grid design, including descriptions of some common difficulties that may be encountered.

Although both positive and negative acting photoresist systems are typically used in solar cell research and development, this discussion will be limited to the use of positive acting photoresist systems. Additionally, although the technologies of electron-beam and x-ray photolithography are quickly becoming more practical, we will consider only UV light (usually Hg arc lamp) as the illumination source. A positive acting photoresist is one in which the exposed areas become soluble in the developer (except in the case of "image reversal," as will be discussed). In general, there are two categories of photolithographic processing used for solar cell metallization definition. The first category includes processes in which the metallization is deposited before photolithographic definition; the excess metal is subsequently etched off everywhere except where it is protected by the patterned photoresist. This type of processing, known as "subtractive," is used extensively in the microelectronics industry and can produce linewidths of about ≥ 2 μm. However, the technique is only as successful as the appropriate selective wet or dry etching process that can be developed (i.e., the etching process which will etch and thus define the metal pattern while not adversely affecting the underlying semiconductor material). Additionally, if a wet etching (isotropic) process is used, the minimum linewidth cannot be narrower than about two times the metal thickness (although in practice, it may need to be much wider). A typical subtractive process is illustrated in Figure 13.16a.

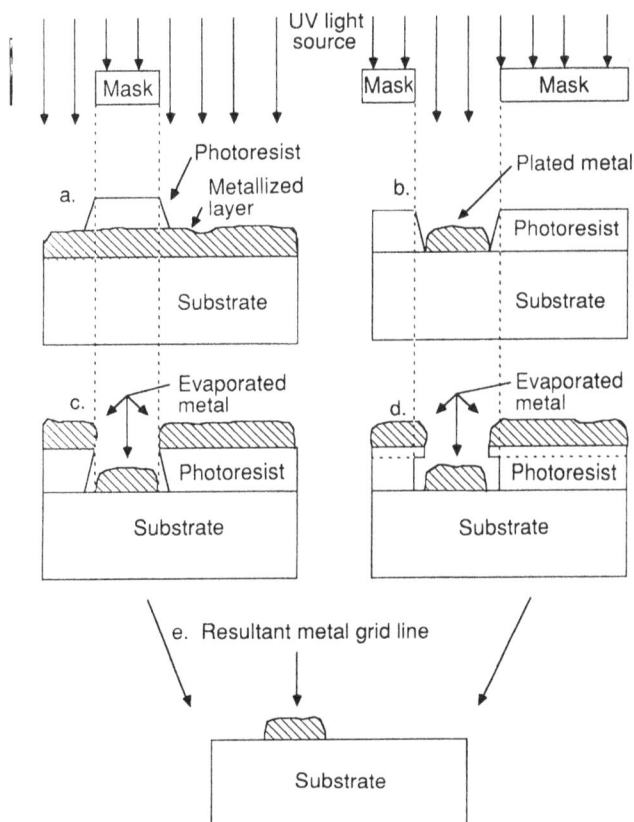

Figure 13.16 Idealized photoresist profiles used for solar cell metallization patterning: (a) Normal profile used in subtractive process; (b) normal profile used in plating process; (c) re-entrant profile produced by image-reversal, shown used in additive "lift-off" process; (d) re-entrant profile produced by altering the solubility of the top portion of the photoresist, shown used in additive "lift-off" process; (e) processed metal grid line after metal etching and/ or photoresist removal of (a–d) above.

In contrast to subtractive processing, the second category includes those processes in which the metallization is performed after the resist patterning; it is therefore distinguished as "additive" processing. Since additive processes (typically known as "lift-off" processes) do not involve isotropic chemical etches for pattern definition, the minimum linewidth is controlled by the much greater resolution of the photolithography process (i.e., minimum linewidth is dependent on the UV exposure source). Additionally, since material-specific selective etching processes do not need to be developed, lift-off processing has been found to be a valuable process tool for metallization studies in a research environment. The key factor in a successful lift-off process is to prepare the photoresist with a slightly undercut profile at the photoresist edge (see Fig. 13.16(c, d)). When the metal is subsequently deposited on the patterned substrate, this undercut, or "re-entrant" profile will produce a discontinuity between the metal on the substrate and the metal on the photoresist; this will facilitate the removal of the overlying metal when the photoresist is dissolved in an appropriate solvent. Thus, the photoresist in a lift-off process is often known as the "release layer" since it aids the release of the unwanted metal.

There are two categories of methods used to produce re-entrant profiles in positive photoresist using a UV exposure source. The first category utilizes the fact that UV light is strongly attenuated by the photoresist, the top surface, absorbing more of the incident light than the region near the substrate. After development, this effect produces a cross-sectional profile in which the top surface of a trench is wider than the bottom (see Fig. 16(a, b)). Although this profile is acceptable for plated contact application, it is the exact opposite of the desired re-entrant profile which is necessary for lift-off processing. Nevertheless, this same artifact of UV attenuation can be exploited to form the proper profile through the process of "image reversal." In this process, the exposed (but not developed) sample is chemically and thermally treated, thereby causing the exposed region to become insoluble in the developer. After a second "flood" exposure and development, a reversed image with the desired profile is produced (see Fig. 13.16(c)) [29,30]. Although image reversal processes are widely used in solar cell metallizations, the added equipment, precautions, and process steps of the predevelopment treatment must be considered and weighed against the probable benefits of the resultant narrower linewidth. Additionally, if a current process is configured to use photomasks appropriate for plated contacts (Fig. 13.16(b)), it should be noted that image reversal photomasks are negatives of these. For these reasons, it is often desirable to employ a lift-off technique which, although not producing as elegant a re-entrant profile, does enable lift-off processing to be used. Shown in Figure 13.16(d) is a re-entrant profile which has been produced by altering the solubility of the top portion of the photoresist [31]. Typically, this is done by soak treating the soft baked photoresist in chlorobenzene or toluene. As illustrated in the figure, the decreased solubility of the top layer produces an "overhang," thereby allowing for the necessary discontinuity in the metal and its subsequent release. As with the image reversal technique, this approach can produce very thin lines. But, since it is difficult to control the laterally dependant solubility of the top layer, edge definition will not be as fine, thus possibly limiting its use to slightly wider line-widths than reliably attainable with image reversal. Nevertheless, since the technique requires no special equipment and can use the same photomasks used to prepare plated metallizations, its use can provide a very cost-effective vehicle to determine if thin-line grid designs would benefit a particular solar cell device.

Although very narrow solar cell grid lines (~ 2 μm wide) can be produced using the lift-off techniques just discussed, lift-off processing is limited by the maximum metal film thickness that can be reliably released. This maximum thickness is normally about two-thirds that of the photoresist thickness. Thus, since normal photoresist thicknesses are about 1.5 to 2.0 μm, the limiting metal thickness will be 1.0 to 1.3 μm. If thicker films are desired, the photoresist viscosity can be increased and/or the spin speed can be reduced, resulting in photoresist thicknesses of about 3.0 μm (allowing for about 2 μm of reliable metal release). As an alternative to thicker photoresist layers, recent process developments incorporating soluble polyimide layers indicate that these techniques may be very well suited for lift-off process applications where thicker metallizations are desired [32,33].

13.4 FUTURE TRENDS AND CONCLUSIONS

Although thus far this discussion has only addressed single-junction solar cells, recent developments in materials and processes indicate that the highest-efficiency cells of the future may likely be multiple-junction devices with as many as six distinct electrical contacts. These multiple-junction cells will enable more of the solar spectrum to be utilized, thereby increasing the electrical conversion efficiency significantly. The highest-efficiency cells, presently nearing 35% (under optical concentration), are two-junction devices relying either on mechanical stacking (four-terminal) or monolithically coupled (three-terminal) designs [2,3]. Additionally, if issues relating to doping uniformity can be addressed, modeling studies indicate that two-junction cells incorporating tunnel-junction interconnects may eventually demonstrate similar performance [34]. In the following section, some of the metallization problems specific to these three- and four-terminal designs will be discussed, illustrating the fact that the high-efficiency, compound semiconductor solar cell grid of the future may be very complex structures.

It is important to realize that the highest efficiency solar cell structures all rely on an IR-sensitive bottom cell. The group IV characteristics of these IR-sensitive cells are such that their performance is much better at high concentrations than at one sun. This is because the band gap is very narrow (low E_g), which makes J_o of Equation (13.2) large and, therefore, V_{oc} low. However, as the concentration increases, the J_{sc}/J_o ratio increases. This raises both the V_{oc} and the FF more significantly, relative to the initial values, than would be the case for higher bandgap devices, thereby leading to greatly improved cell performance. For these reasons, the following discussion will be limited to only concentrator cell designs employing these types of IR-sensitive cells.

13.4.1 Mechanically Stacked, Four-Terminal, Two-Junction Solar Cell Metallization

A mechanically stacked, four-terminal, two-junction cell is one in which two separate solar cell structures, each with its own substrate, are "mechanically" placed on top of each other. The top cell is sensitive to the visible wavelength portion of the spectrum but transparent to the longer wavelength portion (e.g., in the case of InP top cells, those wavelengths >930 nm). The bottom cell, placed below the top cell in the same optical path, absorbs this transmitted near-IR light that has passed through the top cell. The particular metallization issues in this configuration arise in that, if a typical back contact is used on the top cell, the bottom cell will be completely obscured. Indeed, even if the back contact is formed of grids and perfectly aligned with those of the top (this process would require IR mask alignment facilities), the grids of the top cell will obscure the bottom cell unless they also are perfectly aligned with the grids of the bottom cell. In our own work, the back contact of the top cell is formed by contacting the surface of the substrate from the top, using the substrate as a lateral conduction layer and thus eliminating the need for a gridded back contact [2]. Through this process, not only is the need for top and bottom grid alignment eliminated, but also the entire top device can be processed from the top to avoid processing-induced damage to the front surface. Alignment of the top and bottom cell is accomplished by in situ optimization of the J_{sc} of the bottom cell while the two cells are moved relative to each other.

Although the use of Entech covers on these concentrator cells can provide significant benefits, their incorporation must also be assessed carefully; otherwise, non-optimum performance will result. This is primarily because, although the corrugations of the top surface of the (top cell) cover do direct the light away from the grids immediately below, the rays of red and infrared light transmitted through the top cell will strike the bottom cell at non-normal angles of incidence. This will cause the tranmitted light to miss a portion of the bottom cell completely unless a larger bottom cell area is incorporated. Additionally, if a second Entech cover is used on the bottom cell, the cover may not be able to account fully for the deviations from normal incidence. (Note that the rays entering the bottom cell will be optically deviated three times; once by the primary concentrating optics, once by the prismatic cover of the top cell, and finally, the prismatic cover of the bottom cell.)

13.4.2 Monolithic, Three-Terminal, Two-Junction Solar Cell Metallization

In the monolithic, three-terminal, two-junction cell design, several other metallization issues must be addressed. As the name implies, this design uses two cell junctions fabricated on a single, monolithic, substrate. The structure does not employ a tunnel junction; therefore, in addition to the normal front and back contacts, a center contact must be added. This center contact draws current from an epitaxially grown lateral conduction layer located between the top and bottom cells. Since, in this design, all the same issues associated with grid power losses for a two-terminal, single-grid cell are now multiplied by two, the placement and design of the grids becomes an even more difficult matter. To minimize some of these issues, in our own work, extensive use has been made of the Entech prismatic cover. However, in this case, the cover is used to obscure simultaneously both the top and the middle grids [2]. Shown in Figure 13.17 is a cross-sectional view of this structure with the prismatic cover in place. Implied in this figure is the need to fabricate both the top emitter layer and the middle lateral condition layer with the appropriate sheet resistance commensurate with the corrugation spacing of the prismatic cover. In other words, since the finger spacing is fixed by the dimensions of the prismatic cover, the cell must be optimized for the grid (compared to optimizing the grid for the cell!).

By employing these grid procedures, this three-terminal solar cell has demonstrated efficiencies of 31.8% at a concentration ratio of 50 suns. Although efficiency improvements would be expected if the grid metallization were optimized to operate at higher concentrations, the appropriate process improvements are presently constrained somewhat due to finger spacing imposed by the prismatic cover. However, since this device presently relies on plated Au for the contacts, significant improvements may be possible by incorporating evaporated Ag contacts as discussed in Section 13.3.2. Additionally, if emitter losses are found to be the dominant loss mechanism, two-level contact processes, similar to those typically used in VLSI and illustrated in Figure 13.18, may be of considerable benefit. This sort of contact would allow the top and middle finger to be stacked on top of each other, thereby allowing the finger spacing to be decreased to one-half its present value (with the associated reduction in emitter losses).

13.4.3 Conclusions

This chapter has reviewed aspects of metallizations typically encountered in solar cell research and development. These aspects include descriptions of normal operating parameters for one-sun and concentrator solar cells, the related physics of PV devices, and the mathematical analysis used to model the power loss associated with top contact grids. Through related modeling studies, several guidelines have been presented for designing front and back contacts to solar cells, including descriptions of process-related aspects which can affect critical design parameters. Following are some of these guidelines.

For one-sun applications incorporating relatively narrow grid lines, unless it is possible to establish a contact resistance of less than 10^{-2} $\Omega \cdot cm^2$, this term will

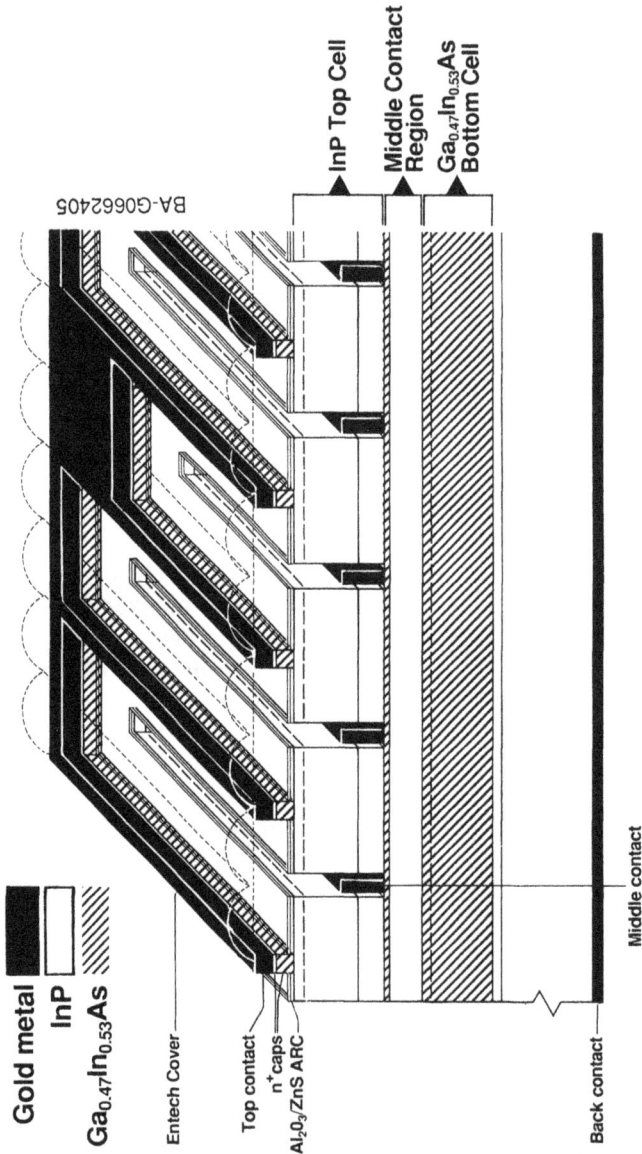

Figure 13.17 Three-dimensional, cross-sectional schematic of the InP/GaInAs monolithic, three-terminal tandem solar cell. Important features include (1) two-level, interdigitated top/middle grid contacts; (2) a middle contact which is common to both subcells; (3) an Entech prismatic cover which eliminates optical losses due to grid obscuration and loss of top cell area. Source: After M.W. Wanlass, J.S. Ward, K.A. Emery, T.A. Gessert, C.R. Osterwald, and T.J. Coutts, "High-Performance Concentrator Tandem Solar Cells Based on Infrared-Sensitive Bottom Cells," to be published in *Solar Cells,* 1990.

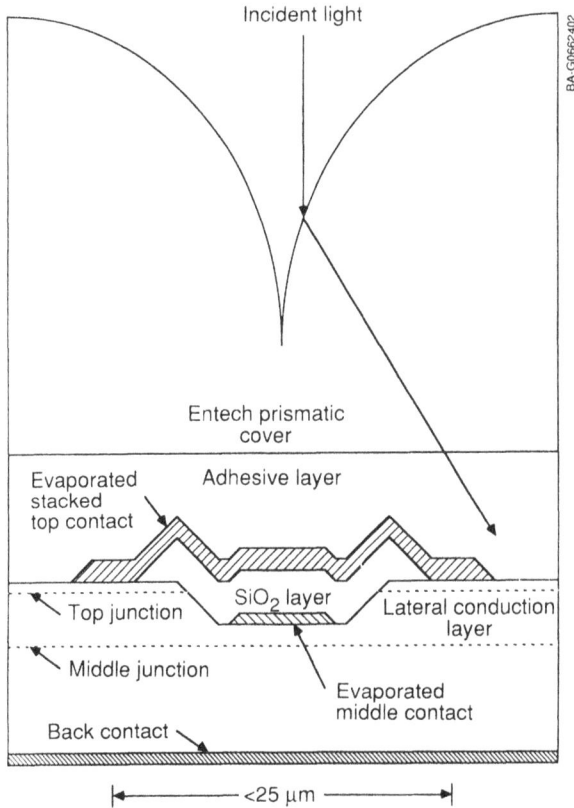

Figure 13.18 Cross-sectional view of proposed stacked contact for future use in three terminal solar cell.

dominate the resistance losses; a significant reduction in power loss would follow if r_c were reduced to 10^{-3} $\Omega \cdot cm^2$. For concentration applications, however, a value of 10^{-4} $\Omega \cdot cm^2$ must be achieved; even lower values, comparable with those of VLSI, would be preferred. Although in practice there may not be complete freedom to lower r_c without damaging the junction or adversely affecting cell performance, a substantial reduction in r_c may be achieved by using techniques such as plasma cleaning and rapid thermal annealing.

The resistivity of the grid-line metal, its thickness, and its aspect ratio are more important parameters in the grid design than is often assumed. It has been indicated that for evaporated metal deposits, variations in resistivity of about two times the ideal value are usual. Similarly, for plated deposits, resistivities up to eight times the ideal value are more usual than may be expected. If not accounted for, these variations in resistivity can dominate the solar cell power loss and possibly affect its stability.

For solar cells operated at one sun using common metallizations, it is always better to design the device with the narrowest grid line that can be fabricated. However, it is necessary to be cautious of the many new problems often encountered when narrow-line designs are fabricated. The most common of these problems is that, as the linewidth narrows, the effect of line broadening becomes much more severe. It is also shown that at a concentration of 1000 suns, not only may the effects of electromigration be present, but the condition of optimum grid linewidth must be met. This condition, which is dependent on the grid-line metal resistivity and its aspect ratio, indicates that the optimum linewidth is usually 5 to 10 μm.

Finally, although the basic procedures for studying solar cell metallization losses were established by previous researchers, the practical application of these techniques is often neglected. Additionally, it is believed that many processes, more typically viewed as being related to the VLSI industry, could be advantageously exploited for PV solar cell metallizations.

ACKNOWLEDGMENTS

The authors wish to thank Dr. Xiaonan Li for very helpful discussions, assistance in sample preparation, and photolithography; Greg Horner for Au/Zn/InP contact resistance measurements; Alice Mason for scanning electron microscopy photographs; and Scott Ward and Mark Wanlass for helpful discussions concerning present process limitations of three- and four-terminal grids. This work was supported by the U.S. Department of Energy under Contract No. DE-ACH02-83CH10093, by NASA Lewis Research Center under Interagency Order No. C-3000-K, and the U.S. Naval Research Laboratories under Interagency No. RU-11-W70-AD.

REFERENCES

1. M.A. Green, S.R. Wenham, J. Zhao, J. Zolper, and A.W. Blakers, "Recent Improvements in Silicon Solar Cell and Module Efficiency," *Proc. 21st IEEE Photovoltaic Specialists Conf.*, Kissimimee, Fla., May 21–25, 1990, New York: IEEE, 1990, p. 207–210.

2. L.M. Frass, J.E. Avery, V.S. Sundaram, V.T. Dinh, T.M. Davenport, J.W. Yerkes, J.M. Gee, and K.A. Emery, "Over 35% Efficient GaAs/GaSb Stacked Concentrator Cell Assemblies for Terrestrial Applications," *Proc. 21st IEEE Photovoltaic Specialists Conf.*, Kissimimee, Fla., May 21–25, 1990, New York: IEEE, 1990, p. 190–195.

3. M.W. Wanlass, J.S. Ward, K.A. Emery, T.A. Gessert, C.R. Osterwald, and T.J. Coutts, "High-Performance Concentrator Tandem Solar Cells Based on Infrared-Sensitive Bottom Cells," *Proc. 10th SERI PVAR&D Meeting*, Lakewood, Colo., Oct. 23–25, 1990, to be published in *Solar Cells*, 1990.

4. I. Weinberg, C.K. Swartz, R.E. Hart, Jr., and T.J. Coutts, "Radiation Resistance and Comparative Performance of ITO/InP and *n/p* InP Homojunction Solar Cells," *Proc. 20th IEEE*

Photovoltaics Specialists Conf., Las Vegas, Nev., Sept. 26–30, 1988, New York: IEEE, 1988, p. 893–897.

5.   M.A. Green, *Solar Cells—Operating Principles, Technology, and Systems Applications*, Englewood Cliffs, N.J.: Prentice-Hall, 1982.

6.   "Standard Test Methods for Electrical Performance of Non-Concentrator Photovoltaic Cells Using Reference Cells," ASTM Standard E948.

7.   D.L. Meier, and D.K. Schroeder, "Contact Resistance: Its Measurement and Relative Importance to Power Loss in a Solar Cell," *IEEE Trans. Electron. Dev.* Vol. ED-31, No. 5, May 1984, pp. 647–653.

8.   H.B. Serreze, "Optimizing Solar Cell Performance by Simultaneous Consideration of Grid Pattern Design and Interconnect Configuration," *Proc. 13th IEEE Photovoltaic Specialists Conf.*, Washington, D.C., June 5–9, 1978, New York: IEEE, 1978, pp. 609–614.

9.   H.H. Berger, "Contact Resistance and Contact Resistivity," *J. Electrochem. Soc., Solid State Sci. Technol.*, Vol. 119, No 4, April 1972, pp. 507–514.

10.   G. SH. Gildenblat, and S.S. Cohen, "Contact Metallization," Chapter 6 in *VLSI Electronics—Microstructure Science*, Vol. 15, *VLSI Metallization*, N.G. Einspruch, S.S. Cohnen, and G. SH. Gildenblat, New York: Academic Press, 1987, pp. 247–292.

11.   T.L. Chu, "Thin Film Cadmium Telluride Solar Cells by Two Chemical Vapor Deposition Techniques," *Solar Cells*, Vol. 23, Jan./Feb. 1988, pp. 31–48.

12.   V.G. Keramidas, "Ohmic Contacts to $Ga_{1-x}Al_xAs$," *Proc. Seventh Int. Symp. on Gallium Arsenide and Related Compounds*, St. Louis, Sept. 1978, Inst. Phys. Conf. Ser. No. 45, Chap. 5, Bristol and London: Inst. of Phys., pp. 396–401.

13.   R.H. Cox, and H. Strack, "Ohmic Contacts for GaAs Devices," *Solid State Electronics*, Vol. 10, Paragomon Press, 1967, pp. 1213–1218.

14.   A.L. Fahrenbruch, and R.H. Bube, *Fundamentals of Solar Cells—Photovoltaic Solar Energy Conversion*, New York: Academic Press, 1983, pp. 197–199.

15.   T.A. Gessert, X. Li, T.J. Coutts, M.W. Wanlass, and A.B. Franz, *Proc. First Int. Conf. on Indium Phosphide and Related Mat. for Adv. Electronic and Optical Devices*, Norman, Okla., March 20–22, 1989, SPIE Proceedings, Vol. 1144, Bellingham, Wash.: SPIE, 1989, pp. 476–487.

16.   W.C. Dautremont-Smith, P.A. Barnes, and J.W. Staylt, Jr., "A Nonalloyed, Low Specific Resistance Ohmic Contact of n-InP," *J. Vac. Sci. Technol. B*, Vol. 2, No. 4, Oct.–Dec. 1984, pp. 620–624.

17.   T.A. Gessert, X. Li, M.W. Wanlass, and T.J. Coutts, "Progress in the ITO/InP Solar Cell," *Proc. Second Int. Conf. on InP and Related Mat.*, Denver: April 23–25, 1990, IEEE Cat. No. 90Ch2895, New York: IEEE, pp. 260–264.

18.   A. Katz, B.E. Weir, D.M. Maher, P.M. Thomas, M. Soler, W.C. Dautermont-Smith, R.F. Karlicek, Jr., J.D. Wynn, and L.C. Kimerling, "Highly Stable W/p-$InP_{0.53}Ga_{0.47}As$ Ohmic Contacts Formed by Rapid Thermal Processing," *Appl. Phys. Lett.*, Vol. 55, No. 21, Nov. 1989, pp. 2220–2222.

19.   A.M. Weisberg, "Why Use Gold," *Gold Plating Technology*, F.H. Reid, and W. Goldie, Ayr, Scotland: Electrochemical Publications Ltd., 1974, p. 14.

20.   I. Ames, F.M. D'Heurle, and R.E. Horstmann, "Reduction of Electromigration in Aluminum Films by Copper Doping," *IBM J. Res. Devel.*, Vol. 14, July 1970, pp. 461–463.

21.   J.E. Sanchez, Jr., J.W. Morris, Jr., and J.R. Lloyd, "Electromigration Failure of Circuit-Level Interconnections," *J. Minerals, Metals and Materials Soc.*, Vol. 24, No. 9, Sept. 1990, pp. 41–45.

22.   M.J. O'Neill, and M.G. Piszczor, "An Advanced Space Photovoltaic Concentrator Array Using Fresnel Lenses, Gallium Arsenide Cells, and Prismatic Cell Covers," *Proc. 20th IEEE Photovoltaic Specialists Conf.*, Las Vegas, Sept. 26–30, 1988, New York: IEEE, 1988, pp. 1007–1012.

23. J.R. Black, "Electromigration—A Brief Survey and Some Recent Results," *IEEE Trans. Electron. Dev.*, Vol. ED-16, No. 4, April 1969, pp. 338–347.

24. K.L. Mittal, "Adhesion Measurement of Thin Films," *Electrocomponent Sci. and Technol.*, Vol. 3, Gordon and Breach Science Pub., 1976, pp. 21–42.

25. K.L. Mittal, "Surface Contamination: An Overview," *Surface Contamination,* Vol. 1, New York: Plenum Press, 1979, pp. 3–45.

26. H. Fischer, and R. Gereth, "New Aspects for the Choice of Contact Materials for Silicon Solar Cells," *Proc. 7th IEEE Photovoltaic Specialists Conf.,* Pasedena, Calif., Nov. 19–21, 1968, New York: IEEE, 1968, pp. 70–76.

27. W.M. Moreau, *Semiconductor Lithography Principles, Practices, and Materials,* New York: Plenum Press, 1988.

28. W.S. DeForest, *Photoresist Materials and Processes,* New York: McGraw-Hill, 1975.

29. H. Moritz, "Optical Single Layer Lift-Off Process," *IEEE Trans. Electron. Dev.,* Vol. ED-32, No. 3, March 1985, pp. 672–676.

30. S.K. Jones, R.C. Chapman, and E.K. Pavelchek, "Image Reversal: A Practical Approach to Lift-Off," *Proc. Advances in Resist Technology and Processing IV,* Santa Clara, Calif., March 2–3, 1987, Proc. of SPIE, Vol 771, pp. 231–241.

31. M. Hatzakis, B.J. Canavello, and J.M. Shaw, "Single-Step Optical Lift-Off Process," *IBM J. Res. Develop.,* Vol. 24, No. 4, July 1980, pp. 452–460.

32. D-Y Shih, E. Galligan, J. Cataldo, J. Paraszczak, S. Nues, R. Serino, W. Graham, and R. McGouey, "In-Plane Solvent Diffusion in a Soluble Polyimid Lift-Off Structure," *J. Vac. Sci. Technol. B,* Vol. 8, No. 5, Sept.–Oct. 1990, pp. 1038–1043.

33. T.D. Flaim, G.A. Barnes, and T. Brewer, "A Novel Release Layer System for IC Processing," *Proc. KTI Microelectronics Seminar—Interface 89,* San Diego, CA, Nov. 6–7, 1989, Sunnyvale, Calif.: KTI Chemicals Inc., 1989, pp. 363–380.

34. M.W. Wanlass, K.A. Emery, T.A. Gessert, G.S. Horner, C.R. Osterwald, and T.J. Coutts, "Practical Considerations in Tandem Cell Modeling," *Solar Cells,* Vol. 27, pp. 191–204.

INDEX

Acceptors
  deactivation of, 164–166
  doping in MOVPE, 135–136
Active channels, 297–299
Additive photolithography, 434
Adhesion, of grid lines, 431–432
AES, *see* Auger electron spectroscopy (AES)
Alloys, quaternary, *see* Quaternary alloys
Alloys, ternary, *see* Ternary alloys
Anisotropic etching, 278
Annealing, rapid thermal, *see* Rapid thermal
  annealing (RTA)
Auger effect, 358–359
Auger electron spectroscopy (AES), 242

Band gap, and heteroepitaxial strain, 29–31
Band offsets, theories of, 5–8
Band structure, of InP and related substances,
  12–22
Barcode readers, 149–150
Base transport, in HBTs, 386–387
Beam flux measurement, 181
Beryllium, 201–202
Boltzmann's constant, 410
Bragg reflector laser, 296–297
Broad-area lasers, 362
Broken-gap lineup, 5
Bulk impurity analysis, 66–67

Cadmium, 166–168
CAIBE, *see* Chemically-assisted ion beam
  etching (CAIBE)
Camel's back structure, 22
Capacitance transient measurements, in DLTS,
  78–80
CBE, *see* Chemical beam epitaxy (CBE)
Cesium ion gun, 49, 58–59
Chemical beam epitaxy (CBE), 107
Chemically-assisted ion beam etching (CAIBE),
  280
Chlorofluorocarbons, in plasma etching, 252
Cobalt, as deep acceptor, 171–172
Collision cascade, 52
Concentration, optical, *see* Optical
  concentration
Concentrator solar cells, 414–415
Confinement factor, in QW lasers, 360–361
Constant capacitance DLTS, 81–82
Contact resistance, 416–420
Contacts, ohmic, *see* Ohmic contacts
Contacts, VLSI, *see* Very-large-scale
  integration (VLSI) contacts

Control system, for MOVPE, 117
Critical point energies, of InP and related
  substances, 14

Damage, etch-induced, 289–293
DCPBH devices, *see* Double–channel planar
  buried heterostructure (DCPBH)
  devices
Deactivation, of acceptors, 164–166
Debye temperature, 10–11
Deep donors and acceptors
  cobalt, 171
  iron, 169–171
  titanium, 171–172
  transition elements, 171–172
Deep levels, 75–99
  early work, 82–86
  evaluation techniques, 75
  radiation effects, 91–96
  recent work, 86–90
  transient spectroscopy, 75–83
    theory of, 76–82
Deep-level transient spectroscopy (DLTS), *see*
  Deep levels, transient spectroscopy
Deformation potential, 28–29
Delta doping, 159
Density of states, in InGaAsP lasers, 356–357
Depth profiling
  high depth resolution, 67, 68–69
  imaging, 69–70
Depth resolution, in SIMS analysis, 47, 61–62
Detection limit, in SIMS analysis, 60
DFB laser, *see* Distributed feedback laser (DFB)
Diagnostics, dry etching, 293–295
Dielectric constants, 35–36
  energy transfer, 341–342
Dielectric deposition technology, 340–348
  guidelines, 343–344
  interfacial considerations, 340–341
  phosphorus loss, 343
  process considerations, 341
  recent advances, 344–348
  substrate temperature, 341
Diffusion, of grid lines, 431–432
Distributed Bragg reflector (DBR) lasers, 371–
  372
Distributed feedback laser (DFB)
  diodes, 277, 367–370
  gratings for, 296–297
DLTS, *see* Deep levels, transient spectroscopy
Dopants